Green Energy Technology

Green Energy Technology

Editors

Wei-Hsin Chen
Hwai Chyuan Ong
Shih-Hsin Ho
Pau Loke Show

MDPI • Basel • Beijing • Wuhan • Barcelona • Belgrade • Manchester • Tokyo • Cluj • Tianjin

Editors
Wei-Hsin Chen
National Cheng Kung University
Taiwan
Hwai Chyuan Ong
University of Malaya
Malaysia
Shih-Hsin Ho
Harbin Institute of Technology
China
Pau Loke Show
University of Nottingham Malaysia
Malaysia

Editorial Office
MDPI
St. Alban-Anlage 66
4052 Basel, Switzerland

This is a reprint of articles from the Special Issue published online in the open access journal *Energies* (ISSN 1996-1073) (available at: https://www.mdpi.com/journal/energies/special_issues/green_energy_technology).

For citation purposes, cite each article independently as indicated on the article page online and as indicated below:

LastName, A.A.; LastName, B.B.; LastName, C.C. Article Title. *Journal Name* **Year**, *Volume Number*, Page Range.

ISBN 978-3-0365-2396-5 (Hbk)
ISBN 978-3-0365-2397-2 (PDF)

Contents

About the Editors

Wei-Hsin Chen is a Distinguished Professor in the Department of Aeronautics and Astronautics, National Cheng Kung University, Taiwan. He was a visiting professor at Princeton University, University of New South Wales, University of Edinburgh, University of British Columbia, and University of Lorraine. He was also invited as an Invited Lecturer at the University of Lorraine in 2019 and 2020. His research topics include bioenergy (combustion, torrefaction, pyrolysis, gasification, liquefaction, bioethanol, etc.), hydrogen production and purification, clean energy (clean coal technology, wind power, thermoelectric generation, carbon capture and utilization, etc.), energy system analysis (optimization, evolutionary computation, machine learning, etc.), and atmospheric science (aerosol absorption and $PM_{2.5}$). He has published around 700 papers with an H-index of 62 (Web of Science). His research provided huge contributions to, and had a large impact on, biomass torrefaction. In this field, his papers, citations, and highly cited papers account for 4.2%, 10.5%, and 25% globally, respectively. He is the associate editor and editorial board member of several prestigious international journals, and the guest editors of over 10 international journals. He has published 10 books and chapters concerning energy science and air pollution. He received 2015 and 2018 Outstanding Research Award (Ministry of Science and Technology, Taiwan), 2015 and 2020 Highly Cited Paper Award (*Applied Energy*, Elsevier), 2016, 2017, 2018, 2019, and 2020 Highly Cited Researcher (Web of Science), 2017 Outstanding Engineering Professor Award (Chinese Institute of Engineers), 2019 Highly Cited Review Article Award (*Bioresource Technology*, Elsevier), 2020 Top Download Paper Award (*International Journal of Energy Research*, Wiley), 2021 TECO Award, etc.

Hwai Chyuan Ong is a faculty member at the Future Technology Research Center, National Yunlin University of Science and Technology. His research interests are wide-ranging, under the general umbrella of renewable energy systems, energy and fuel, green technology and environmental sustainability and solar thermal systems. He has published more than 200 high-impact SCI journal papers with an H-index 50 (WOS). Additionally, he has invited and delivered several keynotes, plenary and invited talks at various international conferences. He was listed as a Highly Cited Researcher (Engineering) by Clarivate Analytics in 2020 and 2019. In 2018 and 2017, he received the Malaysia's Research Star Award (frontier researcher) and, in 2016, he received the Malaysia's Rising Star Award (young researcher) from the Ministry of Higher Education and Clarivate Analytics. In year 2018, he also received the outstanding research award and the most highly cited paper award from the University of Malaya Excellence Award. Currently, he is an Associate Editor of Critical Reviews in Environmental Science and Technology, Alexandria Engineering Journal, Energies and guest editor in FUEL, Sustainable Energy Technologies and Assessment, Environmental Pollution, Biomass Conversion and Biorefinery and Catalyst, as well as an editorial board member for a few other journals. Ong is also a Core Group Member of APRU Sustainable Waste Management (SWM) program.

Shih-Hsin Ho is now a professor at the School of Environment, Harbin Institute of Technology (HIT), China. His research interests mainly focus on microalgae, bioenergy, biochar, wastewater treatment, waste refinery, and bio-based functional materials. He has published over 190 reputed SCI papers (e.g., Nature Biomed. Eng., Trends Biotechnol., Biotechnol. Adv., Appl. Cat. B, Rene. Sus. Energy Rev., Green Chem., Curr. Opi. Biotechnol., Appl. Energy, Water Res., Environ. Sci. and Technol., etc.), and was the first/corresponding author in over 90 of them (including 9 ISI Hi-Ci papers). His SCI total citations are over 10000 (Google Scholar). His H index is 51 (Google Scholar). He has obtained the 2014 Young Top-notch Talents (HIT, China), 2015 National Young Thousand Talents (China), 2017 Minjiang Scholar (Fujian, China), 2019 Young Scientist Studio (HIT, China), etc.

Pau Loke Show is the President of the International Bioprocess Association in Malaysia. He also the director of the Sustainable Food Processing Research Centre and co-director of Future Food Malaysia, Beacon of Excellence, in the University of Nottingham Malaysia. He is a Full Professor of Biochemical Engineering at the Department of Chemical and Environmental Engineering, Faculty of Science and Engineering, University of Nottingham Malaysia. He successfully obtained a PhD two years after obtaining his bachelor's degree from Universiti Putra Malaysia. He is currently a Professional Engineer (PEng), registered with the Board of Engineers Malaysia, Chartered Engineer (CEng) of the Engineering Council UK (MIChemE) and Professional Technologies (PTech), registered with the Malaysia Board of Technologists. Prof Ir. Ts. Dr. Show obtained the Post Graduate Certificate of Higher Education (PGCHE) in 2014, and is now a Fellow of the Higher Education Academy (FHEA) UK. Since he started his career in 2012, he has received numerous prestigious academic awards, including the APEC Science Prize for Innovation, Research and Education ("ASPIRE") Malaysia Award 2020, Malaysia Young Scientist 2019 Award, ASEAN-India Research and Training Fellowship 2019, The DaSilva Award 2018, JSPS Fellowship 2018, Top 100 Asian Scientists 2017, Asia's Rising Scientists Award 2017 and Winner of Young Researcher in IChemE Award 2016. He has graduated more than 20 PhD and MSc students, and is leading a team of 20 members consisting of postdoctoral, PhD and MSc research students. In 2020, he had published more than 550 journal papers in less than 8 years of his career. His publications have been cited over 10000 times in the last 5 years. His current h-index is 50, placing him among the top leaders of his chosen field (Microalgae Technology). He is also the Primary Project leader for more than 35 projects from International, National, and Industry Projects, with a total amount of more than MYR 5.0 million.

energies MDPI

Editorial

Green Energy Technology

Wei-Hsin Chen [1,2,3,*], Hwai Chyuan Ong [4,5,*], Shih-Hsin Ho [6,*] and Pau Loke Show [7,*]

1 Department of Aeronautics and Astronautics, National Cheng Kung University, Tainan 701, Taiwan
2 Research Center for Smart Sustainable Circular Economy, Tunghai University, Taichung 407, Taiwan
3 Department of Mechanical Engineering, National Chin-Yi University of Technology, Taichung 411, Taiwan
4 Centre for Green Technology, Faculty of Engineering and Information Technology, University of Technology Sydney, Sydney, NSW 2007, Australia
5 Future Technology Research Center, National Yunlin University of Science and Technology, 123 University Road, Section 3, Douliou, Yunlin 64002, Taiwan
6 State Key Laboratory of Urban Water Resource and Environment, School of Environment, Harbin Institute of Technology, Harbin 150006, China
7 Department of Chemical and Environmental Engineering, Faculty of Science and Engineering, University of Nottingham Malaysia, Jalan Broga, Semenyih 43500, Malaysia
* Correspondence: chenwh@mail.ncku.edu.tw or weihsinchen@gmail.com (W.-H.C.); ong1983@yahoo.com (H.C.O.); stephen6949@hit.edu.cn (S.-H.H.); PauLoke.Show@nottingham.edu.my (P.L.S.)

Citation: Chen, W.-H.; Ong, H.C.; Ho, S.-H.; Show, P.L. Green Energy Technology. *Energies* **2021**, *14*, 6842. https://doi.org/10.3390/en14206842

Received: 13 October 2021
Accepted: 14 October 2021
Published: 19 October 2021

1. Introduction

Our environment is facing several serious challenges from energy utilization, such as fossil fuel exhaustion, air pollution, deteriorated atmospheric greenhouse effect, global warming, climate change, etc. To solve these problems derived from nonrenewable fuel consumption, a variety of countermeasures, such as the establishment of the RE100 campaign and the target of net zero emissions by 2050, have been launched. To achieve these targets, the development of green energy technology plays a pivotal role. Green Energy Technology (GET) covers technologies, products, equipment, and devices as well as energy services based on software and data protected by patents and/or trademarks. Recent trends underline the principles of a circular economy such as sustainable product design, extending the product lifecycle, reusability, and recycling. Climate change, environmental impact, and limited natural resources require scientific research and novel technical solutions. This Special Issue on Green Energy Technology in *Energies* serves as a publishing platform for scientific and technological approaches to "green"—i.e., environmentally friendly and sustainable—technologies. While a focus lies on energy and bioenergy, it also covers "green" solutions in all aspects of industrial engineering. This Special Issue publishes a comprehensive overview and in-depth technical research paper addressing recent progress in Green Energy Technology. Studies of advanced techniques and methods involving experimental and numerical studies, recent developments, and the current state-of-the-art and emerging technologies in Green Energy Technology are also covered.

This Special Issue of *Energies* on the subject of "Green Energy Technology" contains the successful invited submissions [1–17]. A total of 17 technical papers that cover diversified green energy technology-related research have shown critical results and contributed significant findings in biofuel production [1], energy-saving in buildings [2], wind energy [3], energy policy [4], electricity supply [5], a thermal–hydraulic system [6], fluidized bed reactor [7], green-supply chains [8], SI engine [9], renewable energy grid [10], kinetic energy harvest [11], green hydrogen utilization [12], water conservation [13], hydraulic power [14], Illumination Performance in buildings [15], community renewable energy [16], and heat transfer [17].

The response to our call had the following statistics:

- Submissions (26);
- Publications (17);

- Rejections (9);
- Article types: research articles (17); review articles (0).

Published submissions are related to the most important techniques and analyses applied to green energy technology. In summary, the edition and selections of papers for this Special Issue are very inspiring and rewarding. We thank the editorial staff and reviewers for their efforts and help during the process.

2. Brief Overview of the Contributions to This Special Issue

Most of the published papers focus on providing green energy technology solutions toward environmental sustainability. The following discussion highlights the research findings in accordance with the corresponding research field or work in this Special Issue.

Biofuel is one of the solutions for energy shortage and environmental threats. Nguyen et al. [1] developed microwave-assisted noncatalytic oleic acid esterification for biodiesel production. The highest reaction conversion (97.62%) was achieved by the microwave-assisted esterification process. The study also established a second-order reaction model for the reaction and proven that microwave power greatly influenced the reaction due to nonthermal effects. Moreover, the implementation of poly-ethylene terephthalate (PET)-based waste plastic oil (WPO) at varying engine speed conditions is conducted by Khairil et al. [9]. The properties of WPO were also found to be within the limits mandated in ASTM standard, and the engine performance for different WPO–gasoline blends were close to that of pure gasoline. Pellegrini et al. [12] analyzed the potential of green hydrogen blending in the Italian natural gas network. This work shows how to inject 8100 tons/year of green hydrogen blending in the existing natural gas network with a proper location and sizing of renewable power-to-hydrogen plants. This green hydrogen potential corresponds to an installed capacity of about 78 MW of electrolyzers and about EUR 488 million of investment.

Building energy consumption has risen and will continuously increase with the rise in population, the growth of modern society, and quality of life improvements. The utilization of natural lighting can significantly reduce thermal accumulation in buildings. Thongtha and Boontham [2] designed the application of natural light integrated by using the novel glass units, which provided adequate and efficient daylight illuminance into its interior and decreased heat transmission through the building frames. This led to the conservation of energy consumption from the cooling load of air conditioners and lighting systems. Hence, a significant decline in yearly peak cooling and lighting energy consumption. Jintanawan et al. [11] designed an energy harvesting floor that could convert mechanical energy from people's footsteps to electrical energy. Their outcome showed that the lead-screw model with 45° lead angles coupled with a Genpath prototype-II with a 12-V-DC generator can generate up to 702 mJ which more average energy than others. The efficiency of the EM-generator system is 26% based on the power generation from the heel strike of a human's walk of 2 W per step. The power management and storage circuit were developed to harvested energy into the batteries and to supply other parts to specific loads such as a wireless sensor and Internet of Thing applications. Another building energy analyzed the illumination performance of hollow light pipe which is made from commercial aluminum alloy sheets and commercial zinc alloy sheets [15] aims to reduce electricity consumption for the lighting system. The average illuminance performance of both material types increased with an increase in the incidence angle. The commercial aluminum alloy tube promotes greater light transmission and daylight factor when compared with the commercial zinc alloy tube in each condition. This illuminance measurement demonstrates that the light tube could be included in the lighting systems of some deeper or windowless areas of buildings to decrease the demand for energy consumption in the lighting of buildings.

Many developing countries suffer from high energy-import dependency and inadequate electrification of rural areas. Thus, community renewable energy (CRE) is one of the effective policy tools to provide sustainable energy in a community. Tani and Morone [4] investigated the transition to clean energy technologies in the Boston area, and the outcome showed that the clean energy niche is generally perceived as strong and dynamic in the

Boston area. However, the public de-legitimizing narrative identified gaps such as limited engagement of the local and federal government in breaking through well-established practices and regulatory frameworks, funding, and infrastructure at the policy level. These gaps are likely to delay the market uptake of clean energies in that area. The Philippines' and Singapore's experience in liberalizing their electricity supply industry is presented by Aris et al. [5]. Moreover, Al-Ghussain et al. [10] proposed 100% renewable energy electricity for remote areas in Jordan (Al-Tafilah). The optimal system in Al-Tafilah comprises a 28 MW wind system, 75.4 MW PV, and 1 MW hydropower, with a 259 MWh energy storage system, which is able to provide 99% of energy consumption and reduce up to 47,160 MtCO_2 annually. This model can be easily utilized in other rural cities in Jordan. Luangchosiri et al. [16] analyzed the characteristics of community renewable energy in Thailand to identify the key factors affecting its implementation. They revealed that the primary motivation for implementing CRE is a sustainable development of a community. The internal success factors are intention and vision, human resources, management skills, and community participation, while the external success factors are renewable energy potential, mature technology, financial support, appropriate consultation, and support from the national and local government. It is strongly suggested that external factors, such as financial policy and network support, should be promoted to further facilitate CRE implementation.

Thermal energy also gains much attention recently. Lee and Kim [6] studied the influence of two-phase crossflow for void prediction in bundles using thermal–hydraulic system codes. The outcome of the investigation shown an improvement should be employed a turbulent mixing model based on the Equal Volume and Void Drift to simulate the direct net mass and energy interchanges between channels under two-phase flow conditions. Chang et al. [17] designed the novel printed circuit heat exchanger (PCHE) to enhance the waste heat recovery technology and improve energy efficiency. The PCHE's effectiveness is up to 97.9% for an inlet flow temperature of 95 °C which is better than the others that have fewer layers of PCHE fins.

Other types of green technology such as passive vortex generators, wave energy, floating photovoltaics, liquid-solid circulating fluidized bed, etc. are also presented for this Special Issue. Zhu et al. [3] investigated the effect of passive vortex generators (VGs) on the deep dynamic stall of a wind turbine airfoil. They found that both single-row and double-row VGs effectively suppress the flow separation and reduce the fluctuations in aerodynamic forces. The maximum lift coefficient is therefore increased beyond 40% and suggests that deep dynamic-stall behaviors can be properly controlled by VGs. This study provides the understanding of deep dynamic stall controlled by single-row and double-row VGs. Sureshkumar et al. [7] developed the activated-carbon-coated glass bead adsorbents and applied them for phenol removal from synthetic wastewater which can absorb up to 80% of phenol. Thus, this is an innovative method for effective treatment for toxic phenol contaminant elimination from aqueous solutions. Jusoh et al. [14] applied two major kinds of optimization algorithms (Quadratic Lagrangian (NLPQL) and evolutionary Genetic Algorithm (GA)) on the hydraulic power take-off (HPTO) model for the wave energy converters (WECs). The optimal simulation results showed that the performance of HPTO units has significantly improved up to 96% and 97%, respectively, in regular wave conditions. Both optimal HPTO units were capable of generating electricity up to 62% and 77%, respectively, of their rated capacity in irregular wave circumstances.

Heyibo et al. [13] investigated a new approach of floating photovoltaics or floatovoltaics (FPV) using a flexible crystalline silicon-based photovoltaic module, which is less expensive than conventional pontoon-based FPV. The results show that the foam-backed FPV had a lower operating temperature with a 3.5% higher energy output per unit power. At 50% coverage of Lake Mead, the foam-backed FPV would provide up to127 TWh of clean solar electricity and 633.22 million m^3 of water savings, which would provide enough electricity to retire 11% of the polluting coal-fired plants in the U.S. and provide water for over five million people yearly. The high cost and low profit of green investment are concerns for the government and enterprises deterred from their implementation of green investment. Shi et al. [8] suggested that value co-creation has become a key measure to solve this problem. The green supply chain strategy can promote a high level of firms' value co-creation with their supply chain partners, and different value co-creation modes have different effects on firm performance. Thus, by integrating green supply chain strategies and value co-creation strategies, providing confidence to the firms and their supply chain partners in value co-creation, thus helping them to better implement a green supply chain strategy.

Funding: It received no external funding.

Conflicts of Interest: The authors declare no conflict of interest.

References

1. Nguyen, H.C.; Wang, F.-M.; Dinh, K.K.; Pham, T.T.; Juan, H.-Y.; Nguyen, N.P.; Ong, H.C.; Su, C.-H. Microwave-Assisted Noncatalytic Esterification of Fatty Acid for Biodiesel Production: A Kinetic Study. *Energies* **2020**, *13*, 2167. [CrossRef]
2. Thongtha, A.; Boontham, P. Experimental Investigation of Natural Lighting Systems Using Cylindrical Glass for Energy Saving in Buildings. *Energies* **2020**, *13*, 2528. [CrossRef]
3. Zhu, C.; Wang, T.; Chen, J.; Zhong, W. Effect of Single-Row and Double-Row Passive Vortex Generators on the Deep Dynamic Stall of a Wind Turbine Airfoil. *Energies* **2020**, *13*, 2535. [CrossRef]
4. Tani, A.; Morone, P. Policy Implications for the Clean Energy Transition: The Case of the Boston Area. *Energies* **2020**, *13*, 2615. [CrossRef]
5. Aris, H.; Mohd Zawawi, I.S.; Jørgensen, B.N. The Philippines' and Singapore's Journeys towards Liberalised Electricity Supply Industries—Takeaways for Malaysia. *Energies* **2020**, *13*, 3514. [CrossRef]
6. Lee, Y.; Kim, T. Influence of Two-Phase Crossflow for Void Prediction in Bundles Using Thermal-Hydraulic System Codes. *Energies* **2020**, *13*, 3686. [CrossRef]
7. Sureshkumar, N.; Bhat, S.; Srinivasan, S.; Gnanasundaram, N.; Thanapalan, M.; Krishnamoorthy, R.; Abuhimd, H.; Ahmed, F.; Show, P.L. Continuous Phenol Removal Using a Liquid–Solid Circulating Fluidized Bed. *Energies* **2020**, *13*, 3839. [CrossRef]
8. Shi, X.; Li, G.; Dong, C.; Yang, Y. Value Co-Creation Behavior in Green Supply Chains: An Empirical Study. *Energies* **2020**, *13*, 3902. [CrossRef]
9. Riayatsyah, T.M.I.; Bahri, S.; Sofyan, S.E.; Jalaluddin, J.; Kusumo, F.; Silitonga, A.S.; Padli, Y.; Jihad, M.; Shamsuddin, A.H. Experimental Study on the Performance of an SI Engine Fueled by Waste Plastic Pyrolysis Oil–Gasoline Blends. *Energies* **2020**, *13*, 4196. [CrossRef]
10. Al-Ghussain, L.; Abujubbeh, M.; Darwish Ahmad, A.; Abubaker, A.M.; Taylan, O.; Fahrioglu, M.; Akafuah, N.K. 100% Renewable Energy Grid for Rural Electrification of Remote Areas: A Case Study in Jordan. *Energies* **2020**, *13*, 4908. [CrossRef]
11. Jintanawan, T.; Phanomchoeng, G.; Suwankawin, S.; Kreepoke, P.; Chetchatree, P.; U-viengchai, C. Design of Kinetic-Energy Harvesting Floors. *Energies* **2020**, *13*, 5419. [CrossRef]
12. Pellegrini, M.; Guzzini, A.; Saccani, C. A Preliminary Assessment of the Potential of Low Percentage Green Hydrogen Blending in the Italian Natural Gas Network. *Energies* **2020**, *13*, 5570. [CrossRef]
13. Hayibo, K.S.; Mayville, P.; Kailey, R.K.; Pearce, J.M. Water Conservation Potential of Self-Funded Foam-Based Flexible Surface-Mounted Floatovoltaics. *Energies* **2020**, *13*, 6285. [CrossRef]
14. Jusoh, M.A.; Ibrahim, M.Z.; Daud, M.Z.; Yusop, Z.M.; Albani, A. An Estimation of Hydraulic Power Take-off Unit Parameters for Wave Energy Converter Device Using Non-Evolutionary NLPQL and Evolutionary GA Approaches. *Energies* **2021**, *14*, 79. [CrossRef]
15. Mahawan, J.; Thongtha, A. Experimental Investigation of Illumination Performance of Hollow Light Pipe for Energy Consumption Reduction in Buildings. *Energies* **2021**, *14*, 260. [CrossRef]
16. Luangchosiri, N.; Ogawa, T.; Okumura, H.; Ishihara, K.N. Success Factors for the Implementation of Community Renewable Energy in Thailand. *Energies* **2021**, *14*, 4203. [CrossRef]
17. Chang, C.-Y.; Chen, W.-H.; Saw, L.H.; Arpia, A.A.; Carrera Uribe, M. Performance Analysis of a Printed Circuit Heat Exchanger with a Novel Mirror-Symmetric Channel Design. *Energies* **2021**, *14*, 4252. [CrossRef]

energies

Article

Microwave-Assisted Noncatalytic Esterification of Fatty Acid for Biodiesel Production: A Kinetic Study

Hoang Chinh Nguyen [1], Fu-Ming Wang [2], Kim Khue Dinh [1], Thanh Truc Pham [3], Horng-Yi Juan [3], Nguyen Phuong Nguyen [1], Hwai Chyuan Ong [4,*] and Chia-Hung Su [3,*]

1. Faculty of Applied Sciences, Ton Duc Thang University, Ho Chi Minh City 700000, Vietnam; nguyenhoangchinh@tdtu.edu.vn (H.C.N.); dinhkhue96@gmail.com (K.K.D.); nguyenphuongcass97@gmail.com (N.P.N.)
2. Graduate Institute of Applied Science and Technology, National Taiwan University of Science and Technology, Taipei 10607, Taiwan; mccabe@mail.ntust.edu.tw
3. Graduate School of Biochemical Engineering, Ming Chi University of Technology, New Taipei City 24301, Taiwan; ptttruc1996@gmail.com (T.T.P.); hyjuan@mail.mcut.edu.tw (H.-Y.J.)
4. School of Information, Systems and Modelling, Faculty of Engineering and Information Technology, University of Technology Sydney, Ultimo, NSW 2007, Australia

* Correspondence: ong1983@yahoo.com (H.C.O.); chsu@mail.mcut.edu.tw (C.-H.S.); Tel.: +88-622-908-9899 (ext. 4665) (C.-H.S.)

Received: 22 March 2020; Accepted: 26 April 2020; Published: 1 May 2020

Abstract: This study developed a microwave-mediated noncatalytic esterification of oleic acid for producing ethyl biodiesel. The microwave irradiation process outperformed conventional heating methods for the reaction. A highest reaction conversion, 97.62%, was achieved by performing esterification with microwave irradiation at a microwave power of 150 W, 2:1 ethanol:oleic acid molar ratio, reaction time of 6 h, and temperature of 473 K. A second-order reaction model (R^2 of up to 0.997) was established to describe esterification. The reaction rate constants were promoted with increasing microwave power and temperature. A strong linear relation of microwave power to pre-exponential factors was also established, and microwave power greatly influenced the reaction due to nonthermal effects. This study suggested that microwave-assisted noncatalytic esterification is an efficient approach for biodiesel synthesis.

Keywords: esterification; biodiesel; noncatalytic reaction; kinetic; microwave irradiation; bioenergy

1. Introduction

The energy shortage and environmental threats caused by greenhouse gas emissions have promoted the demand for renewable energy. One of the most promising alternatives is biodiesel, which is known as a biomass-derived fuel produced from the transesterification of triglycerides [1,2]. Biodiesel is a biodegradable, renewable, and green fuel and has superior combustion efficiency [3,4]. Biodiesel use can result in lower hydrocarbon, CO, CO_2, and particulate matter emissions compared with petroleum use [5,6]. Therefore, biodiesel has increasingly been used as a petroleum substitute.

Biodiesel is mainly derived from food oils [7,8], but the biodiesel produced is expensive due to the high cost of these feedstocks. To address this problem, waste and nonedible materials have been proposed as alternatives for producing biodiesel [9–11]. These oils can reduce biodiesel production cost, and they have no competition with the food supply [12,13]. Nevertheless, waste and nonedible oils contain high levels of free fatty acid, which must be reduced via esterification prior to transesterification [14,15]. Commonly, fatty acid is esterified using a liquid acid-catalyzed process [12,16]. Although this method effectively converts fatty acid into biodiesel, the downstream process for catalyst removal is difficult [17]. Notably, liquid acid catalysts cause environmental pollution and corrosive damage to equipment, raising concerns about their use [7]. To overcome the obstacles of liquid acid catalyst use, different solid acid catalysts have been proposed for the esterification such as Propyl-SO3H-SBA-15 [18], Amberlyst-36 [18], WO3-USY zeolite [19], HZSM-5 [20], Amberlyst-15 [21], biocatalysts [21], and ZrO_2-TiO_2 nanorods [22]; however, these solid catalysts exhibit low stability and catalytic activity, thus requiring high catalyst quantities and long reaction times, and they result in a low conversion yield [19,20]. In addition, these catalyst residues can cause a negative impact on the environment. Consequently, the applications of those solid catalysts in the esterification reaction are still limited.

With the growing awareness regarding environmentalism, the efforts have been undertaken to develop green methods for chemical production. For several decades, the enzymatic process has been developed as a substitute for chemically catalyzed esterification for biodiesel synthesis [23,24]. Enzyme-catalyzed esterification is superior to conventional processes in terms of the mild reaction conditions, reduced environmental impact, and selectivity of enzymes toward the substrate [24,25]. However, the high price of enzymes restricts its applications [24,26]. On the other hand, the noncatalytic esterification is considered another eco-friendly process for biodiesel synthesis. This process induced the esterification of fatty acids under thermal conditions without a catalyst [27], eliminating the problems associated with both chemical and enzymatic processes. The supercritical alcohol method has been applied in noncatalytic fatty acid esterification to prepare biodiesel [28,29]. Although this process produces a high yield of esters (up to 97%) within a short reaction time (10–30 min), it proceeds at a high temperature (280–320 °C) and high pressure (10–25 MPa) and consequently requires expensive reactors and extensive safety measures [28,29]. To overcome this, Cho et al. [30] proposed another form of noncatalytic esterification for less-than-supercritical conditions to produce biodiesel. Although their process proceeded at moderate pressure (0.85–1.2 MPa), it still required a high temperature (290 °C) for the reaction [30]. Therefore, another efficient esterification method is required for biodiesel production.

Microwave irradiation has been successfully applied in chemical reactions to enhance the reaction rate [31,32]. This method employs microwaves, which are electromagnetic waves, to transfer energy to the reactants [33]. Through relaxation or resonance generated by microwaves, reactant molecules are induced to vibrate at extremely high frequencies, resulting in fast molecular mobility of reaction species [34,35]. Consequently, microwave irradiation increases reaction rate and conversion yield whilst decreasing energy consumption and reaction time [33,36]. Furthermore, microwave-mediated processes are effective for large-scale operations [37]. Such advantages mean that the microwave approach has been extensively applied in catalytic transesterification and esterification for producing biodiesel [38–40]. Recently, our previous studies have demonstrated the potential use of microwave-mediated noncatalytic/autocatalytic synthesis of fatty acid [41], phytosterol esters [42], and ethyl levulinate [43]. However, its use in noncatalytic esterification to produce biodiesel remains limited.

This study proposed noncatalytic esterification of fatty acid with ethanol using microwave irradiation as an environmentally friendly and energy-efficient method to produce ethyl biodiesel. The esterification was performed under selected microwave powers to enhance the reaction conversion. Ethanol (95%) was used as an acyl acceptor instead of methanol for biodiesel synthesis because ethyl biodiesel is superior to methyl biodiesel due to its oxidation stability, calorific value, cetane number, and cold flow properties [44,45]. Oleic acid was chosen as a model molecule for the esterification, because it is a predominant fatty acid presented in animal fats and vegetable oils [46]. The reaction parameter effect on the reaction conversion was investigated. Furthermore, a mathematical model was proposed for representing the kinetics of noncatalytic esterification reactions.

2. Materials and Methods

2.1. Chemicals

Ethanol (95%), oleic acid (98%), and other reagents were obtained from Tokyo Chemical Industry (Tokyo, Japan).

2.2. Effects of Different Heating Processes

The effect of two heating approaches (conventional heating and microwave irradiation) on the noncatalytic esterification of oleic acid with ethanol was examined. For microwave-assisted esterification, 10-mL glass reactor containing the reaction solution (5 mL, 2:1 molar ratio of ethanol to oleic acid) was sealed and placed in a CEM 908005 microwave oven (Matthews, NC, USA). The microwave was equipped with a gas cooling system to maintain the temperature at a desired level. The reaction was then carried out at a microwave power of 150 W, 433 K, and different reaction times (60–360 min) with stirring. The reaction using conventional heating was undertaken in a 10 mL-sealed stainless-steel reactor placed in an oil bath under the same reaction conditions: 2:1 molar ratio of ethanol to oleic acid, 433 K, and different reaction times (60–360 min). After the reaction was completed, the sample was withdrawn to determine the oleic acid conversion. Each experiment was independently performed in triplicate for each reaction time.

2.3. Analysis

The amount of ethyl oleate synthesized was quantified using a Gas Chromatograph system (GC-2014, Shimadzu, Japan) equipped with a flame ionization detector and Stabilwax column (60 m × 0.25 mm id, 0.25 µm film thickness; Restek, Bellefonte, PA, USA) [1]. Nitrogen was used as the carrier gas and set at 1.0 mL/min. The temperatures of detector and injector were maintained at 250 and 220 °C, respectively. The column temperature was held at 140 °C for 5 min, increased to 240 °C with a ramp rate of 4 °C/min, and maintained for 15 min. Ethyl oleate standard (Sigma-Aldrich, Louis, MO, USA) was used to identify and determine the amount the produced ethyl oleate. One mol of oleic acid could stoichiometrically produce 1 mol of ethyl oleate; the oleic acid conversion (X) was consequently calculated as follows:

$$\begin{aligned} Oleic\ acid\ conversion, X &= Ethyl\ oleate\ conversion \\ &= \frac{Amount\ of\ oleic\ acid\ reacted}{Initial\ amount\ of\ oleic\ acid} \\ &= \frac{282.47 \times amount\ of\ ethyl\ oleate\ produced}{310.51 \times initial\ amount\ of\ oleic\ acid} \end{aligned} \quad (1)$$

2.4. Kinetics of Noncatalytic Esterification Using Microwave Irradiation

Noncatalytic esterification of oleic acid (A) with ethanol (B) to produce ethyl oleate (C) and water (D) is demonstrated as follows:

$$aA + bB \underset{k_2}{\overset{k_1}{\rightleftharpoons}} cC + dD$$

The model established for depicting esterification is considered elementary and reversible. The rate law is therefore as follows:

$$-\frac{dC_A}{dt} = k_1 C_A^a C_B^b - k_2 C_C^c C_D^d \quad (2)$$

where k_1 and k_2 denote forward and reverse reaction rate constants (L mol^{-1} min^{-1}), respectively; C_A, C_B, C_C, and C_D are concentrations of oleic acid, ethanol, ethyl oleate, and water, respectively; a, b, c, and d are the reaction orders of involved species. The reaction is assumed to follow the second-order kinetics ($c = d = a = b = 1$), the rate law is shown in terms of oleic acid conversion (X) as follows Equation (3):

$$\begin{aligned} \frac{dX}{dt} = \frac{1}{C_{A0}}(k_1 C_A C_B & - k_2 C_C C_D) \\ &= \frac{1}{C_{Ao}}\left[k_1 C_{A0}(1-X) C_{A0}(\theta_B - X) - \frac{k_1}{K_e} C_{A0} X C_{A0}(\theta_D + X)\right] \\ &= k_1 C_{A0}\left[(1-X)(\theta_B - X) - \frac{X(\theta_D + X)}{K_e}\right] \end{aligned} \quad (3)$$

where C_{A0} is the initial oleic acid concentration; θ_B and θ_D are the initial ethanol:oleic acid molar ratio and the initial molar ratio of water to oleic acid, respectively; and K_e is the equilibrium rate constant. The reaction rate (dX/dt) will be zero at equilibrium, and K_e is then calculated as follows:

$$K_e = \frac{k_1}{k_2} = \frac{X_e(\theta_D + X_e)}{(1 - X_e)(\theta_B - X_e)} \tag{4}$$

where X_e is the equilibrium conversion of oleic acid. After determination of K_e, Equation (3) can be integrated into Equation (5) using the derivation described by Su [47]. Equation (5) can subsequently be used for determining the rate constant k_1 by linearly plotting $\ln\left[\frac{\left(-1-\theta_B-\frac{\theta_D}{K_e}+\alpha_2\right)X+2\theta_B}{\left(-1-\theta_B-\frac{\theta_D}{K_e}-\alpha_2\right)X+2\theta_B}\right]$ versus $\alpha_2 C_{A0} t$.

$$\ln\left[\frac{\left(-1-\theta_B-\frac{\theta_D}{K_e}+\alpha_2\right)X+2\theta_B}{\left(-1-\theta_B-\frac{\theta_D}{K_e}-\alpha_2\right)X+2\theta_B}\right] = \alpha_2 C_{A0} k_1 t \tag{5}$$

$$\text{where } \alpha_2 = \left[\left(1 + \theta_B + \frac{\theta_D}{K_e}\right)^2 - 4\alpha_1\theta_B\right]^{\frac{1}{2}} \tag{6}$$

$$\text{and } \alpha_1 = 1 - \frac{1}{K_e} \tag{7}$$

The relationship between temperature and the rate constants are represented by the Arrhenius equation:

$$k_1 = A_1 \exp\left(-\frac{E_a}{RT}\right) \tag{8}$$

$$K_e = A_e \exp\left(-\frac{\Delta h}{RT}\right) \tag{9}$$

where A_e and A_1 (L mol^{-1} min^{-1}), respectively, denote pre-exponential factors for the equilibrium and forward rate constants; Δh (J mol^{-1}) and E_a (J mol^{-1}), respectively, represent the molar reaction heat and activation energy of forward reaction; T (K) and R (J mol^{-1} K^{-1}) are the reaction temperature and the ideal gas constant, respectively. These parameters (A_1, A_e, E_a, and Δh) can be obtained from the Arrhenius–Van't Hoff plot [Equations (10) and (11)]:

$$\ln k_1 = \ln A_1 - \frac{E_a}{RT} \tag{10}$$

$$\ln K_e = \ln A_e - \frac{\Delta h}{RT} \tag{11}$$

3. Results and Discussion

3.1. Effect of Various Heating Processes

This study compared noncatalytic esterification using microwave irradiation and conventional heating. As demonstrated in Figure 1, conventional heating resulted in low oleic acid conversions (27.13%–67.13%). This finding is consistent with other works in that conventional heating needed long reaction time to achieve a satisfactory reaction conversion [48]. Conventional heating is an inefficient form of heat transfer, because heat is delivered to the reaction solution by convection, radiation, and conduction from the surface of the reactor, causing heat loss [38,39]. To increase the heat transfer and reaction efficiency, microwave approach was employed as a replacement for the reaction. Results indicated that the oleic acid conversion significantly increased and reached high levels (60.61%–90.18%) with microwave irradiation. Remarkably, at a given reaction time, microwave irradiation exhibited significantly higher oleic acid conversion than conventional heating, signifying that microwave irradiation is preferable to conventional heating for performing the esterification. This result corresponds to the study of Aguilar–Reynosa et al. [49]. Microwave irradiation minimizes loss of heat and additionally provides a nonthermal activation influence on the esterification [50]. Microwaves strongly induce movement, collision, and oscillation of reactant molecules, promoting the reaction efficiency [49,51]. As a result, microwave irradiation is superior than traditional heating regarding to reaction time [43,52]. Therefore, microwave approach was selected for more extensive study.

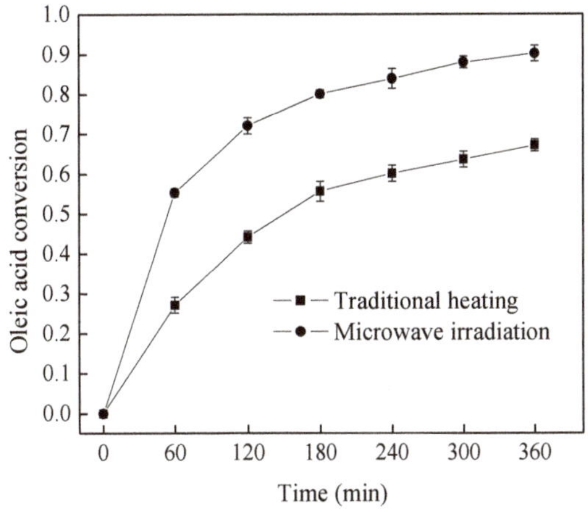

Figure 1. Comparison of esterification using conventional heating and microwave irradiation. Reaction conditions occurred with an ethanol:oleic acid molar ratio of 2:1, temperature of 433 K, and microwave power of 150 W (for microwave irradiation). Vertical bars illustrate the standard deviation of three replicates.

3.2. *Influence of Reactant Molar Ratio*

This work investigated the impact of reactant molar ratio on the reaction by undertaking the reaction at 433 K, at a microwave power of 150 W, and at various ethanol:oleic acid molar ratios (1:1–8:1). As demonstrated in Figure 2, the oleic acid conversion was promoted when molar ratios of the reactants were increased from 1:1 to 2:1. Nevertheless, additional increase in ethanol:oleic acid molar ratio resulted in decreased oleic acid conversion. This may be because large quantities of ethanol used to dilute oleic acid, which plays a role as a catalyst for the esterification reaction, may lower the reaction rate [53,54]. This result corresponds to that of the biodiesel production study by Minami and Saka [54].

3.3. *Temperature Effect on Esterification*

The impact of temperature on reaction was evaluated by conducting the reaction at a microwave power of 150 W, an ethanol:oleic acid molar ratio of 2:1, and various temperatures (413–473 K). Figure 3 presents the progress over time of the oleic acid esterification at different temperatures. Results showed that higher temperature provided a greater conversion of oleic acid, with the highest conversion of 97.62% obtained at 473 K and 6 h. This is because high temperature enhances reaction rate [12], increasing the reaction conversion. This finding is similar to those of other works [12,14,55].

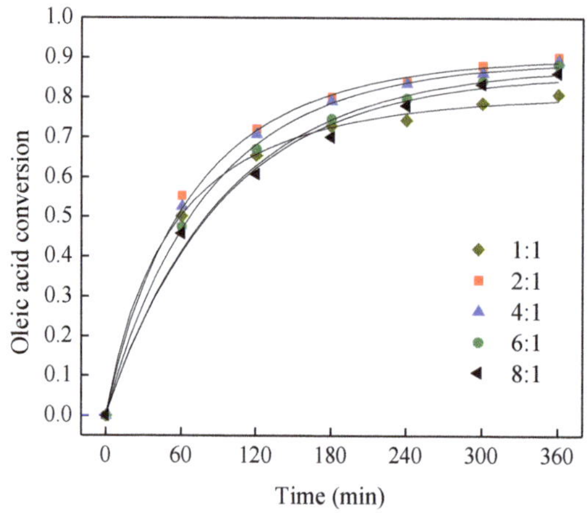

Figure 2. Influence of molar ratio of ethanol to oleic acid on esterification reaction. Reactions occurred at a temperature of 433 K and with microwave power of 150 W.

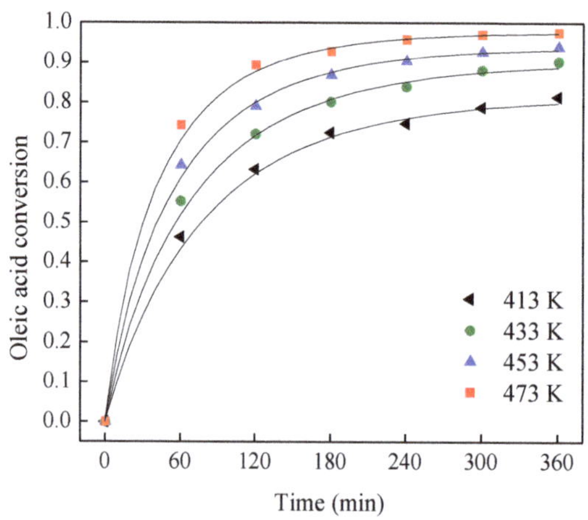

Figure 3. Influence of temperature on esterification reaction. Reactions occurred at a molar ratio of ethanol to oleic acid of 2:1 and microwave power of 150 W.

3.4. Effect of Microwave Power on Reaction

To determine the impact of microwave power on the reaction, the esterification was carried out at different microwave power levels (120–210 W) while maintaining other factors at a constant level. As illustrated in Figure 4, increasing microwave power led to an increase in the oleic acid conversion. This might have been because higher energy generated by higher microwave power resulted in higher activation effects (nonthermal effects), causing faster molecular mobility; therefore, the reaction rate was promoted [50,51]. This result corresponds with that reported for biodiesel production [56] and for pyrolysis of biomass [51].

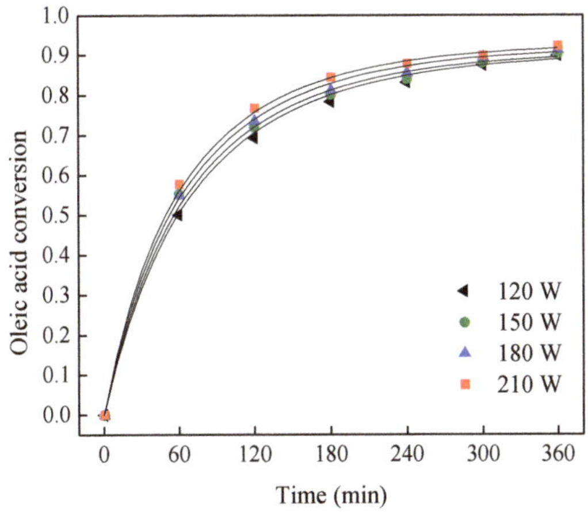

Figure 4. Influence of microwave power on esterification reaction. Reactions occurred at a molar ratio of ethanol to oleic acid of 2:1 and at a temperature of 433 K.

3.5. *Kinetic Model Development*

3.5.1. Determination of Reaction Rate Constant

Figures S1–S3 (in the Supplementary Material), respectively, provide a linear plot of $\ln\left[\frac{\left(-1-\theta_E-\frac{\theta_D}{K_e}+\alpha_2\right)X+2\theta_E}{\left(-1-\theta_E-\frac{\theta_D}{K_e}-\alpha_2\right)X+2\theta_E}\right]$ versus $\alpha_2 C_{A0}t$ at different microwave powers, temperatures, and ethanol:oleic acid molar ratios. As illustrated in Table 1, correlation coefficients (R^2) of the regression lines were in the range of 0.972–0.997, indicating the reliability and correction of the second-order reaction model. Table 1 (runs 1–5) shows that rate constants K_e and k_1 both decreased when the molar ratio of ethanol to oleic acid increased, indicating the negative influence of high ethanol:oleic acid molar ratio on the reaction conversion [53,54]. Nevertheless, the rate constants K_e and k_1 accelerated with increasing temperature (Table 1, runs 6–9), signifying that the reaction is endothermic [47]. Furthermore, increasing microwave power enhanced the rate constants K_e and k_1 (Table 1, runs 10–13), which was similar to the results of other studies [51,56]. This result indicated a positive effect from microwave power on the reaction. Notably, at any ethanol:oleic acid molar ratio and temperature, microwave-based reaction had much higher K_e and k_1 values than the conventional heating-based reaction (see the K_e and k_1 values for the conventional heating-based reaction in the Table S1 in the Supplementary Material). These results indicated that microwave irradiation enhanced the equilibrium rate constant and forward reaction rate constant. Therefore, microwave irradiation is more efficient than conventional heating for conducting the ethyl oleate synthesis.

3.5.2. Determination of Pre-Exponential Factor and Activation Energy

The influence of temperature on reaction rate constants (K_e and k_1) is represented using the Arrhenius equation [Equations (8) and (9)]. The activation energy, pre-exponential factor, and molar reaction heat of the reaction were calculated using the Arrhenius–van't Hoff plots. As illustrated in Figure 5 and in Table 2, high R^2 of the regression lines (0.997 for k_1 and 0.985 for K_e) were achieved; therefore, the model parameters can be determined from the straight line. Results showed that the activation energy of the forward reaction was 23.59 kJ mol^{-1}, and the molar reaction heat was 65.98 kJ mol^{-1}. The pre-exponential factors were calculated to be 2.27 L mol^{-1} min^{-1} and 9.07×10^8 for forward and equilibrium reactions, respectively.

Table 1. Forward reaction rate constant k_1 and equilibrium rate constant K_e for esterification under different reaction conditions.

Run	Molar Ratio of Ethanol to Oleic Acid	Temperature (K)	Microwave Power (W)	Equilibrium Constant, K_e	Forward Reaction Rate Constant, k_1 (L mol^{-1} min^{-1})	R^2
1	1:1	433	150	21.5253	6.50×10^{-3}	0.972
2	2:1	433	150	10.3897	3.18×10^{-3}	0.986
3	4:1	433	150	4.1905	1.59×10^{-3}	0.993
4	6:1	433	150	2.7692	1.06×10^{-3}	0.990
5	8:1	433	150	1.9294	0.92×10^{-3}	0.978
6	2:1	413	150	4.2537	2.41×10^{-3}	0.989
7	2:1	433	150	10.3897	3.18×10^{-3}	0.986
8	2:1	453	150	18.7094	4.23×10^{-3}	0.994
9	2:1	473	150	52.7582	5.75×10^{-3}	0.992
10	2:1	433	120	9.8958	3.02×10^{-3}	0.984
11	2:1	433	150	10.3897	3.18×10^{-3}	0.986
12	2:1	433	180	12.0860	3.39×10^{-3}	0.983
13	2:1	433	210	13.8671	3.59×10^{-3}	0.997

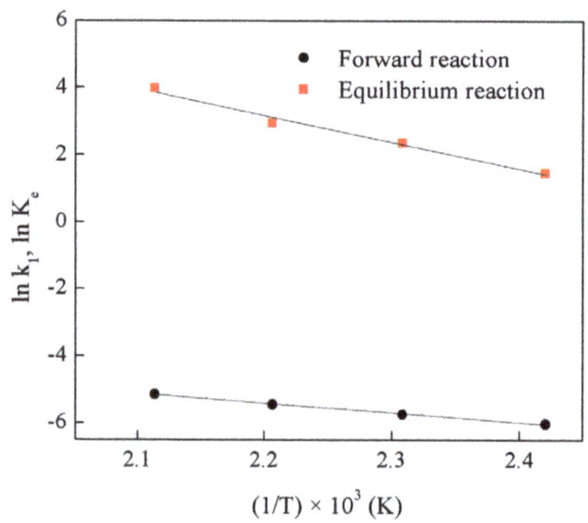

Figure 5. Arrhenius–Van't Hoff plot for equilibrium rate constant and forward reaction rate constant of esterification reaction.

Table 2. Activation energy, molar reaction heat, and pre-exponential factor for oleic acid esterification with ethanol using microwave approach.

Forward Reaction			Equilibrium Reaction		
Activation Energy (kJ mol^{-1})	Pre-Exponential Factor (L mol^{-1} min^{-1})	R^2	Molar Reaction Heat (kJ mol^{-1})	Pre-Exponential Factor	R^2
23.59	2.27	0.997	65.98	9.07×10^8	0.985

3.5.3. Relation of Microwave Power to Pre-Exponential Factor

The modified Arrhenius equation reported by Su et al. [57] was used to determine the influence of microwave heating on the kinetic model:

$$K_e = A_e \exp\left(-\frac{\Delta h}{RT}\right)$$

$$k_1 = A_1 \exp\left(-\frac{E_a}{RT}\right)$$

$$A_e = A_0^e + A_W^e \cdot W \quad (12)$$

$$A_1 = A_0^1 + A_W^1 \cdot W \quad (13)$$

where W is the microwave power; A_W^e and A_W^1, respectively, denote the coefficient for representing the effect of power on the equilibrium and forward reaction; A_0^e and A_0^1, respectively, denote the constant for representing the effect of power on the equilibrium and forward reaction. The pre-exponential factors (A_e and A_1) were assumed to obey linear functions with the microwave power. Both the constants (A_0^1 and A_0^e) and the coefficients (A_W^1 and A_W^e) can therefore be determined by linearly plotting $\frac{K_e}{e^{-\Delta h/RT}}$ versus microwave power W [Equation (14)] or $\frac{k_1}{e^{-E_a/RT}}$ versus microwave power W [Equation (15)]:

$$\frac{K_e}{e^{-\Delta h/RT}} = A_0^e + A_W^e \cdot W \quad (14)$$

$$\frac{k_1}{e^{-E_a/RT}} = A_0^1 + A_W^1 \cdot W \quad (15)$$

Figure 6a,b, respectively, represents a linear relationship between $\frac{K_e}{e^{-\Delta h/RT}}$ and W and $\frac{k_1}{e^{-E_a/RT}}$ and W for the equilibrium reaction and forward reaction. The high R^2 value of the regression line (0.951–0.997) signified that the developed model is reliable, and the assumption is correct. The coefficients (A_W^e and A_W^0) and constants (A_0^e and A_0^1) could therefore be determined from the slope and intercept of each regression line. The value of A_W^e and A_0^e for the equilibrium reaction was 4.11 × 10^6 W^{-1} and 3.69 × 10^8, respectively (Table 3). For the forward reaction, the values of A_W^1 and A_0^1 were respectively determined to be 4.4 × 10^{-3} L mol^{-1} min^{-1} W^{-1} and 1.5282 L mol^{-1} min^{-1}.

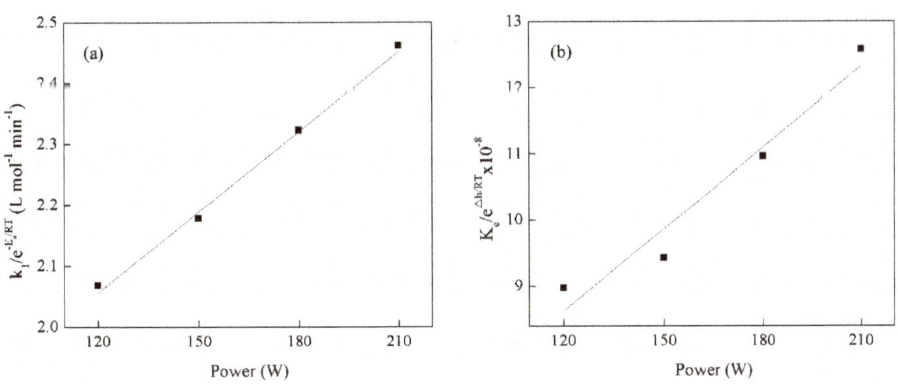

Figure 6. Linear regression of the experimental data for determining microwave power constants and coefficients for forward reaction (**a**) and equilibrium reaction (**b**).

Table 3. Constants and power coefficients for describing the relationship between microwave power and pre-exponential factors.

Forward Reaction			Equilibrium Reaction		
A_0^1 (L mol^{-1} min^{-1})	A_W^1 (L mol^{-1} min^{-1} W^{-1})	R^2	A_0^e	A_W^e (W^{-1})	R^2
1.5282	4.40×10^{-3}	0.997	3.69×10^8	4.11×10^6	0.951

Linear relationship of pre-exponential factors to microwave power: $A_1 = A_0^1 + A_W^1 \cdot W$; $A_e = A_0^e + A_W^e \cdot W$.

3.6. Comparison with Other Works

Table 4 illustrates the comparison of reaction conditions for oleic acid esterification using various processes. The esterification was mainly conducted using a liquid acid catalyst (e.g., H_2SO_4) [16]. This process provides a high conversion yield (99.9%), but it severely corrodes equipment, requires a complicated process for catalyst removal from the reaction solution, and pollutes the environment [16,17]. To avoid the complications associated with homogeneous catalysts, Vieira et al. [20] used HZSM-5 as a solid catalyst for the reaction. Nevertheless, the heterogeneous catalyst demonstrated low catalytic activity, thus requiring large quantities of catalyst (20%) and a long reaction time (7 h), and lowering reaction conversion (80%) [20]. Enzymatic esterification is proposed as a substitute for chemically catalyzed esterification to facilitate environmentally friendly biodiesel production. Nguyen et al. [24] successfully esterified oleic acid with methanol using Eversa Transform lipase for producing biodiesel. Although this process showed high reaction conversion (96.73%), the high cost of enzymes limits its industrial application. Tsai et al. [29] developed another green process called the noncatalytic process using supercritical methanol for oleic acid esterification to eliminate the problems associated with both enzyme- and chemical-catalyzed methods. This efficiently converted oleic acid to esters (97%) within a short reaction time (15 min), but the reaction proceeded at a high temperature (593 K) and high pressure (25 MPa) [29]; it thus required expensive reactors and extensive safety precautions. Melo–Júnior et al. [58] proposed a noncatalytic esterification of fatty acid using microwave irradiation. However, the reaction conversion obtained in their study was low (only 35%), because microwave power was not maintained in their study, leading to lack of microwave-induced nonthermal effect for promotion of the reaction rate [59]. Therefore, the microwave irradiation used in their study is similar to conventional heating. In this study, we proposed another noncatalytic method using microwave irradiation for the oleic acid esterification with ethanol under constant microwave powers to enhance the reaction conversion. The result was a high reaction conversion of 97.62%. The use of microwave irradiation for esterification was more efficient and rapid than conventional heating-based esterification, reducing energy consumption and reaction time. In the reaction mixture, oleic acid and ethanol exist under molecular cage with hydrogen bonding [60]. With the application of microwaves, oleic acid and ethanol absorb the energy from the electromagnetic field, which consequently induces the oscillation of reaction molecule and hydrogen bond variation between ethanol and oleic acid [59]. Resonance is subsequently generated and causes the hydrogen bonds breakage and formation of free small molecules due to reactant molecules escaping from the molecular cage. Consequently, the number of activation molecules is promoted, enhancing the reaction rate [60]. This work signifies that microwave-assisted noncatalytic esterification can be an ecofriendly and efficient process to produce biodiesel.

Table 4. Oleic acid esterification using different methods for producing biodiesel.

Catalyst	Catalyst Loading (%, *w/w*)	Conversion (%)	References
H_2SO_4	1	99.9	[16]
HZSM-5	20	80	[20]
Eversa Transform lipase	11.98	96.73	[24]
Catalyst free (supercritical methanol)	-	97.0	[29]
Catalyst free (microwave irradiation)	-	35	[58]
Catalyst free (traditional heating)	-	67.13	This study
Catalyst free (microwave irradiation)	-	97.62	This study

The potential applications of microwave approach in the large-scale biodiesel production have been widely discussed in literatures [61,62]. Although microwave is superior to conventional heating in the terms of energy consumption and reaction time, the use of microwave for industrial application has several limitations regarding to the control of temperature and safety of the pressurized vessel [61,62]. To address these concerns, the design of microwave reactor is crucial. Studies have showed that the design of microwave reactor for enhancing the temperature and pressure monitoring, safety features, and cooling features is possible [61]. In addition, microwave-based process has been successfully implemented in many industries such as polymers processing, minerals processing, and powder processing [61]. Therefore, the microwave process has a potential application for industrial biodiesel production. However, microwave process has not been used commercially for biodiesel production yet. Further studies are consequently required to evaluate the application of microwave in large-scale biodiesel production.

4. Conclusions

This study developed the noncatalytic oleic acid esterification with ethanol using microwave irradiation for biodiesel production. The microwave-mediated esterification was more efficient than conventional-heating-based esterification. The oleic acid conversion decreased with increasing molar ratio of ethanol to oleic acid but increased by enhancing the temperature, reaction time, and microwave power levels. The high R^2 values of the second-order kinetic model were established in describing the esterification. In addition, a linear relation of microwave power to pre-exponential factors was observed, in which microwave power positively influenced the reaction. This work suggested that microwave-assisted noncatalytic esterification is an ecofriendly and energy-efficient approach for biodiesel synthesis.

Supplementary Materials: The following are available online at http://www.mdpi.com/1996-1073/13/9/2167/s1, Figure S1: Kinetic model for calculating forward reaction rate constant of esterification of oleic acid with ethanol at different reactant molar ratios, Figure S2: Kinetic model for calculating forward reaction rate constant of esterification of oleic acid with ethanol at different temperatures, Figure S3: Kinetic model for calculating forward reaction rate constant of esterification of oleic acid with ethanol at different microwave power levels, Table S1: Forward reaction rate constant k_1 and equilibrium rate constant K_e for esterification using conventional heating.

Author Contributions: Conceptualization, H.C.N., H.C.O., and C.-H.S.; methodology, H.C.N., H.C.O., and C.-H.S.; validation, F.-M.W., K.K.D., and H.-Y.J.; formal analysis, H.C.N., F.-M.W., and K.K.D.; investigation, F.-M.W., K.K.D., T.T.P., H.-Y.J., and N.P.N.; resources, C.H.-S. and H.C.O.; writing—original draft preparation, H.C.N., H.C.O., and C.-H.S.; writing—review and editing, C.-H.S. and H.C.O.; supervision, C.-H.S. and H.C.O.; project administration, C.-H.S.; funding acquisition, C.-H.S. and H.C.O. All authors have read and agreed to the published version of the manuscript.

Funding: This research was funded by the Ministry of Science and Technology of Taiwan, R.O.C.

Conflicts of Interest: The authors declare no conflict of interest.

References

1. Nguyen, H.C.; Liang, S.-H.; Chen, S.-S.; Su, C.-H.; Lin, J.-H.; Chien, C.-C. Enzymatic production of biodiesel from insect fat using methyl acetate as an acyl acceptor: Optimization by using response surface methodology. *Energy Convers. Manag.* **2018**, *158*, 168–175. [CrossRef]

2. Santos, S.; Nobre, L.; Gomes, J.; Puna, J.; Quinta-Ferreira, R.; Bordado, J. Soybean oil transesterification for biodiesel production with micro-structured calcium oxide (CaO) from natural waste materials as a heterogeneous catalyst. *Energies* **2019**, *12*, 4670. [CrossRef]
3. Choi, K.; Park, S.; Roh, H.G.; Lee, C.S. Combustion and emission reduction characteristics of GTL-biodiesel fuel in a single-cylinder diesel engine. *Energies* **2019**, *12*, 2201. [CrossRef]
4. Mahlia, T.M.I.; Syazmi, Z.A.H.S.; Mofijur, M.; Abas, A.E.P.; Bilad, M.R.; Ong, H.C.; Silitonga, A.S. Patent landscape review on biodiesel production: Technology updates. *Renew. Sustain. Energy Rev.* **2020**, *118*, 109526. [CrossRef]
5. Suresh, M.; Jawahar, C.; Richard, A. A review on biodiesel production, combustion, performance, and emission characteristics of non-edible oils in variable compression ratio diesel engine using biodiesel and its blends. *Renew. Sustain. Energy Rev.* **2018**, *92*, 38–49. [CrossRef]
6. Ong, H.C.; Masjuki, H.H.; Mahlia, T.M.I.; Silitonga, A.S.; Chong, W.T.; Yusaf, T. Engine performance and emissions using *Jatropha curcas*, *Ceiba pentandra* and *Calophyllum inophyllum* biodiesel in a CI diesel engine. *Energy* **2014**, *69*, 427–445. [CrossRef]
7. Selvaraj, R.; Praveenkumar, R.; Moorthy, I.G. A comprehensive review of biodiesel production methods from various feedstocks. *Biofuels* **2019**, *10*, 325–333. [CrossRef]
8. Almasi, S.; Ghobadian, B.; Najafi, G.H.; Yusaf, T.; Dehghani Soufi, M.; Hoseini, S.S. Optimization of an ultrasonic-assisted biodiesel production process from one genotype of rapeseed (TERI (OE) R-983) as a novel feedstock using response surface methodology. *Energies* **2019**, *12*, 2656. [CrossRef]
9. Nguyen, H.C.; Liang, S.-H.; Doan, T.T.; Su, C.-H.; Yang, P.-C. Lipase-catalyzed synthesis of biodiesel from black soldier fly (*Hermetica illucens*): Optimization by using response surface methodology. *Energy Convers. Manag.* **2017**, *145*, 335–342. [CrossRef]
10. Khan, I.U.; Yan, Z.; Chen, J. Optimization, transesterification and analytical study of *Rhus typhina* non-edible seed oil as biodiesel production. *Energies* **2019**, *12*, 4290. [CrossRef]
11. Hsiao, M.-C.; Chang, L.-W.; Hou, S.-S. Study of solid calcium diglyceroxide for biodiesel production from waste cooking oil using a high speed homogenizer. *Energies* **2019**, *12*, 3205. [CrossRef]
12. Su, C.-H.; Nguyen, H.C.; Pham, U.K.; Nguyen, M.L.; Juan, H.-Y. Biodiesel production from a novel nonedible feedstock, soursop (*Annona muricata* L.) seed oil. *Energies* **2018**, *11*, 2562. [CrossRef]
13. Silitonga, A.S.; Masjuki, H.H.; Mahlia, T.M.I.; Ong, H.C.; Chong, W.T.; Boosroh, M.H. Overview properties of biodiesel diesel blends from edible and non-edible feedstock. *Renew. Sustain. Energy Rev.* **2013**, *22*, 346–360. [CrossRef]
14. Li, Q.; Zheng, L.; Cai, H.; Garza, E.; Yu, Z.; Zhou, S. From organic waste to biodiesel: Black soldier fly, *Hermetia illucens*, makes it feasible. *Fuel* **2011**, *90*, 1545–1548. [CrossRef]
15. Ong, H.C.; Milano, J.; Silitonga, A.S.; Hassan, M.H.; Shamsuddin, A.H.; Wang, C.-T.; Indra Mahlia, T.M.; Siswantoro, J.; Kusumo, F.; Sutrisno, J. Biodiesel production from *Calophyllum inophyllum-Ceiba pentandra* oil mixture: Optimization and characterization. *J. Clean. Prod.* **2019**, *219*, 183–198. [CrossRef]
16. Lucena, I.L.; Saboya, R.M.; Oliveira, J.F.; Rodrigues, M.L.; Torres, A.E.; Cavalcante, C.L., Jr.; Parente, E.J., Jr.; Silva, G.F.; Fernandes, F.A. Oleic acid esterification with ethanol under continuous water removal conditions. *Fuel* **2011**, *90*, 902–904. [CrossRef]
17. Mardhiah, H.H.; Ong, H.C.; Masjuki, H.; Lim, S.; Lee, H. A review on latest developments and future prospects of heterogeneous catalyst in biodiesel production from non-edible oils. *Renew. Sustain. Energy Rev.* **2017**, *67*, 1225–1236. [CrossRef]
18. Melero, J.A.; Bautista, L.F.; Morales, G.; Iglesias, J.; Sánchez-Vázquez, R. Biodiesel production from crude palm oil using sulfonic acid-modified mesostructured catalysts. *Chem. Eng. J.* **2010**, *161*, 323–331. [CrossRef]
19. Costa, A.A.; Braga, P.R.; de Macedo, J.L.; Dias, J.A.; Dias, S.C. Structural effects of WO_3 incorporation on USY zeolite and application to free fatty acids esterification. *Micropor. Mesopor. Mat.* **2012**, *147*, 142–148. [CrossRef]
20. Vieira, S.S.; Magriotis, Z.M.; Santos, N.A.; Saczk, A.A.; Hori, C.E.; Arroyo, P.A. Biodiesel production by free fatty acid esterification using lanthanum (La^{3+}) and HZSM-5 based catalysts. *Bioresour. Technol.* **2013**, *133*, 248–255. [CrossRef]
21. Álvarez-Mateos, P.; García-Martín, J.; Guerrero-Vacas, F.; Naranjo-Calderón, C.; Barrios-Sánchez, C.; Pérez Camino, M.D.C. Valorization of a high-acidity residual oil generated in the waste cooking oils recycling industries. *Grasas y Aceites* **2019**, *70*, e335. [CrossRef]

22. Fan, M.; Si, Z.; Sun, W.; Zhang, P. Sulfonated ZrO_2-TiO_2 nanorods as efficient solid acid catalysts for heterogeneous esterification of palmitic acid. *Fuel* **2019**, *252*, 254–261. [CrossRef]
23. Selvakumar, P.; Sivashanmugam, P. Ultrasound assisted oleaginous yeast lipid extraction and garbage lipase catalyzed transesterification for enhanced biodiesel production. *Energy Convers. Manag.* **2019**, *179*, 141–151. [CrossRef]
24. Nguyen, H.C.; Huong, D.T.M.; Juan, H.-Y.; Su, C.-H.; Chien, C.-C. Liquid lipase-catalyzed esterification of oleic acid with methanol for biodiesel production in the presence of superabsorbent polymer: Optimization by using response surface methodology. *Energies* **2018**, *11*, 1085. [CrossRef]
25. Amini, Z.; Ilham, Z.; Ong, H.C.; Mazaheri, H.; Chen, W.-H. State of the art and prospective of lipase-catalyzed transesterification reaction for biodiesel production. *Energy Convers. Manag.* **2017**, *141*, 339–353. [CrossRef]
26. Budžaki, S.; Miljić, G.; Sundaram, S.; Tišma, M.; Hessel, V. Cost analysis of enzymatic biodiesel production in small-scaled packed-bed reactors. *Appl. Energy* **2018**, *210*, 268–278. [CrossRef]
27. Rani, K.N.P.; Neeharika, T.S.V.R.; Kumar, T.P.; Satyavathi, B.; Sailu, C. Kinetics of non-catalytic esterification of free fatty acids present in Jatropha oil. *J. Oleo Sci.* **2016**, *65*, 441–445. [CrossRef]
28. Alenezi, R.; Leeke, G.; Winterbottom, J.; Santos, R.; Khan, A. Esterification kinetics of free fatty acids with supercritical methanol for biodiesel production. *Energy Convers. Manag.* **2010**, *51*, 1055–1059. [CrossRef]
29. Tsai, Y.-T.; Lin, H.-M.; Lee, M.-J. Biodiesel production with continuous supercritical process: Non-catalytic transesterification and esterification with or without carbon dioxide. *Bioresour. Technol.* **2013**, *145*, 362–369. [CrossRef]
30. Cho, H.J.; Kim, S.H.; Hong, S.W.; Yeo, Y.-K. A single step non-catalytic esterification of palm fatty acid distillate (PFAD) for biodiesel production. *Fuel* **2012**, *93*, 373–380. [CrossRef]
31. Thakkar, K.; Shah, K.; Kodgire, P.; Kachhwaha, S.S. In-situ reactive extraction of castor seeds for biodiesel production using the coordinated ultrasound–microwave irradiation: Process optimization and kinetic modeling. *Ultrason. Sonochem.* **2019**, *50*, 6–14. [CrossRef] [PubMed]
32. Martinez-Guerra, E.; Howlader, M.S.; Shields-Menard, S.; French, W.T.; Gude, V.G. Optimization of wet microalgal FAME production from *Nannochloropsis* sp. under the synergistic microwave and ultrasound effect. *Int. J. Energy Res.* **2018**, *42*, 1934–1949. [CrossRef]
33. Encinar, J.; González, J.; Martínez, G.; Sánchez, N.; Pardal, A. Soybean oil transesterification by the use of a microwave flow system. *Fuel* **2012**, *95*, 386–393. [CrossRef]
34. Teng, W.K.; Ngoh, G.C.; Yusoff, R.; Aroua, M.K. Microwave-assisted transesterification of industrial grade crude glycerol for the production of glycerol carbonate. *Chem. Eng. J.* **2016**, *284*, 469–477. [CrossRef]
35. Silitonga, A.S.; Shamsuddin, A.H.; Mahlia, T.M.I.; Milano, J.; Kusumo, F.; Siswantoro, J.; Dharma, S.; Sebayang, A.H.; Masjuki, H.H.; Ong, H.C. Biodiesel synthesis from *Ceiba pentandra* oil by microwave irradiation-assisted transesterification: ELM modeling and optimization. *Renew. Energy* **2020**, *146*, 1278–1291. [CrossRef]
36. Milano, J.; Ong, H.C.; Masjuki, H.H.; Silitonga, A.S.; Chen, W.-H.; Kusumo, F.; Dharma, S.; Sebayang, A.H. Optimization of biodiesel production by microwave irradiation-assisted transesterification for waste cooking oil-*Calophyllum inophyllum* oil via response surface methodology. *Energy Convers. Manag.* **2018**, *158*, 400–415. [CrossRef]
37. Cui, Y.; Liang, Y. Direct transesterification of wet *Cryptococcus curvatus* cells to biodiesel through use of microwave irradiation. *Appl. Energy* **2014**, *119*, 438–444. [CrossRef]
38. Ye, W.; Gao, Y.; Ding, H.; Liu, M.; Liu, S.; Han, X.; Qi, J. Kinetics of transesterification of palm oil under conventional heating and microwave irradiation, using CaO as heterogeneous catalyst. *Fuel* **2016**, *180*, 574–579. [CrossRef]
39. Teo, C.L.; Idris, A. Evaluation of direct transesterification of microalgae using microwave irradiation. *Bioresour. Technol.* **2014**, *174*, 281–286. [CrossRef]
40. Ye, J.; Zhu, H.; Yang, Y.; Huang, K.; Raghavan, G.V. Dynamic analysis of a continuous-flow microwave-assisted screw propeller system for biodiesel production. *Chem. Eng. Sci.* **2019**, *202*, 146–156. [CrossRef]
41. Nguyen, H.C.; Lee, H.-Y.; Su, C.-H.; Shih, W.-J.; Chien, C.-C. Green process for fatty acid production from soybean oil through microwave-mediated autocatalytic synthesis. *Chem. Eng. Process.* **2020**, *147*, 107782. [CrossRef]
42. Nguyen, H.C.; Huang, K.-C.; Su, C.-H. Green process for the preparation of phytosterol esters: Microwave-mediated noncatalytic synthesis. *Chem. Eng. J.* **2020**, *382*, 122796. [CrossRef]

43. Nguyen, H.C.; Ong, H.C.; Pham, T.T.T.; Dinh, T.K.K.; Su, C.H. Microwave-mediated noncatalytic synthesis of ethyl levulinate: A green process for fuel additive production. *Int. J. Energy Res.* **2019**, *44*, 1698–1708. [CrossRef]
44. Knothe, G. Dependence of biodiesel fuel properties on the structure of fatty acid alkyl esters. *Fuel Proces. Technol.* **2005**, *86*, 1059–1070. [CrossRef]
45. Li, Q.; Xu, J.; Du, W.; Li, Y.; Liu, D. Ethanol as the acyl acceptor for biodiesel production. *Renew. Sustain. Energy Rev.* **2013**, *25*, 742–748. [CrossRef]
46. Pan, Y.; Alam, M.A.; Wang, Z.; Wu, J.; Zhang, Y.; Yuan, Z. Enhanced esterification of oleic acid and methanol by deep eutectic solvent assisted Amberlyst heterogeneous catalyst. *Bioresour. Technol.* **2016**, *220*, 543–548. [CrossRef] [PubMed]
47. Su, C.-H. Kinetic study of free fatty acid esterification reaction catalyzed by recoverable and reusable hydrochloric acid. *Bioresour. Technol.* **2013**, *130*, 522–528. [CrossRef] [PubMed]
48. Warrand, J.; Janssen, H.-G. Controlled production of oligosaccharides from amylose by acid-hydrolysis under microwave treatment: Comparison with conventional heating. *Carbohydr. Polym.* **2007**, *69*, 353–362. [CrossRef]
49. Aguilar-Reynosa, A.; Romaní, A.; Rodríguez-Jasso, R.M.; Aguilar, C.N.; Garrote, G.; Ruiz, H.A. Comparison of microwave and conduction-convection heating autohydrolysis pretreatment for bioethanol production. *Bioresour. Technol.* **2017**, *243*, 273–283. [CrossRef]
50. Loong, T.C.; Idris, A. Rapid alkali catalyzed transesterification of microalgae lipids to biodiesel using simultaneous cooling and microwave heating and its optimization. *Bioresour. Technol.* **2014**, *174*, 311–315. [CrossRef]
51. Huang, Y.-F.; Chiueh, P.-T.; Kuan, W.-H.; Lo, S.-L. Microwave pyrolysis of lignocellulosic biomass: Heating performance and reaction kinetics. *Energy* **2016**, *100*, 137–144. [CrossRef]
52. Gole, V.L.; Gogate, P.R. Intensification of glycerolysis reaction of higher free fatty acid containing sustainable feedstock using microwave irradiation. *Fuel Process. Technol.* **2014**, *118*, 110–116. [CrossRef]
53. Zanuttini, M.S.; Pisarello, M.; Querini, C.A. *Butia Yatay* coconut oil: Process development for biodiesel production and kinetics of esterification with ethanol. *Energy Convers. Manag.* **2014**, *85*, 407–416. [CrossRef]
54. Minami, E.; Saka, S. Kinetics of hydrolysis and methyl esterification for biodiesel production in two-step supercritical methanol process. *Fuel* **2006**, *85*, 2479–2483. [CrossRef]
55. Ramadhas, A.S.; Jayaraj, S.; Muraleedharan, C. Biodiesel production from high FFA rubber seed oil. *Fuel* **2005**, *84*, 335–340. [CrossRef]
56. Singh, V.; Sharma, Y.C. Low cost guinea fowl bone derived recyclable heterogeneous catalyst for microwave assisted transesterification of *Annona squamosa* L. seed oil. *Energy Convers. Manag.* **2017**, *138*, 627–637. [CrossRef]
57. Su, C.H.; Fu, C.C.; Gomes, J.; Chu, I.; Wu, W.T. A heterogeneous acid-catalyzed process for biodiesel production from enzyme hydrolyzed fatty acids. *AIChE J.* **2008**, *54*, 327–336. [CrossRef]
58. Melo-Júnior, C.A.; Albuquerque, C.E.; Fortuny, M.; Dariva, C.; Egues, S.; Santos, A.F.; Ramos, A.L. Use of microwave irradiation in the noncatalytic esterification of C18 fatty acids. *Energy Fuels* **2009**, *23*, 580–585. [CrossRef]
59. De la Hoz, A.; Diaz-Ortiz, A.; Moreno, A. Microwaves in organic synthesis. Thermal and non-thermal microwave effects. *Chem. Soc. Rev.* **2005**, *34*, 164–178. [CrossRef]
60. Lin, Z.-R.; Zeng, X.-A.; Yu, S.-J.; Sun, D.-W. Enhancement of ethanol–acetic acid esterification under room temperature and non-catalytic condition via pulsed electric field application. *Food Bioprocess Technol.* **2012**, *5*, 2637–2645. [CrossRef]

61. Gude, V.G.; Patil, P.; Martinez-Guerra, E.; Deng, S.; Nirmalakhandan, N. Microwave energy potential for biodiesel production. *Sustain. Chem. Process.* **2013**, *1*, 5. [CrossRef]
62. Mutyala, S.; Fairbridge, C.; Paré, J.J.; Bélanger, J.M.; Ng, S.; Hawkins, R. Microwave applications to oil sands and petroleum: A review. *Fuel Process. Technol.* **2010**, *91*, 127–135. [CrossRef]

energies

MDPI

Article

Experimental Study on the Performance of an SI Engine Fueled by Waste Plastic Pyrolysis Oil–Gasoline Blends

Khairil [1,*], Teuku Meurah Indra Riayatsyah [1], Samsul Bahri [1], Sarwo Edhy Sofyan [1], Jalaluddin Jalaluddin [1], Fitranto Kusumo [2], Arridina Susan Silitonga [2,3], Yanti Padli [4], Muhammad Jihad [1] and Abd Halim Shamsuddin [2]

1 Department of Mechanical Engineering, Universitas Syiah Kuala, Banda Aceh 23111, Indonesia; indraayat@unsyiah.ac.id (T.M.I.R.); samsul.myd@unsyiah.ac.id (S.B.); sarwo.edhy@unsyiah.ac.id (S.E.S.); jalaluddinyunus@unsyiah.ac.id (J.J.); muhammadjihad1994@yahoo.com (M.J.)

2 College Institute of Sustainable Energy, Universiti Tenaga Nasional, Kajang 43000, Malaysia; fitrantokusumo@yahoo.co.id (F.K.); ardinsu@yahoo.co.id (A.S.S.); abdhalim@uniten.edu.my (A.H.S.)

3 Department of Mechanical Engineering, Politeknik Negeri Medan, Medan 20155, Indonesia

4 School of Chemical and Biomolecular Engineering, University of Sydney, Sydney 2006, NSW, Australia; yantipadli@yahoo.com

* Correspondence: khairil@unsyiah.ac.id

† This paper is an extended version of our paper published in International Conference on Sustainable Energy and Green Technology 2019, Bangkok, Thailand, 11–14 December 2019, doi:10.1088/1755-1315/463/1/012002.

Received: 27 May 2020; Accepted: 7 August 2020; Published: 14 August 2020

Abstract: Pyrolyzed waste plastic-based green fuel has been reported to be used as an alternate fuel for diesel engines. Some of the main challenges for implementing this in current automotive technology include evaluating engine performance, emission, noise vibration harshness (NVH), and knock characteristics of this fuel. This study focuses on the engine performance of poly-ethylene terephthalate (PET)-based waste plastic oil (WPO) at varying engine speed conditions. The pyrolysis of mixed-waste plastic was carried out at 300 °C in a fixed-bed reactor. Physicochemical properties such as viscosity, density, calorific value, sulfur, and research octane number (RON) of the plastic fuel and its blends with gasoline were analyzed using ASTM standard test methods. The WPO was blended with two different types of gasoline (RON88 and RON90) at 10, 20, and 30%, and was tested in a spark-ignition (SI) engine. The experimental results showed that different WPO–gasoline blends can be used in an SI engine without any engine modifications, and the performance indicators for different blends were found to be close to that of pure gasoline. The brake power and brake specific fuel consumption (BSFC) were found to be 4.1 kW and 0.309 kg/kW h, respectively. The 10% WPO and 90% RON90 blend produced optimal engine performance at 3500 rpm.

Keywords: waste plastic oil (WPO); engine performance; renewable gasoline; pyrolysis

1. Introduction

Presently, increasing world energy demand is causing an increase in the consumption of fossil fuels, which also causes a negative impact on global warming and the overall environment. Moreover, fossil fuel resources continue to deplete every year due to their non-renewable nature [1,2]. This has prompted researchers to investigate alternative fuels both from waste materials and renewable feedstock to reduce dependence on fossil fuel and ensuing air pollution. Some of the renewable energy sources that have been widely explored include biofuel [3], wind [4], solar [5], geothermal [4], and nuclear energy. Each type of energy sources possesses some advantages and drawbacks. For example, solar and wind energy suffer difficulties in energy storage since they are unstable and available only at certain times [6]. This can be addressed by appropriate energy storage material that can save a large amount of energy, such as phase-change materials that can replace current batteries [7]. On the other hand, energy coming from agricultural waste, or tropical biodiversity of non-edible origin, is still one of the best choices in developing countries, such as Indonesia and Malaysia, as they directly address the issue of energy sustainability [8,9].

Plastic has been widely produced and used since the early 20th century. The use of plastic has risen rapidly and it is presently used in almost every economic sector for packaging products, as it is an easy material to procure as well as to form into an end product. At present, the use of plastic materials in western European countries has reached 60 kg/person/year and 80 kg/person/year in the USA, while in India it is only 2 kg/person/year [10]. High plastic waste is related to population and waste management. Around 80% of plastic waste comes from the mainland. In 2019, waste in Indonesia reached 68 million tons, with 14% of this being plastic waste, at 9.52 million tons. With an estimated 0.7 kg/day of garbage produced by each person, the average daily amount of rubbish dumped in metropolitan cities is approximately 1300 tons/day (the population is more than 1 million) [11]. Furthermore, Indonesia is second highest dumper of plastic waste into world's oceans, after China, depositing approximately 1.29 million tons per year, compared to China's 3.53 million tons per year [11,12]. Managing plastic waste has always been a big problem. Besides being non-biodegradable and difficult to manage, plastic waste can also pollute soil. Standard plastic bags generally made from poly-ethylene are not biodegradable. However, photodegradation can occur in plastics. When exposed to ultraviolet rays from sunlight, the structure becomes fragile and breaks into pieces, which takes a long time. Experts estimate that at least 500 to 1000 years are needed for this decomposition to take place [11]. The negative effects of trash in the ocean has resulted in many marine biota experiencing metabolic disorders, irritation of the digestive system, and death resulting from plastic consumption. According to the Australian Coral Reef Research Center (ACRRC), reefs exposed to plastic waste have about an 89% chance of being affected by diseases, compared to just 4% for reefs not affected by plastic waste. Coral reefs most exposed to plastic wastes appear in Indonesia, which is approximately 26 parts per 100 square meters [13]. Waste management with the theme "3R" (Rescue, Reduce, Recycle) has been considered ineffective.

Several types of plastics commonly used as raw materials are poly-ethylene terephthalate (PET), High-Density Poly-Ethylene (HDPE), Polyvinyl Chloride (PVC), Low-Density Poly-Ethylene (LDPE), Poly Propylene (PP), etc. Among these, PET with the chemical formula $(C_{10}H_8O_4)_n$, is widely available at landfills as it is used as a raw material for mineral-water bottles. It is also very difficult to decompose. One of the advantages of PET plastic waste is that it can be recycled into various types of goods that have economic value such as clothing, bags, furniture, and carpet. Another use of PET plastic is to convert it to fuel oil via pyrolysis [14]. It has been reported in the literature that PET has moderate to high conversion compared to other plastic wastes such as LDPE and HDPE [14,15]. PET plastic possesses a high calorific value, which makes the conversion process using pyrolysis very effective [16].

Based on the literature, it is necessary to find a technique that reduces the amount of plastic waste, by converting plastic waste into an alternative fuel. Many researchers have investigated techniques that would reduce plastic waste in a short time, which is to convert plastic waste into fuel oil via pyrolysis [17], specifically by heating the polymer material without or using only a little oxygen [18,19]. Pyrolysis is the process of thermo-chemical decomposition of organic compounds through the heating process without or in the presence of only a small amount of oxygen and other chemical reagents [20]. Pyrolysis thermally decomposes long-chain polymer molecules into less complex molecules [21]. Pyrolysis has been chosen by many researchers due to its ability to produce a high oil yield of up to 80 wt.% at a temperature of about 500 °C. In many industrial applications, pyrolysis is carried out at atmospheric pressure and operating temperatures above 430 °C. Figure 1 shows the chemical reaction of the pyrolysis process. Pyrolysis is divided into three types: hydrocracking, thermal cracking, and catalytic cracking. According to Cleetus et al. [22], PET undergoes pyrolysis that produces different gases i.e., CO_2 and CO, and other miscellaneous hydrocarbons. Furthermore, condensation occurring in the heat-exchange process produces plastic oil with the chemical formula $C_{13.8}H_{23.56}$ [23]. There are three major products produced during pyrolysis. These are oil, gas, and char, which is valuable for industries, especially production and refineries [24–26].

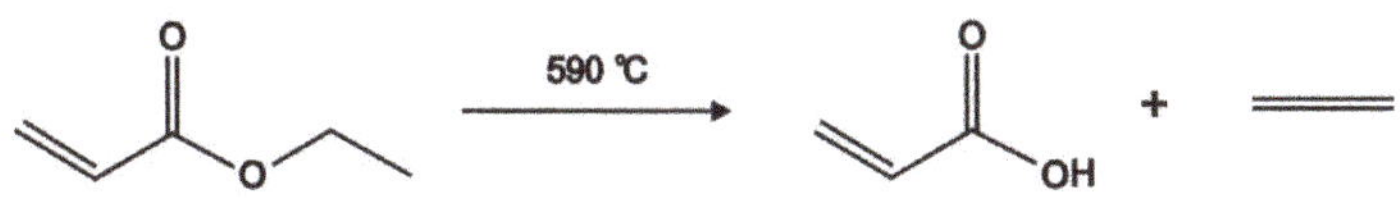

Figure 1. The chemical formula of the pyrolysis process.

Previous research on the pyrolysis of waste plastic oil as has been reported by Ramadhan and Ali [27]. They produced pyrolysis oil from waste LDPE and HDPE at temperatures of 250 °C to 420 °C and found that oil characteristics are similar to that of diesel fuel. Sarker et al. [28] studied the thermal cracking of waste LDPE under temperatures of 150 °C to 420 °C without a catalyst and found that the produced oil properties are similar to those of kerosene.

In Indonesia, the fuel vendorship is distributed by state-owned enterprises (Pertamina) throughout the country. Indonesia has several grades for both diesel and gasoline. Gasoline and diesel fuel standards were updated in 2013. Government regulations stipulate that gasoline fuel should maintain the Euro 2 sulfur limit for RON 88, eliminating lead in the fuel. For diesel fuel, the minimum limit for a cetane number of 48 is mandated in government regulations. There are four types of gasoline grades sold in Indonesia, namely Premium (RON88), Pertalite (RON90), Pertamax (RON91), and Pertamax Plus (RON95) [29]. The Ministry of Energy has since 2008 mandated a minimum limit of biofuel volume to be blended into gasoline and diesel fuels [3,30]. This regulation has undergone several updates. The most recent one, released in 2015, set the bioethanol blend in gasoline to be 5% in 2016 and 10% in 2020. In addition, the biodiesel percentage in diesel for the transportation sector was set to be 20% in 2016 and 30% in 2020.

Previous studies have investigated the use of plastic oil as fuel in gasoline engines. One of them was done by Khan et al. Related to the characteristics of WPO–diesel blends, the study obtained a maximum yield of 77.03% within 2 h. Cleetus et al. [22] studied engine performance using WPO–gasoline blends. However, the type of engine used had a low engine speed of about 1500 rpm. It is important to mention that the effect of WPO–gasoline blends on SI engine performance parameters such as engine power, specific fuel consumption (SFC), and thermal efficiency have not been elucidated properly in previous studies. Thus, the objective of this study is to examine the engine performance of two low-RON gasoline (RON 88 and RON90) blended with WPO at varying engine speeds. In doing so, this study also examines the suitability of gasoline–WPO blend fuel for an SI engine as a potential renewable fuel source for future energy supply.

2. Materials and Methods

2.1. Materials

Plastic waste was collected from the final disposal site on the coast of Banda Aceh, Indonesia. Then, plastic waste was washed several times and dried under sunlight. After that, the waste was cut into small pieces. The experiments were carried out in the Laboratory of Fuel Engines and Propulsion Systems, Department of Mechanical Engineering, Faculty of Engineering, Syiah Kuala University.

2.2. Production of WPO via Pyrolysis

Before the beginning of pyrolysis process, it was ensured that the raw materials were washed thoroughly, cut into pieces, and dried under sunlight. The PET pieces were then weighed up to 1 kg. Figure 2 shows a schematic diagram of the pyrolysis process for plastic waste. The main component consists of a reactor chamber and condenser. The condenser was connected to a reactor chamber by an iron pipe exchanger submerged in cold water so that the condensation process ran smoothly. Digital thermocouples Type K, Length 30cm and diameter 5 mm was directly connected to the reactor chamber to monitor the temperature of the plastic waste gas from the pyrolysis process. The preparation was finalized by installing an iron pipe with heat insulation cladding between the reactor and condenser. The pyrolysis process used Liquefied Petroleum Gas (LPG) as a provider of thermal energy. The produced gas flowed into the reservoir through the heat exchanger pipe using a condensation process as a liquid oil. The pyrolysis process began by turning on a reactor that was set at 300 °C in a vacuum condition. At the same time, the condenser cooler was supplied. The temperature of the reactor was maintained at 300 °C [31] for 2 h 30 min. As the product gases flowed into the condenser, some compounds were condensed. The parts that did not condense remained in the gas phase. The pyrolyzed oil was collected and weighed for analysis. The oil from the plastic waste can be seen in Figure 3. The oil yield from waste plastic through the pyrolysis process at a temperature of 300 °C produced a 53.8% (*w*/*w*) yield of WPO. Sample testing was carried out to determine the chemical content of WPO. The yield, i.e., conversion of plastic to oil, was determined using the following equation [32]:

$$\text{WPO yield (wt, \%)} = \frac{\text{Mass of oil}}{\text{Mass of PET}} \times 100\% \quad (1)$$

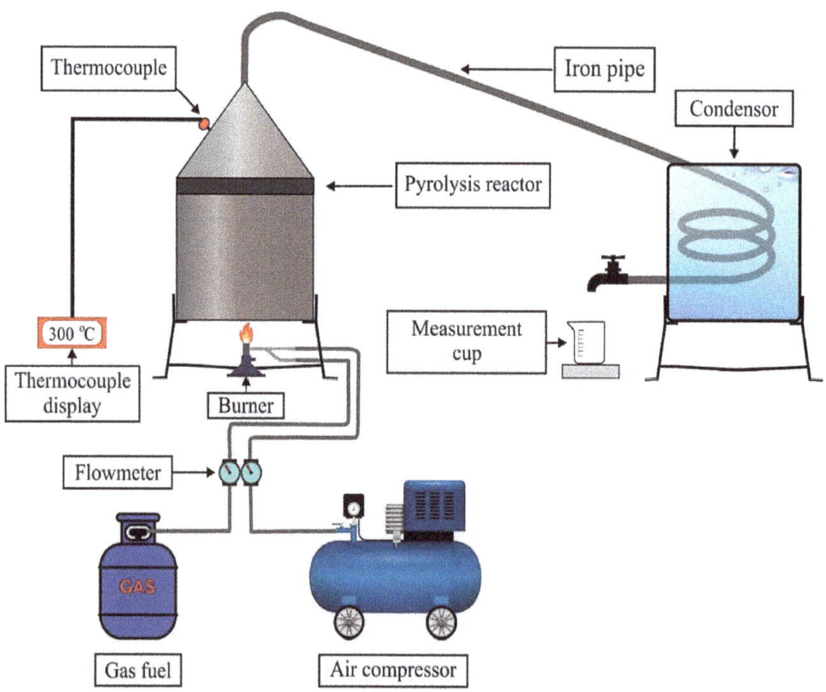

Figure 2. Schematic diagram for converting plastic waste to oil via pyrolysis.

Figure 3. Pictorial view of produced WPO.

2.3. Fuel Preparation

The fuel compositions of WPO–gasoline blends (RON 88 and RON 90) for the three mixture ratios used in the experiment are shown in Table 1. Three mixtures of WPO and gasoline (RON 88), and three mixtures of WPO and gasoline (RON 90), with different percentages of WPO, specifically 10%, 20%, and 30%, were used in the experiment. WPO–gasoline mixtures (RON 88 and RON 90) were placed in a 1 L beaker, then blending was carried out using a magnetic stirrer for 10 min at a speed of 300 rpm at room temperature. The resultant blends were stored in different bottles, separately. To obtain the physicochemical properties of the oil mixture of WPO–gasoline blends, such as viscosity, density, and calorific value, the equation of mixture-oil properties was used as follows:

$$CV_{blend} = X1 \times CV_{WPO} + X2 \times CV_{RON} \quad (2)$$

where:

CV_{blend} = Calorific value of fuel blends; X1 = Percentages volume of WPO; X2 = Percentages volume of pure gasoline; CV_{WPO} = Calorific value of WPO; CV_{RON} = Calorific value of pure gasoline.

Table 1. Preparations of fuel samples of waste oil plastic blend with gasoline (RON88 and RON90).

Fuel Type	WPO (%)	Gasoline Type		Volume
		RON88 (%)	RON90 (%)	
WPO-RON88-10	10	90	-	1 L
WPO-RON88-20	20	80	-	1 L
WPO-RON88-30	30	70	-	1 L
RON88-100	0	100		1 L
WPO-RON90-10	10	-	90	1 L
WPO-RON90-20	20	-	80	1 L
WPO-RON90-30	30	-	70	1 L
RON90-100	0		100	1 L

2.4. Engine Performance Test Procedure

A four-stroke 163 cc overhead valve OHV single-cylinder, horizontal-shaft, water-cooled SI engine was used to perform the tests. The specifications of the engine are shown in Table 2. The engine is mated to a generator with maximum load 3 kW and operated at 220 V–50 Hz and connected to 6 (six) lamps (500 watts each) that used a load bank to control the load of the engine during the performance test. The function of the generator is to supply the voltage and electric current to power the 3000-watt six-lamp configuration when the engine consumes the fuel at varying speeds. A schematic diagram of the engine experimental setup is given in Figure 3. The engine test was conducted under speed variation from 1500 to 3000 rpm with 500 rpm intervals for each type of fuel. A tachometer, voltmeter, and clamp meter was used to determine the engine rotation, the voltage input, and the electrical current of engine load, respectively. The experiment was repeated three times, and the mean value was calculated for each sample. Figure 4 shows a schematic diagram of the engine experimental setup.

Table 2. The specification of the engine.

Brand	Specification (Honda)
Model	GP160H-SD1
Power	5.5 HP
Engine type	Air-Cooled 4 Stroke OHV Single Cylinder, Horizontal Shaft
Cylinder Volume	163 cc
Bore × Stroke	68 × 45 mm
Compression Ratio	8.5:1
Maximum Torque	10.3 Nm at 2500 rpm
Maximum Output	5.5 HP at 3600 rpm
Output Net	3.6 HP at 3600 rpm

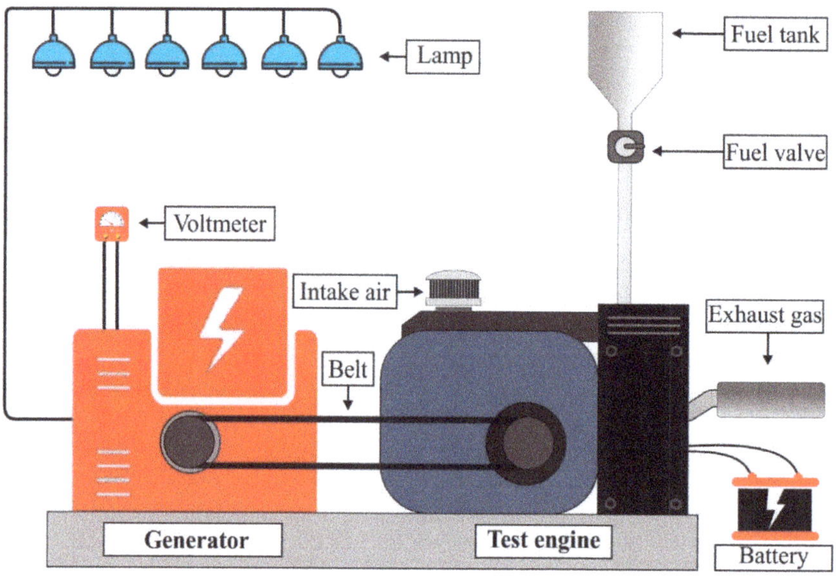

Figure 4. Schematic diagram of engine experimental setup.

2.5. Engine Performance

The gasoline engine performance test included measurement and/or calculation of engine power (P), calculation of SFC, and thermal efficiency (ηth).

2.5.1. Engine Power

The power generated by an engine connected to a single-phase alternating current (AC) generator can be calculated by the equation [33],

$$N_b = \frac{E.\,I.\,pf}{746\ \text{eg}} \tag{3}$$

where:

N_b = Engine Power (HP); E = Voltmeter Reader (Volt); I = Ampere meter Reader (Ampere); pf = Power factor for one single phase = 1; eg = Efficiency of electric generators for engines under 50 kVA = 87–89%, for generators using V belts, the power generated is divided by ηb = 0.96.

2.5.2. Specific Fuel Consumption

The SFC is defined as the amount of fuel consumed by the engine to produce power denoted in kW/h. This SFC of an engine is usually measured by the mass of fuel consumed for the power output. The SFC can be calculated by the following equation [34]:

$$sfc = \frac{mf}{N_b} \quad (4)$$

where:

sfc = Specific fuel consumption (kg/kWh); *mf* = Total fuel consumption (kg/hour); N_b = Engine power (HP).

2.5.3. Thermal Efficiency

Thermal efficiency is defined as the heat-use efficiency of fuel to be converted into mechanical work. The thermal efficiency (ηth) can be calculated using the equation [34]:

$$\eta\text{th} = \frac{N_b \times 632.5}{\text{mf} \times \text{LHV}} \quad (5)$$

where:

ηth = Thermal efficiency (%); N_b = Engine Power (HP); mf = Fuel Consumption (kg/hour); LHV = Fuel Calorific Value (kcal/kg).

3. Results and Discussion

3.1. The Characteristics of Physicochemical Properties of WPO

The characteristic properties of WPO were analyzed and shown in Table 3. The results show that the characteristics of WPO are close to the standards of gasoline fuels that have been set by the Indonesian government. Some of the properties stated in the national standard include viscosity, density, sulfur content, and calorific value. Table 3 shows the measured kinematic viscosity of WPO was 3.5 mm^2/s, which is slightly higher than the reported kinematic viscosity of Ananthakumar et al. [35]. They obtained kinematic viscosity of WPO 2.52 mm^2/s. The density of WPO was 750 Kg/m^3, which is slightly lower than that of gasoline. The sulfur content of WPO was obtained at 0.68% mm. The calorific value of WPO was found to be 45.0196 MJ/kg, which is higher than the calorific value of waste plastic oil of 42.8075 MJ/kg reported by Kumar et al. [36]. Zainuri [37] also reported the calorific value of WPO to be 46.848 MJ/kg.

Table 3. The physicochemical properties of WPO.

Properties	Unit	ASTM D6751 Standard	RON 88 [11]	RON 90 [11,38]	WPO [a]	WPO [b] [36]	Method
Kinematic Viscosity	(mm^2/s)	1.9–6.0	2.0–5.0	2.0–5.0	3.5	2.64	ASTM D4052
Density at 15 °C	(Kg/m^3)	880 max	715–770	715–770	750	830	ASTMD4052 or D1298
Sulfur Content	% mm	Max. 0.05	0.05	0.05	0.68	-	ASTM D664
Calorific Value	MJ/kg	-	47.049	48.20	45.0196	42.8075	ASTM D240
RON	RON (min)	-	88	90	78.8	-	ASTMD2699

[a] Value measured, [b] Vijaya Kumar.

3.2. Engine Performance Fueled by Waste Plastic Oil–Gasoline (RON 88) Blends

Research has been conducted to examine the effect of engine power, SFC, and thermal efficiency on gasoline engines with high-speed motors using waste plastic oil–gasoline (RON 88) blends with different mixture variations.

3.2.1. The Effect of WPO Blends on Brake Power

The engine powers produced using pure gasoline and WPO–gasoline blends are shown in Figure 5. The pure gasoline produced the highest engine power of 1.339 HP. In comparison, the WPO–gasoline blends produced 1.287 HP, 1.244 HP, and 1.229 HP for WPO-RON88-10, WPO-RON88-20, and WPO-RON88-30, respectively. With the increase in engine speed, the brake increased. At 1500 rpm, the pure gasoline fuel gives the highest engine power, which is 1.339 HP, while the WPO–gasoline blends produced 1.287 HP, 1.244 HP, and 1.229 HP for WPO-RON88-10, WPO-RON88-20, and WPO-RON88-30, respectively. At medium speeds of 2000 rpm and 2500 rpm, the engine powers obtained were 1.72 HP and 2.002 HP, respectively, using pure gasoline. On the other hand, when WPO-RON88-30 was used, the engine powers were reduced to 1.392 HP and 1.721 HP, respectively. The RON88 fuel produced the highest engine power of 2.219 HP at 3000 rpm engine speed, while the WPO–gasoline blends obtained 2.19 HP (WPO-RON88-10), 2.05 HP (WPO-RON88-20), and 1.99 HP (WPO-RON88-30). The WPO-RON88-30 blends gave the lowest engine power output. From the figure, it can be seen that the addition of 10% WPO with RON88 does not change the engine power output significantly compared to the pure RON88 result. This can be attributed to a small percentage of low RON and the low calorific value of the WPO in the blend. This is different from the results obtained by Ravinanath et al. [39], which showed an increased power output with increased ethanol content of the blend.

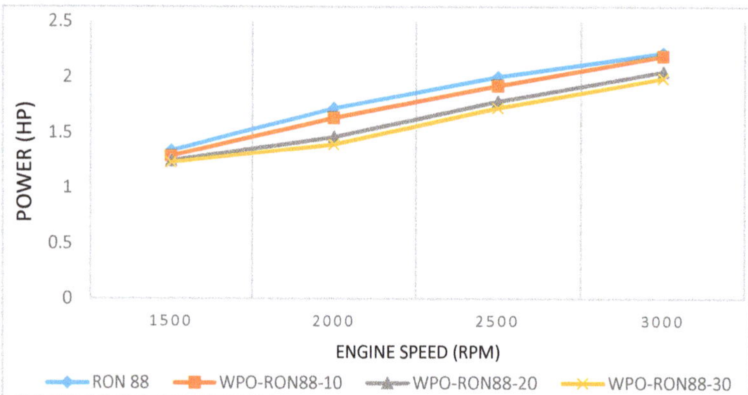

Figure 5. Effect of WPO-RON88 blends on engine power compared to RON88.

3.2.2. The Effect of WPO Blends on SFC

The SFCs of WPO–gasoline blends and pure gasoline are shown in Figure 6. Based on Figure 6, the SFC of WPO-RON88-30 was the highest compared to WPO-10, WPO-20, and RON88. The SFC of WPO-RON88-30 at 1500 rpm was 0.351 kg/kWh compared to 0.3496 kg/kWh, 0.3295 kg/kWh, and 0.3254 kg/kWh for WPO-RON88-10, WPO-RON88-20, and RON88, respectively. Furthermore, the BSFC for WPO-RON88-10 showed identical results to that of RON88 at 1500 rpm and 2500 rpm, which was 0.3295 kg/kWh and 0.404 kg/kWh, respectively. In general, the values of SFC from gasoline–WPO blends were close to gasoline blends, and can be used as a reference fuel for further development with other variations in future research. With the increase in engine speed, the SFC is in line with the existing literature [40]. In addition, with the increase in concentration (%) of WPO in the mixture of RON88-WPO, SFC is increased. This could be due to the lower calorific value of WPO compared to RON88. Cleetus et al. [22] obtained similar results from gasoline–WPO blend when tested in a SI engine.

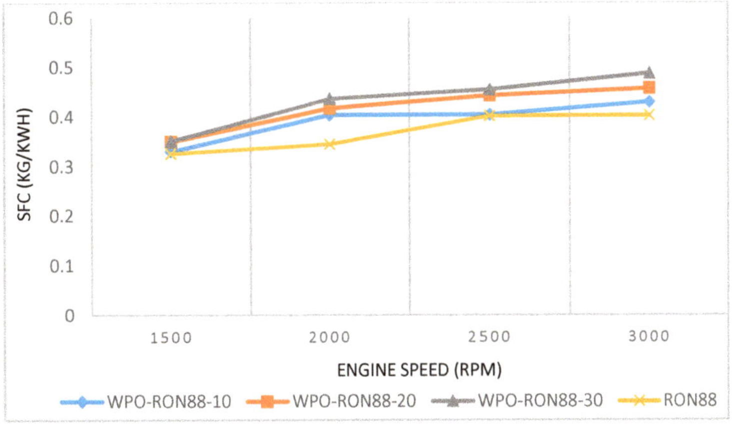

Figure 6. Effect of WPO-RON88 blend on SFC compared to RON88.

3.2.3. The Effect of WPO Blends on Thermal Efficiency

Thermal efficiency is the ratio of work output to energy supplied through fuel injection. The variation of thermal efficiencies using WPO–gasoline blends and pure gasoline is shown in Figure 7. It can be seen that the brake thermal efficiency (BTE) for WPO-RON88-10, WPO-RON88-20, and WPO-RON88-30 were 16.09%, 17.08%, and 17.3%, whereas for RON88, the BTE was about 16.02% at engine speed 1500 rpm. Based on the results of the experiments conducted, the higher the engine speed, the lower the BTE, which can be attributed to increased BSFC at higher engine speeds. It can be seen that the thermal efficiency of content of WPO in WPO–gasoline mixtures was higher than that of pure gasoline, which can be attributed to less frictional power loss than that of lower WPO blends and pure gasoline. Similar results were obtained by Cleetus et al. [22], showing a higher percentage of WPO in WPO–gasoline blends resulting in higher thermal efficiency compared to that of lower WPO containing blends in a CI engine. In addition, higher oxygen content in high WPO blends can result in the presence of more oxygen in the combustion process, therefore achieving better combustion and hence greater thermal efficiency [39].

Figure 7. Effect of WPO-RON88 blend on thermal efficiency (%), compared to RON88.

3.3. *Engine Performance Fueled by WPO–Gasoline (RON 90) Blends*

Research has been conducted to examine the effect of engine power, SFC, and thermal efficiency on gasoline engines with high-speed motor using WPO–gasoline (RON 90) blends with different mixture variations.

3.3.1. The Effect of WPO–Gasoline RON 90 Blends on Engine Power

From Figure 8, it can be seen that engine power increased with increase in engine speed. The rate of increase in power above 3000 rpm was slower compared to other engine speeds, which can be attributed to faster valve operation, reducing the mean effective pressure of the piston in the engine. It is shown in Figure 8 that the power generated from pure RON 90 fuel at 1500 rpm is 1.626 HP, while the WPO-RON90-10, WPO-RON90-20, and WPO-RON90-30 produced powers of 1.573 HP, 1.410 HP, and 1.205 HP, respectively. On the other hand, the engine speeds at 2000 rpm, 2500 rpm, and 3000 rpm for RON 90 produced powers of 2.260 HP, 2.697 HP, and 3.049 HP, respectively. The powers produced by WPO-RON90-10 is close to that of RON 90, which were 2105 HP, 2587 HP, and 2841 HP at 2000 rpm, 2500 rpm, and 3000 rpm, respectively. Furthermore, at 3500 rpm RON90 produced the highest engine power of 3.169 HP. Moreover, the WPO-RON90-30 produced close to the highest powers of pure RON 90, which were 3.078 HP at an engine speed of 3500 rpm, followed by WPO-RON90-10 and WPO-RON90-20, respectively. Ravinanath et al. [39] also reported similar results for higher concentrations (%) of WPO in the blend. As such, blends with higher concentration of WPO was able to approach the engine power output produced by RON90. The WPO-RON90-30 fuel blend produced the highest engine power compared to other blends due to the presence of more fuel-borne oxygen for combustion, therefore producing better combustion.

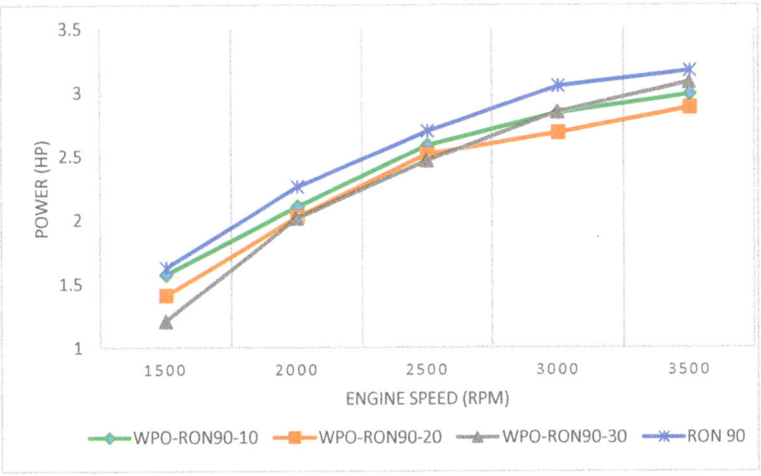

Figure 8. Effect of WPO-RON90 blends on engine power compared to RON90.

3.3.2. The Effect of WPO–Gasoline RON 90 Blends on SFC

The SFC for WPO–gasoline blends as well as pure RON 90 fuels are shown in Figure 9. The SFC increased with an increase in the percentage of WPO in WPO–gasoline blends at the same engine speed. It can be seen in Figure 9 that at an engine speed of 2000 rpm, the SFC of WPO-RON90-10, WPO-RON90-20, WPO-RON90-30, and RON90 were 0.2304 kg/kWh, 0.2665 kg/kWh, 0.2976 kg/kWh, and 0.1388 kg/kWh, respectively. With the increase in engine speed from 2000 rpm to 3500 rpm, an increase in SFC was observed due to higher frictional loss at higher engine speed, as discussed previously. In a previous work, Murni [41] reported that the use of fuels with low kinematic viscosity can increase the amount of fuel being consumed during engine operation. Interestingly, at engine speeds between 1500 rpm to 2000 rpm, the SFC of every fuel is decreased, which is different from the results reported by Cleetus et al. [23]. This can be attributed to the high calorific value and octane number of RON90 fuel.

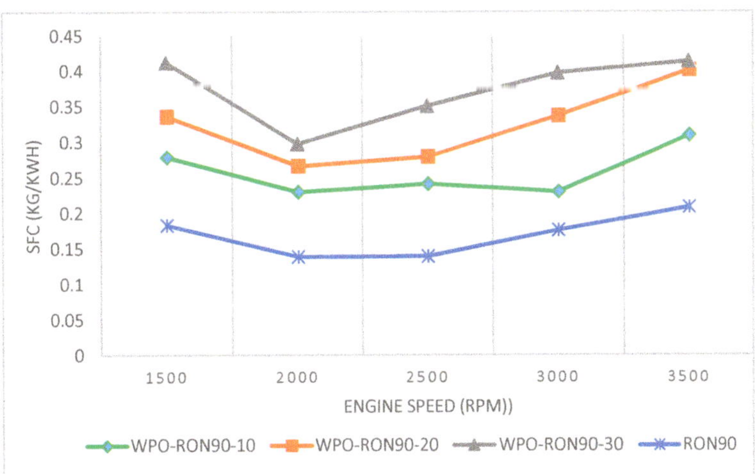

Figure 9. The effect of WPO-RON90 blend on SFC compared to RON90.

3.3.3. The Effect of WPO–Gasoline RON 90 Blends on Thermal Efficiency

Based on experimental results presented in Figure 10, there was a relationship between engine speed and brake thermal efficiency with the use of the different WPO–gasoline blends. Pure RON90 showed higher thermal efficiency compared to WPO blends. The thermal efficiency for WPO-RON90-10, WPO-RON90-20, and WPO-RON90-30 were obtained as 38.65%, 37.5%, and 38.41%, respectively, at an engine speed of 2000 rpm. For the same engine speed, the thermal efficiency of pure RON90 reached 43.05%. The highest brake thermal efficiency for each type of blend was 38.75%, 37.56%, and 38.41% for WPO-RON90-10, for WPO-RON90-20, and WPO-RON90-30, respectively, at 2000 rpm. Overall, the thermal efficiency for all fuels increased at 2000 rpm compared to that of 1500 rpm. However, as the engine speed increased from 2500 rpm to 3500 rpm, the thermal efficiency decreased. The decreased thermal efficiency can be attributed to the lower heating value of WPO. Secondly, after the efficiency increase of the first stage at 1500–2000 rpm engine speed, thermal efficiency continues to decrease for all types of fuel to 3500 rpm engine speed. This is probably due to the high research octane number on the RON90, meaning that fuel can be compressed because of the ability of high-octane fuels to resist auto-ignition. Usually, more power is generated due to increased efficiency and to higher compression.

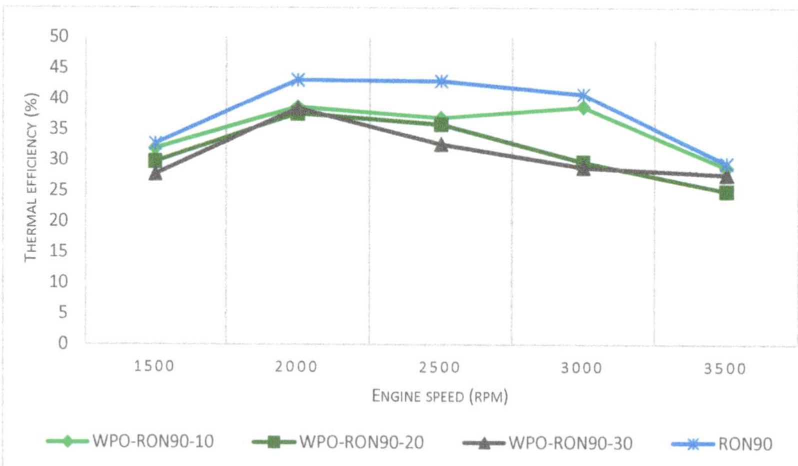

Figure 10. Effect of WPO-RON90 blends on thermal efficiency compared to RON90.

4. Conclusions

Pyrolysis of waste plastic was performed in a batch reactor equipped with temperature-controlled reflux to produce WPO. It was found that the properties of WPO are close to that of gasoline fuel. The properties of WPO were also found to be within the limits mandated in ASTM standard. From the results of this study, the following conclusions can be drawn:

1. The blend ratio of WPO with gasoline (RON88 and RON 90) was varied from 10% to 30% with gasoline. It was found that the engine power outputs of WPO-RON88 and WPO-RON90 blends were close to that of pure RON88 and RON90 fuels.
2. The increasing concentration on the blending of WPO with gasoline (RON88 and RON 90) can reduce engine output power slightly. By contrast, the WPO-RON90-30 obtained satisfactory results at the maximum engine speed of 3500 rpm, which gave results very close to the engine power fueled by pure RON90 fuel.
3. The engine power, SFC, and thermal efficiency improvement due to WPO blending showed more pronounced effects for WPO-RON90 blends than for WPO-RON88 blends.

Author Contributions: Conceptualization, methodology, writing-original draft and supervision by K. T.M.I.R. contribute to validation and writing-review & editing. S.B., S.E.S., J.J. and M.J. contributed to the experiment and investigation. F.K., Y.P., A.S.S. and A.H.S. contributed to analysis the data and improve the manuscript. All authors read and approved the final manuscript.

Funding: This research was funded by the Universitas Syiah Kuala, Minister of national education through Research Professor Program No.: 520/UN11/SPK/PNBP/2019.

Acknowledgments: The authors graciously acknowledge the financial support provided by the Chair of the Renewable Energy Research Fund—Biofuel/Bioenergy Research Program (201801 KETTHA). The authors would like to appreciate many thanks to bachelor students (Razuardi Kumar) and magister students (T.M Hakim Furqan) in the combustion Laboratory of Mechanical Engineering Department who has supported to conducting of the research. The authors wish to express their appreciation to the Direktorat Jenderal Penguatan Riset dan Pengembangan Kementerian Riset dan Teknologi/Badan Riset dan Inovasi Nasional Republik Indonesia and Politeknik Negeri Medan, Medan, Indonesia.

Conflicts of Interest: The authors declare no conflict of interest.

References

1. Silitonga, A.S.; Masjuki, H.H.; Mahlia, T.M.I.; Ong, H.C.; Chong, W.T.; Boosroh, M.H. Overview properties of biodiesel diesel blends from edible and non-edible feedstock. *Renew. Sustain. Energy Rev.* **2013**, *22*, 346–360. [CrossRef]
2. Shahir, S.A.; Masjuki, H.H.; Kalam, M.A.; Imran, A.; Rizwanul Fattah, I.M.; Sanjid, A. Feasibility of diesel–biodiesel–ethanol/bioethanol blend as existing CI engine fuel: An assessment of properties, material compatibility, safety and combustion. *Renew. Sustain. Energy Rev.* **2014**, *32*, 379–395. [CrossRef]
3. Milano, J.; Ong, H.C.; Masjuki, H.H.; Silitonga, A.S.; Chen, W.-H.; Kusumo, F.; Dharma, S.; Sebayang, A.H. Optimization of biodiesel production by microwave irradiation-assisted transesterification for waste cooking oil-Calophyllum inophyllum oil via response surface methodology. *Energy Convers. Manag.* **2018**, *158*, 400–415. [CrossRef]
4. Mason, I.G.; Page, S.C.; Williamson, A.G. A 100% renewable electricity generation system for New Zealand utilising hydro, wind, geothermal and biomass resources. *Energy Policy* **2010**, *38*, 3973–3984. [CrossRef]
5. Ismail, M.S.; Moghavvemi, M.; Mahlia, T.M.I. Characterization of PV panel and global optimization of its model parameters using genetic algorithm. *Energy Convers. Manag.* **2013**, *73*, 10–25. [CrossRef]
6. Mahlia, T.M.I.; Syaheed, H.; Abas, E.P.; Kusumo, F.; Shamsuddin, A.H.; Ong, H.C.; Bilad, M.R. Organic Rankine Cycle (ORC) System Applications for Solar Energy: Recent Technological Advances. *Energies* **2019**, *12*, 2930. [CrossRef]
7. Mofijur, M.; Mahlia, T.M.I.; Silitonga, A.S.; Ong, H.C.; Silakhori, M.; Hasan, M.H.; Putra, N.; Rahman, S.M.A. Phase Change Materials (PCM) for Solar Energy Usages and Storage: An Overview. *Energies* **2019**, *12*, 3167. [CrossRef]
8. Mahlia, T.M.I.; Syazmi, Z.A.H.S.; Mofijur, M.; Abas, A.E.P.; Bilad, M.R.; Ong, H.C.; Silitonga, A.S. Patent landscape review on biodiesel production: Technology updates. *Renew. Sustain. Energy Rev.* **2020**, *118*, 109526. [CrossRef]
9. Ong, H.C.; Masjuki, H.H.; Mahlia, T.M.I.; Silitonga, A.S.; Chong, W.T.; Yusaf, T. Engine performance and emissions using Jatropha curcas, Ceiba pentandra and Calophyllum inophyllum biodiesel in a CI diesel engine. *Energy* **2014**, *69*, 427–445. [CrossRef]
10. Ahvenainen, R. *Modern Plastics Handbook*; Woodhead Publishing Limited: Cambridge, UK, 2003.
11. Jambeck, J.R.; Geyer, R.; Wilcox, C.; Siegler, T.R.; Perryman, M.; Andrady, A.; Narayan, R.; Law, K.L. Plastic waste inputs from land into the ocean. *Science* **2015**, *347*, 768–771. [CrossRef]
12. Finance, M.O. *Bumi Dalam Kantong Plastik*; Kementerian Keuangan: Jakarta, Indonesia, 2019; Volume XIV, No. 144.
13. Lamb, J.B.; Willis, B.L.; Fiorenza, E.A.; Couch, C.S.; Howard, R.; Rader, D.N.; True, J.D.; Kelly, L.A.; Ahmad, A.; Jompa, J.; et al. Plastic waste associated with disease on coral reefs. *Science* **2018**, *359*, 460–462. [CrossRef] [PubMed]
14. Wong, S.L.; Ngadi, N.; Abdullah, T.A.T.; Inuwa, I.M. Current state and future prospects of plastic waste as source of fuel: A review. *Renew. Sustain. Energy Rev.* **2015**, *50*, 1167–1180. [CrossRef]

15. Siddiqui, M.N.; Redhwi, H.H. Pyrolysis of mixed plastics for the recovery of useful products. *Fuel Process. Technol.* **2009**, *90*, 545–552. [CrossRef]
16. Lee, J.; Lee, T.; Tsang, Y.F.; Oh, J.-I.; Kwon, E.E. Enhanced energy recovery from polyethylene terephthalate via pyrolysis in CO_2 atmosphere while suppressing acidic chemical species. *Energy Convers. Manag.* **2017**, *148*, 456–460. [CrossRef]
17. Khairil; Jihad, M.; Indra Riayatsyah, T.M.; Bahri, S.; Sofyan, S.E.; Jalaluddin. The effect of gasoline-waste plastics oil blends on SI engine performance at high-speed rotation. *IOP. Conf. Ser. Earth Environ. Sci.* **2020**, *463*, 012002. [CrossRef]
18. Surono, U.B. Berbagai metode konversi sampah plastik menjadi bahan bakar minyak. *J. Tek.* **2013**, *3*, 32–40.
19. Bezergianni, S.; Dimitriadis, A.; Faussone, G.-C.; Karonis, D. Alternative Diesel from Waste Plastics. *Energies* **2017**, *10*, 1750. [CrossRef]
20. Chong, C.T.; Mong, G.R.; Ng, J.-H.; Chong, W.W.F.; Ani, F.N.; Lam, S.S.; Ong, H.C. Pyrolysis characteristics and kinetic studies of horse manure using thermogravimetric analysis. *Energy Convers. Manag.* **2019**, *180*, 1260–1267. [CrossRef]
21. Ong, H.C.; Chen, W.-H.; Farooq, A.; Gan, Y.Y.; Lee, K.T.; Ashokkumar, V. Catalytic thermochemical conversion of biomass for biofuel production: A comprehensive review. *Renew. Sustain. Energy Rev.* **2019**, *113*, 109266. [CrossRef]
22. Cleetus, C.; Thomas, S.; Varghese, S. Synthesis of Petroleum-Based Fuel from Waste Plastics and Performance Analysis in a CI Engine. *J. Energy* **2013**, *2013*, 608797. [CrossRef]
23. Khan, M.Z.H.; Sultana, M.; Al-Mamun, M.R.; Hasan, M.R. Pyrolytic waste plastic oil and its diesel blend: Fuel characterization. *J. Environ. Public Health* **2016**, *2016*, 7869080. [CrossRef] [PubMed]
24. Soltes, E.J.; Elder, T.J. Pyrolysis. In *Organic Chemicals from Biomass*; CRC Press: Boca Raton, FL, USA, 2018; pp. 63–99.
25. Sultan, M.A.-S.; Yang, Y.; Wang, J.; Gary, A.L. Pyro-Oil and Wax Recovery from Reclaimed Plastic Waste in a Continuous Auger Pyrolysis Reactor. *Energies* **2020**, *13*, 2040.
26. Lee, X.J.; Ong, H.C.; Gan, Y.Y.; Chen, W.-H.; Mahlia, T.M.I. State of art review on conventional and advanced pyrolysis of macroalgae and microalgae for biochar, bio-oil and bio-syngas production. *Energy Convers. Manag.* **2020**, *210*, 112707. [CrossRef]
27. Ramadhan, A.; Ali, M. Pengolahan sampah plastik menjadi minyak menggunakan proses pirolisis. *J. Ilm. Tek. Lingkung.* **2012**, *4*, 44–53.
28. Sarker, M.; Rashid, M.M.; Rahman, M.S.; Molla, M. Environmentally harmful low density waste plastic conversion into kerosene grade fuel. *J. Environ. Prot.* **2012**, *3*, 700. [CrossRef]
29. Indonesia: Fuels: Diesel and Gasoline. Available online: https://www.transportpolicy.net/standard/indonesia-fuels-diesel-and-gasoline/ (accessed on 8 November 2019).
30. Silitonga, A.S.; Atabani, A.E.; Mahlia, T.M.I.; Masjuki, H.H.; Badruddin, I.A.; Mekhilef, S. A review on prospect of Jatropha curcas for biodiesel in Indonesia. *Renew. Sustain. Energy Rev.* **2011**, *15*, 3733–3756. [CrossRef]
31. Mohanraj, C.; Senthilkumar, T.; Chandrasekar, M. A review on conversion techniques of liquid fuel from waste plastic materials. *Int. J. Energy Res.* **2017**, *41*, 1534–1552.
32. Thahir, R.; Altway, A.; Juliastuti, S.R.; Susianto. Production of liquid fuel from plastic waste using integrated pyrolysis method with refinery distillation bubble cap plate column. *Energy Rep.* **2019**, *5*, 70–77. [CrossRef]
33. Priambodo, B.; Maleev, V. *Operasi dan Pemeliharaan Mesin Diesel*; Penerbit Erlangga: Jakarta, Indonesia, 1991.
34. Marthur, S.; Sharma, R. *A Course in Internal Combustion Engine*; JC Kapur: For Dhanpat Rai & Son: Delhi/Julundur, India, 1980.
35. Ananthakumar, S.; Jayabal, S.; Thirumal, P. Investigation on performance, emission and combustion characteristics of variable compression engine fuelled with diesel, waste plastics oil blends. *J. Braz. Soc. Mech. Sci. Eng.* **2017**, *39*, 19–28. [CrossRef]
36. Vijaya Kumar, K.; Puli, R.K. Effects of Waste Plastic Oil Blends on a Multi Cylinder Spark Ignition Engine. *MATEC Web Conf.* **2017**, *108*, 08005. [CrossRef]
37. Zainuri, F. *Pirolisis Sampah Plastik Hingga Suhu 900 °C Sebagai Upaya Menghasilkan Bahan Bakar Ramah Lingkungan*; Universitas Muhammadiyah Surakarta: Surakarta, Indonesia, 2014.

38. Pratama, A.W.; Budi, A. Design of Polypropylene Distillation Fuel Generator (PDFG) as an effort to create alternative energy for standar engine. In Proceedings of the International Conference on Food and Agriculture, Bali, Indonesia, 20–21 October 2018.
39. Rao, R.N.; Silitonga, A.S.; Shamsuddin, A.H.; Milano, J.; Riayatsyah, T.M.I.; Sebayang, A.H.; Nur, T.B.; Sabri, M.; Yulita, M.R.; Sembiring, R.W. Effect of Ethanol and Gasoline Blending on the Performance of a Stationary Small Single Cylinder Engine. *Arab. J. Sci. Eng.* **2020**. [CrossRef]
40. Fattah, I.M.R.; Masjuki, H.H.; Kalam, M.A.; Mofijur, M.; Abedin, M.J. Effect of antioxidant on the performance and emission characteristics of a diesel engine fueled with palm biodiesel blends. *Energy Convers. Manag.* **2014**, *79*, 265–272. [CrossRef]
41. Murni, M. Kaji Eksperimental Pengaruh Temperatur Biodiesel Minyak Sawit Terhadap Performansi Mesin Diesel Direct Injection Putaran Konstan. Master's Thesis, Diponegoro University, Semarang, Indonesia, 2010.

energies

MDPI

Article

Continuous Phenol Removal Using a Liquid–Solid Circulating Fluidized Bed

Nandhini Sureshkumar [1], Samiha Bhat [1], Shwetha Srinivasan [1], Nirmala Gnanasundaram [1,*], Murugesan Thanapalan [2], Rambabu Krishnamoorthy [3], Hatem Abuhimd [4], Faheem Ahmed [5] and Pau Loke Show [6]

1 Mass Transfer Lab, School of Chemical Engineering, Vellore Institute of Technology, Vellore 632014, India; nandhini120698@gmail.com (N.S.); samihabhat17@gmail.com (S.B.); srinivasanshwetha.2009@gmail.com (S.S.)
2 Department of Chemical Engineering, Universiti Tecknologi Petronas, Perak 32610, Malaysia; murugesan@utp.edu.my
3 Department of Chemical Engineering, Khalifa University, Abu Dhabi 127788, UAE; rambabu.krishnamoorthy@ku.ac.ae
4 National Center for Nanotechnology and Semiconductor, King Abdulaziz City for Science and Technology, Riyadh 11564, Saudi Arabia; habuhimd@kacst.edu.sa
5 Department of Physics, College of Science, King Faisal University, P.O. Box-400, Al-Ahsa 31982, Saudi Arabia; fahmed@kfu.edu.sa
6 Department of Chemical and Environmental Engineering, Faculty of Science and Engineering, University of Nottingham Malaysia, Selangor 43500, Malaysia; PauLoke.Show@nottingham.edu.my
* Correspondence: gsnirmala@vit.ac.in

Received: 16 June 2020; Accepted: 24 July 2020; Published: 27 July 2020

Abstract: A liquid-solid circulating fluidized bed (LSCFB) helps to overcome the shortcomings of conventional fluidized beds by using a particle separation and return system as an integral part of the overall reactor configuration. Batch adsorption experiments were carried out for the removal of phenol from a synthetically prepared solution using fresh activated-carbon-coated glass beads. The morphological features and surface chemistry of the adsorbent were analyzed via SEM and FTIR techniques. The adsorbent dosage, contact time and temperature were varied along with solution pH to assess their effects on the adsorbent performance for phenol removal. Isotherm modeling showed that the phenol removal using the activated-carbon glass beads followed the Langmuir model. Effectively, it was observed at an adsorbent loading of 2.5 g/150 mL of feed volume and a contact time of 3 h produced an 80% efficiency in the batch study. Furthermore, on scaling it up to the column, the desired 98% phenol-removal efficiency was obtained with an adsorbent dosage of 250 g and contact time of 25 min. Adsorbent regeneration using 5% (v/v) ethanol showed a 64% desorption of phenol from the sorbent within 20 min in the LSCFB.

Keywords: circulating fluidized bed; adsorption; activated carbon; phenol; glass beads

1. Introduction

Chemical process industries generate tremendous quantities of wastewater that significantly contribute to aquatic environmental pollution. This has demanded the development of sustainable and effective treatment methodologies for industrial effluents [1]. Design and development of innovative process technology to efficiently handle large volumes of polluted water streams in a relatively shorter time are highly desirable and promising for industrial applications. Phenol is the most commonly present contaminant in the industrial effluent stream. It is known to be highly toxic and causes harmful chronic effects on humans and animals alike [2]. This raises concern over the traces of phenolic content present in drinking water and the negative externalities caused by their discharge from the effluent treatment plants on the environment [3]. Many technologies have been developed for the decontamination of phenol-polluted wastewater streams, including membrane separation, advanced oxidation, activated sludge, ion exchange and adsorption [4–6]. Out of the various specified technologies, adsorption proves to be the most viable option due to its economic feasibility, ease of scalability and strategic removal of the target compound(s) [7,8]. It offers a wide variety of highly selective adsorbents with a good regeneration potential [9,10]. Since it does not yield any formation of sludge [11], adsorption does not pose any threat to the environment and achieves a high product quality [12,13]. Batch-mode and column-mode are the two major categories of the adsorption operation.

Column adsorption types consist of four main types: a fixed-bed type; continuous moving-bed type; fluidized-bed type and pulsed-bed type. Of these, the fluidized-bed type is widely adopted in industries due to its ability to handle large feed volumes and its easier process control [9]. The fluidized-bed technology offers the salient advantages of improved heat/mass transfer rates, enhanced interfacial contact area and isothermal operation [10]. Within the fluidized-bed column adsorption, the circulating fluidized-bed (CFB) offers effective liquid-solid contact, uniform temperature, high throughputs and better solid holdup control as compared to the conventional fluidized-bed type [11]. Although gas-solid-type CFBs have been vastly investigated from the 1960s, research studies on a liquid-solid CFB (LSCFB) are very scant. The concept of a LSCFB gained momentum around the 1990s and has been explored widely in the last ten years [10]. The LSCFB technique is very promising for various industries, such as pharmaceuticals, biotechnology catalytic refining, wastewater treatment, etc. [9]. In this work, we primarily focused on phenol adsorption studies via a liquid-solid circulating fluidized bed (LSCFB). The usage of LSCFB for phenol adsorption provides significant advantages over the conventional fluidized beds. These advantages include an enhanced adsorption efficiency, high operational simplicity, better yields, high liquid residence times, low chemical consumption, applicability to non-clarified streams and an increase in the interfacial area between the solid and liquid phases [9,12].

With the use of the appropriate solid particles, the LSCFB system would help overcome the limitations of a conventional fluidized bed. To aid the process of fluidization in the column, the adsorbent used in this study is glass beads [13] coated with commercial activated carbon, primarily because of its high density. Epoxy resin was used for coating these beads with activated carbon due to its strong adhesive properties. Before the column studies, batch investigations were conducted to identify the optimized operating conditions for the LSCFB.

2. Materials and Methods

2.1. Materials

Bituminous coal-based commercial activated carbon (Chemviron CPG-LF) was procured from Calgon Carbon, Brazil, for the preparation of the adsorbent beads. The activated carbon has a surface that is non-polar, which results in an affinity for non-polar adsorbates such as organics. The adsorbent particle sizes were in the range of 1.2–1.4 mm with an iodine number of 950 mg/g. The carbon possessed a specific surface area of 650 m^2/g (based on the Brunauer–Emmett–Teller method) with a mean pore radius of 450 nm and total pore volume of 0.18 cm^3/g. Epoxy resin (Araldite 506), methyl nadic anhydride (MNA hardener) and phenol were obtained from Sigma Aldrich, India. Ethanol (99% pure) was obtained from Fisher scientific, India. Glass beads (1.76 mm diameter) were provided in gift packs from local industry in Vellore SIPCOT, India. All the chemicals and reagents used in the study were of analytic grade and were used directly for experiments. Double-distilled water (Merck Millipore) was utilized for the solution preparation, dilution and wash applications.

2.2. Adsorbent Preparation and Characterizations

Activated carbon was used to coat the glass beads with a dimension 1.76 mm in diameter. This coating was achieved by using epoxy resin. The resin and MNA hardener were mixed in the weight ratio of 1:1. The slurry of the glass beads and epoxy resin was made and mixed, such that the resin was distributed evenly on the glass bead surface to ensure uniform binding of the activated carbon. Following this, a known measure of activated carbon was sprinkled into the slurry, and then rolled and sieved to give the resultant activated-carbon glass bead adsorbent.

The morphology and surface chemistry of the synthesized adsorbent beads were analyzed through scanning electron microscopy (SEM) and Fourier transform infrared spectroscopy (FTIR) studies. SEM micrographs (Evo-18, Carl Zeiss, Dublin, CA, USA) were used to understand the pore structure of the activated carbon glass bead samples. An IR spectrophotometer (IR Affinity-1, Shimazdu, Kyoto, Japan) was used to conduct the FTIR analysis of the adsorbent through the KBr pellet method. The spectral scan was acquired in the range from 4000 to 400 cm^{-1}. SEM and FTIR studies were performed for the fresh, phenol-loaded and regenerated activated-carbon glass bead samples.

2.3. Batch Adsorption Studies

The stock solution prepared was a system of phenol and water. It had an initial concentration of 1000 ppm. The as-prepared phenol water system was stable and non-azeotropic. Batch adsorption studies were conducted for analyzing the effect of four operational parameters, namely adsorbent dosage, contact time, pH and temperature, on the adsorbent performance [14]. Four adsorbent dosage values of 1, 1.5, 2 and 2.5 g were examined using 150 mL of the stock solution. The conical flasks were kept in the orbital shaker at standard conditions of 75 rpm and 30 °C. Samples of 10 mL after a specific time duration was pulled out from the conical flasks to measure their pH and phenol concentration. The solution pH was measured using a digital pH system (HQ411, Hach, Berlin, Germany). The phenol concentration in the test samples were determined using a UV–Vis spectrophotometer (2200, Systronics, Chennai, India) at 270 nm using the calibration graph method. A standard calibration graph was obtained initially using known concentrations of phenol in the water standard samples (x-axis) and the respective absorbance for each of the standard sample (y-axis). The phenol concentration for the test samples were obtained by measuring their absorbance value and using this value to acquire the corresponding concentration from the calibration graph [15].

For each adsorbent dosage, the contact time was varied from 0 to 3 h. Test samples were obtained at 0.25, 0.5, 0.75, 1, 2 and 3 h of contact time. For analyzing the effect of temperature, the given adsorbent dosage was subjected to four different temperatures (20, 30, 40, 50 and 60 °C) for 1 h operational time and subjected to spectrophotometry to calculate the final phenol concentrations. To investigate the regeneration capacity of the adsorbent, 5% (v/v) ethanol was used to desorb phenol from the adsorbent while varying the contact time from 0 to 3 h. Test samples were acquired at 0.25, 0.5, 0.75, 1, 2 and 3 h of the regeneration step. The phenol removal percentage was calculated by Equation (1) [16]:

$$\%R = \frac{C_i - C_e}{C_i} \times 100 \quad (1)$$

where C_i and C_e are the phenol concentrations (ppm) of the initial and equilibrium states of adsorption.

The adsorption capacity of the activated-carbon glass beads after time t of adsorption for the adsorbent (q_t) was calculated using Equation (2) [17]:

$$q_t = \frac{(C_i - C_t) \times V_a}{W_a} \quad (2)$$

where C_t (ppm), V_a (L) and W_a (g) are the phenol concentrations after time t, the volume of feed solution taken and the weight of the adsorbent used, respectively. The equilibrium adsorption capacity (q_e) for the adsorbent was calculated from Equation (2) using C_e instead of C_t.

2.4. LSCFB Study

The LSCFB system consisted of three main components—A riser, downcomer and liquid-solid separator, as shown in Figure 1. The proportionate design of the LSCFB system was based on our previous reported work [18]. The riser was of dimensions 1.5 m height and 4 cm diameter. The downcomer measured a height of 1.8 m and a width of 8 cm in diameter. The riser was fitted with two inlet feed lines—A primary feed line and a secondary feed line. The primary feed line was regulated using a rotameter with a flow rate of 3000 L/h and another flow meter of 2400 L/h regulated the secondary pipe inlet. The capacity of the inlet feed tank to the riser for adsorption was 100 L. The riser was provided with two distributors—A primary distributor and a secondary distributor at its bottom. The primary distributor occupied 20% of the total bed area while the secondary distributor was 5% open of the total bed area. The riser had two pressure tapings, one near the lower end of the column above the distributors and the other one at the upper end just before the elbow bend into the liquid-solid separator. These two pressure tapings were connected to a manometer filled by the manometric fluid to record the pressure drop in the column. The top end of the riser was connected to the liquid-solid separator just after the elbow bend. The liquid that overflowed was circulated back to the feed tank while the rest of the contents paved their way to the downcomer through the top dynamic seal. The capacity of the inlet feed tank to the downcomer for desorption was 100 L. A rotameter of 2400 L/h was used to regulate the inlet feed line to the downcomer. The downcomer had a diffuser that uniformly provides an inlet for the desorption liquid. There was a valve provided at the bottom dynamic seal to regulate the solid holdup between the riser and the downcomer. There were provisions for wash water provided at the bottom of the liquid-solid separator and the downcomer. This was operated after every adsorption and desorption cycle.

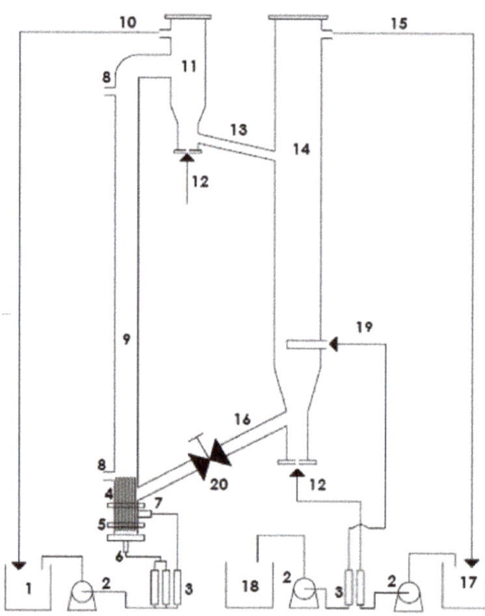

Figure 1. Schematic diagram of the experimental setup used. (1) Feed Tank; (2) Pump; (3) Rotameter; (4) Primary Distributor; (5) Secondary Distributor; (6) Primary Feed Line; (7) Secondary Feed Line; (8) Pressure Tapings; (9) Riser; (10) Liquid-Solid Separator Outlet; (11) Liquid-Solid Separator; (12) Wash Water; (13) Top Dynamic Seal; (14) Downcomer; (15) Downcomer Outlet; (16) Bottom Dynamic Seal; (17) Wash Water Storage Tank; (18) Desorption Liquid Storage Tank; (19) Desorption Liquid Inlet; (20) Check Valve.

2.4.1. Adsorption Cycle

The column was washed with water before operating it at the required conditions. The adsorbent was fed into the column to fill 35% of the riser height. The feed tank was filled with 1000 ppm stock solution. Primary liquid was pumped into the LSCFB through calibrated flow meters at a rate of 1100 L/h. The secondary feed was pumped at 750 L/h. The combined velocity offered by the primary and secondary feed streams was higher, which enabled the particles to move up at a velocity higher than the terminal velocity and less than the critical velocity. At this flow condition, the adsorbents got entrained by the liquid flowing vertically up in the riser and was passed to the liquid-solid separator. A solid hold up developed in the liquid-solid separator allowing for more interactions between the adsorbent and the adsorbate. Subsequently, the bulk and solid phases entered into the downcomer section of the LSCFB. The downcomer facilitated further adsorption due to a higher residence time offered by its larger diameter. The bulk phase flowed back into the riser through the bottom dynamic seal that was regulated using a valve. The column was run for 50 min and 10 mL solutions were withdrawn from the feed tank after proper mixing at 5 min intervals up to 50 min. The solution samples were then analyzed using a UV-V is spectrophotometer for its absorbance. The feed was then drained from the column to begin the desorption cycle after the water wash.

2.4.2. Desorption Cycle

The column was water-washed again, before its next run. The desorption liquid feed tank was supplied with a 5% (*v*/*v*) ethanol-water solution. It was pumped into the downcomer at a liquid flow rate of 1200 L/h. The direction of flow was reversed and the contents flowed backward into the liquid-solid separator and into the riser. The column ran for 50 min and samples of 10 mL volume were withdrawn from the desorption feed tank after proper mixing at 5 min intervals. The obtained samples were examined using a UV-Vis spectrophotometer for its absorbance. The bulk phase was then removed from the column and the adsorbent was analyzed through SEM.

3. Results

3.1. Batch Study

3.1.1. Effect of Adsorbent Dosage

The performance of the activated-carbon glass beads as a phenol removal adsorbent was mainly assessed by two factors: (i) the phenol-removal efficiency and (ii) the adsorption capacity [19]. The effect of the adsorbent dosage on these performance factors of the adsorbent was analyzed through batch adsorption experiments. For all the adsorbent dosages considered, the Langmuir isotherm produced a better correlation coefficient (R^2 = 0.9431) as compared to the Freundlich isotherm (R^2 = 0.9073) and, hence, the q_m value was analyzed for the factor of adsorption capacity for the various adsorbent dosages.

The phenol-removal efficiency is linked to the availability of active sites for the phenol molecules to get adsorbed [20,21]. On the other hand, the adsorption capacity is associated with the saturation of the binding spots on the adsorbent for the adsorbate to get adsorbed [22]. On analyzing the graph as shown in Figure 2a, it was seen that 1 g of the adsorbent produced the same percentage of adsorption for the various contact times and hence it was considered to be an ineffective dosage. For the case of 1.5 g of adsorbent dosage, better results were produced than 1 g in terms of the phenol-removal efficiency for a relatively lower value of adsorption capacity, but reached saturation at 65% itself, which was undesirable. Additionally, 2 and 2.5 g of adsorbent dosage produced approximately the same but much better results compared to 1 g and 1.5 g, resulting in a 78% and 80% phenol-removal efficiency, respectively, thus making it redundant to continue our experimentation of the batch studies with higher adsorbent loadings. Hence, it was concluded that 2.5 g of activated carbon yielded a desirable and effective phenol removal of 80% from the feed solution. This showed that despite having the least adsorption capacity among the various dosage runs, the adsorbent still possessed considerable amount of vacant active sites for phenol adsorption and this would lead to better results (>80% phenol-removal efficiency) in the continuous LSCFB system.

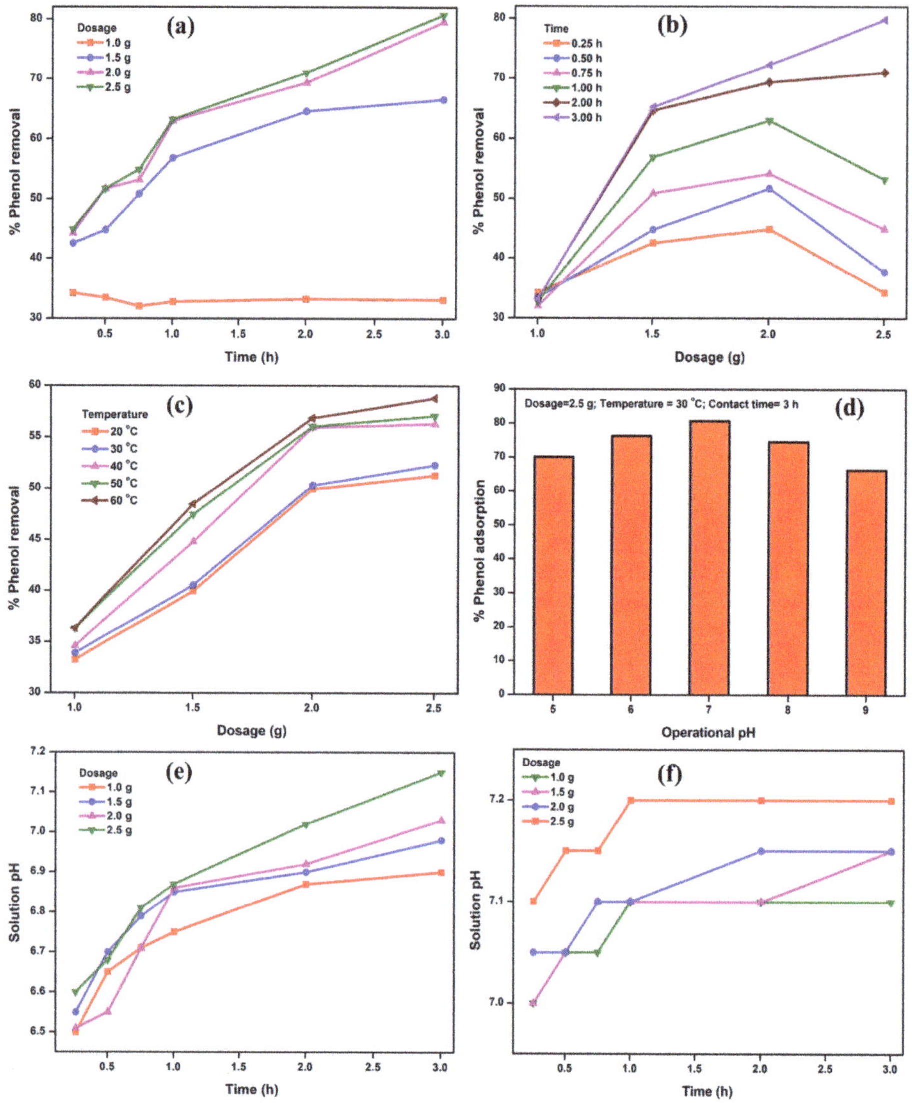

Figure 2. Effect of (**a**) adsorbent dosage; (**b**) contact time; (**c**) temperature and (**d**) operational pH on % phenol removal. Solution pH studies for (**e**) the adsorption system (phenol-activated carbon) and (**f**) control system (distilled-water-activated carbon).

3.1.2. Effect of Contact Time

As shown in Figure 2b, with an increase in adsorbent loading, it took more time for the bulk of the activated-carbon-coated glass beads to come into contact with the phenol solution, thus promoting increased interactions between the adsorbate and adsorbent [23]. This can be explained from the results portrayed in the figure. For instance, 1 g being the lowest adsorbent loading, it took less time for the bulk of the beads to come in contact with the phenolic solution and hence all contact times had approximately the same phenol-removal efficiency. Moreover, as the adsorbent loading increased, it was evident that the larger adsorbent loadings took a longer time to reach an effective phenol-removal efficiency. For example, a 2.5 g loading showed variation in the adsorption percentage from a minimum of 35% adsorption in the first 0.25 h to a maximum of 80% adsorption within the next 3 h. Additionally, another reason for the trend observed is linked to the excessive contact time to which the system was subjected to. As the contact time increased, the system reached a point wherein the binding sites on the adsorbent become saturated and no more adsorption was practically possible. On closely observing the 3 h timeline for the four adsorbent loadings, it was evident that the phenol-removal efficiency increased with incremental levels of adsorbent dosage. This shows that the beads with an adsorbent loading of 2.5 g would produce the best results of >80% phenol removal for an optimal time of 3 h. With respect to the LSCFB, running the column at an appropriate adsorbent loading for a lesser amount of time was the viable option to maintain the surface morphology of the activated-carbon-coated glass beads.

3.1.3. Effect of Temperature

Generally, the behavior of the phenol–water system for temperature variations is very similar to that of a normal aqueous system till the attainment of the critical temperature of the binary system [4]. In this case, with the initial phenol solution taken being very dilute (1000 ppm) and the critical temperature of the system being ~70 °C, it was safe to increase the temperature of the solution till 60 °C. It is a very well-known and understood fact that on increasing the temperature of the solution, the kinetic energy of the molecules increases, which in turn increases the interaction between the adsorbate and the adsorbent, leading to more binding of the phenol on the adsorbent and hence an increased percentage adsorption [19].

From the graph as depicted in Figure 2c, it was observable that, at higher temperatures, all the four adsorbent dosages followed the same trend as explained earlier and displayed better percentages of adsorption for increased temperatures. For 1 g of adsorbent dosage, it was observed that the phenol-removal percentage was very less, at 35% for all the temperatures. However, higher dosage values of 1.5, 2 and 2.5 g showed better percentages of adsorption with an increase in operational temperature. For the optimal loading of 2.5 g, the phenol-removal efficiency was increased from 51% at 20 °C to 58.5% at 60 °C. Additionally, no significant increase in the phenol adsorption percentage for the temperature increment from 20 to 60 °C was observed. This in turn indicated that a relatively low operational temperature is more preferable for the phenol removal studies in the LSCFB. For the continuous column study, an operational temperature of 30 °C was selected as this was very close to the average room temperature of the current research work environment.

3.1.4. pH Variation Studies

The operational pH of the adsorption system is an important parameter that impacts the phenol-removal efficiency from the feed wastewater. In this study, the pH dependency examinations were performed by varying the operational pH in the range of 5 (acidic condition) to 9 (basic condition). Other parameters, such as dosage, contact time and temperature for the experimental tests, were fixed at conditions of 2.5 g, 3 h and 30 °C, respectively. Results for the influence of pH on the % phenol removal is presented in Figure 2d. A maximum phenol-removal efficiency of 80% was observed at a neutral pH of 7. Both the acidic (pH < 7) and basic (pH > 7) operational pH values resulted in a lower phenol-removal performance. Furthermore, the % phenol removal was comparatively lower in the basic environment than in the acidic environment. This was mainly due to the interference of the basic OH ions that hindered the diffusional effects of the phenol molecules into the pores of the activated carbon [4].

Further, in the batch experiments, continuous removal of phenol by the activated-carbon glass beads altered the pH value of the bulk solution [24]. The change in the pH value of the bulk phase would alter the instantaneous adsorption phenomenon for a given adsorbent dosage value. Figure 2e presents the variation of the solution pH for the four adsorbent dosages at different contact times. It was seen that for lower adsorbent dosages and for smaller intervals of time, the pH values were closer to 6.5. While, on the contrary, for higher adsorbent loadings kept for a longer time, say until 3 h, the pH values reached a neutral value of 7. In order to ensure the attainment of a neutral pH by the bulk phase for higher adsorbent dosage and contact time, control runs were performed using distilled-water-activated carbon beads to understand the variation of solution pH. Results for the control run are presented in Figure 2f, which show a slight increment in the solution pH with increased time and adsorbent dosage. The trivial increase in the solution pH could be ascribed to the basic functional groups present on the surface of the activated carbon. This confirmed that the attainment of a neutral pH for the investigated phenol-activated carbon glass beads system was predominately due to the adsorptive removal of phenol from the bulk phase by the activated-carbon glass beads.

In general, the pH of a dilute phenolic solution having a concentration of 1000 ppm (0.01 M or 1 N) or less possessed a weak acidic character with its pH ranging from 6 to 7. The pH of the initial solution of 1000 ppm concentration before subjecting to adsorption study was 6.2, as calculated using a pH meter. The ability of phenol to exhibit a weak acidic character despite the presence of an −OH group is attributed to the stability of the benzene ring. Phenol loses an H^+ ion-producing phenoxide ion that stabilizes itself by delocalizing the negative ion with the pie bonds throughout the ring [25,26].

3.1.5. Desorption Study

As shown in Figure 3, the regeneration of the phenol-adsorbed activated-carbon glass beads using a 5% (*v*/*v*) ethanol solution was performed by assessing the desorption potential of the phenol from the activated-carbon-coated glass beads in presence of an ethanol medium [27]. From the obtained results, it was clear that using a 5% (*v*/*v*) ethanol solution resulted in a 56% phenol desorption efficiency for the activated-carbon-coated glass beads for a regeneration period of 2 h. Hence, the viable option while working with the downcomer of the LSCFB would be increasing the concentration of the ethanol solution that can bring about a satisfactory desorption or removal of phenol from the beads within a lesser amount of time.

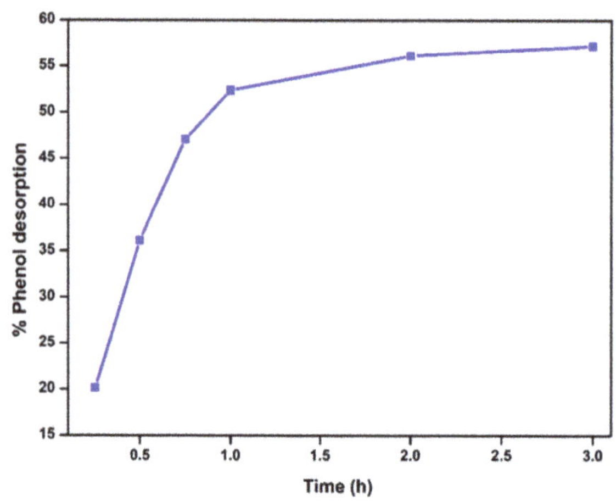

Figure 3. The % desorption of phenol over time with the use of 5% (*v*/*v*) ethanol.

3.1.6. Adsorption Isotherms

The Langmuir and Freundlich isotherm models were examined to predict the interactive nature between the phenol and activated-carbon glass beads. The Langmuir isotherm postulates the theory of energetically equivalent active sites of the sorbent. Accordingly, the adsorption would result in a single layer formation of adsorbate on the adsorbent. The isotherm advocates a characteristic theoretical maximum of adsorption capacity (q_{max}). According to Langmuir, the relation between C_e and q_e in a linearized form is shown in Equation (3) [28]:

$$\frac{C_e}{q_e} = \frac{1}{q_m K_L} + \frac{C_e}{q_m} \tag{3}$$

where q_m (mg/g) and K_L (L/mg) are the monolayer maximum adsorption capacity and the adsorption constant for Langmuir isotherm, respectively. The separation factor (R_L) explicates the favor of adsorption if $0 < R_L < 1$ and is given by Equation (4) [29]:

$$R_L = \frac{1}{1 + K_L C_i} \tag{4}$$

The Freundlich isotherm supports the heterogeneous and rough surface nature of the solid phase. The isotherm highlights the existence of interactions between the surface-bonded and free molecules of the liquid phase, which results in multilayer formation of the sorbate molecules. The Freundlich isotherm relates C_e and q_e linearly, as presented in Equation (5) [30]:

$$\ln(q_e) = \ln\left(K_f\right) + \frac{1}{n}\ln(C_e) \tag{5}$$

where K_f ($mg^{1-1/n}$ $L^{1/n}$ g^{-1}) and n are the adsorption constant for the Freundlich isotherm and the adsorption intensity, respectively.

Results of the equilibrium modeling with experimental data are shown in Figure 4. The values of the Langmuir (K_L) and Freundlich (K_f) constants were evaluated as 0.0062 L/mg and 1.9051 $mg^{1-1/n}$ $L^{1/n}$/g, respectively. Various other isotherm parameters are tabulated in Table 1. The higher R^2 value of the Langmuir isotherm showed that the phenol adsorption on the activated carbon glass beads was homogenous with monolayer formation. The separation factor (R_L) was calculated from the Langmuir constant (K_L) and the initial concentration of the solution (C_0). The ideal range for R_L is theoretically estimated to be between 0 and 1. When the R_L value is in the range of 0 to 1, the resultant adsorption process is said to be a favorable process. On the contrary, any value of $R_L > 1$ indicates that the adsorption process is reversible. The separation factor for the reported adsorption batch study using a 2.5 g adsorbent dosage was estimated to be 0.1398, which substantiated the fact that the adsorption of phenol on activated-carbon-coated glass beads was physisorption [31].

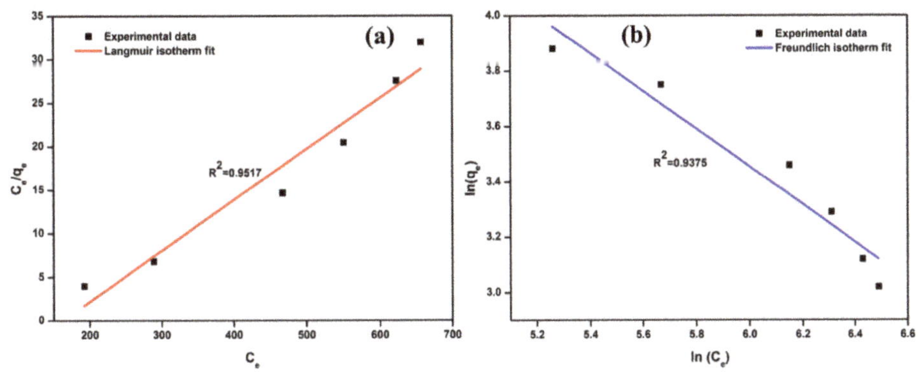

Figure 4. Adsorption isotherms: (**a**) The Langmuir isotherm for an adsorbent dosage of 2.5 g; and (**b**) the Freundlich isotherm for an adsorbent dosage of 2.5 g.

Table 1. Equilibrium isotherm parameters.

Isotherm	Parameters	Values
Langmuir	R^2	0.9571
	q_m (mg/g)	17.12
	K_L (L/mg)	0.0062
	R_L	0.1398
Freundlich	R^2	0.9375
	n	1.4643
	K_f (mg$^{1-1/n}$ L$^{1/n}$/g)	1.9051

3.2. Column Study

3.2.1. Operational Parameters

The LSCFB was designed and fabricated to run for two cycles, namely, the adsorption and desorption cycle, operated for a total duration of 50 min per cycle and maintaining a primary flow rate of 1100 L/h and a secondary flow rate of 750 L/h. The operating feed volume was maintained at 35% of the total volume of the riser, whose dimensions were 1.5 m in height and 0.04 m in diameter. All the experiments were carried out at a room temperature of 30 °C. The solid holdup was estimated by the pressure gradient in the riser. On eliminating any kind of effects due to wall friction, the average solid holdup (ε_s) was determined using Equations (6) and (7).

$$-\frac{\Delta P}{\Delta L} = (\varepsilon_s \rho_s + \varepsilon_l \rho_l) g \tag{6}$$

$$\varepsilon_s + \varepsilon_l = 1 \tag{7}$$

where $\frac{\Delta P}{\Delta L}$, ε_s ε_l, ρ_s, ρ_l and g are the pressure drop gradient (N/m^3), solid holdup, liquid holdup, density of the solid (kg/m^3), density of the liquid (kg/m^3) and acceleration due to gravity (m/s^2), respectively.

The value of the terminal velocity (u_{tr}) of the particle is theorized to be estimated from Equation (8). This value tends to be greater than the minimum fluidization velocity of the particle for the liquid.

$$u_{tr} = \left[\frac{4(\rho_s - \rho_l)^2 g^2}{225 \rho_l \mu}\right]^{\frac{1}{3}} d_p \tag{8}$$

where d_p and μ are the solid diameter (m) and the liquid viscosity (kg/m s).

The cumulative flow rate of the primary and the secondary inlet pipes provided the total net liquid flow rate. The critical velocity being greater than the terminal velocity of the particle is given by Equation (9) [32,33]:

$$u_{cr} = 1.2 * u_{tr} \tag{9}$$

The column was operated above the terminal velocity and below the critical transition velocity. The operational conditions of the LSCFB are enlisted in Table 2 [18].

3.2.2. Adsorption Cycle

From the experimental data achieved, it was evident that the LSCFB facilitated a better adsorbate-adsorbent interaction. This was corroborated from the results presented in Figure 5a that showed that the phenol-removal efficiency in the LSCFB was 98%, which was 18% more than the phenol-removal efficiency achieved by 2.5 g of adsorbent dosage in the batch study; also, this enhancement in the phenol-removal efficiency was achieved in a very short operation time of 25 min, as compared with the 3 h contact time of the batch mode.

Table 2. Operational parameters for the liquid-solid circulating fluidized bed (LSCFB) column.

Primary Liquid Flow Rate (L/h)	1100
Secondary Liquid Flow Rate (L/h)	750
Desorption Liquid Flow Rate (L/h)	1200
Archimedes Number	56,108
Terminal Velocity of Adsorbent (m/s)	0.2783
Minimum Fluidization Velocity (m/s)	0.0180
Critical Velocity (m/s)	0.3340
Pressure Drop (N/m^2)	701.19
Solid Holdup	0.5193
Liquid Holdup	0.4087

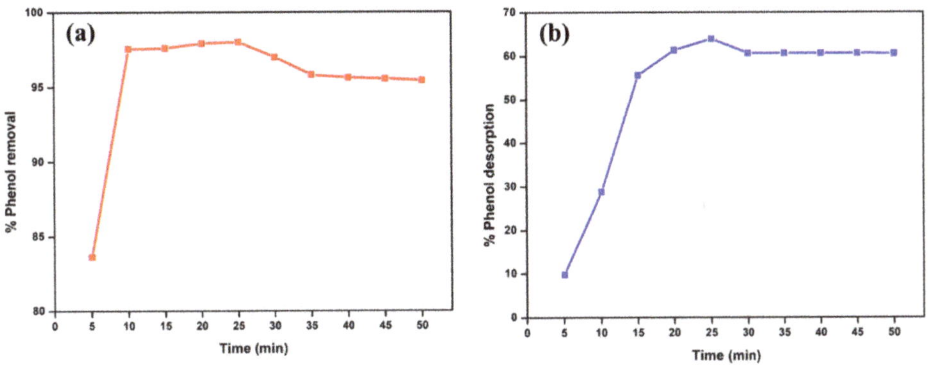

Figure 5. (**a**) The % phenol removal vs. time in an LSCFB; and (**b**) % phenol desorption vs. time in an LSCFB.

3.2.3. Desorption Cycle

An adsorbent dosage of 2.5 g in the batch study produced a maximum of 56% desorption. Under the same conditions, the continuous column study using the LSCFB resulted in a 64% desorption in a lesser time of 20 min, as shown in Figure 5b. Hence, it was primarily evident that a continuous flow system enriched the desorption level of the adsorbent compared to the batch mode. Additionally, it was also interpreted that on increasing the ethanol concentration, better desorption rates can be achieved in a lesser period.

3.2.4. Column Adsorption Models

On plotting the breakthrough curve for the column [34], a skewed S-shaped curve was obtained, as shown in Figure 6a, which signified the different packing densities in the bed provided by the solid and liquid voidage, thus maintaining the heterogeneity of the bed.

Further modeling was carried out using the Yoon-Nelson, Adam-Bohart and modified-dose response models. The results for the analysis of various column models are presented in Figure 6b–d. The Yoon-Nelson model for column adsorption is described by Equations (10,11) [35]:

$$\ln\left[\frac{C_t}{C_0 - C_t}\right] = k_{YN}t - k_{YN}\tau \tag{10}$$

$$q_{YN} = \frac{C_0\, Q\tau}{1000\, w} \tag{11}$$

where, k_{YN}, τ, Q, w and q_{YN} are the rate constant (1/min), time required for 50% adsorbate breakthrough (min), flow rate (L/min), weight of the adsorbent (g) and Yoon-Nelson adsorptive capacity of the bed (mg/g).

The Adam-Bohart model is based on the theory that the concentration of the feed solution is weak with the speed of adsorption limited by external mass transfer [36]. The model is explained by Equation (12), which relates the value of C_0/C_t with time (t) in an open system. It has been formulated as follows:

$$\ln\left[\frac{C_0}{C_t} - 1\right] = \left(\frac{k_{AB}N_0 Z}{u}\right) - k_{AB}C_0 t \tag{12}$$

where, k_{AB}, Z, N_0 and u are the kinetic constant (L/g/min), bed depth in the column (m), Adam-Bohart adsorptive capacity of the bed (mg/g) and outlet velocity of the bed (m/min).

The modified dose-response (MDR) model is a topical fit variant for the column adsorption studies. It is a non-linear logistic fit mostly used to describe the column adsorption behavior of heavy metals. The model is described by the non-linear correlation as specified in Equations (13) and (14) [37].

$$\frac{C_t}{C_0} = 1 - \frac{1}{\left(\frac{ut}{b} + 1\right)^a} \tag{13}$$

$$q_{MDR} = \frac{C_0\, b}{w_a} \tag{14}$$

where a and b (mL) are the characteristic factors of the MDR model.

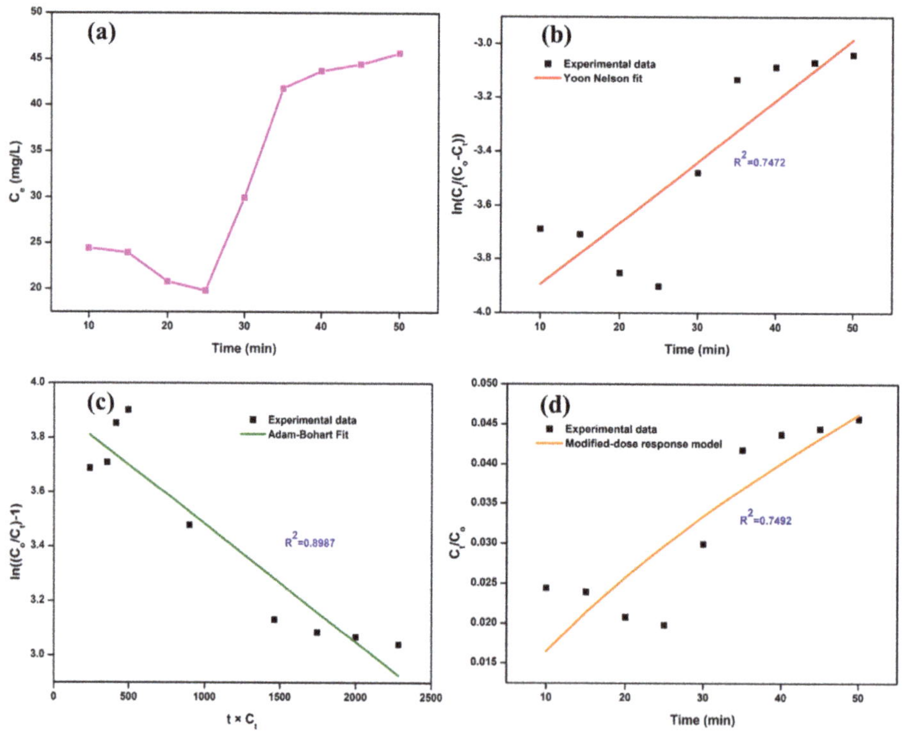

Figure 6. (**a**) Phenol-activated carbon breakthrough curve analysis; (**b**) Yoon-Nelson fit model; (**c**) Adam–Bohart fit model; and (**d**) modified-dose response model.

Table 3 presents the characteristic column adsorption capacity obtained for the experimental studies as well as using various theoretical models examined for the column adsorption. The higher correlation coefficient of R^2 = 0.8987 indicated that the column adsorption studies obeyed the Adam-Bohart model. The adsorbent capacity predicted by this model was also very close to that of the experimental values. This showed that the column adsorption of phenol onto the activated-carbon-coated glass beads involved a quasi-chemical interactions along with the absence of the internal diffusion effects of the bulk phase on the solid phase [36].

Table 3. Column study—Adsorption modeling.

Model	Characteristic Parameter	Parameter Value	R^2
Breakthrough curve	Bead Capacity (BC)	250 mg/g	-
Yoon-Nelson	Adsorbent Capacity (q_{0YN})	163 mg/g	0.7472
Adam-Bohart	Adsorbent Capacity (N_0)	279 mg/g	0.8987
Modified-dose response	Adsorbent Capacity (q_{MDR})	291 mg/g	0.8492

3.3. SEM Analysis

The morphology of the fresh, spent and regenerated activated-carbon glass beads was studied by the SEM technique. The micrographs obtained for the adsorbent beads are presented in Figure 7. As shown in Figure 7a, the activated carbon possessed a cellular and honeycomb-like structure, despite the process undergone in the preparation of the adsorbent using the glass beads, epoxy resin and activated carbon. The thin porous sheet-like structure of the adsorbent was homogenously distributed, which elevated the pore density of the activated carbon beads. These numerous pores hosted the necessary active sites for phenol adsorption [38].

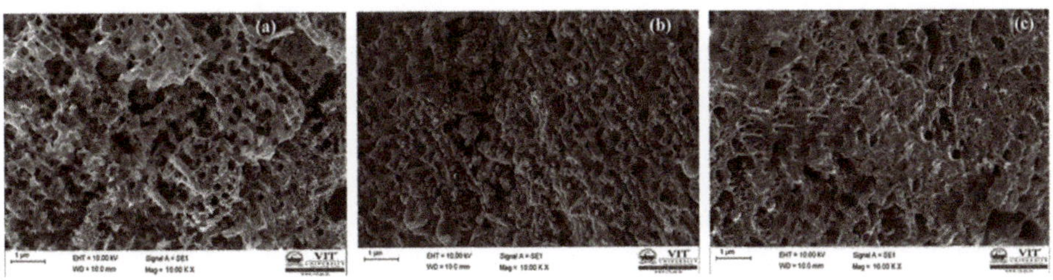

Figure 7. SEM micrographs of the activated-carbon-coated glass beads: (**a**) before adsorption; (**b**) after adsorption; and (**c**) after desorption.

The SEM image for the spent adsorbent (Figure 7b) showed the successful adsorption of the phenol molecules on the activated-carbon-coated glass beads. The densified nature of the adsorbent with complete coverage of the porous honeycomb structure confirmed the phenol binding to the adsorbent surface. The micrograph also showed a uniform coverage of the active sites by the phenol molecules, which indicated the energetically homogenous nature of the activated-carbon glass beads [39].

Figure 7c depicts the regenerated surface of the activated-carbon-coated glass beads. The quasi-porous nature of the regenerated adsorbent clearly showed that the ethanol medium efficiently desorbed the phenol molecules from the pores of the adsorbent, allowing the reuse of the adsorbent in the LSCFB.

3.4. FTIR Analysis

The surface chemistry of the activated-carbon glass beads at different stages of the experiment was analyzed using the FTIR technique. Figure 8 and Table 4 contain the results for the pristine adsorbent. The characteristic stretch observed at 3626.17 cm^{-1} was due to the –OH band of the adsorbed water whereas the peaks at 3095.75 cm^{-1} and 3066.82 cm^{-1} were due to the –OH vibrations of the carboxylic acids group present in the activated carbon. The spectra showed the presence of peaks in the fingerprint region, which corresponded to the presence of SiO_2 and C–C bonds, thus confirming the presence of glass beads and activated carbon, respectively. The epoxy resin ($C_{21}H_{25}ClO$) showed a strong C–Cl presence in the infrared region between 800 and 600 cm^{-1}. The large number of teeth-like structures was attributed to the occurrence of vibrations of the C–C bonds when subjected to infra-red radiation. The absence of other peaks throughout the entire wavelength region confirmed that the adsorbent contained only the aforementioned compounds [40].

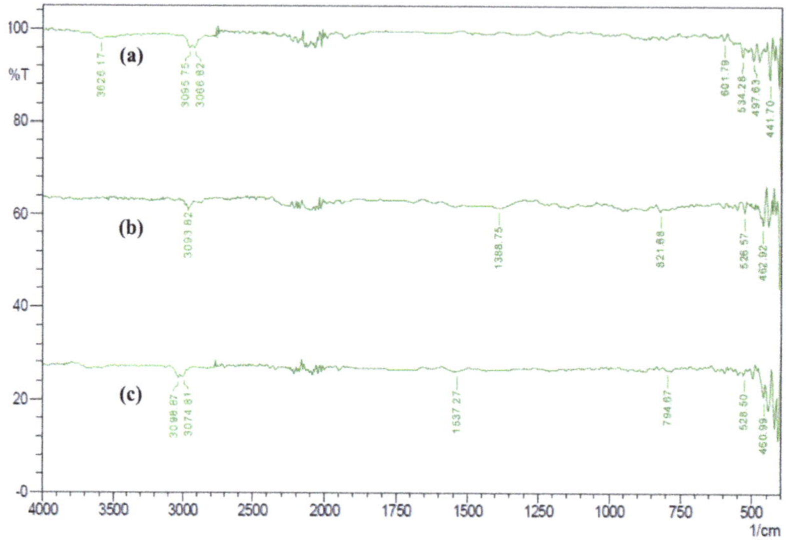

Figure 8. FTIR analysis of the adsorbent: (**a**) before adsorption; (**b**) after adsorption; and (**c**) after desorption.

Table 4. FTIR functional group analysis for the activated-carbon beads.

Wavelength Region (cm^{-1})	Functional Group	Compound
Fresh activated-carbon beads		
600–400	-	SiO_2 (Glass Beads)
Teeth-like structures between 1500 and 600	-	C–C (Activated Carbon)
800–600	C–Cl	Epoxy Resin
Spent activated-carbon beads		
1410–1310	–OH	Phenol
Teeth-like structures between 2500 and 2000	Aromatic (Benzene)	C = C (Strong)
900–700	Aromatic (Benzene)	C = C (sp^2 C–C)
Regenerated activated-carbon beads		
1600–1400	Aromatic (Benzene)	C = C (Weak)
Teeth-like structures between 2500 and 2000	Aromatic (Benzene)	C = C
800–590	–OH	Ethanol (Weak)

FTIR spectra for the phenol-adsorbed activated-carbon glass beads are presented in Figure 8 and Table 4. The results indicated the presence of phenol at a stretch between 1410 and 1310 cm^{-1} and at an appropriate value of 1388.75 cm^{-1}. Secondly, the prevalence of teeth-like structures between 2500 and 2000 cm^{-1} showed strong C = C functionality due to the presence of an aromatic ring. Moreover, the existence of a benzene ring in phenol adsorbed by the adsorbent can be confirmed by the peak at 821.68 cm^{-1}. Given the fact that the solutions taken were dilute, peaks were not obtained between 3200 and 3100 cm^{-1}, which would have indicated an alcohol-water solution in the system. The red shift in the –OH band from 3095.75 cm^{-1} (pertaining to fresh adsorbent) to 3093.82 cm^{-1} indicated the interactions of the carboxylic groups present in the adsorbent for phenol adsorption. Additionally, the peaks obtained in the fingerprint region adhere to those previously obtained, thus proving that no other compounds were formed during adsorption [41].

The results for surface functionality of the ethanol-washed adsorbent beads are presented in Figure 8 and Table 4. The presence of a peak at 794.67 cm^{-1} indicated the weak presence of an alcohol group, which was possibly ethanol. Secondly, the broad stretch between 1600 and 1400^{-1} with a peak at 1537.27 cm^{-1} showed a weak aromatic character, which indicated the reduced aromatic tendencies as compared to Figure 8b. This confirmed the successful desorption of phenol in the regenerated adsorbent beads; also, the restoration of the characteristic bands at positions 3098.87 cm^{-1} and 3074.81 cm^{-1} confirmed the successful regeneration of the adsorbent beads. The presence of no other peaks highlighted that no other side compounds were formed that led to any changes in the system and both adsorption and desorption occurred successfully. Additionally, the fingerprint region remained the same as before.

4. Conclusions

In this study, activated-carbon-coated glass bead adsorbents were developed and applied for phenol removal from synthetic wastewater. Batch adsorption studies indicated a maximum phenol removal of 80% for the optimal operational conditions of an adsorbent dosage 2.5 g, contact time 3 h and temperature 30 °C. Reusability studies of the adsorbent beads using a 5% (*v*/*v*) ethanol solution showed a good regeneration percentage of 56%, which were associated with the activated-carbon beads for phenol removal studies. To further enhance the performance of the phenol-removal process using the as-developed activated-carbon beads, a column study was performed in an LSCFB. A very high phenol-removal efficiency of 98% was achieved in the LSCFB with an enhanced regeneration efficiency of 64% for the adsorbent beads, within a shorter duration of 20 min (as compared to batch adsorption results). The results analyzed from the batch and column study indicated that the Langmuir and Adam-Bohart models provided the best fit, respectively. This elucidated the monolayer deposition of the phenol on the adsorbent beads through a quasi-chemical reaction phenomenon. Furthermore, the morphological analysis of the adsorbent beads at various stages of adsorption and regeneration studies exhibited their intact structural stability. Furthermore, surface chemistry studies of the adsorbent using the FTIR technique showed that the chemical functionality of the adsorbent was well maintained, and that successful phenol adsorption/desorption occurred on the surface of the activated-carbon glass beads. The results highlighted that the continuous mode of phenol removal and adsorbent regeneration using the LSCFB was more advantageous and significant than the batch mode. Thus, phenol removal using activated-carbon-coated glass beads in an LSCFB is reported as a novel approach for effective treatment of phenol-polluted wastewater streams. The promising results of the study indicate the possible industrial potential of the presented technique, especially towards toxic phenol contaminant elimination from aqueous solutions.

Author Contributions: Conceptualization, N.G. and R.K.; methodology, N.S., S.B. and S.S.; software, N.G. and R.K.; data curation, N.S., S.B. and S.S.; data validation, N.G., R.K. and P.L.S.; writing—original draft, N.S., S.B., S.S. and R.K.; writing—review and editing, N.G., M.T., F.A. and H.A.; project administration, N.G., M.T. and R.K.; funding acquisition, N.G. and H.A. All authors have read and agreed to the published version of the manuscript.

Funding: This research received no external funding.

Acknowledgments: The authors are grateful to the Vellore Institute of Technology, Vellore, for providing SEED money and an efficient infrastructure facility to carry out the present study, as well as for providing a scanning electron microscopy analysis facility aided by DST-FIST-SBST-VIT (SEM EVO 18, CARL ZEISS).

Conflicts of Interest: The authors declare no conflict of interest.

References

1. Bharath, G.; Hai, A.; Rambabu, K.; Banat, F.; Ashraf, M.T.; Schmidt, J.E. Systematic production and characterization of pyrolysis oil from date tree wastes for bio-fuel applications. *Biomass Bioenergy* **2020**, *135*, 105523. [CrossRef]
2. Pernyeszi, T.; Farkas, V.; Felinger, A.; Boros, B.; Dékány, I. Use of non-living lyophilized Phanerochaete chrysosporium cultivated in various media for phenol removal. *Environ. Sci. Pollut. Res.* **2018**, *25*, 8550–8562. [CrossRef] [PubMed]
3. Ali, A.; Bilal, M.; Khan, R.; Farooq, R.; Siddique, M. Ultrasound-assisted adsorption of phenol from aqueous solution by using spent black tea leaves. *Environ. Sci. Pollut. Res.* **2018**, *25*, 22920–22930. [CrossRef] [PubMed]
4. Nirmala, G.; Murugesan, T.; Rambabu, K.; Sathiyanarayanan, K.; Show, P.L. Adsorptive removal of phenol using banyan root activated carbon. *Chem. Eng. Commun.* **2019**, 1–12. [CrossRef]
5. Bharath, G.; Hai, A.; Rambabu, K.; Savariraj, D.; Ibrahim, Y.; Banat, F. The fabrication of activated carbon and metal-carbide 2D framework-based asymmetric electrodes for the capacitive deionization of Cr(vi) ions toward industrial wastewater remediation. *Environ. Sci. Water Res. Technol.* **2020**, *6*, 351–361. [CrossRef]
6. Evangeline, C.; Pragasam, V.; Rambabu, K.; Velu, S.; Monash, P.; Arthanareeswaran, G.; Banat, F. Iron oxide modified polyethersulfone/cellulose acetate blend membrane for enhanced defluoridation application. *Desalin. Water Treat.* **2019**, *156*, 177–188. [CrossRef]
7. Krishnamoorthy, R.; Govindan, B.; Banat, F.; Sagadevan, V.; Purushothaman, M.; Show, P.L. Date pits activated carbon for divalent lead ions removal. *J. Biosci. Bioeng.* **2019**, *128*, 88–97. [CrossRef]
8. Hai, A.; Bharath, G.; Babu, K.R.; Taher, H.; Naushad, M.; Banat, F. Date seeds biomass-derived activated carbon for efficient removal of NaCl from saline solution. *Process Saf. Environ. Prot.* **2019**, *129*, 103–111. [CrossRef]
9. Wang, J.; Shao, Y.; Yan, X.; Zhu, J. Review of (gas)-liquid-solid circulating fluidized beds as biochemical and environmental reactors. *Chem. Eng. J.* **2019**, *386*, 121951. [CrossRef]
10. Zhu, J.X.; Zheng, Y.; Karamanev, D.G.; Bassi, A.S. (Gas-)liquid-solid circulating fluidized beds and their potential applications to bioreactor engineering. *Can. J. Chem. Eng.* **2000**, *78*, 82–94. [CrossRef]
11. Zheng, Y.; Zhu, J.X.; Marwaha, N.S.; Bassi, A.S. Radial solids flow structure in a liquid-solids circulating fluidized bed. *Chem. Eng. J.* **2002**, *88*, 141–150. [CrossRef]
12. Lan, Q.; Bassi, A.S.; Zhu, J.X.; Margaritis, A. Continuous Protein Recovery with a Liquid-Solid Circulating Fluidized-Bed Ion Exchanger. *AIChE J.* **2002**, *48*, 252–261. [CrossRef]
13. Rao, V.V.B.; Rao, S.R.M. Adsorption studies on treatment of textile dyeing industrial effluent by flyash. *Chem. Eng. J.* **2006**, *116*, 77–84.
14. Foo, K.Y.; Hameed, B.H. Insights into the modeling of adsorption isotherm systems. *Chem. Eng. J.* **2010**, *156*, 2–10. [CrossRef]
15. Rambabu, K.; Velu, S. Improved performance of $CaCl_2$ incorporated polyethersulfone ultrafiltration membranes. *Period. Polytech. Chem. Eng.* **2016**, *60*, 181–191. [CrossRef]
16. Bharath, G.; Rambabu, K.; Hai, A.; Anwer, S.; Banat, F.; Ponpandian, N. Synthesis of one-dimensional magnetite hydroxyapatite nanorods on reduced graphene oxide sheets for selective separation and controlled delivery of hemoglobin. *Appl. Surf. Sci.* **2020**, *501*, 144215. [CrossRef]
17. Rambabu, K.; Bharath, G.; Banat, F.; Show, P.L. Biosorption performance of date palm empty fruit bunch wastes for toxic hexavalent chromium removal. *Environ. Res.* **2020**, *187*, 109694. [CrossRef] [PubMed]
18. Gnanasundaram, N.; Loganathan, M.; Perumal, K. Solid holdup in liquid solid circulating fluidized bed with viscous liquid medium. *Alexandria Eng. J.* **2014**, *53*, 959–968. [CrossRef]
19. Anjum, H.; Gnanasundaram, N.; Johari, K.; Murugesan, T. Decontamination of benzene from aqueous solution by green functionalization of activated carbon. In Proceedings of the IOP Conference Series: Materials Science and Engineering, Kuala Lumpur, Malaysia, 13–14 August 2018.
20. Oladipupo Kareem, M.; Edathil, A.A.; Rambabu, K.; Bharath, G.; Banat, F.; Nirmala, G.S.; Sathiyanarayanan, K. Extraction, characterization and optimization of high quality bio-oil derived from waste date seeds. *Chem. Eng. Commun.* **2019**, 1–11. [CrossRef]

21. Rambabu, K.; Hai, A.; Bharath, G.; Banat, F.; Show, P.L. Molybdenum disulfide decorated palm oil waste activated carbon as an efficient catalyst for hydrogen generation by sodium borohydride hydrolysis. *Int. J. Hydrogen Energy* **2019**, *44*, 14406–14415. [CrossRef]
22. Rambabu, K.; Show, P.-L.; Bharath, G.; Banat, F.; Naushad, M.; Chang, J.-S. Enhanced biohydrogen production from date seeds by Clostridium thermocellum ATCC 27405. *Int. J. Hydrogen Energy* **2019**, in press. [CrossRef]
23. Padmavathy, K.S.; Madhu, G.; Haseena, P.V. A study on Effects of pH, Adsorbent Dosage, Time, Initial Concentration and Adsorption Isotherm Study for the Removal of Hexavalent Chromium (Cr (VI)) from Wastewater by Magnetite Nanoparticles. *Procedia Technol.* **2016**, *24*, 585–594. [CrossRef]
24. Bharath, G.; Rambabu, K.; Banat, F.; Anwer, S.; Lee, S.; Binsaleh, N.; Latha, S.; Ponpandian, N. Mesoporous hydroxyapatite nanoplate arrays as pH-sensitive drug carrier for cancer therapy. *Mater. Res. Express* **2019**, *6*, 085409. [CrossRef]
25. Gaber, D.; Abu Haija, M.; Eskhan, A.; Banat, F. Graphene as an Efficient and Reusable Adsorbent Compared to Activated Carbons for the Removal of Phenol from Aqueous Solutions. *Water. Air. Soil Pollut.* **2017**, *228*, 320. [CrossRef]
26. Banat, F.; Al-Asheh, S.; Al-Makhadmeh, L. Utilization of raw and activated date pits for the removal of phenol from aqueous solutions. *Chem. Eng. Technol.* **2004**, *27*, 80–86. [CrossRef]
27. Delgado, J.A.; Uguina, M.A.; Sotelo, J.L.; Águeda, V.I.; Gómez, P.; Hernández, V. Modelling the desorption of ethanol from an externally heated saturated activated carbon column by purging with air. *Adsorpt. Sci. Technol.* **2010**, *28*, 689–704. [CrossRef]
28. Anjum, H.; Johari, K.; Gnanasundaram, N.; Appusamy, A.; Thanabalan, M. Investigation of green functionalization of multiwall carbon nanotubes and its application in adsorption of benzene, toluene & p-xylene from aqueous solution. *J. Clean. Prod.* **2019**, *221*, 323–338.
29. Anjum, H.; Johari, K.; Gnanasundaram, N.; Appusamy, A.; Thanabalan, M. Impact of surface modification on adsorptive removal of BTX onto activated carbon. *J. Mol. Liq.* **2019**, *280*, 238–251. [CrossRef]
30. Gnanasundaram, N.; Loganathan, M.; Singh, A. Optimization and Performance parameters for adsorption of Cr6+ by microwave assisted carbon from Sterculia foetida shells. In Proceedings of the IOP Conference Series: Materials Science and Engineering, Miri, Sarawak, Malaysia, 1–3 December 2016.
31. Sarma, G.K.; Sen Gupta, S.; Bhattacharyya, K.G. Nanomaterials as versatile adsorbents for heavy metal ions in water: A review. *Environ. Sci. Pollut. Res.* **2019**, *26*, 6245–6278. [CrossRef]
32. Gnanasundaram, N.; Venugopal, A.; Ullas, G.; Katragadda, Y. Effect of Liquid Viscosity and Solid Inventory on Hydrodynamics in a Liquid-solid Circulating Fluidized Bed. *J. Appl. Fluid Mech.* **2017**, *10*, 267–274.
33. Trivedi, U.; Bassi, A.; Zhu, J.X. (Jesse) Continuous enzymatic polymerization of phenol in a liquid-solid circulating fluidized bed. *Powder Technol.* **2006**, *169*, 61–70. [CrossRef]
34. Edathil, A.A.; Pal, P.; Kannan, P.; Banat, F. Total organic acid adsorption using alginate/clay hybrid composite for industrial lean amine reclamation using fixed-bed: Parametric study coupled with foaming. *Int. J. Greenh. Gas Control* **2020**, *94*, 102907. [CrossRef]
35. Patel, H. Fixed-bed column adsorption study: A comprehensive review. *Appl. Water Sci.* **2019**, *9*, 45. [CrossRef]
36. Shi, S.L.; Lv, J.P.; Liu, Q.; Nan, F.R.; Jiao, X.Y.; Feng, J.; Xie, S.L. Application of Phragmites australis to remove phenol from aqueous solutions by chemical activation in batch and fixed-bed columns. *Environ. Sci. Pollut. Res.* **2018**, *25*, 23917–23928. [CrossRef] [PubMed]
37. Song, J.; Zou, W.; Bian, Y.; Su, F.; Han, R. Adsorption characteristics of methylene blue by peanut husk in batch and column modes. *Desalination* **2011**, *265*, 119–125. [CrossRef]
38. Sriram, A.; Swaminathan, G. Utilization of dye-loaded activated carbon as a potential alternative fuel source: A feasibility study through calorific and thermo-gravimetric analysis. *Environ. Sci. Pollut. Res.* **2018**, *25*, 33140–33152. [CrossRef]
39. Omidi Khaniabadi, Y.; Jafari, A.; Nourmoradi, H.; Taheri, F.; Saeedi, S. Adsorption of 4-chlorophenol from aqueous solution using activated carbon synthesized from aloe vera green wastes. *J. Adv. Environ. Heal. Res.* **2015**, *3*, 120–129.

40. Pal, P.; Edathil, A.A.; Chaurasia, L.; Rambabu, K.; Banat, F. Removal of sulfide from aqueous solutions using novel alginate–iron oxide magnetic hydrogel composites. *Polym. Bull.* **2018**, *75*, 5455–5475. [CrossRef]
41. Rambabu, K.; Srivatsan, N.; Gurumoorthy, A.V.P. Polyethersulfone-barium chloride blend ultrafiltration membranes for dye removal studies. In Proceedings of the IOP Conference Series: Materials Science and Engineering, Vellore, Tamil Nadu, India, 2–3 May 2017; Volume 263.

energies

MDPI

Article

Experimental Investigation of Natural Lighting Systems Using Cylindrical Glass for Energy Saving in Buildings

Atthakorn Thongtha * and Pirompom Boontham

Department of Physics, Faculty of Science, Naresuan University, Phitsanulok 65000, Thailand; Pirompornb57@email.nu.ac.th

* Correspondence: atthakornt@nu.ac.th; Tel.: +665-596-3550

Received: 21 April 2020; Accepted: 13 May 2020; Published: 16 May 2020

Abstract: This research focuses on the use of natural lighting integrated into buildings. Cylindrical glass was fitted into the top of our test model, which was 1 m × 1 m × 1 m, which enhanced the light inside it. The glass fitted comprised a single layer (G), two layers (2G), or two layers of glass filled with distilled water (2GW). Each combination of glass increased the number of glass cylinders from two to six. The nine formats were tested indoors using a light intensity of 1000 W/m^2 and the temperature was controlled at 25 °C. The lowest temperature averaged 34.4 °C, which was recorded using only two glass cylinders that had two layers of glass filled with distilled water. The average internal illumination was 549 lux, which agreed with the CIE standard. Then, the two layers of glass filled with water were examined under natural conditions. It was found that the highest average inside temperature was 40.4 °C at 1:30 p.m. The average illuminant values for three days were in the range of 300–500–750 lux, which concurred with the CIE standard. Additionally, the use of the 2S-2GW resulted in the conservation of electrical energy consumed by the cooling load and the illumination of the building between 9:00 a.m. and 3:00 p.m.

Keywords: energy saving; daylight; heat flux reduction; illumination; CIE standard

1. Introduction

Building energy consumption figures have risen from 26% in 1980 to 54% in 2010 and are predicted to rise to 84% in 2050. They will continuously increase day by day because of the rise in population, the growth of modern society, and quality of life improvements [1,2]. Lighting in offices contributes to approximately 30% of the total energy consumption in buildings [3–6]. In recent years, buildings have been appropriately developed, constructed, and maintained so as to supply their inhabitants with a better environment quality and electrical energy conservation through optimal design and functional practices [6,7].

Natural daylight is a very important source of illumination in buildings because it is free and helps reduce energy consumption [8–10]. Sufficient and effective daylight utilization results in energy saving. The use of natural daylight in buildings will also be significant for the visual and physical comfort and health of the people working within them [6,11–14]. The advance of daylight investigation contributes to energy conservation, improves human welfare, enhances physiological capacities, and avoids disease. One of the consequences of the adjustment of the human eyes to light over time is visual ability, and subsequently, the significant function of the psychophysical levels of daylight for numerous activities is affected. Undoubtedly, daylight conditions can impact the inhabitants of buildings both mentally and physically. Many studies have proven that the hormone melatonin is suppressed by daylighting, which enables us to improve regular light–dark rhythms that can help individuals gain satisfactory rest [15–18]. Daylighting will be used for various applications and it has outstanding solutions to reduce energy consumption for internal building areas. The regular daylight system comprises a top skylight and a window. Windows allow natural light to illuminate the interior of a building while also allowing for the transfer of heat. Some areas of a building cannot be reached by natural light, so they do not heat up as much as the external parts of the structure [19]. The effects of inadequate daylight deeper in a building necessitate extra use of electricity from lighting, which contributes to approximately 30% of total energy consumption [20].

The conservation of artificial lighting during the daytime allows for great energy saving [21,22]. Architectural structures can be designed to provide adequate and efficient daylight for the internal areas of a building, which will result in energy conservation [19,23–26]. Daylight illuminates the inside of a building and transfers infrared radiation, which can be significantly absorbed by water vapor in the air, which is an important cause of thermal gain in buildings [27,28]. As a result of excessive heat gain in the building, ventilating fans and air conditioners are essential for providing a cool, controlled environment, removing the hot air from the building, or cooling the interior air. This is necessary for providing a comfortable environment for the inhabitants. This requires significant power consumption. To significantly reduce thermal accumulation in buildings and appropriately utilize natural lighting, the top of the test box was designed and constructed using wood and glass cylinders filled with distilled water. Our experiment concentrated on testing both thermal performance and the amount of natural light inside the box by using various glass cylinders that were constructed to meet the required heat transfer reduction and energy conservation levels within buildings.

2. Materials and Methods

2.1. Testing Model Design, Temperature, and Illuminant Measurements under Controlled Conditions

The thermal and illuminant performance of our model, which was constructed from wood on five sides and insulated using polyethylene sheets with different configurations of glass inserted into the top of the box, were investigated using nine different cylindrical glass designs. The dimensions of our model had an area of 1 m^2, and a volume of 1 m^3, as shown in Figure 1. The top of the testing unit was interchanged using two glass cylinders (2S), four glass cylinders (4S), and six glass cylinders (6S) to allow light to enter the box, creating illumination within it. Each glass design was constructed as either a single layer of glass (G), two layers of glass (2G) or two layers of glass filled with distilled water (2GW), as shown in Figure 2. Insolation was provided by nine 500 W halogen lamps placed 0.25 m away from the top of the unit. The nine lamps were calibrated to deliver a light intensity of 1000 W/m^2, which was regulated using a basic voltage control device. K-type thermocouples with an accuracy of ±0.5 °C were employed to measure the temperature changes, and were attached to both the exterior and interior surfaces of the glass and the top of the unit using thermal paste to ensure good thermal contact, and were insulated using aluminum foil tape. The ambient temperature in the laboratory was set at approximately 25 °C using an air conditioner. The temperature inside the box was measured by suspending a thermocouple in the center of it. All data were recorded at 2 min intervals continuously for 3 h using a data logger, and the illuminance was measured using a lux meter (DIGICON LX-70).

(a) (b)

Figure 1. (**a**) View of the test unit in controlled conditions. (**b**) Fixed locations of the thermocouples for temperature testing.

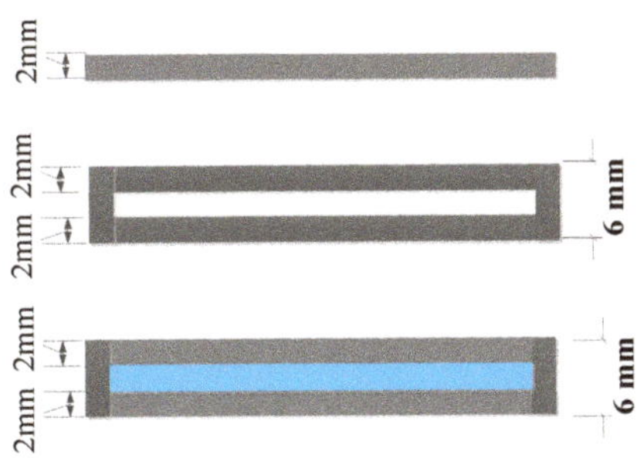

Figure 2. Views of the glass units.

2.2. Testing the Boxes' Thermal Behaviors and Illuminant Measurements under Natural Light and Weather Conditions

The following experiment was conducted under natural light and weather conditions. The surrounding solar radiation intensity swung from 0 to 0.840 kW/m^2 throughout the day between 6:00 a.m. to 6:00 p.m. The surrounding temperature fluctuated between ~24 °C in the morning and the maximum value of approximately 35.8 °C at around 1:00 p.m. Wind speed surrounding the testing area was between 0.10 and 3.50 m/s. The thermal and illuminant performance of the optimum model was further studied using two experimental boxes, as shown in Figure 3. Both test boxes were built to the same specifications using identical materials. The ambient temperature, internal temperature, and exterior and interior surface temperatures of the top of the units were measured and recorded at 5 min intervals continuously for 12 h in actual ambient conditions. A pyranometer was set in an outdoor area to measure the solar radiation intensity, and the wind speed was recorded using an anemometer. The illuminance measurement was recorded using a lux meter (DIGICON LX-70).

Figure 3. View of the testing boxes in actual weather conditions.

3. Results and Discussion

3.1. Surface Temperatures on the Top and Internal Temperatures of Each Model under Controlled Conditions

The exterior (T_{go}) and interior (T_{gi}) glass surface temperatures on the top of the box, the exterior (T_{wo}) and interior (T_{wi}) surface temperatures of the wooden top, and the internal temperature (T_r) and ambient temperature (T_a) of the testing model, using two glass units constructed from two layers of glass sealed with air inside (2S-2G), are shown in Figure 4. While controlling the artificial light intensity at 1000 W/m^2, the temperature variations in each position were observed for 180 min. When the testing time exceeded 40 min, the temperature in each location remained stable. The temperature values for longer than 40 min were calculated to determine the average temperatures in each position, as exhibited in Figures 5 and 6. For the 2S-2G, the average temperatures of T_{go}, T_{gi}, T_{wo}, T_{wi}, T_r, and T_a were approximately 58.7 °C, 63.5 °C, 66.8 °C, 77.7 °C, 37.2 °C, and 24.9 °C, respectively, as shown in Figures 4–6. The average temperatures of the other glass units were similar to those of the 2S-2G, as exhibited in Figures 5 and 6.

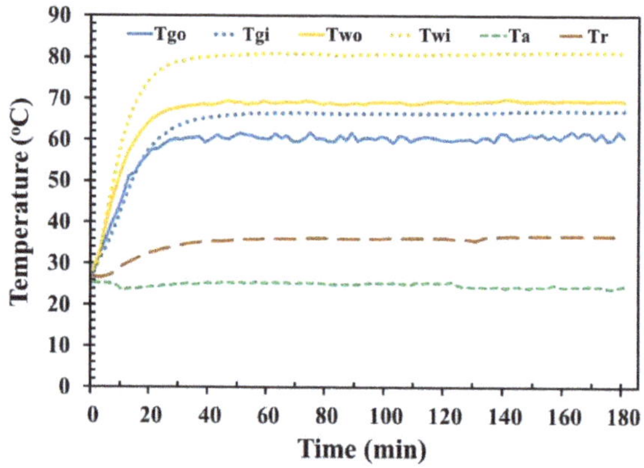

Figure 4. Temperature variations at different locations of the 2S-2G glass units.

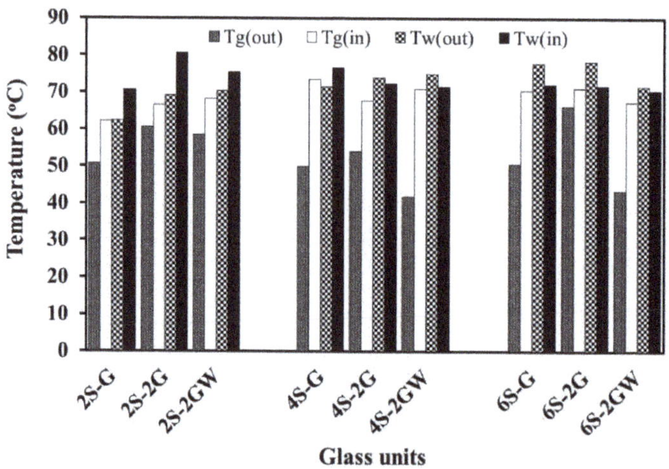

Figure 5. Comparison of the internal and external temperatures of the top surface of the box and the nine glass units.

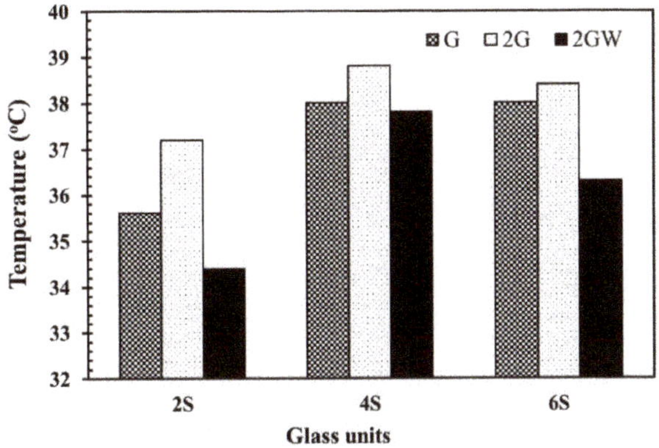

Figure 6. Comparison of internal unit temperatures of each configuration of the glass units.

The internal temperatures of the 2S-G, 2S-2G, 2S-2GW, 4S-G, 4S-2G, 4S-2GW, 6S-G, 6S-2G, and 6S-2GW, increased until they reached a steady value at a testing time of longer than 40 min, as shown in Figure 4. The average internal unit temperatures of the models were approximately 35.6 °C, 37.2 °C, 34.4 °C, 38.0 °C, 38.8 °C, 37.8 °C, 38.0 °C, 38.4 °C, and 36.3 °C, respectively, as shown in Figure 6. It was observed that the internal temperature of the 2S-2GW was lower than those of the 2S-G, 2S-2G, 4S-G, 4S-2G, 4S-2GW, 6S-G, 6S-2G, and 6S-2GW units, which were 1.2 °C, 2.8 °C, 3.6 °C, 4.4 °C, 3.4 °C, 3.6 °C, 4.0 °C and 1.9 °C, respectively. The two glass units constructed from two layers of glass sealed with distilled water inside (2S-2GW) reduced the inside unit temperature by more than 3% when compared with the other glass units. This demonstrated that a decrease in heat propagation from the exterior surface to the interior was achieved in the test unit. This clearly indicates that the optimum glass units, which were filled with distilled water, demonstrated optimal insulating properties, which led to a reduction in the heat transmission load. The lower interior temperature of the box created by the 2S-2GW unit could lead to energy savings in buildings and a significant reduction in the yearly peak cooling requirement [29–31].

3.2. Indoor Illuminance under Controlled Conditions

The average illuminances of the nine different glass designs (2S-G, 2S-2G, 2S-2GW, 4S-G, 4S-2G, 4S-2GW, 6S-G, 6S-2G, and 6S-2GW) are shown in Figure 7. When using the 2S-G, 2S-2G, and 2S-2GW in a two-unit configuration, the average interior illuminances decreased from 710 lux to 549 lux, respectively. The average illuminances of the 4S and 6S were similar to those of the 2S, as displayed in Figure 7. Although the average interior illuminances decreased when the model of the two layers of glass filled with distilled water was used instead of the single glass unit, the illuminance level of this configuration was in the range of 300–500–750 lux, which is also based on the International Commission on Illumination standard. Additionally, this unit reduced the room temperature by more than 3% when compared with the single glass unit. When using the single layer of glass in a configuration of two, four, or six units, the average illuminance values increased from 710 lux to 748 lux, respectively, as shown in Figure 7. For the cases of the two-layer glass (2G) and the one filled with distilled water (2GW), the average illuminances were similar to that of a single layer of glass (G), as shown in Figure 7. This demonstrated that an increase in the number of glass units directly increased the illumination. The average interior illuminance of each of the nine configurations was in the range of 300–500–750 lux, which achieved the International Commission on Illumination standard (CIE standard).

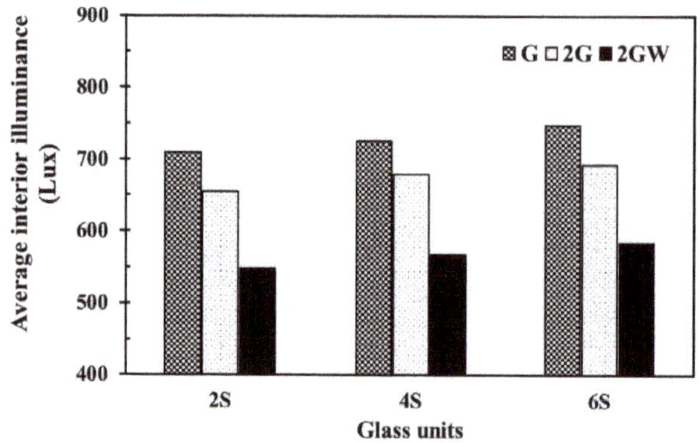

Figure 7. Comparison of illuminance of each of the configurations of the glass units.

3.3. Comparison of Thermal Behavior and Inside Illuminance under Natural Weather Conditions

The difference between the thermal behaviors and interior illuminance of two of the models was investigated using two glass units, with two layers of glass filled with distilled water (2S-2GW) in only one of the models under natural weather conditions, which were concurrently tested from 6:00 a.m. to 6:00 p.m. daily for three days, giving a 12 h test cycle (29 October, 31 October, and 4 November 2019). The fluctuations in the solar radiation, ambient temperature, interior surface temperature, exterior surface temperature, and room temperature of the two models were examined and compared, as illustrated in Figures 8–11 and Tables 1 and 2.

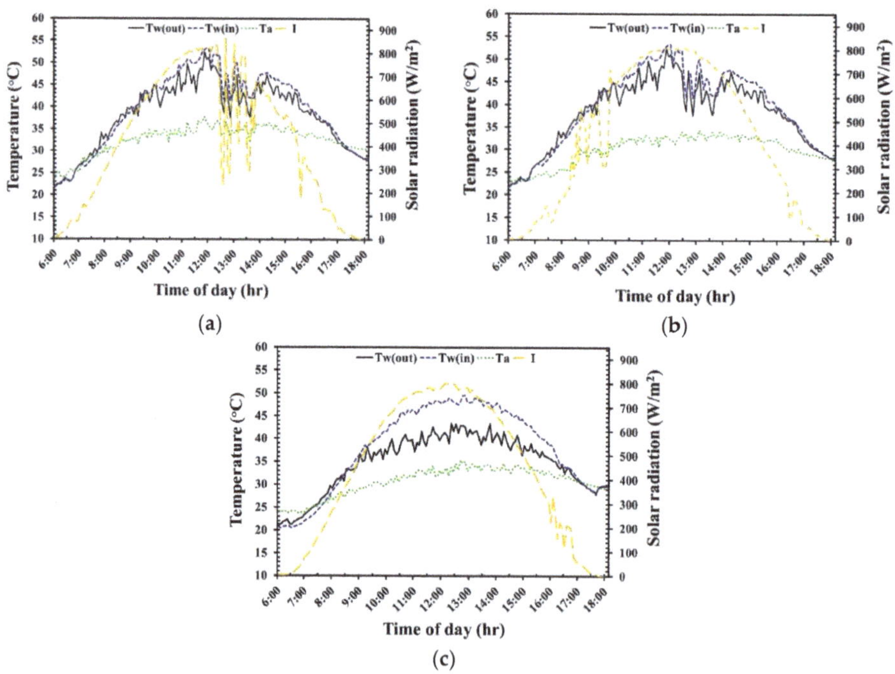

Figure 8. Reference testing of the exterior and interior surface temperatures of the box under natural weather conditions: (**a**) 29 October 2019, (**b**) 31 October 2019, and (**c**) 4 November 2019.

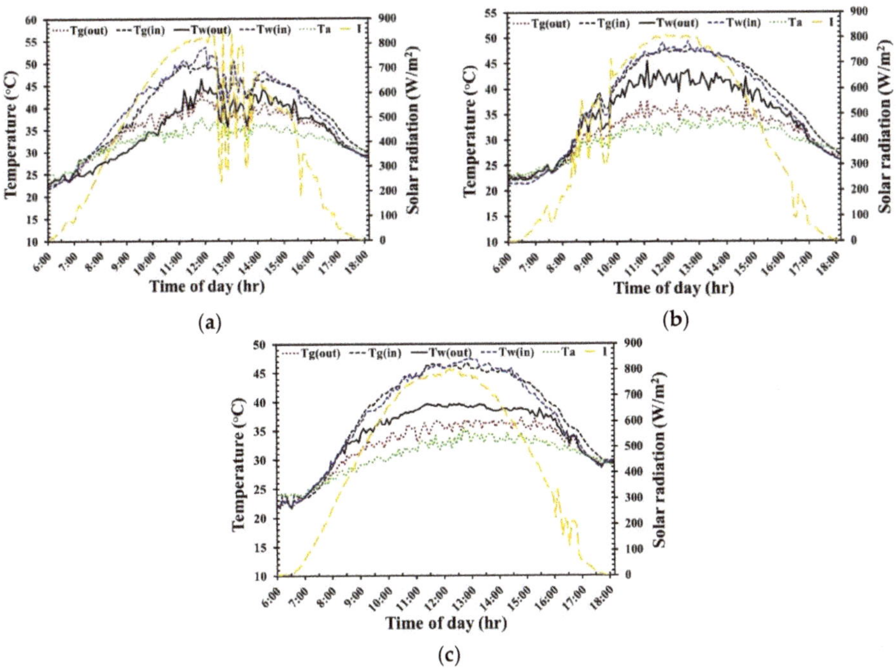

Figure 9. Testing the exterior and interior surface temperatures of the 2S-2G glass unit under natural weather conditions: (**a**) 29 October 2019, (**b**) 31 October 2019, and (**c**) 4 November 2019.

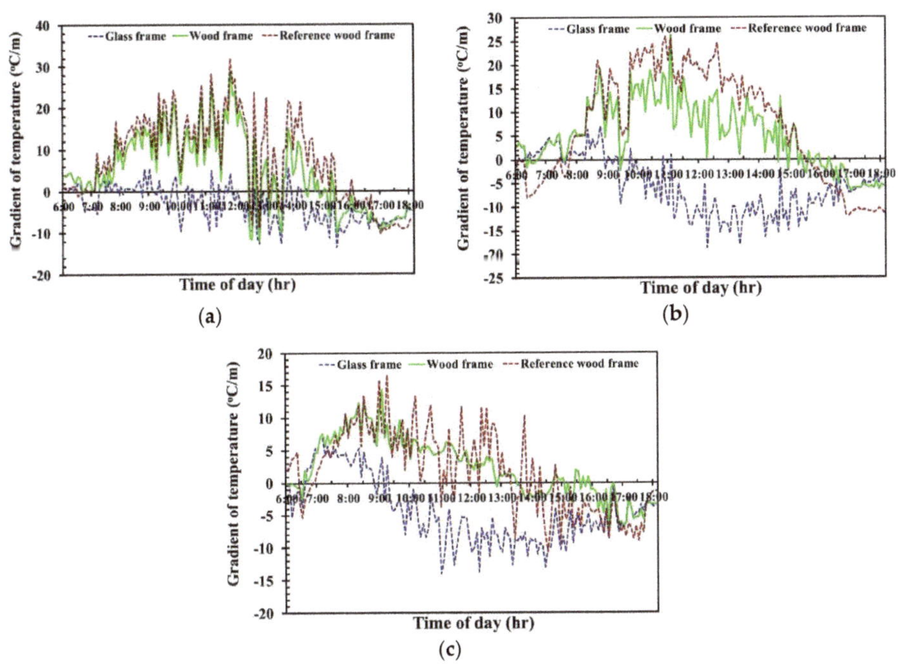

Figure 10. Comparison of the temperature gradients of both boxes under natural weather conditions: (**a**) 29 October 2019, (**b**) 31 October 2019, and (**c**) 4 November 2019.

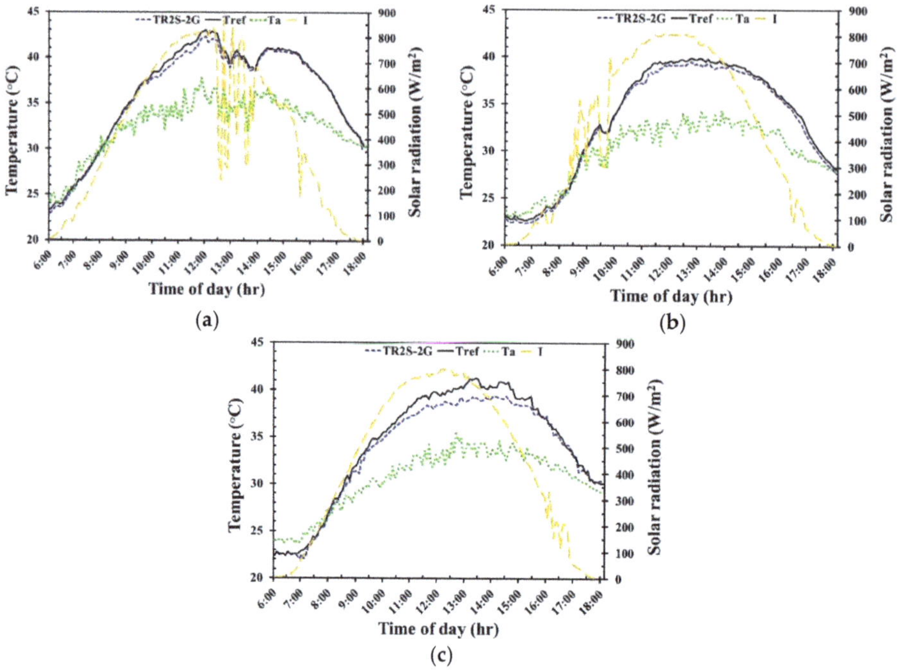

Figure 11. Inside temperatures of the two testing units under natural weather conditions: (**a**) 29 October 2019, (**b**) 31 October 2019, and (**c**) 4 November 2019.

Table 1. Average highest temperatures at different locations of the reference and 2S-2GW glass models under natural weather conditions for three days.

Day	Type of Model									
	Reference				2S-2GW					
	T_{wo} (°C)	T_{wi} (°C)	T_r (°C)	T_a (°C)	T_{wo} (°C)	T_{wi} (°C)	T_{go} (°C)	T_{gi} (°C)	T_r (°C)	T_a (°C)
1st	52.4	53.3	43.0	37.8	46.4	53.6	43.4	49.8	42.3	37.8
2nd	47.0	49.3	39.9	34.4	45.4	49.3	37.8	47.8	39.5	34.4
3rd	43.4	49.6	41.3	35.3	39.7	47.4	37.3	46.7	39.4	35.3
Avg.	47.6	50.7	41.4	35.8	43.8	50.1	39.5	48.1	40.4	35.8

Table 2. Average temperatures at various positions of the reference and 2S-2GW glass units under natural weather conditions.

Day	Type of Model									
	Reference				2S-2GW					
	T_{wo} (°C)	T_{wi} (°C)	T_r (°C)	T_a (°C)	T_{wo} (°C)	T_{wi} (°C)	T_{go} (°C)	T_{gi} (°C)	T_r (°C)	T_a (°C)
1st	39.4	40.6	37.1	33.2	35.7	41.0	35.7	40.5	36.7	33.2
2nd	34.9	36.1	33.1	29.8	34.1	36.4	31.1	36.8	32.7	29.8
3rd	34.7	37.7	34.3	30.6	34.3	37.3	32.3	37.7	33.6	30.6
Avg.	36.3	38.1	34.8	31.2	34.7	38.2	33.0	38.3	34.3	31.2

3.3.1. Surface Temperature Fluctuation

Periodically, clouds altered the intensity of solar radiation throughout the day from 6:00 a.m. to 6:00 p.m. with a maximum intensity of approximately 0.840 kW/m^2 at around midday, as shown in Figures 8, 9 and 11. The surrounding temperature was affected by both the weather conditions and the wind speed which was recorded at 0.10 to 3.50 m/s. The surrounding temperature fluctuated between ~24 °C in the morning, around 6:00 a.m., and approximately 35.8 °C at around 1:00 p.m.

The surface temperature fluctuation of the reference model is shown in Figure 8 and Table 1. The average maximum exterior and interior temperatures of the wooden top, and the ambient temperatures for three days reached as high as 47.6 °C, 50.7 °C, and 35.8 °C, at around 12:00, 12:30, and 1:00 p.m., respectively, and then decreased in value. The average exterior and interior surface temperatures of the wooden top, and the ambient temperatures for three days, were approximately 36.3 °C, 38.1 °C and 31.2 °C, as listed in Table 2, which were observed consecutively from 6:00 a.m. to 6:00 p.m. each day.

The surface temperature variation of the 2S-2GW model is exhibited in Figure 9 and Table 1. The average highest exterior and interior temperatures of the wooden top, and the glass top, including the ambient temperatures for three days, reached as high as 43.8 °C, 50.1 °C, 39.5 °C, 48.1 °C, and 35.8 °C, at around 12:00 p.m., 1:00 p.m., 12:00 p.m., 1:00 p.m., and 1:00 p.m., respectively, and then decreased in value, while the overall averages were approximately 34.7 °C, 38.2 °C, 33.0 °C, 38.3 °C, and 31.2 °C, respectively, as shown in Table 2, which were observed consecutively from 6:00 a.m. to 6:00 p.m. each day.

3.3.2. Temperature Gradient

The temperature gradients on both the interior and exterior surfaces of the top of the two test units were investigated for three days, as shown in Figure 10. The temperature gradients on the top surface of the units were positive when the outer surface temperatures were higher than those inside, while the temperature gradients on the top surfaces were negative when the outer surface temperatures were less than those of the interior. The temperature gradients of the 2S-2GW and reference models had a positive gradient of temperature for three days, with the highest average values being approximately 23 °C/m and 26 °C/m around midday, which then decreased in value during the afternoon. The temperature gradients of the top of the 2S-2GW glass units after 9:00 a.m. were negative when the temperature on the outer surface was lower than that of the inside of the model. The average maximum negative value on the top of the 2S-2GW glass unit surfaces was approximately 15 °C/m. This shows that the heat from the surrounding weather can obviously transfer to the inside of the box during the daytime, while part of the heat can flow through the glass and back into the ambient weather at the same time. This relates to a decrease in heat accumulation within the box when using the 2S-2GW glass unit, and implies a conservation of energy consumption from cooling loads in buildings, which is clearly a significant result.

3.3.3. Indoor Space Temperature and Illuminance

Figure 11 presents the variations of the inside space temperatures of the 2S-2GW and reference models. The natural fluctuation in solar radiation, surrounding temperature, and exterior and interior surface temperatures on the top of the boxes affected the inside space temperatures, which varied throughout the day, and were recorded from 6:00 a.m. to 6:00 p.m. for three days. It was observed that the inside temperatures of the two test units were approximately equal between 6:00 a.m. and 10:00 a.m., and then after 10:30 a.m. their temperatures started to diverge, which was due to different peak temperatures being reached at various times. The interior temperatures of the 2S-2GW and reference units rose more rapidly, and achieved the highest peak value of 40.4 °C and 41.4 °C, as shown in Table 1, which averaged at 12:30 p.m. and 12:00 p.m., respectively. The average inside temperatures of the 2S-2GW and reference test unit were 34.3 °C and 34.8 °C, respectively, over a period of three days, as shown in Table 2. It was observed that the peak and average temperatures inside the 2S-2GW were lower than the reference model by around 1.0 °C and 0.5 °C, which demonstrates a decrease in heat propagation from the exterior surface to the interior. Using the 2S-2GW glass units to reduce the temperature in a room allows for energy saving in buildings and a significant decline in the yearly peak cooling requirement [29–31].

The average indoor illuminance of the 2S-2GW is shown in Figure 12. The average indoor illuminance increased and reached its highest at around midday, and then dropped in value, which was recorded between 300 and 750 lux from 9:00 a.m. to 3:00 p.m. The interior illuminance of the 2S-2GW glass unit was in the range 300–500–750 lux, which adhered to the International Commission on Illumination standard (CIE standard). If this glass unit was to be adopted into buildings according to geographical locations in the tropics, the energy consumption reduction in both the cooling load of air conditioners and artificial lighting would be achieved, creating a significant decline in the yearly peak energy consumption requirement. Although a life cycle assessment of the 2S-2GW roof window costs has not been investigated in this specific context, we may analyze a more widespread context in the next work. In view of long-term benefits, the design using the two glass units that had two layers of glass filled with distilled water can reduce the internal space temperature at the highest peak and provide adequate and efficient daylight illuminance, which leads to this new form of knowledge.

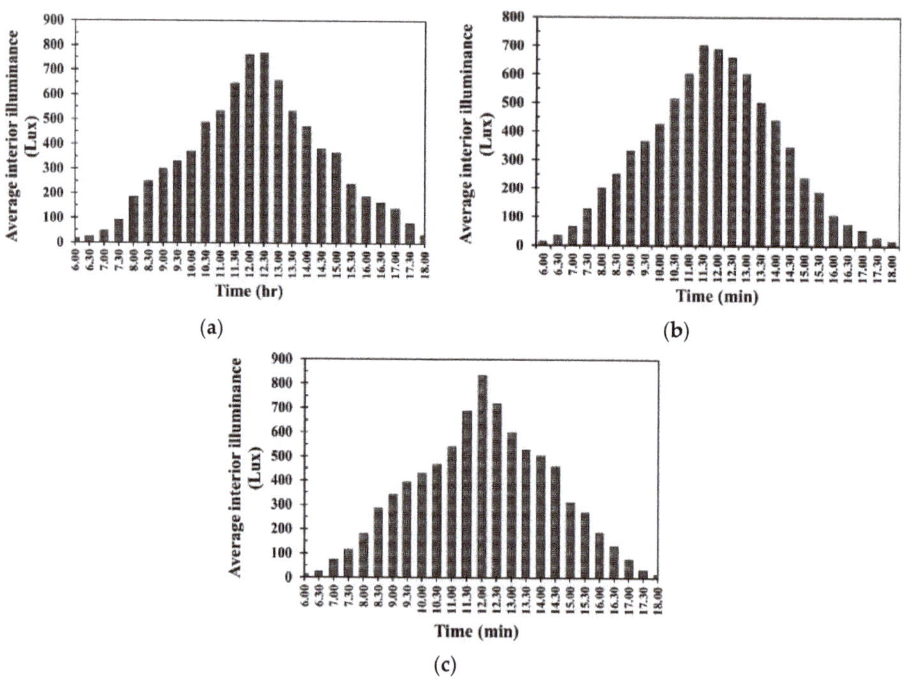

Figure 12. Variability of the average indoor illuminance under natural weather conditions: (**a**) 29 October 2019, (**b**) 31 October 2019, and (**c**) 4 November 2019.

4. Conclusions

This paper contained sufficient contributions which were based on sound scientific knowledge. The application of natural light integrated into the test box was successfully investigated by using the novel glass units, which provided adequate and efficient daylight illuminance into its interior and decreased heat transmission through the building frames. The average highest room temperature of the two glass units that had two layers of glass filled with distilled water decreased the value at the highest peak of the day by 1.0 °C when compared with that of the reference testing model under natural weather conditions. The average internal illuminant values for three days were in the range of 300 to 750 lux from 9:00 a.m. to 3:00 p.m., which agreed with the standard of the International Commission on Illumination. Lower room temperatures were achieved by utilizing the integration of natural light into buildings which led to the conservation of energy consumption from the cooling load of air conditioners and lighting systems. This demonstrates a significant decline in yearly peak cooling and lighting energy consumption.

Author Contributions: Conceptualization, A.T.; methodology, A.T. and P.B.; formal analysis, A.T. and P.B.; investigation, A.T. and P.B.; resources, A.T.; data curation, A.T.; writing—original draft preparation, A.T. and P.B.; writing—review and editing, A.T.; visualization, A.T.; supervision, A.T.; project administration, A.T.; funding acquisition, A.T. All authors have read and agreed to the published version of the manuscript.

Funding: This research was funded by the Faculty of Science, Naresuan University, grant number R2563E026.

Acknowledgments: The authors would like to thank the Thailand Research Fund (TRF), the Faculty of Science, Naresuan University for providing financial support to this research work, and their research center. Thanks are given to Peter Barton from the Division of International Affairs and Language Development (DIALD), Naresuan University, for editing the English in this document.

Conflicts of Interest: The authors declare no conflict of interest.

References

1. Ürge-Vorsatz, D.; Cabeza, L.F.; Serrano, S.; Barreneche, C.; Petrichenko, K. Heating and cooling energy trends and drivers in buildings. *Renew. Sustain. Energy Rev.* **2015**, *41*, 85–98. [CrossRef]
2. Ürge-Vorsatz, D.; Petrichenko, K.; Staniec, M.; Eom, J. Energy use in buildings in a long-term perspective. *Curr. Opin. Environ. Sustain.* **2013**, *5*, 141–151. [CrossRef]
3. Caicedo, D.; Pandharipande, A.; Willems, F.M. Daylight-adaptive lighting control using light sensor calibration prior-information. *Energy Build.* **2014**, *73*, 105–114. [CrossRef]
4. Doulos, L.T.; Sioutis, I.; Tsangrassoulis, A.; Canale, L.; Faidas, K. Revision of threshold luminance levels in tunnels aiming to minimize energy consumption at no cost: Methodology and case studies. *Energies* **2020**, *13*, 1707. [CrossRef]
5. Alrubaih, M.S.; Zain, M.F.M.; Alghoul, M.A.; Ibrahim, N.L.N.; Shameri, M.A.; Elayeb, O. Research and development on aspects of daylighting fundamentals. *Renew. Sustain. Energy Rev.* **2013**, *21*, 494–505. [CrossRef]
6. Yoo, S.; Kim, J.; Jang, C.Y.; Jeong, H. A sensor-less LED dimming system based on daylight harvesting with BIPV systems. *Opt. Express* **2014**, *22*, A132–A143. [CrossRef]
7. Saxena, R.; Biplab, K.; Rakshit, D. Quantitative assessment of phase change material utilization for building cooling load abatement in composite climatic condition. *J. Sol. Energy Eng.* **2017**, *140*, 1–15. [CrossRef]
8. Kim, I.T.; Kim, Y.S.; Cho, M.; Nam, H.; Choi, A.; Hwang, T. High-performance accuracy of daylight-responsive dimming systems with illuminance by distant luminaires for energy-saving buildings. *Energies* **2019**, *12*, 731. [CrossRef]
9. Kim, C.H.; Kim, K.S. Development of sky luminance and daylight illuminance prediction methods for lighting energy saving in office buildings. *Energies* **2019**, *12*, 592. [CrossRef]
10. Freewan, A.A.Y. Impact of external shading devices on thermal and daylighting performance of offices in hot climate regions. *Sol. Energy* **2014**, *102*, 14–30. [CrossRef]
11. Berardi, U.; Wang, W. Daylighting in an atrium-type high performance house. *Build. Environ.* **2014**, *76*, 92–104. [CrossRef]
12. Kwon, C.W.; Lee, K.J. Integrated daylighting design by combining passive method with daysim in a classroom. *Energies* **2018**, *11*, 3168. [CrossRef]
13. Kaminska, A.; Ożadowicz, A. Lighting control including daylight and energy efficiency improvements analysis. *Energies* **2018**, *11*, 2166. [CrossRef]
14. Lee, K.S.; Han, K.J.; Lee, J.W. The impact of shading type and azimuth orientation on the daylighting in a classroom–focusing on effectiveness of façade shading, comparing the results of DA and UDI. *Energies* **2017**, *10*, 635. [CrossRef]

15. Van der Rhee, H.J.; de Vries, E.; Coebergh, J.W. Regular sun exposure benefits health. *Med. Hypotheses* **2016**, *97*, 34–37. [CrossRef]
16. Leslie, R.P. Capturing the daylight dividend in buildings: Why and how? *Build. Environ.* **2003**, *38*, 381–385. [CrossRef]
17. Kechichian, E.; Ezzedine, K. Vitamin D and the Skin: An Update for Dermatologists. *Am. J. Clin. Dermatol.* **2018**, *19*, 223–235. [CrossRef]
18. Valles, X.; Alonso, M.H.; Lopez-Caleya, J.F.; Diez-Obrero, V.; Dierssen-Sotos, T.; Lope, V.; Molina-Barcelo, A.; Chirlaque, M.D.; Jimenez-Moleon, J.J.; Tardon, G.F.; et al. Colorectal cancer, sun exposure and dietary vitamin D and calcium intake in the MCC-Spain study. *Environ. Int.* **2018**, *121*, 428–434. [CrossRef]
19. Kim, J.T.; Kim, G. Overview and new developments in optical daylighting systems for building a healthy indoor environment. *Build. Environ.* **2010**, *45*, 256–269. [CrossRef]
20. Pandharipande, A.; Caicedo, D. Daylight integrated illumination control of LED systems based on enhanced presence sensing. *Energy Build.* **2011**, *43*, 944–950. [CrossRef]
21. Musumeci, S.; Bojoi, R.; Armando, E.; Borlo, S.; Mandrile, F. Three-legs interleaved boost power factor corrector for high-power LED lighting application. *Energies* **2020**, *13*, 1728. [CrossRef]
22. Cannavale, A.; Ayr, U.; Fiorito, F.; Martellotta, F. Smart electrochromic windows to enhance building energy efficiency and visual comfort. *Energies* **2020**, *13*, 1449. [CrossRef]
23. Acosta, I.; Campano, M.Á.; Amarillo, S.D.; Muñoz, C. Dynamic daylight metrics for electricity savings in offices: Window size and climate smart lighting management. *Energies* **2018**, *11*, 3143. [CrossRef]
24. Galatioto, A.; Beccali, M. Aspects and issues of daylighting assessment: A review study. *Renew. Sustain. Energy Rev.* **2016**, *66*, 852–860. [CrossRef]
25. Sibley, M. Let There Be Light! Investigating Vernacular Daylighting in Moroccan Heritage Hammams for Rehabilitation, Benchmarking and Energy Saving. *Sustainability* **2018**, *10*, 3984. [CrossRef]
26. Baglivo, C.; Bonomolo, M.; Congedo, P.M. Modeling of light pipes for the optimal disposition in buildings. *Energies* **2019**, *12*, 4323. [CrossRef]
27. Chylek, P.; Lesins, G.B.; Videen, G.; Wong, J.G.D.; Pinnick, R.G.; Ngo, D.; Klett, J.D. Black carbon and absorption of solar radiation by clouds. *J. Geophys. Res. Atmos.* **1996**, *101*, 23365–23371. [CrossRef]
28. Zhang, Q.; Thompson, J.E. A model for absorption of solar radiation by mineral dust within liquid cloud drops. *J. Atmos. Solar Terres. Phys.* **2015**, *133*, 121–128. [CrossRef]
29. Maneewan, S.; Janyoosuk, K.; Chan, H.Y.; Thongtha, A. Incorporating black dust into autoclaved aerated concrete wall for heat transfer reduction. *J. Metals Mater. Miner.* **2019**, *29*, 82–87.
30. Thongtha, A.; Khongthon, A.; Boonsri, T.; Chan, H.Y. Thermal effectiveness enhancement of autoclaved aerated concrete wall with PCM-contained conical holes to reduce the cooling load. *Materials* **2019**, *12*, 2170. [CrossRef]
31. Punlek, C.; Maneewan, S.; Thongtha, A. Phase change material coating on autoclaved aerated lightweight concrete for cooling load reduction. *Mater. Sci. (MEDŽIAGOTYRA)* **2017**, *23*, 145–149. [CrossRef]

energies

MDPI

Article

Experimental Investigation of Illumination Performance of Hollow Light Pipe for Energy Consumption Reduction in Buildings

Jiraphorn Mahawan and Atthakorn Thongtha *

Department of Physics, Faculty of Science, Naresuan University, Phitsanulok 65000, Thailand; mjiraphorn@yahoo.com

* Correspondence: atthakornt@nu.ac.th; Tel.: +665-596-3550

† This paper is an extended and revised article presented at the International Conference on Sustainable Energy and Green Technology 2019 (SEGT 2019) on 11–14 December 2019 in Bangkok, Thailand.

Abstract: This work investigates the light illumination intensity, light transmission performance, light distribution on the floor, and daylight factor of vertical light hollow tubes at various incident elevation angles of a light source. The light tubes were made from commercial aluminum alloy sheets and commercial zinc alloy sheets to investigate internal illuminance for buildings and reducing the demand of electrical energy from artificial lighting. The vertical light tubes with a constant length of 0.5 m and diameters of 0.20, 0.25, and 0.30 m were designed in a testing room model, with dimensions of 1 m × 1 m × 1 m. A 20-W light-emitting diode (LED) lamp was used as the light source for the lighting simulations, which was placed away from the top of the light tube. The incident elevation angle of the light source was changed between 0° and 80° with 5° increments. It was found that the elevation angle of the incidence light had an influence on the light intensity distribution on both ends of light tube. The average illuminance performance of both material types increased with an increase of the incidence angle from 0° to 80° and an increase of the tube diameter from 0.20 m to 0.30 m. The commercial aluminum alloy tube promotes greater light transmission and daylight factor when compared with the commercial zinc alloy tube in each condition. This illuminance measurement demonstrates that the light tube could be included in the lighting systems of some deeper or windowless areas of buildings to decrease the demand of energy consumption in the lighting of buildings.

Keywords: daylight; light pipe; light transmission; daylight factor; illumination

Citation: Mahawan, J.; Thongtha, A. Experimental Investigation of Illumination Performance of Hollow Light Pipe for Energy Consumption Reduction in Buildings . *Energies* **2021**, *14*, 260. https://doi.org/10.3390/en14020260

Received: 19 November 2020
Accepted: 4 January 2021
Published: 6 January 2021

1. Introduction

A high level of solar radiation is experienced throughout most of Thailand throughout the year. The average daily value of solar radiation is around 18.2 MJ/m^2-day with the highest levels of solar radiation experienced in the months of April and May, between 20 and 24 MJ/m^2-day [1–3]. These conditions cause significant thermal accumulation in buildings. Therefore, energy consumption in buildings is due mainly to the air conditioning system and lighting, which has been increasing due to the energy demands of modern society to improve the comfort of the occupants. Approximately 30% of total energy consumption in residential and commercial buildings is demand for artificial lighting. Nowadays, buildings are designed and constructed to provide their occupants with a better-quality environment and to improve energy conservation with optimal designs and functional practices [4,5].

Daylight is one solution to save energy consumption in buildings because it is free and a valuable light source for internal building areas throughout the day [6–8]. Effective daylight utilization can result in energy savings. The use of natural daylight in buildings also significantly improves the visual and physical comfort of the building. Individuals spend most of their time inside [9–13]. Windows permit daylight to illuminate interior building spaces, but parts of the deeper internal building areas do not obtain daylight, and heat transfers through windows as well [14]. Effects of insufficient daylight within deeper building areas and thermal accumulation in buildings due to daylight illumination mean that the artificial lighting system can be an option for illumination and contributes to reducing approximately 30% of total building energy consumption [15].

Reducing the artificial lighting energy consumption during the daytime is an issue for energy savings [13,16,17]. The design of architectural structures that can carry adequate daylight into the internal building areas results in energy conservation for both illumination and the air conditioning system [14,15,18–23]. Light pipes are an alternative way to provide daylight into indoor spaces of buildings, which is useful for spaces with or without glazing opening [24,25]. Light pipe systems may be straight or have bends. Commercial light pipes can be defined as a hollow tube to allow the illumination into deeper parts of buildings that do not receive sufficient daylight. Light pipes can be responsible for reducing the electricity consumption of artificial lighting systems [5].

Many studies have investigated various components of light pipe systems to improve the performance, such as integrating other functions such as ventilation. The integration of a light pipe system into a natural stack ventilation and solar heating was investigated by Shao and Riffat [26]. A horizontal light pipe with a trapezoidal shape was designed by Canziani et al. who used an active reflector to track the solar rays and further improved illumination into deeper areas by increasing the uniformity [27]. Light tubes with apertures attached have also been used to transmit daylight [28]. Uniform lighting levels on different levels of buildings can be provided by vertical light pipes, lighting multiple areas [29]. An illuminance proportion of 14% under cloudy sky and 7% for sunny conditions in winter and cloudy conditions was investigated using the light pipes [30]. Mohelnikova evaluated the efficiency of light pipes to be between 0.2 and 0.5, when different diameters of light pipes were studied [31]. Apart from these, the term "daylight penetration factor" (DPF) of light pipes was used to determine the performance of a daylight system, which describes the relation of the internal illuminance due to a light pipe against the total external illuminance [32–34]. A good lighting system requires 1–2% daylight factor (DF) for activity in residence and 2–4% DF for activities in office buildings under International Commission on Illumination (CIE) standard for overcast conditions [35]. The relationship of the average inside illuminance and different diameters of light pipes was investigated by Vasilakopoulou et al. [36]. The performance of light pipes was experimentally investigated under subtropical climates in Instanbul [25], Hong-Kong [37], Korea [38], Beijing [39], and Jordan [40] which demonstrated that good results and a uniform light distribution can be provided to buildings using light pipes.

In Thailand, aluminum and zinc alloy sheet metal were generally used as roofing and siding material of buildings because of its lightweight, superior corrosion resistance and higher chemical durability than steel sheets [41–43]. Our previous presented work [44] designed and constructed the light tubes with a fixed length of 0.5 m and the different diameters of 0.20, 0.25, and 0.30 m, which were made from commercial aluminum and zinc alloy sheets. That work demonstrated only the reflection performance of hollow light tubes at each condition. To appropriately consider and utilize the illumination of the vertical light hollow tubes for transmitting light into buildings, this current investigation was focused on examining the improved illumination distribution at the top and bottom ends, light transmission performance, the internal illuminance distribution on the floor plane, and the daylight factor at various incident angles of the light source. Furthermore, the correlation to these obtained experimental values of the vertical light hollow tubes in each material type at different diameters and incident elevation angle was also investigated and compared.

2. Materials and Methods

The light tubes were produced with either a commercial aluminum alloy sheet or a zinc alloy sheet. Both were designed as tubes with a length of 0.5 m and different diameters of 0.20 m, 0.25 m, and 0.30 m. Each light tube was installed on the top of a testing room to allow light to transmit into the interior space. The six sides of the testing room were built using wood. The model room's walls each had an area of 1 m^2 and a volume of 1 m^3, as shown in Figure 1a. A 20-W artificial LED lamp was used as the light source. The illumination changes, as the elevation angle was varied between 0° and 80° with a step size of 5°, are exhibited in Figure 1. The illumination at nine positions—at the top and bottom ends—of each light tube was measured by using an illuminance lux meter (DIGICON LX-70) which is also based on the International Commission on Illumination standard (CIE standard) as shown in Figure 2. All values at the top and bottom side ends were calculated to determine the average illuminance of vertical aluminum and zinc alloy tubes with the diameters of 0.20 m, 0.25 m, and 0.30 m. The average luminous intensity at the top and bottom side ends was evaluated to identify the improved light transmission efficiency. The illuminance distribution in the model area was also tested and measured at 25 locations using an illuminance lux meter according to the CIE standard, as exhibited in Figure 3. The daylight factor was defined as the proportion of average internal illumination on the floor plane and the related luminance at ambient areas on horizontal and unshaded areas.

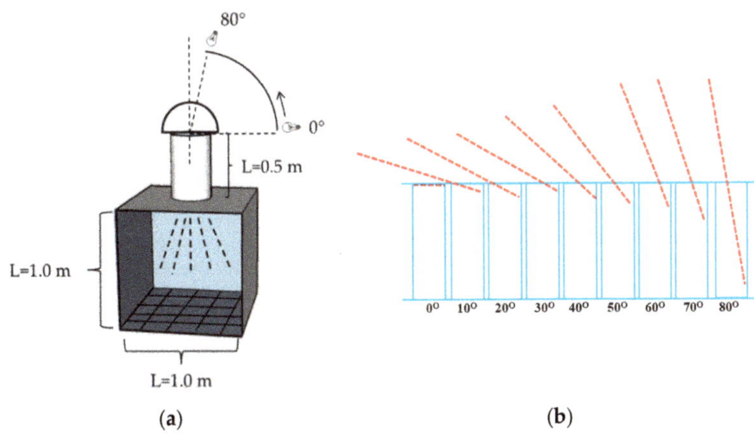

Figure 1. (**a**) View of the testing room. (**b**) Incident angle.

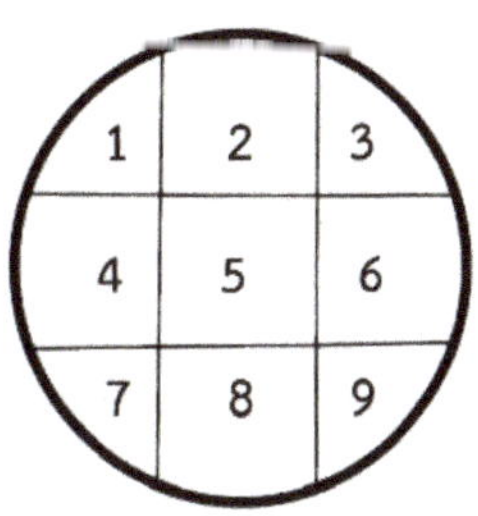

Figure 2. Fixed location of the nine illuminance measurements at the top and bottom end positions of light tube.

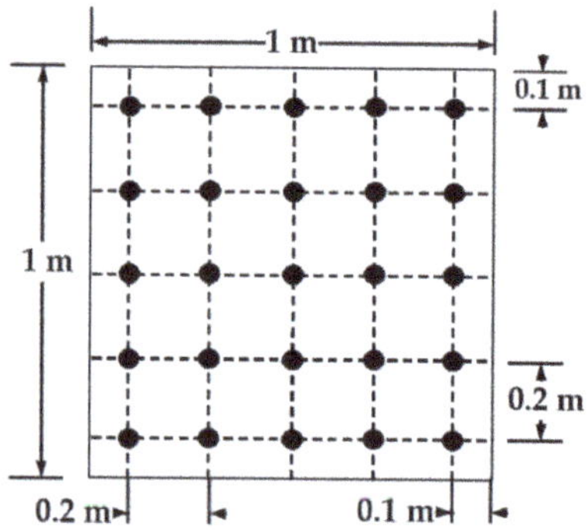

Figure 3. Fixed positions of the 25 illumination measurements on the floor plane.

3. Results and Discussion

Figure 4 presents the internal illumination contours of the top positions of the aluminum alloy light tube, with a diameter of 0.25 m, at various incident light angles. A color map on the two-dimensional horizontal plane demonstrates the illuminance distribution at the top position of the hollow light pipe. The illumination of the top position was between 248 lux and 282 lux at the incident light angle of 0°, which was uniformly distributed. While increasing the incident light angle from 0° to 30°, the illuminance distribution at the top position was nearly uniform in value. When the incident light angle increased from 30° to 80°, the internal illuminance distribution became nonuniform in some positions, as exhibited in Figure 4d–f. As the incident light angle increased from 0° to 80°, the internal illumination distribution at the top end changed in each condition. The illuminance value at each position on the top increased in value when the incident angle increased, as displayed in Figure 4.

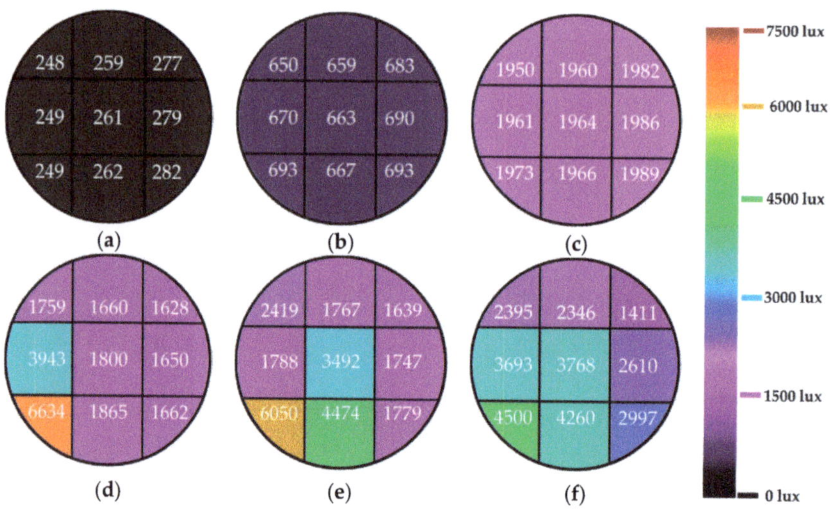

Figure 4. Internal illumination contours of the aluminum alloy tube on the top end for the incident light angles of (**a**) 0°, (**b**) 15°, (**c**) 30°, (**d**) 45°, (**e**) 60°, and (**f**) 75°.

Figure 5 illustrates illumination distribution of the top end of the zinc alloy hollow light pipe with a diameter of 0.25 m with various angles of incidents. At the incident angle of 0°, the illumination of the end was between 165 lux and 199 lux, which uniformly illuminated the area, as shown in the color map. Increasing the incident angle from 0° to 30° led to nearly uniform illuminance for all angles. When the incident light angle increased from 30° to 80°, the illumination became nonuniformly distributed, as exhibited in Figure 5d–f. When increasing the incident angle from 0° to 80°, the illumination distinctly increased.

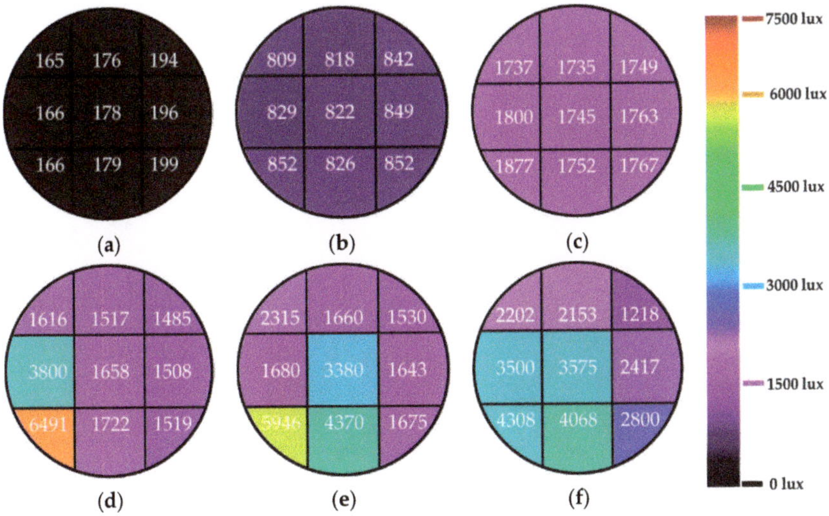

Figure 5. Illumination contours from zinc alloy light tube on the top end at the incident angles of (**a**) 0°, (**b**) 15°, (**c**) 30°, (**d**) 45°, (**e**) 60°, and (**f**) 75°.

Figure 6 demonstrates the illumination levels on the bottom of both aluminum alloy and zinc alloy hollow light pipes, with the same diameter of 0.25 m and tube length of 0.5 m at different incident angles. When using the aluminum alloy hollow light pipe, the illumination was between 12 lux and 52 lux at the incident light angle of 0°, which was uniformly distributed. The zinc alloy tube had illumination values between 2 lux and 34 lux with a nonuniform illuminance distribution at the incident angle of 0°. When the incident angle was increased from 0° to 30°, the illuminance distribution at the bottom of both types was nearly uniform. When the incident light angle increased from 30° to 80°, the illuminance distribution became nonuniformly distributed, as exhibited in Figure 6d–f. The illumination distribution for both aluminum alloy and zinc alloy tubes at the bottom varied when the incident light angle was changed from 0° to 80°. The illuminance at the bottom increased with the increase of the incident angle from 0° to 80°, as illustrated in Figure 6.

The nine measurements of illumination at the top and bottom were used to calculate the average illuminance for both aluminum alloy and zinc alloy hollow light pipes with a length of 0.5 m and different diameters of 0.20 m, 0.25 m, and 0.30 m. When using the aluminum alloy hollow light pipe with diameters of 0.20 m, 0.25 m, and 0.30 m, the average illuminance at different incident angles is shown in Figure 7. When the incident angle was varied, a change of luminous intensity at the top and bottom of the tube was achieved. At the top position of the tube with the diameter of 0.20 m, the average luminous intensity increased from 243 lux to 3022 lux; at the bottom end of the tube with a diameter of 0.20 m, the average luminous intensity increased from 22 lux to 1702 lux, when the incident light angle to the tube increased from 0° and 80°, as displayed in Figure 7a. For the diameters of 0.25 m and 0.30 m, the trend of the average illuminance at the top and bottom positions was similar to that of the tube with a 0.20 m diameter, as exhibited in Figure 7b,c. For the tube with a diameter of 0.25 m, the average luminous intensity increased from 261 lux to 3142 lux at the top end and from 31 lux to 2304 lux at the bottom end position of the tube with an increase in the incident light angle to the light tube between 0° and 80°, as illustrated in Figure 7b. For the diameter of 0.30 m, the average luminous intensity increased from 220 lux to 3549 lux at the top end and from 40 lux to 2591 lux at the bottom end when there was an increase of the incident angle to the light tube from 0° to 80°, as illustrated in Figure 7c.

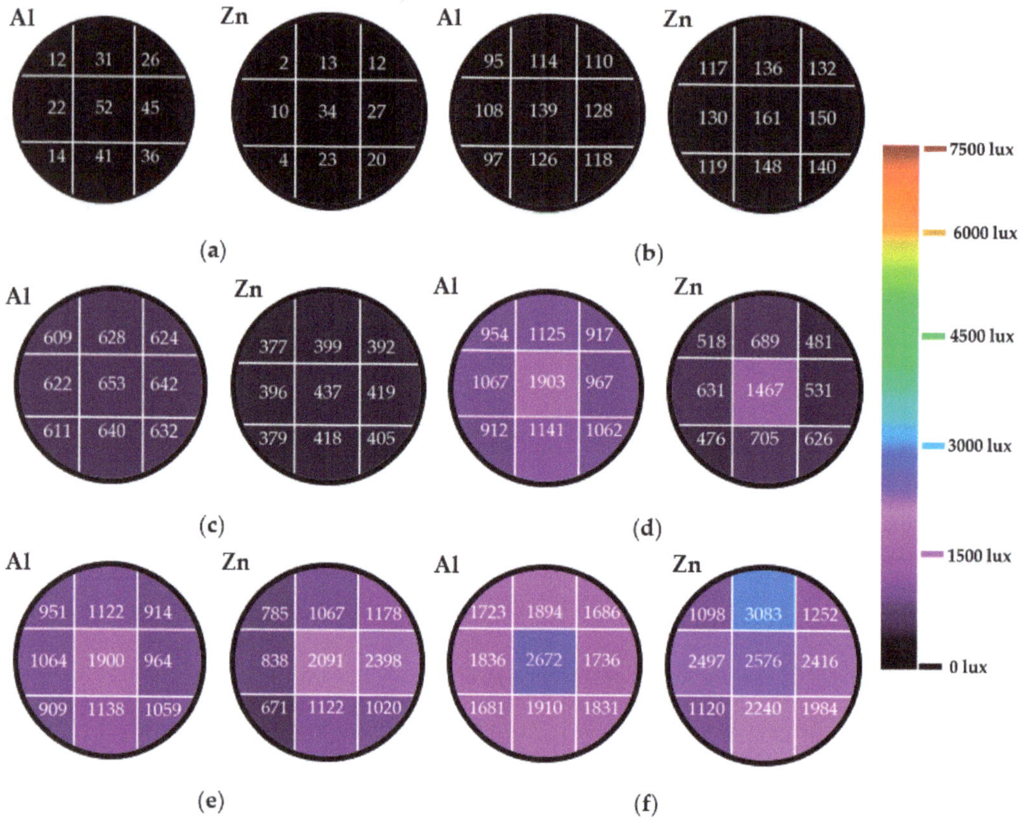

Figure 6. Illumination contours of aluminum alloy and zinc alloy hollow light pipe on the bottom end positions at the incident light angles of (**a**) 0°, (**b**) 15°, (**c**) 30°, (**d**) 45°, (**e**) 60°, and (**f**) 75°.

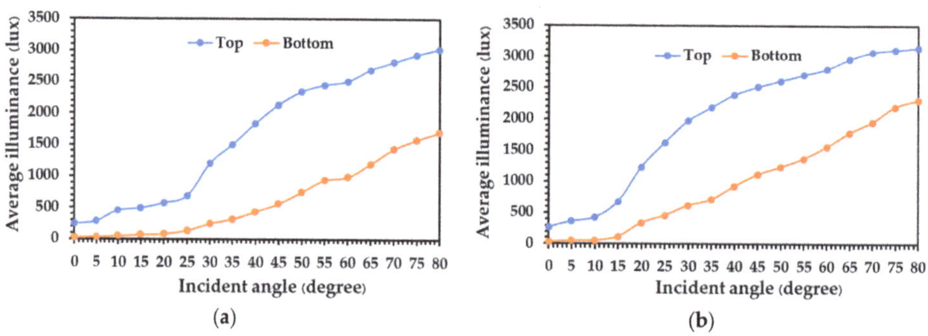

Figure 7. *Cont.*

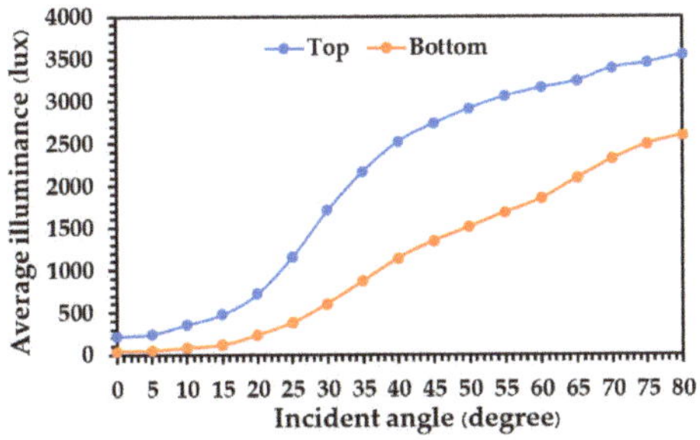

(c)

Figure 7. Average illuminance at the top and bottom end positions of aluminum alloy tube with a diameter of (**a**) 0.20 m, (**b**) 0.25 m, and (**c**) 0.30 m.

The illuminance of zinc alloy tubes with different diameters of 0.20 m, 0.25 m, and 0.30 m, and height of 0.5 m, at various incident angles are shown in Figure 8. A variation in the incident light angle to the tube led to the changes in luminous intensity at the top and bottom of the zinc alloy tube. The trend of the illuminance of zinc alloy tubes with the diameters of 0.20 m, 0.25 m, and 0.30 m the at the top and bottom was similar to that of the aluminum alloy tube, as illustrated in Figure 8. At the top end of zinc alloy light tubes, the illuminance intensity increased from 200 lux to 2889 lux for 0.20 m, from 180 lux to 3052 lux for 0.25 m, and from 190 lux to 3469 lux for the 0.30 m diameter pipes, respectively, with an increase in the incident light angle to the light tube between 0° and 80°. At the bottom of the zinc alloy light tubes, the illuminance intensity increased from 12 lux to 1502 lux for 0.20 m, from 16 lux to 2133 lux for 0.25 m, and from 17 lux to 2448 lux for 0.30 m, respectively, when there was an increase of the incident light angle to the light tube from 0° to 80°. It was observed that the luminous intensity increased at the top and bottom of both aluminum alloy and zinc alloy vertical light tube with an increase of the incident angle to the light source. When considering the low incident angles into the light tube, a major part of the light beam was reflected and little was directly transmitted through the vertical tube, which could lead to the lower illuminance intensity at the top and bottom end positions at the low incident angles of the light source. While the higher incident angles of light source into the light tube was examined, an increase of the illuminance intensity at the top and bottom of the light tubes was achieved, which was a result of more direct light penetration and less incidental light reflection.

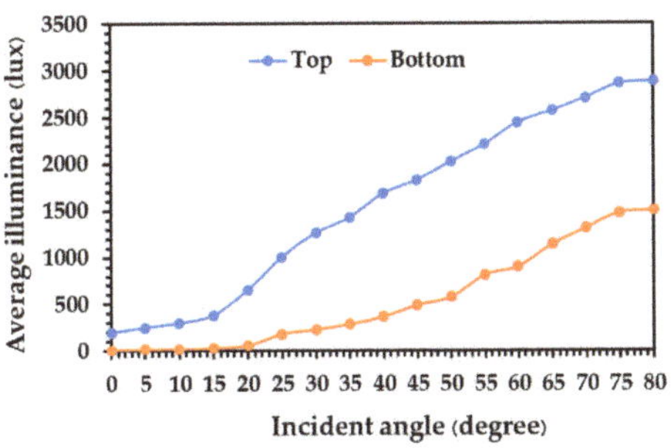

(a)

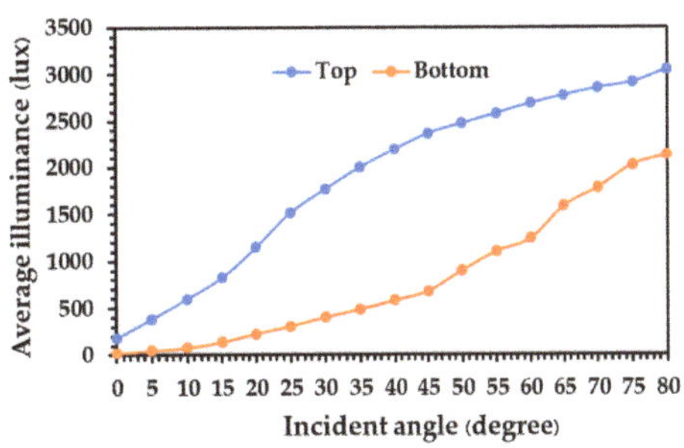

(b)

Figure 8. *Cont.*

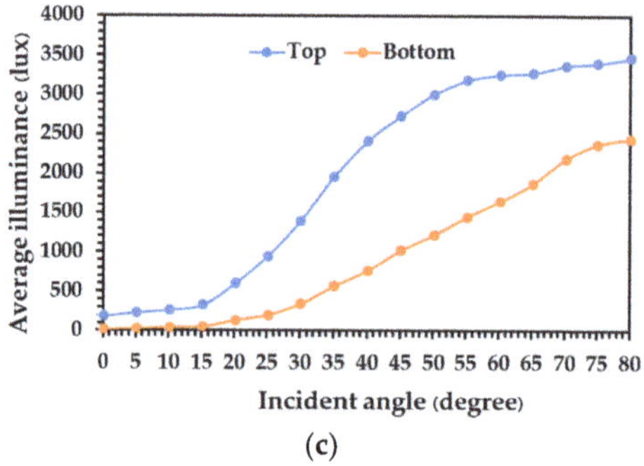

Figure 8. Average illuminance at the top and bottom tube position of zinc alloy tube with diameters of (**a**) 0.20 m, (**b**) 0.25 m, and (**c**) 0.30 m.

When considering the different types of tube materials, it was observed that the average luminous intensity at the bottom end of the aluminum alloy tubes having a diameter 0.20 m was higher than that of the zinc alloy tube for each incident light angle, as demonstrated in Figure 9a. For the diameters of 0.25 m and 0.30 m, the average illuminance of the aluminum alloy tube and zinc alloy tube was similar to that of the light tube with a diameter of 0.20 m, as shown in Figure 9b,c. This indicates that the internal surface of the aluminum alloy tube light showed better reflection, leading to an improvement in the light transmission performance within the tube.

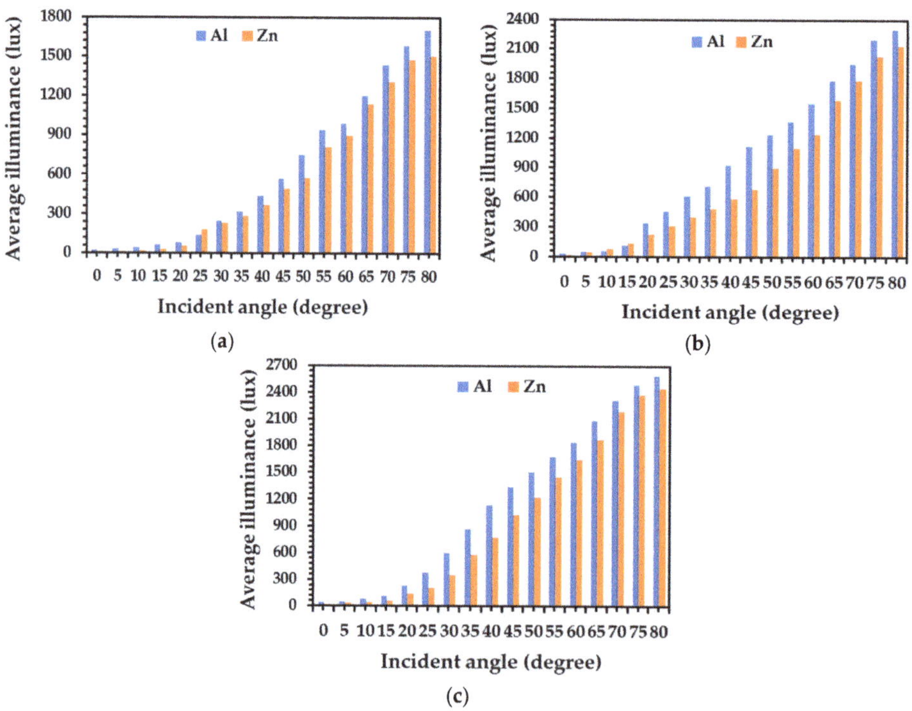

Figure 9. Average illuminance at the bottom end positions of aluminum alloy tube and zinc alloy tube with the diameters of (**a**) 0.20 m, (**b**) 0.25 m, and (**c**) 0.30 m.

All of the luminous intensity values at the top and bottom ends of both aluminum alloy and zinc alloy tubes were considered to calculate the light transmission performance of the tubes, as exhibited in Figure 10. Light transmission of the vertical aluminum alloy tube with a diameter of 0.20 m increased from 9.1% to 56.3% when the incident light angle to the tube increased from 0° to 80°, as shown in Figure 10a. When considering the diameters of 0.25 m and 0.30 m, the light transmission performance was similar to those of the 0.20 m diameter tube, as illustrated in Figure 10a. Light transmission performance of the aluminum alloy tubes with a diameter of 0.25 m increased between 11.8% and 73.3%, and that of the 0.30 m tube increased from 18.3% to 73.1% when the incident light angle increased from 0° to 80°.

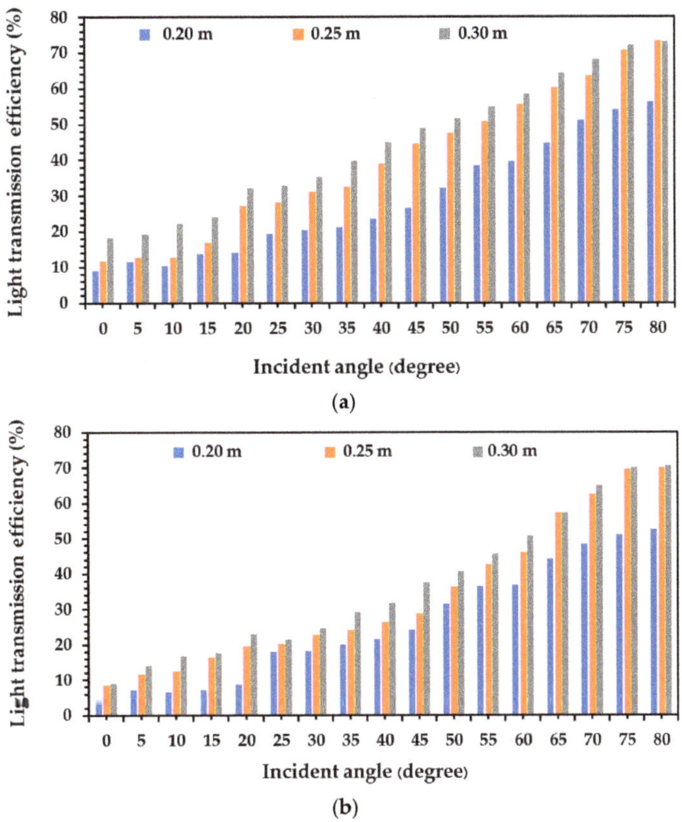

Figure 10. Light transmission performance of (**a**) aluminum alloy tube and (**b**) zinc alloy tube.

When using the zinc alloy tube with a diameter 0.20 m and a height of 0.50 m, light transmission increased from 4.7% to 52.5% when the incident light angle increased from 0° to 80°. For the diameters of 0.25 m and 0.30 m, the light transmission performance was similar to those of the 0.20 m diameter tube, as illustrated in Figure 10b. Light transmission performance of the zinc alloy tube with a diameter of 0.25 m increased between 8.7% and 69.9%, and that of 0.30 m tube increased between 9.1% and 70.6% with an increase of the incident light angle from 0° to 80°. This increase of light transmission performance is achieved from an increase of the incident light angle, demonstrating the reduced number of reflections within the light tube surfaces, which leads to minimal losses of luminous intensity.

Additionally, when using the aluminum alloy tube with a height of 0.5 m and diameters of 0.20 m, 0.25 m, and 0.30 m, the light transmission increased from 9.1% to 18.3% with an increase of the tube diameter from 0.20 m to 0.30 m at the incident light angle of 0°. The light transmission performance at the incident light angles of 5°, 10°, 15°, 20°, 25°, 30°, 35°, 40°, 45°, 50°, 55°, 60°, 65°, 70°, and 80° were similar to those of the incident angle of 0°, as illustrated in Figure 10a. For the case of the zinc alloy tube, the light transmission performance also increased from 4.7% to 9.1% when the tube diameter increased from 0.20 m to 0.30 m at the incident angle of 0°. At the incident light angles of 5°, 10°, 15°, 20°, 25°, 30°, 35°, 40°, 45°, 50°, 55°, 60°, 65°, 70°, and 80°, the light transmission performance also increased from 7.3% to 14.2%, 6.7% to 16.8%, 7.1% to 17.7%, 8.8% to 23.1%, 18.0% to 21.5%, 18.2% to 24.7%, 20.0% to 29.2%, 21.7% to 31.8%, 24.2% to 37.4%, 31.5% to 40.6%, 36.5% to 45.6%, 36.8% to 50.7%, 44.2% to 57.2%, 48.3% to 65.1%, 51.1% to 69.9%, and 52.5% to 70.6%, respectively, as exhibited in Figure 10b. This demonstrates that higher diameters of aluminum alloy and zinc alloy light tubes lead to an increase in light transmission performance, relating to less losses of luminous intensity that occur from the decrease in the number of reflections in the inner tube surfaces.

With comparing the aluminum alloy tube and zinc alloy tube of the same diameter, average light transmission performance of the aluminum alloy tube was higher than that of zinc alloy tube in each incident light angle and diameter of light tube, as shown in Figure 11. It was observed that the average light transmission performance of the aluminum alloy tube can transmit the luminous intensity more than 4% when compared with that of zinc alloy tube in all conditions. This demonstrates that the aluminum alloy tube shows more reflectivity than zinc alloy tubes. This indicates that this technology could be considered as an alternative daylight system in deeper rooms or could substitute artificial lighting in windowless spaces, leading to greater energy conservation in buildings.

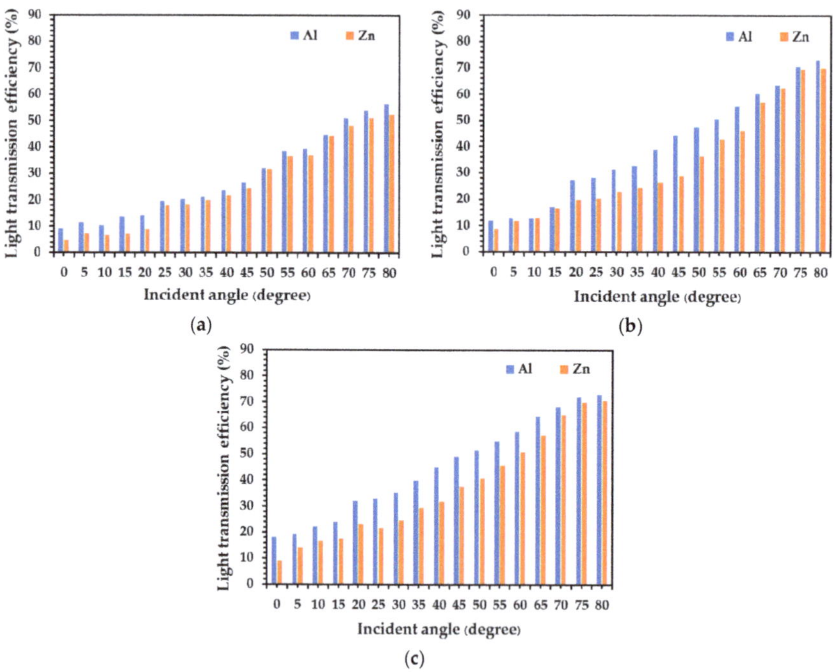

Figure 11. Light transmission performance of light tubes with different materials with diameters of (**a**) 0.20 m, (**b**) 0.25 m, and (**c**) 0.30 m.

The illumination distribution on the floor of the model room from the aluminum alloy and zinc alloy light tubes, with a diameter of 0.25 m and a tube height of 0.5 m, are presented in a 2-D color map as exhibited in Figures 12 and 13. The illuminance distribution from the aluminum alloy was between 12 lux and 31 lux, and the zinc alloy light tube was between 12 lux and 19 lux, at the incident light angle of 0°. When the incident angle increased from 0° to 45°, the illuminance distribution on the floor from both types was uniformly distributed at the same value throughout. When the incident light angle increased from 45° to 80°, the internal illuminance distribution was nonuniformly distributed, as shown in Figures 12d–f and 13d–f. When the incident angle increased from 0° to 80°, a higher illuminance in each position was achieved, as shown in Figures 12 and 13.

The 25 positions of illumination measurements on the floor were averaged for both aluminum and zinc alloy light tube with a length of 0.5 m and different diameters (0.20 m, 0.25 m, and 0.30 m) as exhibited in Figures 14 and 15. When considering the aluminum alloy tube, with a diameter of 0.20 m, the average illuminance on the floor increased from 11 lux to 236 lux when the incident angle increased from 0° to 80°. For the cases of the diameters of 0.25 m and 0.30 m, the average illuminance values on the floor were similar to those of the 0.20 m diameter tube, as illustrated in Figure 14a. The average illuminance on the floor with the aluminum alloy tube with a diameter of 0.25 m increased between 16 lux and 311 lux, and that of 0.30 m increased between 9 lux and 418 lux when there was an increase in the incident light angle from 0° to 80°.

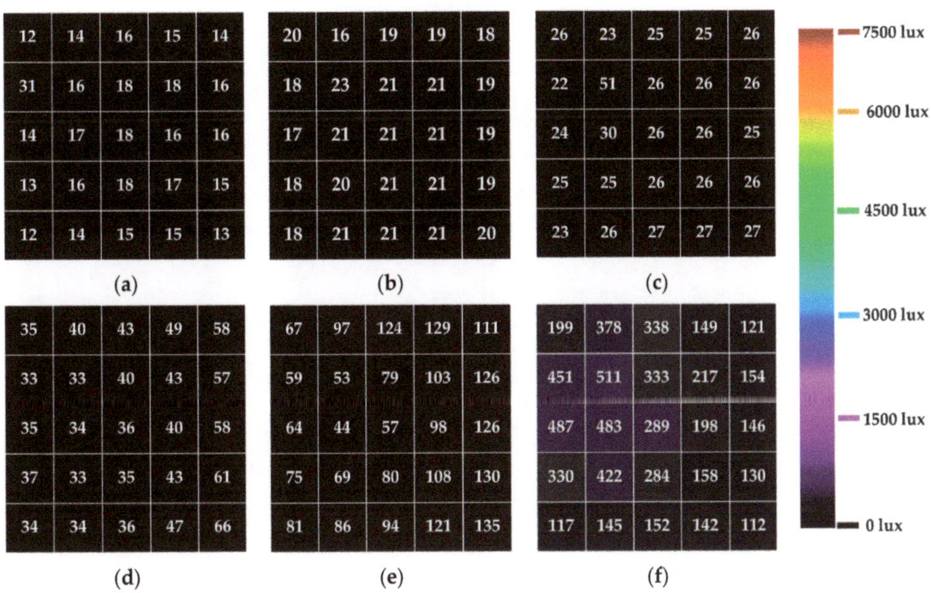

Figure 12. Illuminance distribution on the floor from the aluminum alloy light tube at the incident angles of (**a**) 0°, (**b**) 15°, (**c**) 30°, (**d**) 45°, (**e**) 60°, and (**f**) 75°.

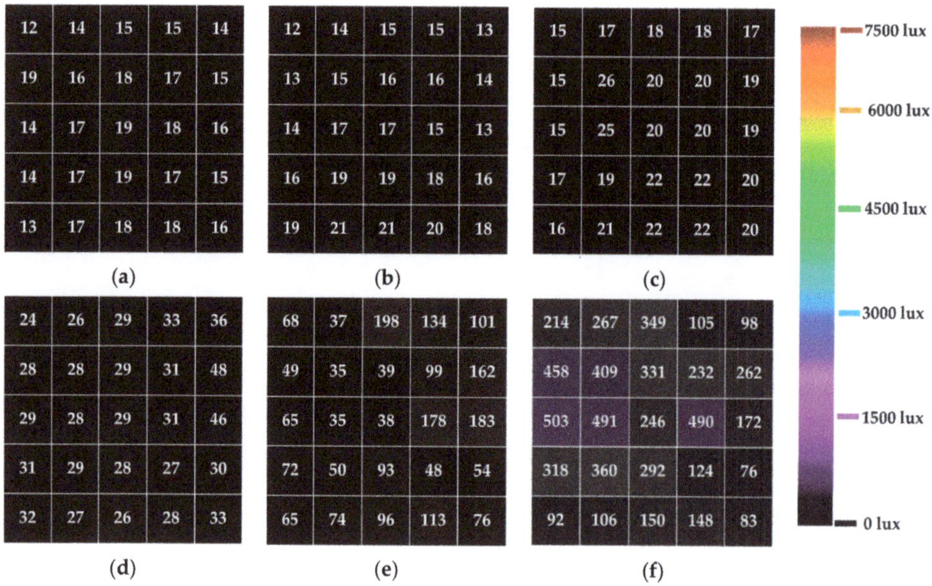

Figure 13. Illuminance distribution on the floor from the zinc alloy light tube at the incident angles of (**a**) 0°, (**b**) 15°, (**c**) 30°, (**d**) 45°, (**e**) 60°, and (**f**) 75°.

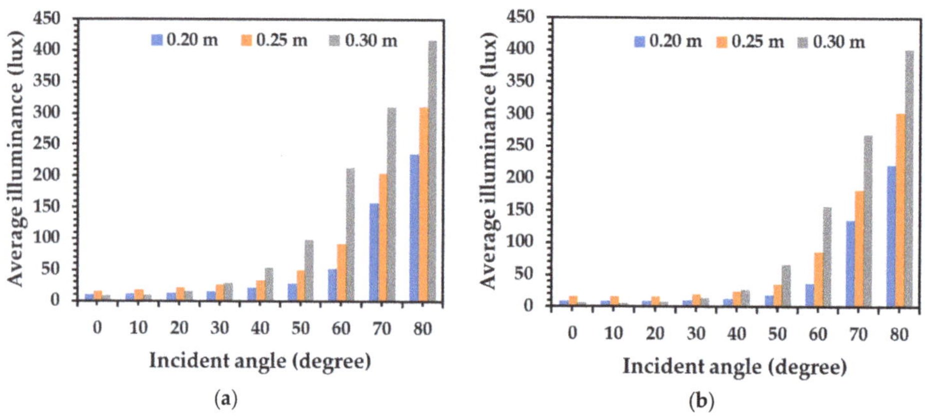

Figure 14. Average internal illuminance contours on the floor for the (**a**) aluminum alloy and (**b**) zinc alloy light tubes.

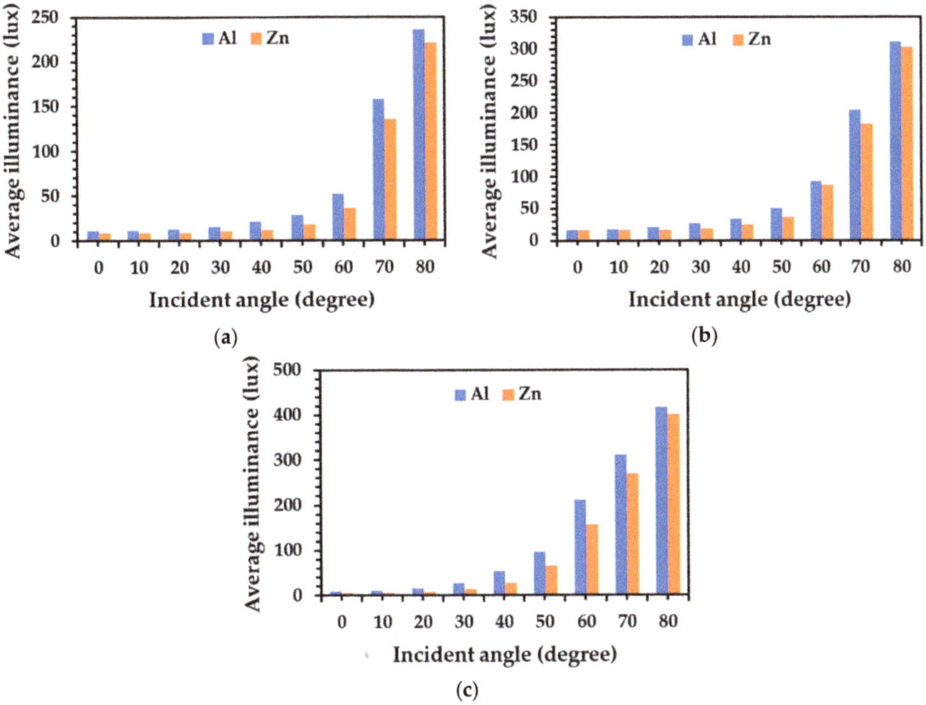

Figure 15. Average internal illuminance of light tube with different materials and diameters (**a**) 0.20 m, (**b**) 0.25 m, and (**c**) 0.30 m.

When considering the zinc alloy tube with a diameter of 0.20 m, the average illuminance on the floor increased from 9 lux to 20 lux when the incident light angle increased from 0° to 80°. For the cases of the diameters of 0.25 m and 0.30 m, the average illuminance values on the floor were similar to those of the 0.20 m diameter tube, as illustrated in Figure 14b. The average illuminance on the floor with the zinc alloy tube with a diameter of 0.25 m increased between 16 lux and 302 lux, and that of the 0.30 m diameter tube increased between 6 lux and 400 lux when there was an increase in the incident angle from 0° to 80°.

Comparing the diameter of both aluminum alloy tubes and zinc alloy tubes, the average illuminance on the floor was between 9 lux to 28 lux when the tube diameter increased from 0.20 m to 0.30 m at the incident light angles of 0°, 5°, 10°, 15°, 20°, 25°, and 30°. For the incident light angles of 35°, 40°, 45°, 50°, 55°, 60°, 65°, 70°, and 80°, the average illuminance on the floor of both aluminum alloy tube and zinc alloy tube increased with an increase in diameter from 0.20 m to 0.30 m, as illustrated in Figure 14.

Comparing the aluminum and zinc alloy tubes at the same diameter of 0.20 m, the average illuminance on the floor from the aluminum alloy tube was higher than that of zinc alloy tube at all incident light angles, as shown in Figure 15a. The aluminum alloy tube was more than 7% brighter when compared with zinc alloy tubes at incident light angles between 0° and 80°. For the diameters of 0.25 m and 0.30 m, an increase of the average illuminance on the floor from the aluminum alloy tube was more than 3% when compared with that of zinc alloy tube at all incident light angles, as shown in Figure 15b,c. This indicates that the aluminum alloy tube has more illuminance efficiency than zinc alloy tubes at each diameter size and incident light angle.

The daylight factor is defined as the average internal illumination on the floor plane and illuminance on the external testing model in an unshaded area. The exterior illuminance was measured to be between 890 lux and 1140 lux and obtained an average luminous intensity of 950 lux. The daylight factor for both aluminum alloy and zinc alloy light tubes in all patterns is demonstrated in Figure 16. When considering the aluminum alloy tube with a diameter of 0.20 m, the average daylight factor increased from 1.2% to 22.7% when the incident light angle increased from 0° to 80°. For the cases of the diameters of 0.25 m and 0.30 m, the average daylight factor was similar to those of the diameter of 0.20 m as displayed in Figure 16a. The average daylight factor from the aluminum alloy tube with a diameter of 0.25 m increased between 1.6% and 32.7%, and that of 0.30 m increased between 0.9% and 43.9% when there was an increase of incident light angle from 0° to 80°.

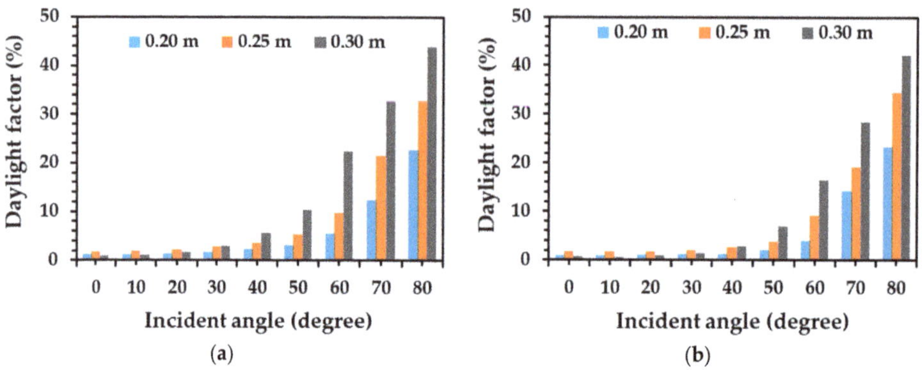

Figure 16. Daylight factor of the vertical light tube using (**a**) aluminum alloy and (**b**) zinc alloy.

Considering the case of the zinc alloy tube with a diameter 0.20 m, the average daylight factor increased from 0.9% to 23.2% with an increase in the incident light angle from 0° to 80°. For the cases of the diameters of 0.25 m and 0.30 m, the trend of average daylight factor was similar to those of the diameter of 0.20 m, as exhibited in Figure 16b. The average daylight factor from zinc alloy tube with a diameter of 0.25 m increased between 1.7% and 34.5%, and that of 0.30 m increased between 0.7% and 42.2% when there was an increase of incident light angle from 0° to 80°. When comparing the incident light angle to the tube of both aluminum alloy tube and zinc alloy tube, the average maximum daylight factor was found at the incident light angle of 80° from both aluminum alloy tubes and zinc alloy tubes.

Comparing the diameter variation of both aluminum alloy tubes and zinc alloy tubes, the daylight factor was similar and between 0.7% and 2.9% when the tube diameter increased from 0.20 m to 0.30 m at the incident light angles of 0°, 5°, 10°, 15°, 20°, 25°, and 30°. For the incident light angles to the tube of 35°, 40°, 45°, 50°, 55°, 60°, 65°, 70°, and 80°, the daylight factor of both aluminum alloy tube and zinc alloy tube increased with an increase in diameter from 0.20 m to 0.30 m, as illustrated in Figure 16. This observation demonstrated that the aluminum alloy tubes can provide a better daylight factor compared with zinc alloy tubes for each condition.

This daylight factor obtained from the aluminum alloy tubes and zinc alloy tubes in this work corresponded to those of previous works [29,31], which demonstrated the daylight contribution in areas with 1–2% of daylight factor for activities in residence and 2–4% of daylight factor for activities in office buildings [29,31]. For the case of window, the occupants visualize the outdoor environment and use natural ventilation, while the shading from adjacent obstacles and barriers appeared in many cases and insufficient daylight within deeper building areas was achieved. Skylights can provide high light intensity, natural ventilation, and uniform daylighting, but immoderate solar gains and overheating in the internal spaces of rooms was obtained. The distant areas in buildings achieved insufficient light transmission. Hollow light tubes in this work can transport daylight into some deeper or windowless areas of buildings and uniformly distribute light. Importantly, the investment price of this system was cheap for inventing the aluminum alloy tubes and zinc alloy tubes. Although an investigation of thermal behavior and inside illuminance of the models under natural weather conditions, and an analysis of the reduction of energy consumption using systems on a full scale could be led to more sound scientific knowledge for integrating the vertical light tubes into buildings; a more widespread context will be studied and analyzed under natural weather conditions in the future work. In view of the benefits, the invented light tubes can use commercial aluminum alloy sheets and commercial zinc alloy sheets to provide illumination for activities in residences and offices for some incident light angles for each diameter of light tube, which leads to an alternative inexpensive material for producing a lighting system. This investigation could be considered as an alternative daylight system in deeper parts of buildings or the windowless spaces to conserve the energy consumption for lighting buildings.

4. Conclusions

This investigation evaluated the light transmission performance of aluminum alloy and zinc alloy hollow light tubes as a function of both incident angle and the diameter of the light tubes. The luminous intensity at the bottom of the aluminum alloy and zinc alloy light tubes with a diameter of 0.25 m increased from 31 lux to 2304 lux and 16 lux to 2133 lux, respectively, when the incident angle of light was increased from 0° to 80°. This is a result of the proportion of reflected light and direct light being transmitted through the vertical tube. The light transmission performance of both light tube types increased when the incident light angle increased from 0° to 80° in each diameter and the diameter of tube increased from 0.20 m to 0.30 m for each incident angle. The type of light tube material has an effect on the reflective performance of the light tube. The aluminum alloy pipe can improve the light transmission performance by approximately 4% and achieved a better daylight factor when compared with the zinc alloy tube for each condition. This work could be the initial investigation to contribute to enhancing daylight in some deeper areas of buildings or replacing artificial lighting in windowless areas. Light tubes can conserve energy consumption for lighting in buildings.

Author Contributions: Conceptualization, A.T.; methodology, A.T. and J.M.; formal analysis, A.T. and J.M.; investigation, A.T. and J.M.; resources, A.T.; data curation, A.T.; writing—original draft preparation, A.T. and J.M.; writing—review and editing, A.T.; visualization, A.T.; supervision, A.T.; project administration, A.T.; funding acquisition, A.T. All authors have read and agreed to the published version of the manuscript.

Funding: This research was funded by Faculty of Science, Naresuan University, grant number P2564C010.

Institutional Review Board Statement: Not applicable.

Informed Consent Statement: Not applicable.

Acknowledgments: The authors would like to thank the Thailand Science Research and Innovation (TSRI), Faculty of Science, Naresuan University for providing financial support to this research work, and our research center. Thanks are given to Kyle V. Lopin for editing this document.

Conflicts of Interest: The authors declare no conflict of interest.

References

1. Khedari, J.; Sangprajak, A.; Hirunlabh, J. Investigation of thermal performance by applying a solar chimney with PCM towards the natural ventilation of model house under Climate of Thailand. *J. Renew. Energy* **2001**, *25*, 267–280. [CrossRef]
2. Thongtha, A.; Khongthon, A.; Boonsri, T.; Chan, H.Y. Thermal effectiveness enhancement of autoclaved aerated concrete wall with PCM-contained conical holes to reduce the cooling load. *Materials* **2019**, *12*, 2170. [CrossRef]
3. Kunchornrat, A.; Namprakai, P.; Du, P.T. The impacts of climate zones on the energy performance of existing Thai buildings. *Resour. Conser. Recycl.* **2019**, *53*, 545–551. [CrossRef]
4. Yan, T.; Sun, Z.W.; Gao, J.J.; Xu, X.H.; Yu, J.H.; Gang, W.J. Simulation study of a pipe-encapsulated PCM wall system with self-activated heat removal by nocturnal sky radiation. *Renew. Energy* **2020**, *146*, 1451–1464. [CrossRef]
5. Kim, J.T.; Kim, G. Overview and new developments in optical daylighting systems for building a healthy indoor environment. *Build. Environ.* **2010**, *45*, 256–269. [CrossRef]

6. Kim, I.T.; Kim, Y.S.; Cho, M.; Nam, H.; Choi, A.; Hwang, T. High-performance accuracy of daylight-responsive dimming systems with illuminance by distant luminaires for energy-saving buildings. *Energies* **2019**, *12*, 731. [CrossRef]
7. Kim, C.H.; Kim, K.S. Development of sky luminance and daylight illuminance prediction methods for lighting energy saving in office buildings. *Energies* **2019**, *12*, 592. [CrossRef]
8. Freewan, A.A.Y. Impact of external shading devices on thermal and daylighting performance of offices in hot climate regions. *Sol. Energy* **2014**, *102*, 14–30.
9. Berardi, U.; Wang, W. Daylighting in an atrium-type high performance house. *Build. Environ.* **2014**, *76*, 92–104. [CrossRef]
10. Azad, A.S.; Rakshit, D.; Patil, K.N. Model development and evaluation of global and diffuse luminous efficacy for humid sub-tropical region. *Renew. Energy* **2018**, *119*, 375–387. [CrossRef]
11. Rezaei, S.D.; Shannigrahi, S.; Ramakrishna, S. A review of conventional, advanced, and smart glazing technologies and materials for improving indoor environment. *Sol. Energy Mater. Sol. Cells* **2017**, *159*, 26–51. [CrossRef]
12. Sharma, P.; Rakshit, D. Financial viability of energy conservation using natural light in an academic building in temperate zone. *J. Sol. Energy Eng.* **2016**, *138*, 1–10. [CrossRef]
13. Valles, X.; Alonso, M.H.; Lopez-Caleya, J.F.; Diez-Obrero, V.; Dierssen-Sotos, T.; Lope, V.; Molina-Barcelo, A.; Chirlaque, M.D.; Jimenez-Moleon, J.J.; Tardon, G.F.; et al. Colorectal cancer, sun exposure and dietary vitamin D and calcium intake in the MCC-Spain study. *Environ. Int.* **2018**, *121*, 428–434. [CrossRef] [PubMed]
14. Thongtha, A.; Boontham, P. Experimental investigation of natural lighting systems using cylindrical glass for energy saving in buildings. *Energies* **2020**, *13*, 2528. [CrossRef]
15. Pandharipande, A.; Caicedo, D. Daylight integrated illumination control of LED systems based on enhanced presence sensing. *Energy Build.* **2011**, *43*, 944–950. [CrossRef]
16. Li, D.H.W.; Lam, J.C. Evaluation of lighting performance in office buildings with daylighting controls. *Energy Build.* **2001**, *33*, 793–803.
17. Li, D.H.W.; Lam, T.N.T.; Wong, S.L. Lighting and energy performance for an office using high frequency dimming controls. *Energy Convers. Manag.* **2006**, *47*, 1133–1145.
18. Vasilakopoulou, K.; Kolokotsa, D.; Santamouris, M.; Kousis, I.; Asproulias, H.; Giannarakis, I. Analysis of the experimental performance of light pipes. *Energy Build.* **2017**, *151*, 242–249. [CrossRef]
19. Ghosh, A.; Mallick, T.K. Evaluation of colour properties due to switching behaviour of a PDLC glazing for adaptive building integration. *Renew. Energy* **2018**, *120*, 126–133.
20. Chirarattananon, S.; Chaiwiwatworakul, P.; Pattanasethanon, S. Daylight availability and models for global and diffuse horizontal illuminance and irradiance for Bangkok. *Renew. Energy* **2002**, *26*, 69–89. [CrossRef]
21. Krarti, M.; Erickson, P.M.; Hillman, T.C. A simplified method to estimate energy savings of artificial lighting use from daylighting. *Build. Environ.* **2005**, *40*, 747–754. [CrossRef]
22. Li, D.H.W.; Lam, T.N.T.; Wong, S.L.; Tsang, E.K.W. Lighting and cooling energy consumption in an open plan office using solar film coating. *Energy* **2008**, *33*, 1288–1297. [CrossRef]
23. Park, K.W.; Athienitis, A.K. Work plane illuminance prediction method for daylighting control systems. *Sol. Energy* **2003**, *75*, 277–284. [CrossRef]
24. Lopin, M.K. An overview of daylighting systems. *Sol. Energy* **2002**, *73*, 77–82. [CrossRef]
25. Ekren, N.; Gorgulu, S. An investigation into the usability of straight light-pipes in Istanbul. *Energy Educ. Sci. Tech. Part A Energy Sci. Res.* **2012**, *30*, 637–644.
26. Shao, L.; Riffat, S.B. Daylighting using light pipes and its integration with solar heating and natural ventilation. *Int. J. Lighting Res. Technol.* **2000**, *32*, 133–139. [CrossRef]
27. Canziani, R.; Peron, F.; Rossi, G. Daylight and energy performances of a new type of light pipe. *Energy Build.* **2004**, *36*, 1163–1176. [CrossRef]
28. Kennedy, D.M.; O'Rourke, F. Experimental analysis of a scaled, multi-aperture, light-pipe, daylighting system. *Sol. Energy* **2015**, *122*, 181–190. [CrossRef]
29. Hansen, V.G.; Edmonds, I. Methods for the illumination of multilevel buildings with vertical light pipes. *Sol. Energy* **2015**, *117*, 74–88. [CrossRef]
30. Shao, L.; Riffat, S.; Icks, W.H. A study of performance of light pipes under cloudy and sunny conditions in the UK. *Right Light* **1997**, *1*, 155–159.
31. Mohelnikova, J. Daylighting and energy savings with tubular light guides. *WSEAS Trans. Environ. Dev.* **2008**, *4*, 200–209.
32. Alrubaih, M.S.; Zain, M.F.M.; Alghoul, M.A.; Ibrahim, N.L.N.; Shameri, M.A.; Elayeb, O. Research and development on aspects of daylighting fundamentals. *Renew. Sustain. Energy Rev.* **2013**, *21*, 494–505. [CrossRef]
33. Smith, B.J.; Phillips, G.M.; Sweeney, M. Daylighting. In *Environmental Science*; Routledge: London, UK, 2014; Chapter 9.
34. Zhang, X.; Muneer, T. Mathematical model for the performance of light pipes. *Light. Res. Technol.* **2000**, *32*, 141–146. [CrossRef]
35. Reinhart, C.F.; Mardaljevic, J.; Rogers, Z. Dynamic daylight performance metrics for sustainable building design. *Leukos* **2006**, *3*, 7–31. [CrossRef]
36. Vasilakopoulou, K.; Synnefa, A.; Kolokotsa, D.; Karlessi, T.; Santamouris, M. Performance prediction and design optimisation of an integrated light pipe and artificial lighting system. *Int. J. Sustain. Energy* **2014**, *35*, 675–685. [CrossRef]
37. Li, D.H.W.; Tsang, E.K.W.; Cheung, K.L.; Tam, C.O. An analysis of light-pipe system via full-scale measurements. *Appl. Energy* **2010**, *87*, 799–805. [CrossRef]
38. Yun, G.Y.; Shin, H.Y.; Kim, J.T. Monitoring and evaluation of a light-pipe system used in Korea. *Indoor Built Environ.* **2010**, *9*, 129–136.

39. Yanpeng, W.; Rendong, J.; Deying, L.; Wenming, Z.; Chongfang, M. Experimental Investigation of Top Lighting and Side Lighting Solar Light Pipes under Sunny Conditions in Winter in Beijing. In Proceedings of the SPIE—The International Society for Optical Engineering, Beijing, China, 2 February 2009.
40. Mushtaha, E.; Kana'an, B.A.; Al-Jawazneh, R.A.; Hammad, R.S. Effect of using different light pipe parameters on the daylight quality in buildings: The case of Jordan. *Int. J. Green Energy* **2016**, *3*, 1590–1598. [CrossRef]
41. Gago, E.J.; Muneer, T.; Knez, M.; Köster, H. Natural light controls and guides in buildings. Energy saving for electrical lighting, reduction of cooling load. *Renew. Sustain. Energy Rev.* **2015**, *41*, 1–13. [CrossRef]
42. Pleshkov, S.; Brakale, G.; Vedishcheva, I. A Project aimed to increase energy efficiency of the object swimming pool universitetsky by application of hollow mirrored tubular light guides under trade mark. *Mater. Sci. Eng.* **2018**, *463*, 042050. [CrossRef]
43. Ahsan, M.d.; Ahsan, R.U.; Kim, Y.R.; Ashiri, R.; Cho, Y.J.; Jeong, C.; Park, Y.D. Cold metal transfer (CMT) MAW of zinc coated steel. *Weld. J.* **2016**, *95*, 120–132.
44. Mahawan, J.; Thongtha, A.; Promphak, K.; Chansomsak, S. Application of solar tube integrating with roof for energy consumption reduction in building. In Proceedings of the International Conference on Sustainable Energy and Green Technology 2019 (SEGT 2019), Bangkok, Thailand, 11–14 December 2019; The Joint Graduate School of Energy and Environment—Center of Excellence on Energy Technology and Environment (JGSEE-CEE), Centre for Energy Sciences (University of Malaya, Malaysia), and Centre for Vehicular Technology (UTAR, Malaysia): Bangkok, Thailand, 2019.

energies

MDPI

Article

Design of Kinetic-Energy Harvesting Floors

Thitima Jintanawan [1], Gridsada Phanomchoeng [1,2,*], Surapong Suwankawin [3], Phatsakorn Kreepoke [1], Pimsalisa Chetchatree [1] and Chanut U-viengchai [3]

1 Department of Mechanical Engineering, Chulalongkorn University, Bangkok 10330, Thailand; thitima.j@chula.ac.th (T.J.); prai_extremely@hotmail.co.th (P.K.); pimsalisa.ch@gmail.com (P.C.)
2 Smart Mobility Research Unit, Chulalongkorn University, Bangkok 10300, Thailand
3 Department of Electrical Engineering, Chulalongkorn University, Bangkok 10330, Thailand; surapong.su@chula.ac.th (S.S.); chanut0@gmail.com (C.U.-v.)
* Correspondence: gridsada.phanomchoeng@gmail.com; Tel.: +66-2-218-6630

† This paper is an extended and revised article presented at the International Conference on Sustainable Energy and Green Technology 2019 (SEGT 2019) on 11–14 December 2019 in Bangkok, Thailand.

Received: 18 September 2020; Accepted: 12 October 2020; Published: 16 October 2020

Abstract: Alternative energy generated from people's footsteps in a crowded area is sufficient to power smart electronic devices with low consumption. This paper aims to present the development of an energy harvesting floor—called Genpath—using a rotational electromagnetic (EM) technique to generate electricity from human footsteps. The dynamic models of the electro-mechanical systems were developed using MATLAB®/Simulink to predict the energy performances of Genpath and help fine-tune the design parameters. The system in Genpath comprises two main parts: the EM generator and the Power Management and Storage (PMS) circuit. For the EM generator, the conversion mechanism for linear translation to rotation was designed by using the rack-pinion and lead-screw mechanism. Based on the simulation analysis, the averaged energy of the lead-screw model is greater than that of the rack-pinion model. Thus, prototype-II of Genpath with 12-V-DC generator, lead-screw mechanism was recently built. It shows better performance when compared to the previous prototype-I of Genpath with 24-V-DC-generator, rack-pinion mechanism. Both prototypes have an allowable displacement of 15 mm. The Genpath prototype-II produces an average energy of up to 702 mJ (or average power of 520 mW) per footstep. The energy provided by Genpath prototype-II is increased by approximately 184% when compared to that of the prototype-I. The efficiency of the EM-generator system is ~26% based on the 2-W power generation from the heel strike of a human's walk in one step. Then, the PMS circuit was developed to harvest energy into the batteries and to supply the other part to specific loads. The experiment showed that the designed PMS circuit has the overall efficiency of 74.72%. The benefit of the design system is for a lot of applications, such as a wireless sensor and Internet of Thing applications.

Keywords: energy harvesting; electromagnetic generator; energy floor tile; power management system; footstep energy harvesting; piezoelectric; energy harvesting paver

1. Introduction

The smart internet of things (IoT) has been brought to the world's attention lately. The smart IoT-devices, e.g., Radio-frequency identification (RFID) sensor, Global Positioning System (GPS) tracking, etc., are widely used in the crowded areas such as airports and public stations, commercial buildings, and department stores [1,2]. Seeking an alternative energy for such devices, it was noticeable that a large population can generate power from their footsteps, grabbing the authors' attention. Therefore, the ultimate goal was to develop an affordable Vibration Energy Harvesting (VEH) system—called Genpath—the smart floor capable of conversing kinetic energy from thousands of footsteps into electrical energy. Genpath can be installed in places with a crowded population for harvesting energy.

Energy harvesting from human motions is an interesting and applicable issue, thanks to the ultra-low power consumption of electronic devices lately [3]. Focused on walking, the energy produced by the heel strike of a person's walk is 1–5 J or 2–20 W per step [4]. There are various commercial products which harvest energy from people's walks, such as energy storage shoes [5] and the energy floor [6–11]. Pavegen and Energy Floors have produced a commercial system that generates power from footsteps [6,7]. The Pavegen system, using electromagnetic generators, can produce 2 to 4 joules, or around 5 watts of power of off-grid electrical energy per step. The Energy Floors focuses on harvesting energy from humans dancing and playing games. Dutch Railways built a novel phone charger for Utrecht Central Station using a swing set called Play for Power [8]. The system turns kinetic energy from the swings into power dispensed through charging cables. However, few technical details of those products are published thus far.

Kinetic energy, among practical energy-harvesting sources, comes in various forms, e.g., seismic noise, vibration of rotating machinery, motions of vehicles and humans, etc. Vibrational energy harvesting (VEH) is the concept of converting kinetic energies present in the environment into electrical energy. There are two main techniques of kinetic-to-electrical energy conversion applicable to the design of the VEH floor: i.e., piezoelectric generators and electromagnetic generators [12]. The piezoelectric generator produces an electrical charge and energy when deformed under mechanical stress. Piezoelectric generators can provide high voltage levels up to several volts. With its compact size, the piezoelectric generator gives high-power density per unit of volume [13]. The electromagnetic (EM) generator generates an electromotive force (EMF) and thus energy when the permanent magnet relatively moves through the coil. Both techniques of VEH discussed are suitable for generating energy for low-power electronic instruments. However, the piezoelectric generator effectively gives out maximum energy at around its natural frequencies, i.e., higher than 200 Hz, whereas the bandwidth of human motion is between 1–10 Hz. In contrast, the electromagnetic generator is effective in the frequency range of 2–20 Hz. Hence, it is more suitable for harvesting energy from human movements. In addition, the electromagnetic generators also yield higher power density and their costs are much lower than the piezoelectric generators [14]. For the piezoelectric generator, it needs a structural design not only to optimize the power density but also to protect the piezo element itself due to its fragility, while the structural design of the EM generator is much simpler for the application of a VEH floor.

There are also two types of the EM generator in terms of energy-conversion mechanisms: linear oscillation to electrical energy and rotation to electrical energy [15]. The mechanism of linear oscillation conversion is simpler, but it requires the larger amplitude of excitation to produce electricity, whereas human pedals are random in vibration with low amplitude. Consequently, the rotational EM generator was chosen for the VEH floor to convert kinetic energy from people's footsteps into electrical energy.

The objective of the paper is to design a simple but efficient VEH floor embedded with the rotational electromagnetic (EM) generator. To achieve this purpose, the dynamic models of the electro-mechanical systems were developed using MATLAB®/Simulink for predicting the energy performances of the VEH floors and fine-tuning the design parameters. The entire system consists of two main parts of (1) the EM generator, including the translation-to-rotation conversion mechanism, and (2) the Power Management and Storage (PMS) circuit. For simplicity, a direct-current (DC) generator was used in the design to produce electricity. The rack-pinion and lead-screw mechanisms were adopted to converse a linear motion from a human's pedal to a rotation of the generator's rotor. The PMS circuit with extra low energy consumption was designed to simultaneously convert and store electrical energy. The paper is organized into the following sections. In Section 2, the design of each sub-system is described in detail. Then, the installation and demonstration of application are presented in Section 3. Finally, the conclusion is stated in Section 4.

2. Design of Subsystems

The development of the VEH floor—called Genpath—capable of harvesting kinetic energy from people's footsteps and converting it to electrical energy is presented. In the process of energy generation, the rotational EM generator is deployed as it is independent from the resonant frequency and achieves higher energy density when compared to the linear EM generators [15]. Figure 1 shows the diagram of the complete system in Genpath. In Figure 1, when the force from a footstep is applied on the floor-tile, the mechanism for a movement-converter changes the translation of the floor-tile to the rotation of the EM generator to induce voltage. With the connected PMS circuit, the electrical voltage and power generated by the EM generator is processed. The harvested power is then stored in the rechargeable batteries so it can be supplied to the smart IoT-devices with low energy consumption. Two main parts, i.e., (1) the system of EM generator, including the movement-converter and (2) the system of PMS circuit, are presented in detail in the following subsections.

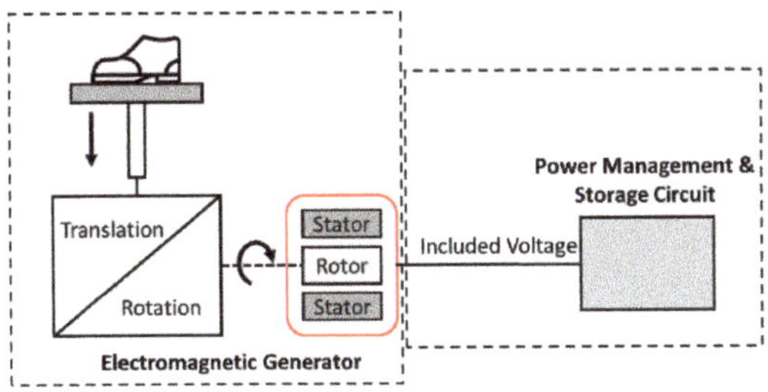

Figure 1. Genpath Concept Diagram.

2.1. The System of EM Generator

2.1.1. Conceptual Design

Figure 2 shows the two proposed designs of the generator system in Genpath, with the only difference in the mechanisms for the movement converter. A set of Genpath comprises a floor-tile block embedded with the translation-to-rotation conversion mechanism which is connected to the DC generator and the PMS circuit board. The entire dimension is 40 × 40 × 10 cm^3 with the maximum allowable displacement of 15–20 mm. Two types of mechanisms, rack pinion and lead screw, are used to convert the translation from a footstep to the rotation of the generator. For the rack-pinion mechanism in Figure 2a, the pinion connected to the floor-tile drives the DC-generator shaft through the additional pinion gears which help transform low-speed power to high-speed power. For the lead-screw mechanism in Figure 2b, the nut fixed to the floor-tile's center moves up and down and drives the lead screw to rotate about its axis. The set of bevel gears transmits the rotation from the lead screw to the DC generator and changes the direction of rotation by 90°. The rotation of the DC generator in both designs then induces the voltage. The springs with the maximum displacement of 15–20 mm connected to the four corners of the block help restore the top floor-tile back to the equilibrium position. Considering the limits of the dimension and the displacement, thus, the small size of 12/24-V-DC motor was decided for the DC generator.

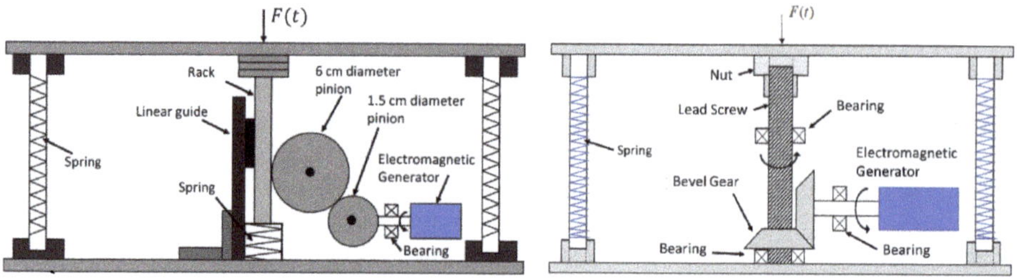

(**a**) Rack-Pinion Design (**b**) Lead-Screw Design

Figure 2. (**a**) Rack-Pinion Design; (**b**) Lead-Screw Design.

2.1.2. Analysis

To predict the energy performances of the EM-generator designs, the dynamic models of the two electro-mechanical systems, shown in Figure 2, were developed using MATLAB®/Simulink. Figure 3a,b show the physical models corresponding to the two systems in Figure 2a,b, respectively. In Figure 3a, the system consists of the elements of rack, pinion and gear on the mechanical side, and the DC generator (with its own resistance, R_G, and inductance, L) connected to the load R_L on the electrical side. Contrastively, the elements of rack, pinion and gear are replaced by the nut, lead screw, and bevel gears for the system in Figure 3b. The footstep force $F(t)$ is modeled as the arbitrary function reported in [9] and presented in Figure 4. The spring with a maximum compression of 15–20 mm provides the restoring force F_s to restore the floor-tile back to the equilibrium position. The dynamic equations governing the electro-mechanical models of both designs are formulated as follows.

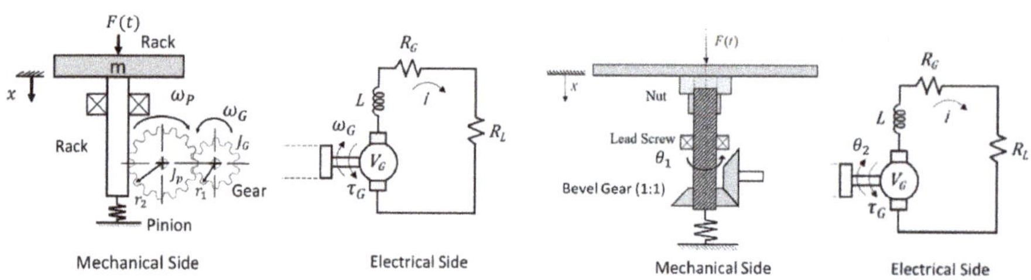

(**a**) Rack and Pinion Model (**b**) Lead-Screw Model

Figure 3. Physical models: (**a**) Rack and Pinion Model; (**b**) Lead-Screw Model.

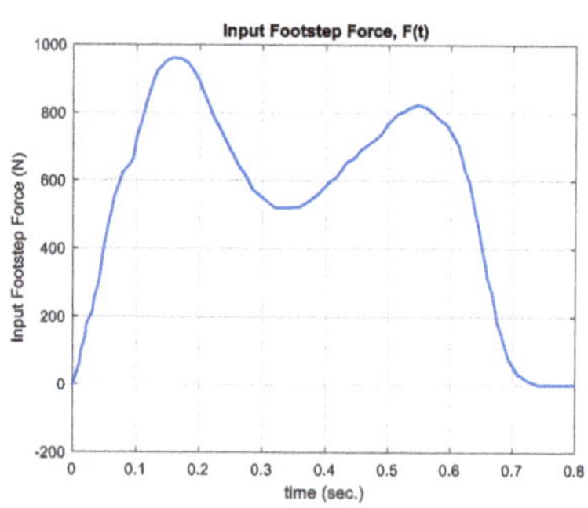

Figure 4. Input Footstep Force.

For the electrical side in Figure 3, Kirchhoff's voltage law yields

$$V_L + V_{RG} + V_{RL} - V_G = 0, \tag{1}$$

where V_G is the back emf of the generator; i.e., $V_G = K_t\omega_G$ where K_t is the back emf (torque) constant. In addition, V_L, V_{RG} and V_{RL} are the voltages across the generator's inductor, generator's resistor and load's resistor, respectively. With the voltage-current relations, (1) becomes

$$\dot{i} + \left(\frac{R_G + R_L}{L}\right)i - \left(\frac{K_t}{L}\right)\omega_G = 0, \tag{2}$$

where i is the current, K_t is the back emf (torque) constant, ω_G is the generator rotational speed, L is the inductance of the generator, R_G is the resistance of the generator and R_L is the resistance of the load. Equation (2) is the differential equation governing the armature winding of the generator.

From the free body diagram (FBD) of the mechanical system with the rack and pinion in Figure 5, Newton's second law and the law of angular momentum describe the translation of the rack, and the rotations of the pinion and the generator rotor, respectively, as

$$m\ddot{x} = F(t) - F_r - F_s, \tag{3}$$

$$J_p\dot{\omega}_p = J_p\frac{\ddot{x}}{r_2} = (F_r - f_r)r_2, \tag{4}$$

$$J_G\dot{\omega}_G = J_G\frac{\ddot{x}}{r_1} = (f_r - \tau_G)r_1, \tag{5}$$

where m is the mass of the plate and rack, and J_p and J_G are the moments of inertia of the pinion and the gear. x is the displacement of the rack, ω_p is the angular velocity of the pinion, ω_G is the angular velocity of the gear. In (3)–(5), $F(t)$ is the input force, F_s is the restoring spring and damper force, F_r is the friction force between the rack and pinion, f_r is the friction force between the pinion and gear. In addition, τ_G is the electromagnetic torque of the DC generator, where $\tau_G = K_t i$, and r_1 and r_2 are the radius of the gear and pinion, respectively.

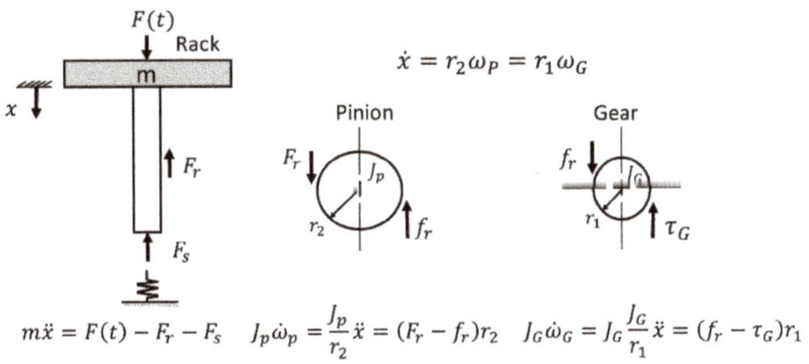

Figure 5. Rack and Pinion Free Body Diagram.

Eliminating F_r and f_r from (3)–(5), the differential equation governing dynamics of the mechanical system with the rack and pinion is obtained as

$$M\ddot{x} + \frac{K_t}{r_1}i + F_s = F(t), \tag{6}$$

where $M = \left(m + \frac{J_p}{r_2^2} + \frac{J_G}{r_1^2}\right)$. Then the governing equations of the electro-mechanical system from combining (2) and (6) are

$$\begin{bmatrix} \dot{i} \\ \dot{x} \\ \ddot{x} \end{bmatrix} = \begin{bmatrix} -\left(\frac{R_G+R_L}{L}\right) & 0 & \frac{K_t}{Lr_1} \\ 0 & 0 & 1 \\ -\frac{K_t}{Mr_1} & 0 & 0 \end{bmatrix} \begin{bmatrix} i \\ x \\ \dot{x} \end{bmatrix} + \begin{bmatrix} 0 \\ 0 \\ \frac{F(t)-F_s}{M} \end{bmatrix}, \tag{7}$$

Similarly, from the FBD of the mechanical system with lead and screw as shown in Figure 6, the equations govern the translation of the nut and the rotations of the lead screw and the generator rotor, respectively, are

$$m\ddot{x} = F(t) - F_a - F_s, \tag{8}$$

$$J_1\ddot{\theta}_1 = \frac{2\pi J_1}{l}\ddot{x} = \Gamma - T_B, \tag{9}$$

$$J_G\ddot{\theta}_2 = \frac{2\pi J_G}{l}\ddot{x} = T_B - \tau_G, \tag{10}$$

where m is the mass of the floor-tile and the nut, J_1 is the moment of inertia of the lead screw and J_G is the moment of inertia of the bevel gear. l is the pitch of the lead screw, x is the displacement of the plate and nut, θ_1 is the angular position of the lead screw and θ_2 is the angular position of the bevel gear. In addition, $F(t)$ is the applied force from the footstep, F_s is the restoring spring and damper forces, F_a is the friction force between nut and lead screw, T_B is the friction torque of the bevel gear, τ_G is the electromagnetic torque of the DC generator; where $\tau_G = K_t i$. Note that Γ in (9) is the transmitted torque from the nut to the lead screw which is proportional to F_a as

$$\Gamma = \Delta F_a,$$

where $\Delta = \frac{l}{2\pi\eta_{thread}\eta_{thrust}}$, with η_{thread} and η_{thrust} are the efficiencies of the *thread* and the *thrust* bearing, respectively.

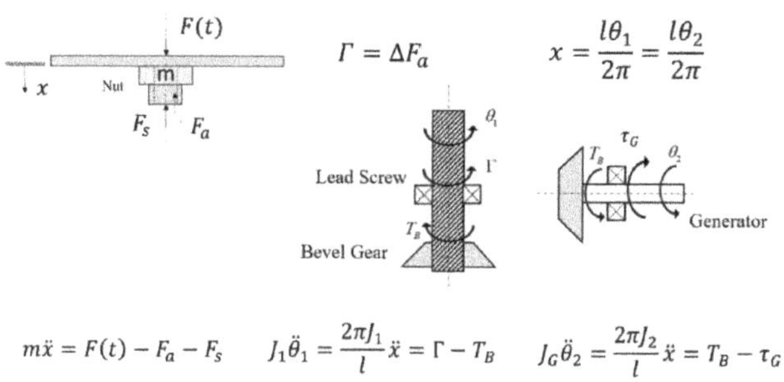

Figure 6. Lead-Screw Free Body Diagram.

By rearranging (8)–(10), the equations become

$$J_{eq}\ddot{\theta}_1 + \frac{K_t}{\Delta}i + F_s = F(t), \tag{11}$$

$$\frac{2\pi J_{eq}}{l}\ddot{x} + \frac{K_t}{\Delta}i + F_s = F(t), \tag{12}$$

where $J_{eq} = \frac{ml}{2\pi} + \frac{(J_1+J_G)}{\Delta}$ is the equivalent moment of inertia corresponding to the mass of the plate and nut m, and the mass moments of inertia of the lead screw and bevel gear J_1 and J_G, respectively.

By combining (2) and (11)–(12), the equations governing the electro-mechanical system with the lead and screw design are obtained as

$$\begin{bmatrix} \dot{i} \\ \dot{\theta}_1 \\ \ddot{\theta}_1 \end{bmatrix} = \begin{bmatrix} -\left(\frac{R_G+R_L}{L}\right) & 0 & \frac{K_t}{L} \\ 0 & 0 & 1 \\ -\frac{K_t}{J_{eq}\Delta} & 0 & 0 \end{bmatrix} \begin{bmatrix} i \\ \theta_1 \\ \dot{\theta}_1 \end{bmatrix} + \begin{bmatrix} 0 \\ 0 \\ \frac{F(t)-F_s}{J_{eg}} \end{bmatrix}, \tag{13}$$

or

$$\begin{bmatrix} \dot{i} \\ \dot{x} \\ \ddot{x} \end{bmatrix} = \begin{bmatrix} -\left(\frac{R_G+R_L}{L}\right) & 0 & \frac{K_t}{L} \\ 0 & 0 & 1 \\ -\frac{lK_t}{2\pi J_{eq}\Delta} & 0 & 0 \end{bmatrix} \begin{bmatrix} i \\ x \\ \dot{x} \end{bmatrix} + \begin{bmatrix} 0 \\ 0 \\ \frac{l(F(t)-F_s)}{2\pi J_{eg}} \end{bmatrix}, \tag{14}$$

The MATLAB®/Simulink models corresponding to (7) and (14) were developed to predict the voltages and currents for various load resistance R_L that both the EM generator systems with rack pinion and lead screw could generate. The Simulink model of the system with lead screw is presented in Figure 7. With the selected parameters shown in Tables 1 and 2, the simulation results were compared to the corresponding test results as illustrated in Figures 8 and 9 (The test procedure will be described in Section 2.1.4.). Figures 8 and 9 show that the analytical models accurately predict the magnitudes of the voltages and currents generated by the EM generator. The voltage and current signals in Figures 8 and 9 can be divided into two stages according to the movement, i.e., forward and return stages. In the forward stage, ~0.2–0.8 s, the floor-tile moves downwards when the footstep-force applied, causing the generator to rotate in one direction and hence induce negative voltage and current as shown in Figures 8 and 9. During 0.2–0.8 s, the floor-tile might reach the lowest position, causing the generator to stop the rotation and induce no voltage and current before the return stage begins. In the return stage, ~0.8–1.4 s, the floor-tile moves upwards to the equilibrium position according to the restoring spring forces and drives the generator to rotate backward and induce positive voltage and current as seen in Figures 8 and 9. Note that in Figure 9, there exists the small humps of the predicted voltage and current during 0.6–0.8 s. These signals correspond to the applied force at the same interval when the floor-tile is moving downwards. The discrepancy of this analytical prediction and the test result might be because of the difference between the actual force and the force function in Figure 4.

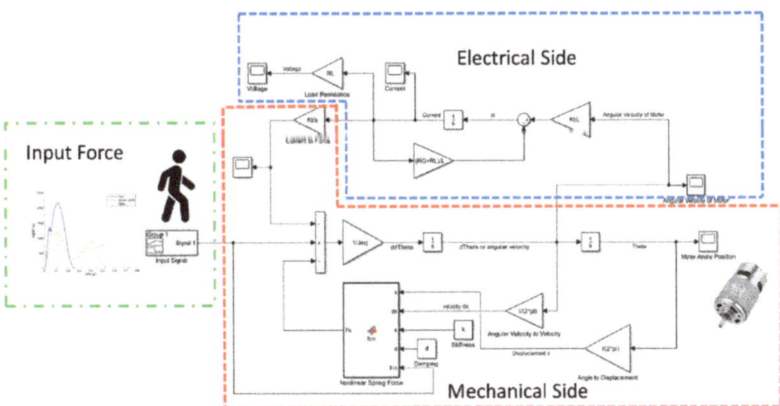

Figure 7. Simulink model for the electromagnetic (EM) generator with lead-screw design.

Table 1. Rack-Pinion Parameters.

Parameters	Value
Mass of Rack and Plate (m)	3.016 kg
Radius of Pinion (r_1)	0.72×10^{-2} m
Radius of Gear (r_2)	3×10^{-2} m
Moment of inertia of bevel gear (J_G)	8.6756×10^{-7}
Moment of Inertia of Pinion (J_p)	1×10^{-5} kg m
Spring Coefficient (k)	20,500 N/m
Damping Coefficient (d)	900 N·s/m
Resistance of Generator (R_G)	42 Ohm
Inductance (L)	19.6×10^{-3} H
Generator constant (K_t)	0.5854 Vs/rad
Resistance of Load (R_L)	30 Ohm

Table 2. Lead-Screw Parameters.

Parameters	Value
Pitch of Lead Screw (l)	8 mm
Mass of Nut and Plate (m)	2.16 kg
Moment of inertia of bevel gear (J_G)	8.6756×10^{-7}
Moment of Inertia of lead screw (J_l)	2.5536×10^{-6} kg m^2
Lead angle	45 degree
Spring Coefficient (k)	40,000 N/m
Damping Coefficient (d)	13,600 N·s/m
Resistance of Generator (R_G)	37 Ohm
Inductance (L)	19.6×10^{-3} H
Generator constant (K_t)	0.392 Vs/rad
Resistance of Load (R_L)	30 Ohm
Friction coefficient (μ)	0.21
Efficient of *thrust* bearing η_{thrust}	0.6529
Efficient of *thread* η_{thread}	0.8132

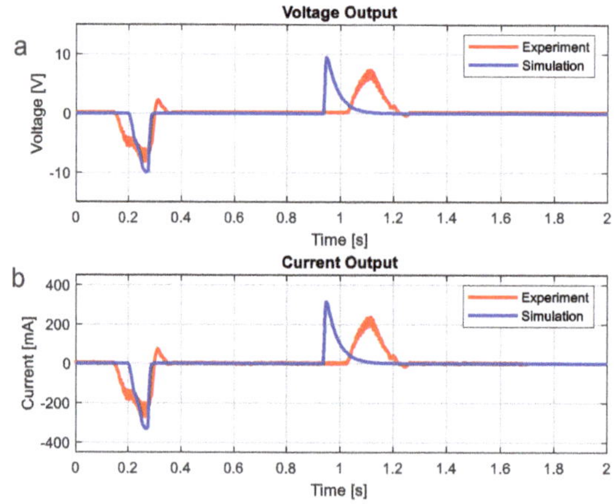

Figure 8. Voltage and Current of Rack-Pinion Model from Experiment and Simulation. (**a**) Voltage of Rack-Pinion Model. (**b**) Current of Rack-Pinion Model.

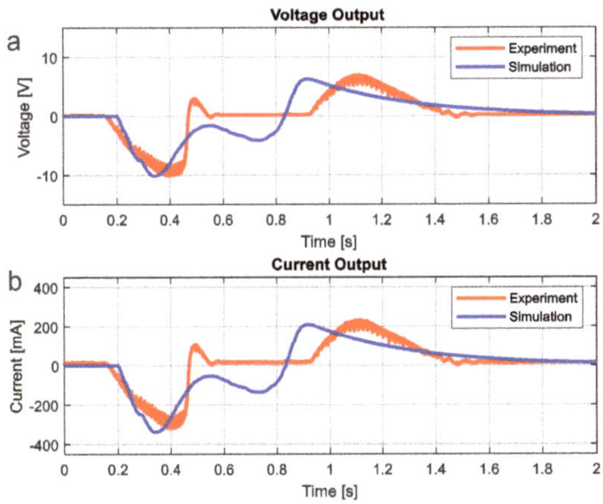

Figure 9. Voltage and Current of Lead-Screw Model from Experiment and Simulation. (**a**) Voltage of Lead-Screw Model. (**b**) Current of Lead-Screw Model.

In summary, the induced voltages and currents for both EM-generator designs predicted by the analytical models were compared in Figure 10. Then the performances of the two EM-generators were predicted and summarized in Table 3. The results mainly show that the designed systems generate an averaged 216–886 mJ of electrical energy per footstep, or the averaged power of 216–590 mW. It is sufficient to power electronic devices with low power consumption in the vicinity, such as sensors and communication instruments. This finding assures the possibility of building both of Genpath's prototypes. Moreover, the verified analytical models were used in the parametric design as presented in Section 2.1.3.

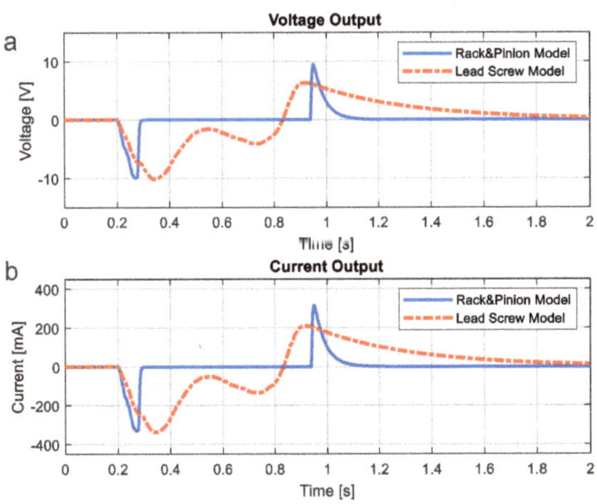

Figure 10. Simulation of Rack-Pinion and Lead-Screw Model. (**a**) Voltage of Rack-Pinion and Lead-Screw Model. (**b**) Current of Rack-Pinion and Lead-Screw Model.

Table 3. Performances of the EM-generator designs.

Variables	Rack-Pinion Design	Lead-Screw Design
	Values per Footstep	Values per Footstep
Maximum voltage	9.92 V	10.13 V
Average voltage	0.99 V	4.16 V
Maximum current	330.9 mA	337.9 mA
Average current	33.15 mA	138.8 mA
Maximum power	3.28 W	3.42 W
Average power	216.4 mW	590.3 mW
Wave duration	1.00 s	1.50 s
Average energy	216.4 mJ	885.8 mJ

2.1.3. Design of Elements

The critical elements of the EM generator systems in Figures 2 and 3 were decided with the use of the analytical models to tune for the optimized parameters. Design of the elements such as the rack pinion and lead screw, the springs, the transmitted gears and the DC generator is summarized as follows.

The rack pinion and lead screw shown in Figures 2 and 3 are used to change the translation to rotation. A mechanism of the rack-pinion was first adopted in the first prototype [16] because of its availability and economy cost. The drawback of the rack-pinion mechanism arises from its coarse tolerance, resulting in large friction loss. In addition, only the rough-pitch models were found. Thus, for the rack's allowable displacement of 15 mm, the angular displacement of the pinion is very limited. Therefore, the EM generator system using the rack and pinion is inefficient for harvesting energy from the footstep with limited displacement. To improve the design of the movement converter, the rack-pinion mechanism is replaced by the lead screw in the second prototype. The lead screw has more variety in dimensions for the selection. With finer pitch or smaller lead angles, the angular displacement of the lead can be extended with the limited stroke of the nut. Two sets of the lead angles, i.e., 45° and 60°, were selected and installed to the 24-V-DC-generator system for the comparison. Table 4 presents the energy produced by the generator, when connected to 49-Ω resistance load, for three different designs of the movement converters: the rack pinion, the lead screw with 60° lead angles and the lead screw with 45° lead angles. It was found that the EM generator system with the 45° lead screw produces the highest level of energy among the three designs. With the smaller value of the lead angles, the lead proceeds to larger angular displacement within the same limited stroke of 15 mm, resulting in the greater period for the generator to spin. Therefore, the accumulative energy the generator provides is higher.

Table 4. Comparison of Rach-Pinion, Lead-Screw (60° lead angles) and Lead-Screw (45° lead angles) Average Energy.

Design	Averaged Energy (mJ)
Rack pinion	319
60° lead angles Lead screw	353
45° lead angles Lead screw	488

The spring and transmitted gears are also the parts critical to harvest the energy. The softer spring is preferable in the design. For the explanation, Figure 11 shows the predicted voltages and currents for the two EM-generator systems varying in the stiffness coefficients. The softer spring, with less value of stiffness coefficient, results in the higher levels of the voltage and current in the forward stage, and hence yields the greater power in harvesting. The softer spring inserts less restoring forces and causes the floor-tile to move down with higher speed that is converted to a higher rotational speed of the generator. With more speed, the generator can produce more power. Although the softer springs are theoretically desirable, they should be sufficiently hard enough to restore the system back to equilibrium due to the friction of the system. To satisfy such conditions, the optimized springs with the wire diameter of 2.2 mm or the stiffness of 40 kN/m were selected.

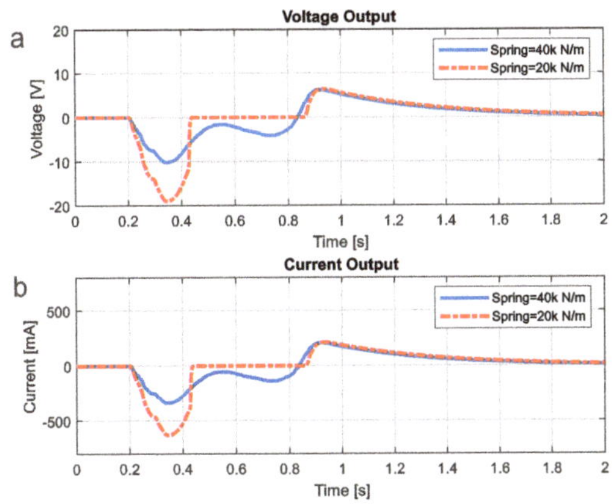

Figure 11. Simulation of Lead-Screw Model with Soft and Hard Springs. (**a**) Voltage of Lead-Screw Model with Soft and Hard Springs. (**b**) Current of Lead-Screw Model with Soft and Hard Springs.

Moreover, the sets of gear train and bevel gears in Figures 2 and 3, respectively, are used to transmit the rotation from the movement converters to the generator's rotor. The gear ratio larger than 1:1 can help increase the speed of the generator. However, the increase of the gear ratio is limited by the amount of the resistance force in the system. The increase of the gear ratio leads to the greater frictions and the greater resistance torque provided by the generator. If the resistance force exceeds the applied force from the footstep, the floor-tile will not move. Consequently, the maximum gear ratio of 4:1 is designed for the rack-pinion system and originated from a pinion's 6 cm diameter and transmitted to a gear's 1.5 cm diameter as shown in Figure 2. In addition, the maximum gear ratio of the bevel gears in Figure 3 for the lead-screw system is set to 1:1.

The DC generator was used in the design for simplicity. In order to generate at least 3.3 V for operating the micro-controller, the typical 12-V or 24-V-DC-motor generator was selected to ensure such criteria. To choose a proper DC generator's speed, the kinematic relation of the transmission system was analyzed. With the maximum value of 20 mm displacement for safely walking, the maximum value of the angular velocity is obtained at 210 rpm. Hence, the model of a motor generator with a 300 rpm rated speed was selected. Two types of the DC generators, 12 V and 24 V with the properties shown in Table 5, were installed in the second prototype of Genpath for comparison of their performance. The induced voltage and current for both types of the DC generators are compared in Figure 12. The energy per footstep produced by the DC generators for various rated load resistances are shown in Table 6. The 12-V motor provides the energy as much as 2.5 times that of the 24-V motor because of the resistance during the transience.

Table 5. Comparison of 12- and 24-V-DC-Generator Parameters.

Voltage (V)	Resistance RG (Ω)	Inductance L (mH)	Kt (Vs/rad)
12	37	3.6	0.2903
24	42	19.6	0.5854

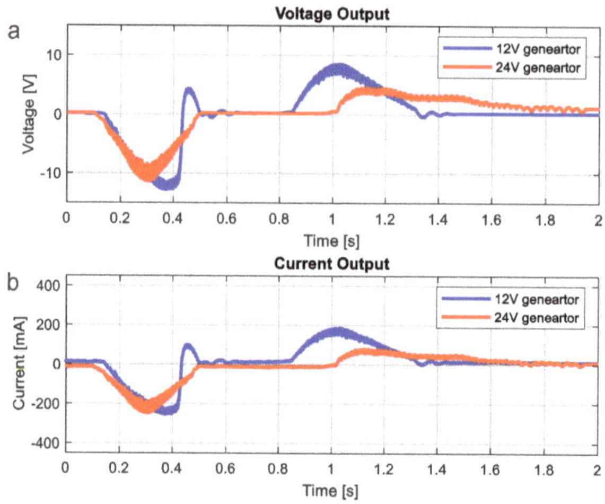

Figure 12. Comparison Voltage and Current of Lead-Screw Model with 12– and 24–V-DC Generator. (**a**) Voltage of Lead-Screw Model. (**b**) Current of Lead-Screw Model.

Table 6. Energy per footstep produced by the DC generators for various rated load resistances.

Load (Ω)	Average Energy (mJ)	
	12 V	**24 V**
30	798.2	321.5
39	750.0	313.2
49	745.5	488.2

In conclusion, the analytical models developed in the previous section were utilized to obtain the best-fit design. First, the simulation results show that both the rack-pinion and lead-screw models yield about the same maximum power, according to the same levels of both voltage and current magnitudes as seen in Figure 10. However, the simulation results in Figure 10 also indicate that the lead screw provides the longer time in movement and yields the larger angular displacement of the rotor within the limited stroke. This results in more time for the generator in the lead-screw design to generate power. It was also found that the lead-screw design with the finer pitch or, i.e., 45° lead angles, provides the highest energy per step. Second, although the simulation results show that the softer spring could provide higher power in the forward stage, the harder spring with the optimum stiffness value of 40 kN/m was selected to enable to restore the system back to equilibrium. Finally, the 12-V-DC generator, compared to the 24-V-DC generator, gives a better performance probably because of the lower resistance during the transient, as the properties show in Table 5.

2.1.4. Development of the Prototypes

Figures 13 and 14 show the two prototypes of Genpath built with the key components as listed in Table 7. Prototype-I [16] was installed with the 24-V-DC generator and uses the rack pinion for the movement converter. Prototype II is the improved prototype built with the lead screw for the movement converter and the 12-V-DC generator. The experiment was then performed to test the prototypes' performances as shown in Figure 15. First, each prototype was connected to the rated resistor R_L to provide the maximum power output. Then the voltage across R_L, the current i and the corresponding electrical power when a normal footstep is applied were measured using an oscilloscope and a current probe. The test results are shown in Figure 16 and summarized in Table 8. Genpath prototype-II with 12-V-DC generator and lead-screw mechanism was significantly improved when compared to the prototype-I [16]. It stated in Table 8 that the latest Genpath prototype produces an average energy of 702 mJ (or average power of 520 mW), the maximum voltage of 9.5 V and the maximum current of 285 mA per footstep in the duration of 1.35 s. The energy provided by the EM-generator in Genpath's prototype-II was increased by approximately 184% when compared to that of the prototype-I [16]. The efficiency of the EM-generator system is 26% based on the power generation from the heel strike of a human's walk of 2 W per step. This amount of energy could sufficiently power typical low-power electrical devices, as previously described.

Figure 13. Photograph of Rack and Pinion Prototype.

Figure 14. Photograph of Lead-Screw Prototype.

Table 7. Components of rack and pinion and lead-screw prototypes.

Prototype I (Rack and Pinion)			**Prototype II (Lead Screw)**		
Item	Dimensions	#	Item	Dimensions	#
Acrylic plate	400 × 400 × 10 mm	2	Wood plate	400 × 400 × 5 mm	2
Linear guide	Dia 12 Length 90 mm	4	Linear guide	Dia 12 Length 90 mm	4
Linear bearing	Inner dia 12 mm	4	Linear bearing	Inner dia 12 mm	4
Shaft coupling	Inner dia 12 mm	4	Shaft coupling	Inner dia 12 mm	4
Coil spring	Length 60 mm Dia 1.6 mm	4	Coil spring	Length 60 mm Dia 2.2 mm	4
Shaft to generator	Dia 8 mm Length 60 mm	1	Shaft to generator	Dia 8 mm Length 60 mm	1
Rack and pinion	Pinion radius 3 cm	1	Nut and lead screw	Dia 8 mm Pitch 2 mm	1
Flexible coupling	8 mm	1	Flexible coupling	8 mm	1
Gear	Radius 0.75 cm	1	Bevel gear	Inner dia 8 mm	2
Ball bearing	Inner dia 8 mm	3	Ball bearing	Inner dia 8 mm	3
Generator	ZGA37RG 24V 300 rpm	1	Generator	ZGA37RG 12V 300 rpm	1

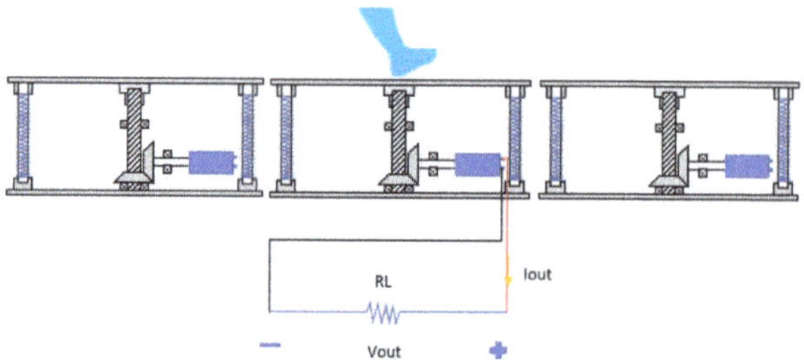

Figure 15. Test Set up Diagram for the Prototypes.

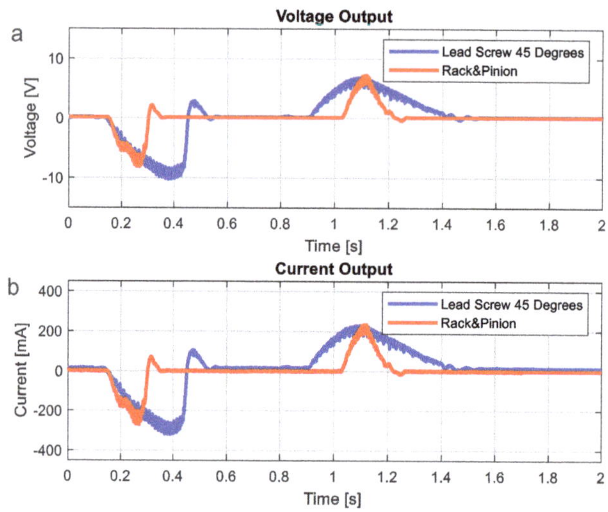

Figure 16. Comparison Voltage and Current of Prototype I and Prototype II. (**a**) Voltage of Prototype I and Prototype II. (**b**) Current of Prototype I and Prototype II.

Table 8. Performances of Genpath prototypes I and II.

Variables	Prototype I (Rack-Pinion)	Prototype II (Lead-Screw)
	Values per Footstep	Values per Footstep
Maximum voltage	7.5 V	9.5 V
Average voltage	1.26 V	2.88 V
Maximum current	246 mA	285 mA
Average current	42.5 mA	88 mA
Maximum power	1.85 W	2.71 W
Average power	216 mW	520 mW
Wave duration	1.14 s	1.35 s
Average energy	247 mJ	702 mJ

2.2. *The system of Power Management and Storage Circuit*

The power management and storage (PMS) circuit was designed to convert and store electrical energy at the same time. Figure 17 shows the circuit diagram and the real-world circuit is depicted in Figure 18. From the performance test of Genpath as shown in Figure 16, it is clearly seen that the generated voltage and current waveform are AC signals. Negative portions occur when a footstep is applied, causing the generator to rotate in one direction. Positive portions occur during the restoration period, resulting in the opposite direction of the generator's rotation.

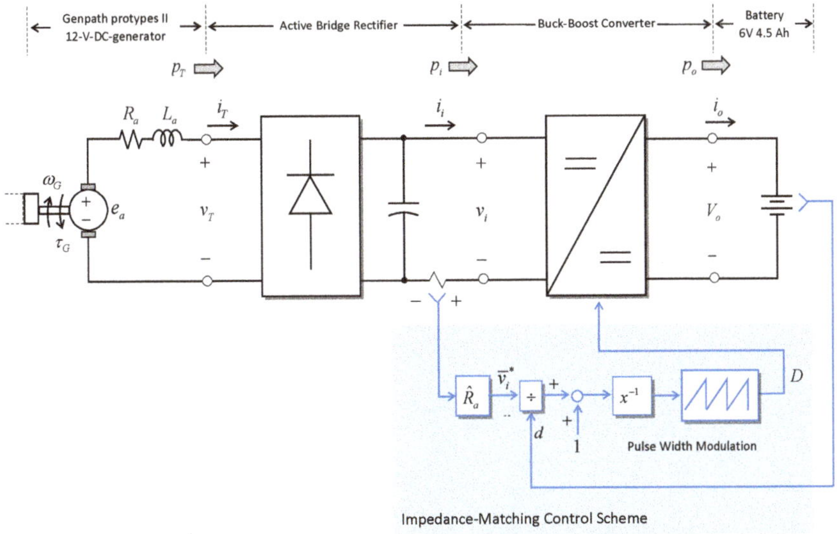

Figure 17. Circuit Diagram of Power Management and Storage System.

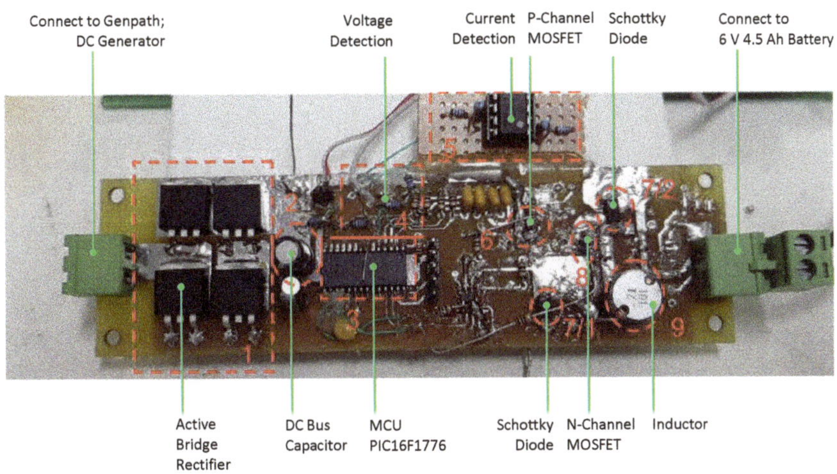

Figure 18. Real-World Circuit of Power Management and Storage System.

The energy from the generator is stored to a 6-V, 4.5-Ah battery through a two-stage power converter. First, the active bridge rectifier converts the AC voltage to the DC voltage. Since the metal–oxide–semiconductor field-effect transistors (MOSFETs) used in the rectifier have extremely low turn-on resistances and junction voltage drops, this kind of rectifier can perform with high efficiency. Second, the buck-boost converter helps convert the variable DC voltage from the rectifier to the battery. This buck-boost converter is operated in accordance with a matching-impedance control scheme which paves the way for the maximum power transfer. In this scheme, the reactance term by inductance is assumed to be small and is neglected from the calculated impedance. This assumption is valid by investigating the value of inductance in Table 5 and the low-frequency AC voltage exhibited in Figure 19b. (Figure 19a shows the output voltage and current without power management and a storage circuit.) The reactance is, therefore, insignificant in comparison with the resistance and it is neglected in the impedance-matching control scheme for the sake of simplicity. This control scheme is implemented with the microcontroller PIC16F1776 which includes the extreme low-power consumption feature. (Note: the input force to the prototype II in this section is different from the previous section due to the environment setup.)

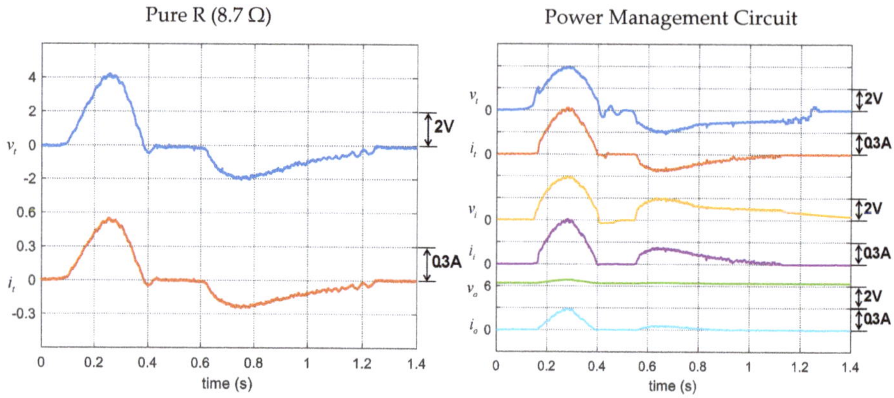

(**a**) Output Voltage and Current without Power Management and Storage Circuit

(**b**) Output Voltage and Current with Power Management and Storage Circuit

Figure 19. Experimental result showing the operation of power management and storage circuit; v_t and i_t are outputs of generator; v_i and i_i are outputs of the rectifier circuit; v_o and i_o are outputs of the buck-boost circuit.

The experiment is conducted to evaluate the performance of the two-stage converter. The AC voltage is efficiently rectified; the efficiency of active bridge rectifier is about 95.78%. Figures 19b and 20b show the operation of the buck-boost converter along with the impedance matching control scheme. The converter helps charge the power into the battery and the control scheme can match the impedance to achieve the maximum power transfer. The power management system is capable of gaining the averaged power of 280 mW from each footstep and storing the accumulative energy of 302 mJ into the battery at the output stage. The efficiency of the buck-boost converter is about 78.00%, thus the overall efficiency of the power management system is 74.72%. Table 9 gives the detailed performances of power management and storage system for each footstep. If comparing the Genpath prototype II with the commercial product such as the Pavegen's system, the Pavegen's system which has three generators per tile can generate the energy approximately 2 J per step and the Genpath prototype II which has one generator per tile can generate the energy approximately 0.3 J per step. The Genpath prototype II generates the energy approximately 6 times less than that of the Pavegen's system. Even though it cannot generate as much energy as the commercial one, the Genpath prototype II is developed based on open hardware which is easy to access and build. The mathematical model of the system exists. Thus, it is possible to develop Genpath's system in any community to generate more energy in the future.

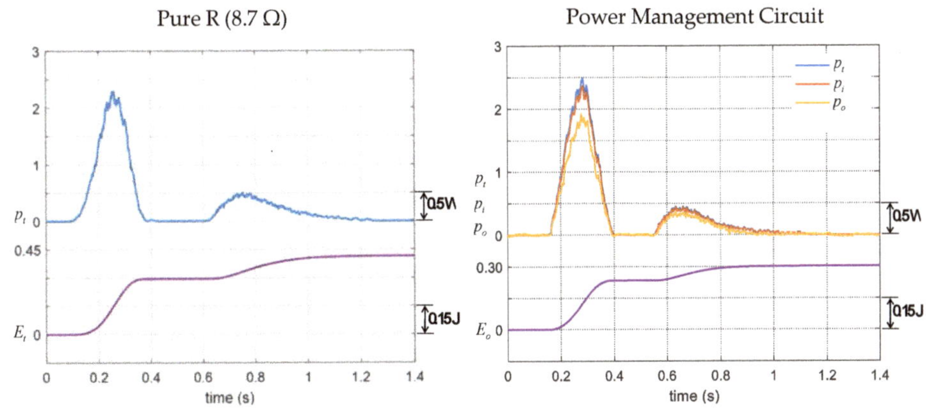

(a) Power and Energy without Power and Storage Management Circuit.

(b) Power and Energy with Power Management and Storage Circuit

Figure 20. Experimental result showing the operation of the power management system regarding the process of power conversion and energy storage; P_t and E_t are power and energy outputs of the generator; P_i is power output of the rectifier circuit; P_o and E_o are power and energy outputs of the buck-boost circuit.

Table 9. Performances of power management and storage circuit with Genpath prototypes II and 12-V-DC generator.

Variables	Values per Footstep	
	Pure R (8.7 Ω)	Power Management and Storage Circuit
Maximum voltage; max (v_t)	4.22 V	3.97 V
Maximum current; max (i_t)	548 mA	635 mA
Maximum power; max (p_t)	2.29 W	2.48 W
Average power; $\overline{p}_t$	351 mW	374 mW
Average power; $\overline{p}_i$	-	359 mW
Average power; $\overline{p}_o$	-	280 mW
Wave duration	1.18 s	1.08 s
Stored energy at 1.4 s; E_o (t = 1.4 s)	415 mJ	302 mJ
Efficiency of Active Rectifier	-	95.78%
Efficiency of Buck-Boost Converter	-	78.00%
Overall Efficiency	-	74.72%

3. Installation and Demonstration

To demonstrate the application of the developed system, the system was assembled as a floor tile with the dimension of 40 × 40 × 15 cm and installed by three sets of the floor tiles side-by-side without wiring at an exhibition hall of a 100-year-old engineering building at Chulalongkorn University, on the mechanical engineering project exhibition day. Figure 21 shows the photo of the system. Then the exhibitors were invited to walk on the floor tiles. Once they stepped on the floor tiles, the system generated power to store in the battery and lit up the array of LEDs, as seen in Figure 21. This demonstration made exhibitors relive the importance of green energy and energy harvesting in their daily lives. The benefit of the designed system can apply for a lot of applications, such as a wireless sensor and Internet of Thing applications.

Figure 21. Installation and demonstration.

4. Conclusions

The paper presented a design of an energy harvesting floor capable of converting mechanical energy from people's footsteps to electrical energy. The system, comprising the translation-to-rotation conversion mechanism, the EM generator, and the power management circuit system, generates electricity from people's footsteps. For the EM generator, the conversion mechanism for linear translation to rotation was designed by using the rack-pinion and lead-screw mechanism. Based on simulation analysis, the averaged energy of the lead-screw model is more than that of the rack-pinion model. Moreover, the design of lead-screw model elements was studied. The results show that the lead-screw model with 45° lead angles can generate more averaged energy than others, and the 12-V-DC generator can provide more energy than the 24-generator due to the resistance during the transience, and the softer spring can convert translation motion to higher rotational speed of the generator which can produce more power. Although the softer springs are theoretically desirable, they should be sufficiently hard enough to restore the system back to equilibrium due to the friction of the system. Then, Genpath prototype-II with 12-V-DC generator, lead-screw mechanism, and allowable displacement of 15 mm was built. It was significantly improved when compared to the prototype-I [16]. This Genpath prototype produces an average energy of up to 702 mJ (or average power of 520 mW). The energy provided by Genpath prototype-II is increased by approximately 184% when compared to that of the prototype-I [16]. The efficiency of the EM-generator system is 26% based on the power generation from the heel strike of a human's walk of 2 W per step. Next, the power management and storage circuit were developed to harvested energy into the batteries and to supply other parts to specific loads. The experiment showed that the circuit has the overall efficiency of 74.72%. The benefit of the designed system can apply for a lot of applications, such as a wireless sensor and Internet of Thing applications.

Author Contributions: Conceptualization, T.J., S.S., and G.P.; methodology, T.J., S.S., and G.P.; software, T.J., S.S., and G.P.; validation, P.K., P.C. and C.U.-v.); formal analysis, T.J.; investigation, P.K., P.C. and C.U.-v.; resources, T.J.; data curation, G.P.; writing—original draft preparation, T.J., S.S., and G.P.; writing—review and editing, T.J., S.S., and G.P.; visualization, T.J.; supervision, T.J.; project administration, G.P. All authors have read and agreed to the published version of the manuscript.

Funding: This research was funded by Ratchadaphiseksomphot Endowment Fund Chulalongkorn University, grant number CU_GI_63_07_21_01.

Acknowledgments: We would like to thank W. Lowattanamart, V. Suttisung, and S. Sintragoonchai who initiated and supported the idea on the project and built up Genpath prototype-I.

Conflicts of Interest: The authors declare no conflict of interest. Also, the funders had no role in the design of the study; in the collection, analyses, or interpretation of data; in the writing of the manuscript, or in the decision to publish the results.

References

1. Lazaro, A.; Villarino, R.; Girbau, D. A Survey of NFC Sensors Based on Energy Harvesting for IoT Applications. *Sensors* **2018**, *18*, 3746. [CrossRef] [PubMed]
2. Cottone, F. Energy Harvesting: Introduction. In Proceedings of the NiPS Summer School, Fiuggi, Italy, 7–12 July 2015; p. 50.
3. Panwar, N.L.; Kaushik, S.C.; Kothari, S. Role of Renewable Energy Sources in Environmental Protection: A Review. Renew. *Renew. Sustain. Energy Rev.* **2011**, *15*, 1513–1524. [CrossRef]
4. Beeby, S.P.; Torah, R.N.; Tudor, M.J.; Glynne-Jones, P.; O'Donnell, T.; Saha, C.R.; Roy, S. A Micro Electromagnetic Generator for Vibration Energy Harvesting. *J. Micromech. Microeng.* **2007**, *17*, 1257–1265. [CrossRef]
5. Riemer, R.; Shapiro, A. Biomechanical Energy Harvesting from Human Motion: Theory, State of the Art, Design Guidelines, and Future Directions. *J. Neuroeng. Rehab.* **2011**, *8*, 22. [CrossRef] [PubMed]
6. Energy Floors 2019. Available online: https://energy-floors.com (accessed on 1 March 2019).
7. Pavegen 2020. Available online: https://pavegen.com/ (accessed on 7 October 2020).
8. Rain-Noe. Swingset-Powered Phone Chargers. *Ieyenews*. 2020. Available online: https://www.ieyenews.com/swingset-powered-phone-chargers/ (accessed on 7 October 2020).
9. Liu, M.; Lin, R.; Zhou, S.; Yu, Y.; Ishida, A.; McGrath, M.; Kennedy, B.; Hajj, M.; Zuo, L. Design, Simulation and Experiment of a Novel High Efficiency Energy Harvesting Paver. *Appl. Energy* **2018**, *212*, 966–975. [CrossRef]
10. Hwang, S.J.; Jung, H.J.; Kim, J.H.; Ahn, J.H.; Song, D.; Song, Y.; Lee, H.L.; Moon, S.P.; Park, H.; Sung, T.H. Designing and Manufacturing a Piezoelectric Tile for Harvesting Energy from Footsteps. *Curr. Appl. Phys.* **2015**, *15*, 669–674. [CrossRef]
11. Kim, K.B.; Cho, J.Y.; Jabbar, H.; Ahn, J.H.; Hong, S.D.; Woo, S.B.; Sung, T.H. Optimized Composite Piezoelectric Energy Harvesting Floor Tile for Smart Home Energy Management. *Energy Convers. Manag.* **2018**, *171*, 31–37. [CrossRef]
12. Vocca, H.; Cottone, F. Kinetic Energy Harvesting. In *ICT-Energy-Concepts Towards Zero–Power Information and Communication Technology*; Intechopen: London, UK, 2014. [CrossRef]
13. Yang, Z.; Zhou, S.; Zu, J.; Inman, D. High-Performance Piezoelectric Energy Harvesters and Their Applications. *Joule* **2018**, *2*, 642–697. [CrossRef]
14. Larkin, M.; Tadesse, Y. HM-EH-RT: Hybrid Multimodal Energy Harvesting from Rotational and Translational Motions. *Int. J. Smart Nano Mater.* **2013**, *4*, 257–285. [CrossRef]
15. Arnold, D.P. Review of Microscale Magnetic Power Generation. *IEEE Trans. Magn.* **2007**, *43*, 3940–3951. [CrossRef]
16. Lowattanamart, W.; Suttisung, V.; Sintragoonchai, S.; Phanomchoeng, G.; Jintanawan, T. Feasibility on Development of Kinetic-Energy Harvesting Floors. *IOP Conf. Ser. Earth Environ. Sci.* **2020**, *463*, 12107. [CrossRef]

energies

MDPI

Article

Effect of Single-Row and Double-Row Passive Vortex Generators on the Deep Dynamic Stall of a Wind Turbine Airfoil †

Chengyong Zhu, Tongguang Wang *, Jie Chen and Wei Zhong

Jiangsu Key Laboratory of Hi-Tech Research for Wind Turbine Design, Nanjing University of Aeronautics and Astronautics, Nanjing 210016, China; rejoicezcy@nuaa.edu.cn (C.Z.); chenjie0320@nuaa.edu.cn (J.C.); zhongwei@nuaa.edu.cn (W.Z.)

* Correspondence: tgwang@nuaa.edu.cn; Tel.: +86-25-84896138

† This paper is a revised and extended version of our paper published in the International Conference on Sustainable Energy and Green Technology 2019 (SEGT 2019), Bangkok, Thailand, 11–14 December 2019. IOP Conf. Series: Earth and Environmental Science. 2020; 463:012118.

Received: 8 April 2020; Accepted: 13 May 2020; Published: 16 May 2020

Abstract: Passive vortex generators (VGs) have been widely applied on wind turbines to boost the aerodynamic performance. Although VGs can delay the onset of static stall, the effect of VGs on dynamic stall is still incompletely understood. Therefore, this paper aims at investigating the deep dynamic stall of NREL S809 airfoil controlled by single-row and double-row VGs. The URANS method with VGs fully resolved is used to simulate the unsteady airfoil flow. Firstly, both single-row and double-row VGs effectively suppress the flow separation and reduce the fluctuations in aerodynamic forces when the airfoil pitches up. The maximum lift coefficient is therefore increased beyond 40%, and the onset of deep dynamic stall is also delayed. This suggests that deep dynamic-stall behaviors can be properly controlled by VGs. Secondly, there is a great difference in aerodynamic performance between single-row and double-row VGs when the airfoil pitches down. Single-row VGs severely reduce the aerodynamic pitch damping by 64%, thereby undermining the torsional aeroelastic stability of airfoil. Double-row VGs quickly restore the decreased aerodynamic efficiency near the maximum angle of attack, and also significantly accelerate the flow reattachment. The second-row VGs can help the near-wall flow to withstand the adverse pressure gradient and then suppress the trailing-edge flow separation, particularly during the downstroke process. Generally, double-row VGs are better than single-row VGs concerning controlling deep dynamic stall. This work also gives a performance assessment of VGs in controlling the highly unsteady aerodynamic forces of a wind turbine airfoil.

Keywords: deep dynamic stall; passive vortex generators; wind turbine airfoil; URANS simulations

1. Introduction

Although passive vortex generators (VGs) are very simple, they have been proven to suppress the flow separation effectively and then boost the aerodynamic performance of horizontal axis wind turbines (HAWTs) [1]. Conventional VGs are composed of some pairs of vanes sticking out from the surface, angled to the incoming flow [2]. The height of VGs is close to the boundary-layer thickness. The fundamental principle of VGs is to produce streamwise vortices. These vortices can reenergize the boundary layer to resist the adverse pressure gradient.

The effectiveness of VG designs is primarily determined by the evolution of streamwise vortices. This vortex evolution is further impacted by various VG parameters. Godard and Stanislas [3] measured the boundary layer flow of a two-dimensional bump with VGs, using stereo particle image velocimetry (PIV) and hot-film probes. They found the triangular VGs better than rectangular VGs in decreasing the drag penalty at low angles of attack (AOAs). The counter-rotating configuration was also found better than the co-rotating one in generating upwash and downwash wake regions. Mueller-Vahl et al. [4] carried out wind-tunnel measurements of the NACA $63_{(3)}$-618 airfoil equipped with triangular VGs. Their research implied that decreasing the spanwise spacing of VGs could not only delay the onset of static stall, but also cause a high drag penalty. Baldacchino et al. [5] systematically studied the effect of VG parameters on the aerodynamic performance of DU97-W-300 airfoil using wind-tunnel experiments. They found that the vane height and chordwise location of VGs are the main factors in the airfoil performance. The chordwise location plays a significant role in the post-stall behavior of airfoil. Positioning VGs too downstream can result in an early abrupt stall, because VGs become very prone to be submerged in the separation zones [4,5]. Wang et al. [6] studied the effect of rectangular VGs on the performance of NREL S809 airfoil by URANS simulations. Compared to single-row VGs, they found double-row VGs to further suppress the flow separation and then further delay the static stall.

RANS-based simulations of airfoil flow with VGs are by far the most common, although some researches were performed by highly expensive DNS/LES-type simulations [7,8]. Nevertheless, RANS methods also need a large quantity of computational cost, because the boundary layer with VGs requires adequately fine resolution. To reduce the cost of fully resolved RANS method, the VG modelling is often simplified [9]. The idea is to add a flow-dependent forcing term to the momentum equations based on the thin airfoil theory. This method can successfully predict the aerodynamic performances of both the airfoil and blade with VGs [10,11]. However, this simplified modelling has to calibrate the key coefficient first.

VGs have succeeded in aerospace engineering and have been practically applied in wind turbine engineering (Figure 1). However, the HAWT blade flow controlled by VGs remains unclear. Most studies have focused on the two-dimensional steady airfoil flow controlled by VGs. In contrast, the blade flow is three-dimensional, rotational, and often becomes unsteady. The unsteady operating conditions are attributed to complicated environmental effects such as wind gust, turbulent inflow, and yaw misalignment [12,13]. The blade sections therefore undergo a time-varying AOA. If the AOA variation is dramatic enough, dynamic stall of the rotating blade will occur [14].

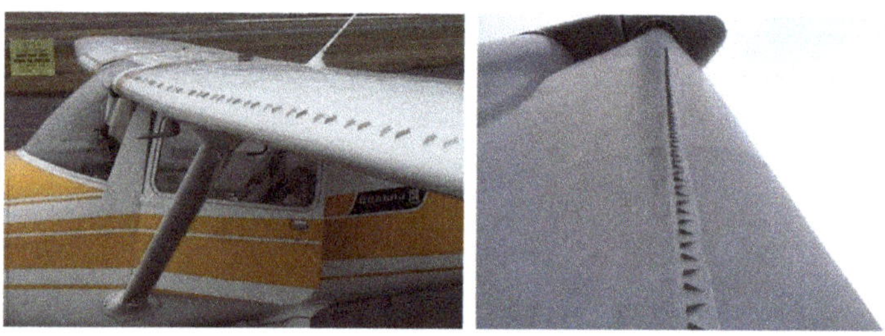

Figure 1. Applications of passive vortex generators (VGs) on the aircraft wings and wind turbine blades [15,16].

Dynamic stall is characterized by the shedding and passage of a strong vortical disturbance over the suction surface, thereby causing a highly nonlinear fluctuating pressure field [17]. Dynamic stall often means the unsteady blade loads and there is noticeable aerodynamic hysteresis. These unsteady aerodynamic forces are directly linked to the structure failures, reduced turbine life, and increased operating maintenance.

Therefore, some methods were proposed to control dynamic stall, including aerodynamic blowing [18], trailing-edge flap [19], co-flow jet [20], and plasma actuator [21]. These existing ways can be classified as active control techniques, which will introduce auxiliary power equipment. This leads to a more complicated design of blades. Consequently, active control techniques are often limited to wind turbine blades [22]. In contrast, passive control techniques can also improve the wind turbine performance without external energy expenditure, among which VGs are very cost-effective.

Nevertheless, the effect of VGs on dynamic stall has been rarely investigated and hence is still poorly understood. Our previous works [23,24] demonstrated that VGs could effectively suppress the flow separation of oscillating wind turbine airfoil, thereby attenuating the aerodynamic hysteresis. In this regard, double-row VGs are more effective than single-row VGs. Our previous works only focused on the light dynamic stall controlled by VGs. However, the deep dynamic stall is often accompanied by stronger vortex motions and severer flow separation.

This work aims to investigate the effect of single-row and double-row VGs on the deep dynamic stall. The URANS method is used to identify the unsteady airfoil flow characteristics with and without VGs. The aerodynamic hysteresis loops, flow structures, and boundary-layer velocity profiles are analyzed in detail to reveal the effect of VGs on deep dynamic stall.

2. Numerical Modelling

2.1. Geometry and Mesh Generation

Figure 2 illustrates the geometry of VGs used on the NREL S809 airfoil. The main VG parameters are h = 5 mm, d/h = 3.5, D/h = 7, L/h = 3, and β = 18°, based on the VG design methodology [4,5]. Two chordwise locations are considered: x_{VG}/c = 15% (single-row); x_{VG}/c = 15% and 40% (double-row). x_{VG} is measured between the leading edges of airfoil and VGs, and c is the airfoil chord length.

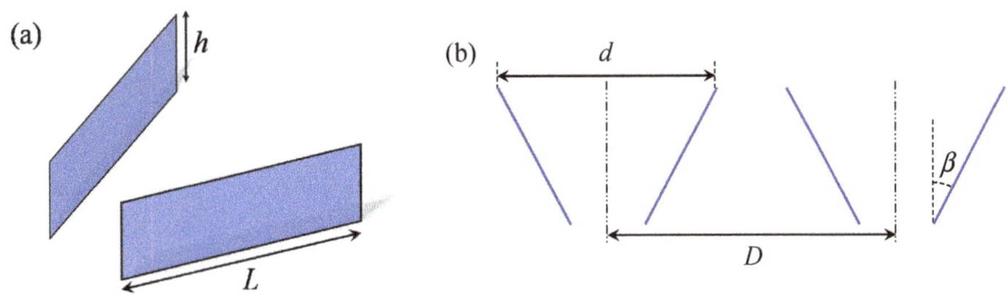

Figure 2. Schematic of the rectangular vane-type vortex generators in a counter-rotating configuration. (**a**) Isometric view; (**b**) planar view.

The present numerical modelling is essentially the same as that in our previous works [23,25]. Table 1 gives the main features of the mesh. The computational mesh includes only one pair of VGs (Figure 3), and the translational periodic boundary condition is used on spanwise boundaries. The mesh dependency study has been done by the General Richardson Extrapolation method [23]. The selected mesh with single-row VGs has about 1000 cells on each VG vane. There are 200 × 190 × 80 points in the wrap-around, normal, and spanwise directions, respectively. The Reynolds number is 1×10^6 (i.e., c = 0.457 m and U_0 = 33.68 m/s), where U_0 is the freestream velocity.

Table 1. Main features of the computational mesh.

Mesh Configuration	Structured O-Type
y^+	Always < 1
Normal growth ratio	1.08
Far-field Distance	20 c
Mesh size (million)	2.9 (single-row VGs) 3.5 (double-row VGs)

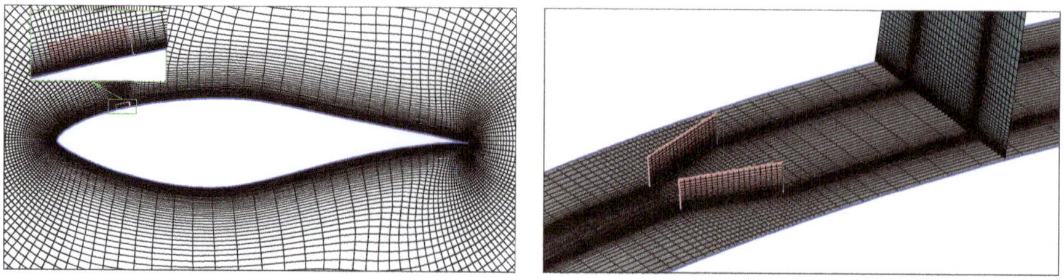

Figure 3. Boundary-layer and surface mesh of the airfoil with single-row VGs.

2.2. URANS Settings

Dynamic stall of the airfoil is obtained by sinusoidal pitch oscillation about the quarter-chord axis. The sliding mesh method [26] is used to simulate the dynamic motion of airfoil. The instantaneous AOA follows the sinusoidal variation:

$$\alpha = \alpha_{\mathrm{m}} + A\sin(2\pi ft) \tag{1}$$

The reduced frequency is defined as $k = \pi fc/U_0$. In this work, the deep dynamic stall condition of $\alpha_m = 18.75°$, $A = 10.3°$, and $k = 0.078$ is simulated following the wind-tunnel experiments [27].

In the mesh motion simulated by sliding mesh method, nodes rigidly move in a given dynamic zone, but the cells defined by these nodes will not deform. A sliding interface is also introduced to connect multiple cell zones. The sliding interface is updated and synchronized with the mesh motion to reflect the new positions. Therefore, the computational domain is divided into two subdomains. The inner is a rotating region, and the outer a stationary region. The interaction between these two regions is made through a cylindrical sliding interface at radius of 4.4c in this work.

The commercial software ANSYS/FLUENT 16.0 [26] is used to numerically solve the URANS equations. Table 2 provides the main URANS settings. Ekaterinaris and Platzer [28] found that the proper consideration of transitional flow effect can improve the predictive accuracy of aerodynamic hysteresis. Therefore, the turbulence is simulated by the SST k-ω eddy viscosity model [29] incorporated with the γ-Re_θ transition model [30]. This turbulence modelling has been proven to be reliable in simulating the dynamic stall of wind turbine airfoils [31]. The time step is set to assure 540 steps over each cycle and 20 inner iterations per time step, based on our previous works [13]. Iterative convergence criterion is met by assuring the cycle-to-cycle force variations negligible.

Table 2. Main settings of the URANS simulations.

Spatial Discretization	Third-order MUSCL convection scheme
Temporal Discretization	Bounded second-order implicit scheme
Pressure-Velocity Coupling	Coupled algorithm
Time Steps Per Cycle	540
Inner Iterations	20
Turbulence Model	SST k-ω model
Transition Model	γ-Re_θ model

2.3. Validation of Numerical Modelling

Due to the lack of experimental data of dynamic stall with VGs (unsteady-controlled), present numerical modelling has been validated against two sets of available experimental data: steady-controlled and unsteady-uncontrolled.

For steady-controlled data, the numerical modelling can reliably predict the pressure distributions of the DU97-W-300 airfoil with and without triangular VGs (Figure 4). Figure 4 also demonstrates that VGs are effective in suppressing the trailing-edge separated flow at $\alpha = 15°$, thereby leading to a high leading-edge suction and greatly increasing the lift coefficient C_l. VGs, however, have a marginal effect on the pressure distribution when the flow is fully attached at $\alpha = 10°$.

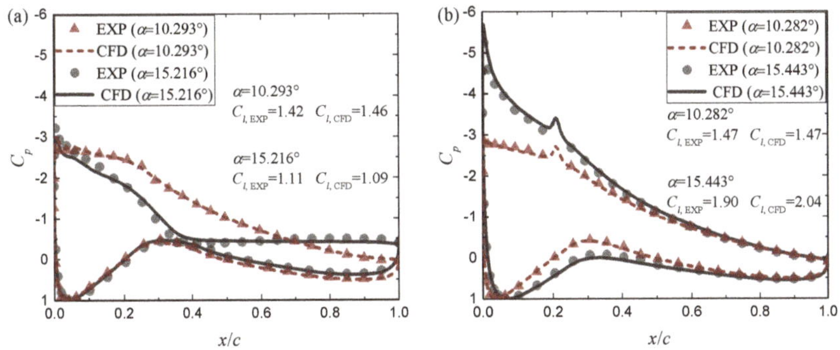

Figure 4. Calculated and measured pressure distributions of the DU97-W-300 airfoil with and without VGs. (**a**) Without VGs; (**b**) with VGs ($x_{VG}/c = 20\%$).

For unsteady-uncontrolled data, the obtained results also show a good agreement with the experimental data of the NREL S809 airfoil in light dynamic stall [23]. Moreover, Figure 5 suggests that the calculated aerodynamic hysteresis loops generally agree with the experimental data [27] and Johansen's CFD results [31] in deep dynamic stall. The hysteresis loops also show noticeable fluctuations at the high AOAs. This is due to the severe vortex shedding and passage over the suction surface. The unsteady aerodynamic forces are accurately predicted during the flow separation and flow reattachment processes. Consequently, present numerical modelling of the deep dynamic stall of the NREL S809 airfoil with VGs should be adequately correct.

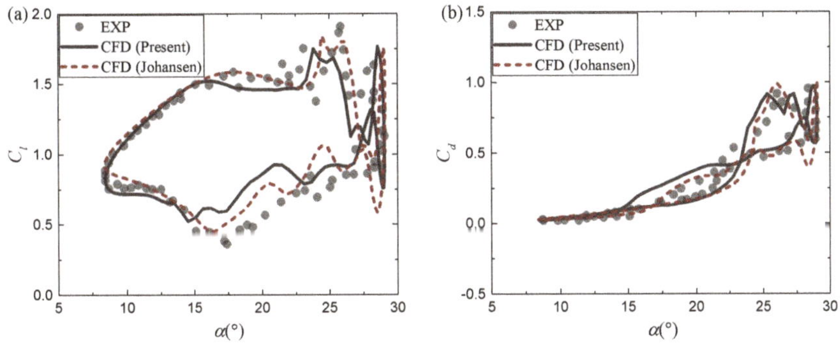

Figure 5. C_l-α and C_d-α hysteresis loops of the NREL S809 airfoil undergoing deep dynamic stall ($\alpha_m = 18.75°$, $A = 10.3°$, and $k = 0.078$). (**a**) C_l; (**b**) C_d.

3. Results and Discussion

3.1. Aerodynamic Hysteresis Loops

Figure 6 shows the calculated aerodynamic hysteresis loops with and without VGs. During the upstroke process, the aerodynamic coefficients of the airfoil with single-row and double-row VGs were relatively close. VGs significantly delayed the onset of dynamic stall. The C_l of clean airfoil began to diverge from the linear regime at $\alpha = 16°$, implying the start of flow separation (Figure 6a). However, the C_l with VGs well followed the linear theory until $\alpha = 22°$. At the high AOAs, the strong dynamic stall vortex motion caused large fluctuations in the aerodynamic coefficients of the clean airfoil. Figure 6 also indicates that the degree of these fluctuations can be decreased by VGs.

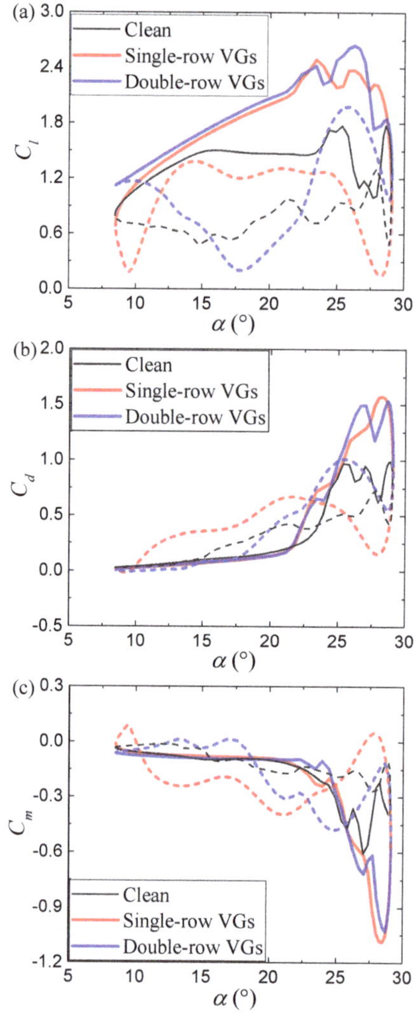

Figure 6. Aerodynamic hysteresis loops of the NREL S809 airfoil with and without VGs. Solid lines denote increasing angles of attack (AOA), and dashed lines indicate decreasing AOA. (**a**) C_l; (**b**) C_d; (**c**) C_m.

During the downstroke process, the aerodynamic coefficients showed a clear difference between single-row and double-row VGs. The downstroke process from $\alpha_{\max}$ to $\alpha_{\min}$ can be further divided into three parts:

- From $\alpha_{\max}$ to $\alpha = 25°$, double-row VGs quickly restored the decreases in C_l and C_d in comparison with single-row VGs. This suggests that the second-row VGs impacted greatly on the massive flow separation when the airfoil began to pitch down.

- From α = 25° to α = 13°, the C_l with single-row VGs kept high but was accompanied by high hysteresis intensities of the C_d and C_m. In contrast, the C_l with double-row VGs decreased gradually at first and then increased slowly.
- From α = 13° to α_{min}, single-row VGs produced considerable increases in hysteresis intensities. The second-row VGs significantly helped the C_l readjust to the linear regime, so that double-row VGs led to low hysteresis intensities.

Table 3 provides the dynamic-stall parameters extracted from the hysteresis loops in Figure 6. The definition of aerodynamic pitch damping ζ_{Cm} is given by:

$$\zeta_{Cm} = -\oint C_m d\alpha / \left(4A^2\right) \tag{2}$$

A high ζ_{Cm} also implies a high torsional aeroelastic stability. If the ζ_{Cm} decreases to a negative value, the amplitude of airfoil pitch will increase rapidly, and then the flutter occurs unless restrained [17].

Table 3. Dynamic-stall parameters of the airfoil with and without VGs.

Case Name	$C_{l,max}$	α_{Clmax} (°)	$C_{l,dec}$	$C_{m,inc}$	$C_{m,dec}$	$C_{m,min}$	ζ_{Cm}
clean	1.78	28.62	0.67	−0.34	−0.120	−0.605	0.153
singe-row VGs	2.49	23.37	1.26	−0.23	−0.280	−1.086	0.055
double-row VGs	2.65	26.24	1.95	−0.60	−0.399	−1.028	0.124

Single-row and double-row VGs increased the $C_{l,max}$ of NREL S809 airfoil by 40% and 49%, respectively. This is because double-row VGs could further delay the onset of dynamic stall. Both single-row and double-row VGs dramatically reduced the $C_{m,min}$ by almost 70%. The reason is that VGs hindered the forward motion of the center of pressure with the trailing-edge flow separation effectively suppressed.

Table 3 also indicates a large decrease of 64% in the ζ_{Cm} of the airfoil with single-row VGs. Therefore, single-row VGs can reduce the torsional aeroelastic stability, thereby likely causing the airfoil flutter. In this regard, double-row VGs are better to only reduce the ζ_{Cm} from 0.153 to 0.124.

3.2. Flow Field Developments

Figure 7 illustrates the flow field developments around the airfoil with and without VGs. Three AOAs of 9.83°, 18.75°, and 27.67° were chosen to represent the three typical degrees of flow separation: fully attached flow, trailing-edge (TE) separated flow, and massively separated flow, respectively. The clean airfoil flow also showed wider separation zones during the downstroke process than during the upstroke process. This manifests that the aerodynamic hysteresis was attributed to the retarded flow reattachment when the airfoil pitched down.

During the upstroke process, both single-row and double-row VGs eliminated the TE separation vortex at α = 18.75°. The C_l of airfoil with VGs was therefore dramatically increased (Figure 6a). At α = 27.67° (↑), there were three separation vortices on the upper surface of the clean airfoil. Two small separation vortices were located on the first half chord, and one large separation vortex was near the trailing edge. Furthermore, single-row and double-row VGs produced a fourth small TE separation vortex. This small vortex crowded out the large separation vortex, thereby leading to a high TE suction peak (Figure 8a). Surprisingly, the second-row VGs seemed to bring about an undesirable effect to reduce the TE suction peak.

During the downstroke process, the leading-edge (LE) and TE separation vortices shed into the wake alternately. At α = 27.67° (↓), the airfoil flow field with single-row VGs was highly distorted, because the LE separation vortex was hardly attached to the surface. Consequently, the suction value with single-row VGs was greatly decreased, even lower than that of the clean airfoil (Figure 8b). Double-row VGs, however, suppressed the LE flow separation effectively, and hence kept a high suction on the first half chord. Additionally, both single-row and double-row VGs avoided the secondary separation vortex near the trailing edge.

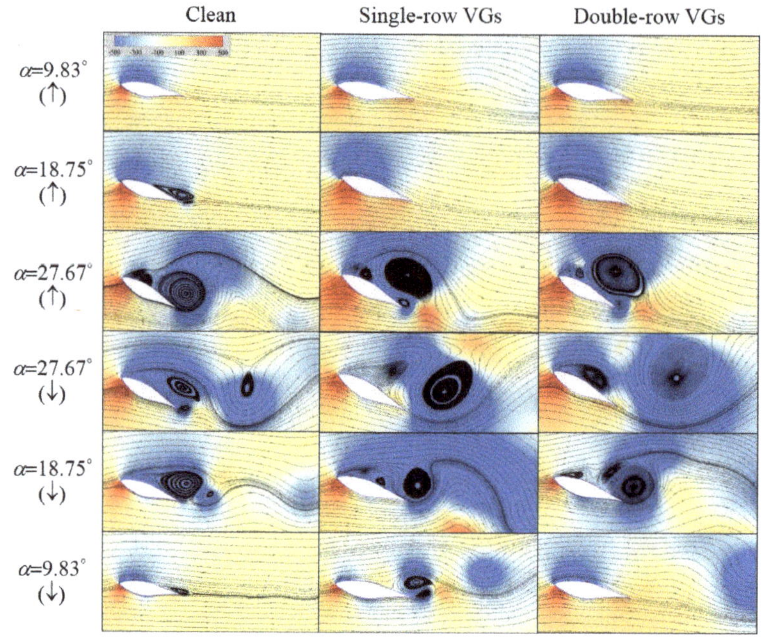

Figure 7. Streamlines and pressure fields around the airfoil with and without VGs.

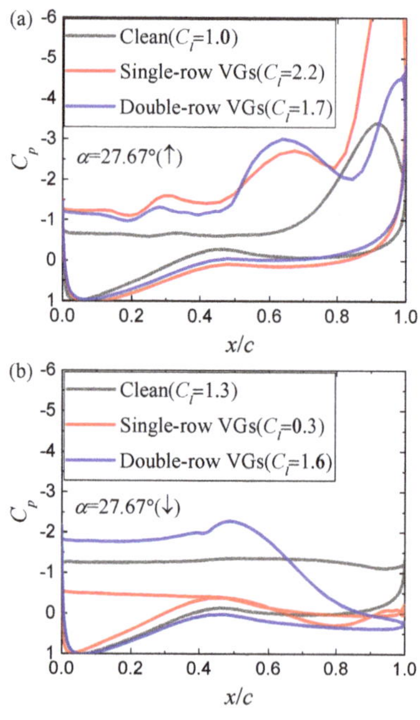

Figure 8. *Cont.*

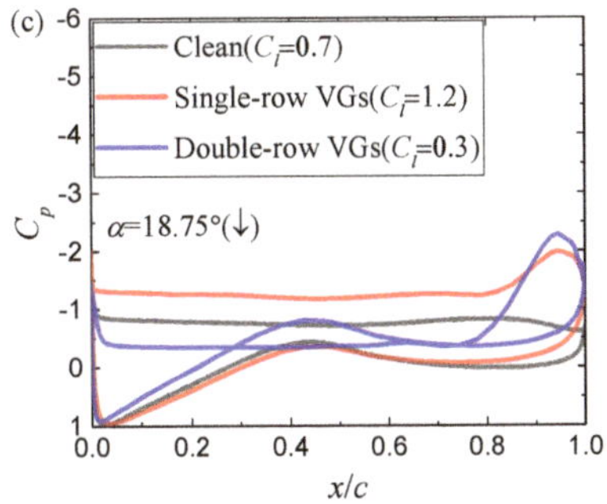

Figure 8. Pressure distributions of the airfoil with and without VGs. (**a**) $\alpha = 27.67°$ (↑); (**b**) $\alpha = 27.67°$ (↓); (**c**) $\alpha = 18.75°$ (↓).

At $\alpha = 18.75°$ (↓), a large LE separation vortex appeared in the clean airfoil flow. The decreased height of separation vortex and the downstream movement of the vortex core suggest that the clean airfoil flow began to reattach gradually. However, both single-row and double-row VGs caused a tertiary vortex. This separation vortex was even detached from the upper surface due to double-row VGs, so that the suction on the upper surface and the C_l of airfoil with double-row VGs were vastly reduced (Figure 8c). Interestingly, although double-row VGs were positioned on the upper side, they significantly affected the C_p distribution on the lower side. This could decrease the pressure difference between the upper and lower sides. At $\alpha = 9.83°$ (↓), Figure 7 also implies that the second-row VGs further accelerated the flow reattachment and hence resulted in a high LE suction peak.

3.3. Boundary-Layer Velocity Profiles

Figures 9 and 10 show the non-dimensionalized streamwise velocity profiles when the AOA reached 27.67° during the upstroke and downstroke processes, respectively. Note that $\alpha = 27.67°$ means the airfoil flow fell into the deep dynamic-stall process. Boundary-layer velocity profiles at $x/c = 10\%$ and $x/c = 75\%$ can represent the LE and TE separation vortices, respectively.

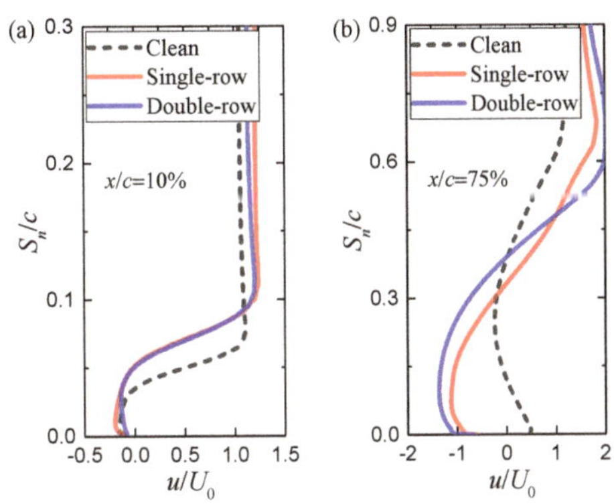

Figure 9. Streamwise velocity profiles at $\alpha = 27.67°$ (↑). S_n denotes the normal distance away from the wall surface, and u the streamwise velocity. (**a**) $x/c = 10\%$; (**b**) $x/c = 75\%$.

During the upstroke process, the streamwise velocity profiles with single-row and double-row VGs were quite close (Figure 9). This is attributed to the similar flow fields around the airfoil with single-row and double-row VGs at $\alpha = 27.67°$ (↑) (Figure 7). Figure 9 also indicates that VGs increased the height of LE separation vortex and the severity of TE reverse flow. Nevertheless, the external flow was effectively accelerated by VGs, thereby producing a high suction value (Figure 8a).

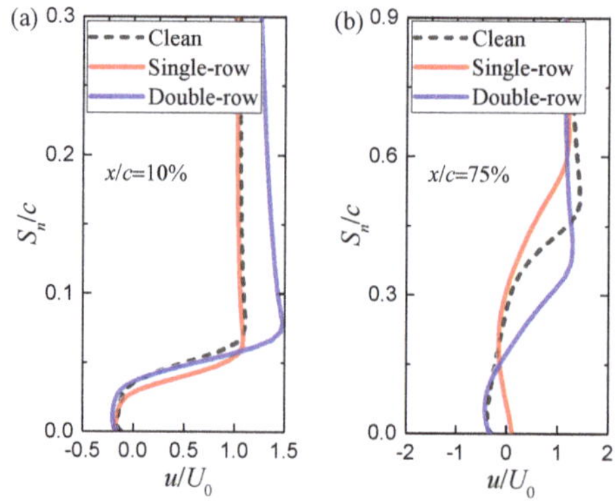

Figure 10. Streamwise velocity profiles at $\alpha = 27.67°$ (↓). (**a**) $x/c = 10\%$; (**b**) $x/c = 75\%$.

During the downstroke process, the streamwise velocity profiles also showed a clear difference between single-row and double-row VGs (Figure 10). At $x/c = 10\%$, although the boundary-layer thicknesses with double-row VGs and without VGs were close, the external flow was effectively accelerated due to double-row VGs (Figure 10a). This suggests that double-row VGs made the LE flow withstand a higher adverse pressure gradient. However, the boundary-layer thickness with single-row VGs was decreased, because the LE separation vortex seemed to be detached from the wall surface (Figure 7). At $x/c = 75\%$, the boundary-layer thicknesses from high to low was in the sequence of single-row VGs, clean, and double-row VGs. This sequence also determined the severity of TE flow separation. Interestingly, double-row VGs could effectively counteract the adverse pressure gradient and then suppress the TE flow separation, but single-row VGs could not (Figure 8b). This finding highlights the great impact of the second-row VGs during the downstroke process.

4. Conclusions

This paper gives a flow analysis of deep dynamic stall of the NREL S809 airfoil controlled by single-row and double-row VGs. VGs were fully resolved, and URANS simulations were conducted with the transitional SST k-ω eddy viscosity model. Based on this study, several conclusions were reached as follows.

- Present numerical modelling can accurately predict the aerodynamic loads of both an airfoil with VGs and an airfoil undergoing deep dynamic stall.
- Both single-row and double-row VGs postpone the flow separation from $\alpha = 16°$ to $\alpha = 22°$, when the airfoil pitches up. Then, the $C_{l,\max}$ of airfoil with VGs is considerably increased beyond 40%.
- Although single-row and double-row VGs produce an additional TE separation vortex, they can reduce the fluctuations in aerodynamic coefficients near the $\alpha_{\max}$.
- Single-row VGs bring about a vast decrease in the C_l from 2.2 to 0.3 near the $\alpha_{\max}$ when the airfoil begins to pitch down, implying severe dynamic-stall behaviors. Single-row VGs also seriously retard the flow reattachment near the $\alpha_{\min}$. Therefore, single-row VGs considerably reduce the ζ_{Cm} by 67% and hence undermine the torsional aeroelastic stability of airfoil.
- Double-row VGs can quickly restore the decrease in C_l and C_d near the $\alpha_{\max}$ in comparison with single-row VGs. Double-row VGs also significantly help the C_l readjust to the linear regime with the flow reattachment effectively accelerated.
- Double-row VGs can effectively counteract the adverse pressure gradient and hence suppress the TE flow separation during the downstroke process, but single-row VGs cannot. This explains the clear difference in aerodynamic responses between single-row and double-row VGs.

This paper also provides a performance assessment of VGs in controlling highly unsteady aerodynamic loads of a wind turbine airfoil. This study may contribute to understanding the deep dynamic stall controlled by single-row and double-row VGs. Future work should concentrate on the effect of passive VGs on a rotating blade undergoing dynamic stall.

Author Contributions: C.Z. conceived of the research and wrote the manuscript. C.Z. and J.C. conducted the data collection. T.W. and W.Z. contributed technical guidance and revised the manuscript. All authors have read and agreed to the published version of the manuscript.

Funding: This work is funded by the National Key Research and Development Project under grant No. 2019YFB1503701-02, CAS Key Laboratory of Wind Energy Utilization under grant No. KLWEU-2016-0102, and the Priority Academic Program Development of Jiangsu Higher Education Institutions.

Acknowledgments: The authors would like to express their gratitude to the conference chairs of SEGT 2019 for recommendation of this publication in *Energies*.

Conflicts of Interest: The authors declare no conflict of interest.

Nomenclature

↑	upstroke process
↓	downstroke process
α	angle of attack (AOA)
α_{Clmax}	AOA of $C_{l,\max}$
α_m	mean AOA
$\alpha_{\max}$	maximum AOA
$\alpha_{\min}$	minimum AOA
β	geometric vane inflow angle
ζ_{Cm}	aerodynamic pitch damping
A	AOA amplitude
c	chord length
C_d	drag coefficient
C_l	lift coefficient
$C_{l,\mathrm{dec}}$	C_l at α_{Clmax} (↓)
$C_{l,\max}$	maximum C_l
C_m	pitching moment coefficient
$C_{m,\mathrm{dec}}$	C_m at α_{Clmax} (↓)
$C_{m,\mathrm{inc}}$	C_m at α_{Clmax} (↑)
$C_{m,\min}$	minimum (maximum nose-down) C_m
C_p	pressure coefficient
D	inter-vane spacing
d	intra-vane spacing
f	frequency of oscillation
h	vane height
k	reduced frequency
L	vane length
S_n	normal distance away from the wall surface
u	streamwise velocity
U_0	freestream velocity
x	chordwise location
x_{VG}	chordwise location measured between the airfoil and VG leading edges

References

1. Barlas, A.; van Kuik, G. Review of state of the art in smart rotor control research for wind turbines. *Prog. Aerosp. Sci.* **2010**, *46*, 1–27. [CrossRef]
2. Lin, J.C. Review of research on low-profile vortex generators to control boundary-layer separation. *Prog. Aerosp. Sci.* **2002**, *38*, 389–420. [CrossRef]

3. Godard, G.; Stanislas, M. Control of a decelerating boundary layer. Part 1: Optimization of passive vortex generators. *Aerosp. Sci. Technol.* **2006**, *10*, 181–191. [CrossRef]
4. Mueller-Vahl, H.; Pechlivanoglou, G.; Nayeri, C.N.; Paschereit, C.O. Vortex generators for wind turbine blades: A combined wind tunnel and wind turbine parametric study. In Proceedings of the Asme Turbo Expo 2012, New York, NY, USA, 11–15 June 2012; pp. 899–914.
5. Baldacchino, D.; Ferreira, C.S.; De Tavernier, D.; Timmer, W.; Van Bussel, G.J.W. Experimental parameter study for passive vortex generators on a 30% thick airfoil. *Wind Energy* **2018**, *21*, 745–765. [CrossRef]
6. Wang, H.; Zhang, B.; Qiu, Q.; Xu, X. Flow control on the NREL S809 wind turbine airfoil using vortex generators. *Energy* **2017**, *118*, 1210–1221. [CrossRef]
7. Velte, C.; Hansen, M.O.L.; Cavar, D. Flow analysis of vortex generators on wing sections by stereoscopic particle image velocimetry measurements. *Environ. Res. Lett.* **2008**, *3*, 015006. [CrossRef]
8. Spalart, P.R.; Shur, M.; Strelets, M.K.; Travin, A.K. Direct Simulation and RANS Modelling of a Vortex Generator Flow. *Flow Turbul. Combust.* **2015**, *95*, 335–350. [CrossRef]
9. Jirasek, A. Vortex-Generator Model and Its Application to Flow Control. *J. Aircr.* **2005**, *42*, 1486–1491. [CrossRef]
10. Baldacchino, D.; Manolesos, M.; Ferreira, C.S.; Salcedo, Á.G.; Aparicio, M.; Chaviaropoulos, T.; Diakakis, K.; Florentie, L.; García, N.R.; Papadakis, G.; et al. Experimental benchmark and code validation for airfoils equipped with passive vortex generators. In *Science of Making Torque from Wind*; Iop Publishing Ltd.: Bristol, UK, 2016; Volume 753.
11. Troldborg, N.; Zahle, F.; Sørensen, N.N. Simulations of wind turbine rotor with vortex generators. In *Science of Making Torque from Wind*; Institute of Physics Publishing: Munich, Germany, 2016; Volume 753.
12. Leishman, J.G. Challenges in modelling the unsteady aerodynamics of wind turbines. *Wind Energy* **2002**, *5*, 85–132. [CrossRef]
13. Zhu, C.; Wang, T. Comparative Study of Dynamic Stall under Pitch Oscillation and Oscillating Freestream on Wind Turbine Airfoil and Blade. *Appl. Sci.* **2018**, *8*, 1242. [CrossRef]
14. Zhu, C.; Wang, T.; Zhong, W. Combined Effect of Rotational Augmentation and Dynamic Stall on a Horizontal Axis Wind Turbine. *Energies* **2019**, *12*, 1434. [CrossRef]
15. Solar Thermal Magazine. Shark-Inspired Vortex Generators Improve Wind Turbine Blade Performance. Available online: https://solarthermalmagazine.com (accessed on 8 February 2018).
16. Micro AeroDynamics Inc. Available online: https://microaero.com/cessna_150/ (accessed on 14 May 2020).
17. McCroskey, W.J. *The Phenomenon of Dynamic Stall*; National Aeronautics and Space Administration: Washington, DC, USA, 1981.
18. Müller-Vahl, H.F.; Strangfeld, C.; Nayeri, C.; Paschereit, C.O.; Greenblatt, D. Control of Thick Airfoil, Deep Dynamic Stall Using Steady Blowing. *AIAA J.* **2015**, *53*, 277–295. [CrossRef]
19. Gerontakos, P.; Lee, T. Dynamic Stall Flow Control via a Trailing-Edge Flap. *AIAA J.* **2006**, *44*, 469–480. [CrossRef]
20. Xu, H.-Y.; Qiao, C.-L.; Ye, Z. Dynamic Stall Control on the Wind Turbine Airfoil via a Co-Flow Jet. *Energies* **2016**, *9*, 429. [CrossRef]
21. Li, G.Q.; Zhang, W.G.; Jiang, Y.B.; Yang, P.Y. Experimental investigation of dynamic stall flow control for wind turbine airfoils using a plasma actuator. *Energy* **2019**, *185*, 90–101. [CrossRef]
22. Johnson, S.J.; van Dam, C.P. *Active Load Control Techniques for Wind Turbines*; Sandia National Laboratories: Springfield, VA, USA, 2008.
23. Zhu, C.; Chen, J.; Wu, J.; Wang, T. Dynamic stall control of the wind turbine airfoil via single-row and double-row passive vortex generators. *Energy* **2019**, *189*, 116272. [CrossRef]
24. Zhu, C.; Wang, T.; Wu, J. Numerical Investigation of Passive Vortex Generators on a Wind Turbine Airfoil Undergoing Pitch Oscillations. *Energies* **2019**, *12*, 654. [CrossRef]
25. Zhu, C.; Wang, T.; Chen, J.; Zhong, W. Flow analysis of the deep dynamic stall of wind turbine airfoil with single-row and double-row passive vortex generators. In Proceedings of the International Conference on Sustainable Energy and Green Technology, Bangkok, Thailand, 11–14 December 2019; IOP Conference Series: Earth and Environmental Science. 2020; Volume 463, p. 012118. [CrossRef]
26. ANSYS Inc. *FLUENT Theory Guide, Release 16.0*; ANSYS Inc.: Canonsburg, PA, USA, 2015.
27. Ramsay, R.F.; Hoffman, M.J.; Gregorek, G.M. *Effects of Grit Roughness and Pitch Oscillation on the S809 Airfoil*; National Renewable Energy Laboratory: Golden, CO, USA, 1995.

28. Ekaterinaris, J.A.; Platzer, M. Computational prediction of airfoil dynamic stall. *Prog. Aerosp. Sci.* **1998**, *33*, 759–846. [CrossRef]
29. Menter, F.R. Two-Equation Eddy-Viscosity Transport Turbulence Model for Engineering Applications. *AIAA J.* **1994**, *32*, 1598–1605. [CrossRef]
30. Menter, F.R.; Langtry, R.B.; Likki, S.R.; Suzen, Y.B.; Huang, P.G.; Völker, S. A Correlation-Based Transition Model Using Local Variables—Part I: Model Formulation. *J. Turbomach.* **2006**, *128*, 413–422. [CrossRef]
31. Johansen, J. *Unsteady Airfoil Flows with Application to Aeroelastic Stability*; Risø National laboratory: Roskilde, Denmark, 1999.

energies

MDPI

Article

The Philippines' and Singapore's Journeys Towards Liberalised Electricity Supply Industries—Takeaways for Malaysia †

Hazleen Aris [1,*], Iskandar Shah Mohd Zawawi [2] and Bo Nørregaard Jørgensen [3,*]

1. Institute of Informatics and Computing in Energy, Universiti Tenaga Nasional, Kajang 43000, Malaysia
2. Faculty of Computer and Mathematical Sciences, Universiti Teknologi MARA, Shah Alam 40450, Malaysia; iskandarshah@uitm.edu.my
3. Center for Energy Informatics, Mærsk Mc-Kinney Moller Institute, University of Southern Denmark, 5230 Odense M, Denmark

* Correspondence: hazleen@uniten.edu.my (H.A.); bnj@mmmi.sdu.dk (B.N.J.); Tel.: +60-3-8928-7370 (H.A.)

† This paper is an extended version of our paper published in International Conference on Sustainable Energy and Green Technology 2019, SEGT 2019, 9 January 2020, Volume 463, 2020.

Received: 1 May 2020; Accepted: 30 June 2020; Published: 8 July 2020

Abstract: Malaysia is in the process of liberalising its electricity supply industry (ESI) further, with the second reform series announced in September 2018. If everything goes as planned, Malaysia would be the third country in the Association of Southeast Asia Nations (ASEAN) to have a fully liberalised ESI after the Philippines and Singapore. A number of initiatives have been in the pipeline to be executed and a lot more will be planned. At this juncture, it is important for Malaysia to look for the best practices and lessons that can be learnt from the experience of other countries that have successfully liberalised their ESIs. Being in the same region, it is believed that there is a lot that Malaysia can learn from the Philippines and Singapore. This paper therefore presents and deliberates on the chronological development of the countries' progressive journeys in liberalising their ESIs. The aim is to discern the good practices, the challenges as well as the lessons learnt from these transformations. Analysis is being made and discussed from the following four perspectives; legislative framework, implementation phases, market components and impact on renewable energy penetration. Findings from this study would provide useful insight for Malaysia in determining the course of actions to be taken to reform its ESI. Beyond Malaysia, the findings can also serve as the reference for the other ASEAN countries in moving towards liberalising their ESIs.

Keywords: regional electricity market; energy security; energy efficiency; energy sustainability; ASEAN; electricity liberalisation; electricity unbundling

1. Introduction

Like most other countries, the electricity supply industry (ESI) in Malaysia started rather spontaneously, driven by the need of individual enterprises to electrify their businesses. Natural monopoly began in 1982 after the private and state owned electricity supply companies are consolidated under the National Electricity Board (NEB). Subsequently, NEB became the sole state-owned body that 'monopolised' the supply of electricity to the whole of peninsular Malaysia, having the transmission lines that stretched over 6300 km and close to two million customers at the time [1]. In a monopolised or vertically integrated ESI, the operations of generation, transmission and distribution of electricity are owned and managed by one giant utility company as an entity or together with a number of other companies related to it. Thus, competition is almost non-existent. On the contrary, a liberalised or 'open' ESI provides opportunities for other players in the industry to participate in the business. It therefore promotes competition, which often translates into efficiency that results from improved operation, management, customer services et cetera. A liberalised ESI also facilitates the creation of regional electricity market where electricity is traded between a number of nearby countries. Such regional electricity market does not exist yet in the Association of Southeast Asia Nations (ASEAN) region, although the intention is there and is not new. The ASEAN Power Grid (APG) initiative was mandated in December 1997 with the aim to enhance regional energy sustainability, security and reliability [2]. According to the plan, the construction of the APG would begin with a regional power interconnection on cross-border bilateral terms, which then expands to sub-regional basis towards a totally integrated South East Asia power grid system. More than twenty years after the mandate, the envisioned APG is still very much a work-in-progress [2]. While the physical interconnection of the grids is mostly ready, the electricity trading between the countries is not [3]. The latter is currently done through bilateral power purchase agreements between the bordering countries. In comparison, the Nord Pool, which started the similar initiative at about the same time, has been thriving successfully and even regarded as the most robust, efficient and harmonised regional electricity market [4,5]. The success is largely attributed to, and made possible by, the region's move to liberalise their ESIs. With ESI liberalisation, efficiency is attained through means such as increased competition in the market that leads to improved operations [6], which can result in the elimination of unnecessary overhead supply (reserve margin) in the monopolised markets [7]. This allows for the capital resources to be utilised more effectively. Liberalised ESIs also promote the creation of pooled energy sources that allows for more efficient use of electricity through increased consumers' response to prices [8]. Greater interconnections have also allowed for increased energy security [9]. In terms of governance, liberalisation has significantly improved the governance of monopoly utilities, the prospects for competition and innovation, and the quality of policy instruments for environmental emissions control [10].

Thus far, there are only two ASEAN countries that have liberalised their ESIs fully and have their own electricity markets; the Philippines and Singapore. Bearing differences in terms of geographical condition, economic development status, total population et cetera, both countries have undergone different experience in liberalising the ESIs. While it has been a 'bumpy ride' for the Philippines, Singapore experiences a smoother transition. Malaysia is expected to be the next ASEAN country to go for full ESI liberalisation. In September 2018 [11], the Malaysia electricity supply industry (MESI) started the second series of its reform, MESI 2.0, where opening the electricity market further is put forth as one of the main agenda. To address the stated MESI 2.0 objectives, recommendations have been made for Malaysia to consider unbundling of its ESI, using the achievements of the countries in the Nordic as comparison [12]. However, emulating the Nordic countries might not be the best option due to the vast differences between the two regions in many aspects. The two countries in the Southeast Asia region that have already attained full ESI liberalisation would be more relevant for Malaysia to look at. This article therefore presents the chronological review of the Philippines' and Singapore's ESIs transformation towards liberalisation with the aim to identify the good practices, the challenges as well as the lessons learnt from their experience. In particular, the review focuses on the following perspectives.

1. The regulatory framework and institutional bodies to institute reform.
2. The phases of reform implementation.
3. The components that constitute the wholesale electricity market.
4. The impact of ESI liberalisation on the renewable energy (RE) penetration.

The first three are essential to be considered right from the beginning of a reform move while the RE penetration is included in the scope because Malaysia has clearly set its RE targets. Thus, it is relevant to identify the extent to which ESI liberalisation is able to help achieve the targets. The required information is obtained from the published research work, technical reports, newspaper clips and online communication with the relevant authorities. Analysis results conclude on the significant role played by the enacted legislative framework that is subsequently followed by the formation of the institutional bodies and agencies with sufficient authorities to instate the ESI reform. It is also found that gradual yet clearly defined phases of reform with explicit activities and targets are important to ensure its smooth implementation. Despite the differences in market operations, common components that constitute an electricity market can be identified. However, no significant impact is seen on the amount of RE penetration into the grids as a result of the liberalisation. Beyond Malaysia, the recommendations can potentially be considered for use by the other ASEAN countries too. This is especially necessary when a review of the reform experiences of the countries found a significant disparity between the expected and actual outcomes [13].

The rest of this article is organised into the following sections. Section 2 and Section 3 elaborate on the Philippines' and Singapore's ESI reform journeys respectively. Analysis, lessons learnt and policy implications are presented in Section 4. Section 5 concludes the article and points to potential future work and outlook.

2. ESI Liberalisation in the Philippines

Liberalisation of the Philippines' ESI began in 2001. Prior to the liberalisation, the state-owned monopoly National Power Corporation (NPC) was the vertically integrated power utility engaged in the production, transmission and distribution of electricity and used to be the largest provider and generator of electricity in the Philippines. The efforts towards energy restructuring in the Philippines started to materialise with the change in government in 1986. The Energy Regulatory Board (ERB) constituted in 1987 assumed the responsibility to regulate the energy sector, overseeing the power rates and services of the private electric utilities in the country. The year 1988 saw the beginning of independent power producers (IPPs) in response to an executive order from the then President, driven by the need to address the looming electricity shortages in the island [14]. By 1994, more than 40 IPP contracts were accumulated, a number that is more than any other developing countries [15]. Table 1 shows the current generation market share in the Philippines. The Asian financial crisis in 1997 led to the Philippines having the second highest electricity prices in the world that was putting more pressure for reform. Thus, a more serious move towards deregulation of the industry began in 2001, when the Republic Act 9136, known as the Electric Power Industry Reform Act (EPIRA) was enacted after almost eight years on the table. It was regarded as the most comprehensive piece of legislation of its kind in Asia by the Asian Development Bank. Being comprehensive, it is also the most complicated with the involvement of many parties [16]. The EPIRA called for the following [17].

- Creation of the National Transmission Company (TransCo) to assume the transmission function of the NPC.
- Privatisation of NPC assets, including the newly created TransCo.
- Creation of an independent, quasi-judicial entity called the Energy Regulatory Commission (ERC) to ensure a transparent, competitive, and reliable electricity market.

Table 1. Key generation players' share in the Philippines [18].

Player	Market Share (%)
San Miguel Energy Corp	20
Aboitiz Power Corp.	17
First Gas/First Gen.	17
PSALM	18
AES Transpower	4
SEM Calaca	4
NPC	6
Global Business Power Corp.	4
Salcon Phils./Atlas	2
K-Water	2
Others	6

TransCo was created immediately in the same year and started operating and managing the power transmission system that links power plants to the electricity distribution utilities nationwide in 2003. The ERC, an independent, quasi-judicial regulatory body that promotes competition, encourages market development, ensures customer choice and penalises abuse of market power, replaced the ERB. The EPIRA also resulted in the creation of the Power Sector Assets and Liabilities Management (PSALM) Corporation in 2001, a wholly-owned and -controlled government entity to take over the ownership of all existing generation assets of the NPC, IPP contracts, real estate, and all other disposable assets including the transmission assets of the TransCo. By the same token, PSALM assumed all outstanding obligations of NPC arising from loans, issuance of bonds, securities, and other instruments of indebtedness. The principal purpose of PSALM is to manage the orderly sale and privatisation of these assets with the objective of optimally liquidating all of the NPC's financial obligations. The Small Power Utilities Group (SPUG), which provides electricity to the off-grid customers and has been in existence since 1996, was put under the NPC and will absorb the remaining unsold assets of the NPC. Under this new structure, the Philippines' ESI is constituted of the following key players [19].

- DOE: Government agency in charge of planning and policy making for the electricity sector. EPIRA mandated DOE to organise and establish the appropriate market design and governance structure of the Wholesale Electricity Spot Market (WESM) and supervise the restructuring of the electricity industry.
- ERC: Regulates natural monopolies (distribution and transmission wires business), issues generation and supply licenses, oversees competition in the power market and enforces the implementing rules and regulations of the EPIRA.
- PSALM: Government-owned and controlled corporation that manages the orderly sale, disposition, and privatisation of NPC generation and other disposable assets, and its IPP contracts; as well as optimally liquidate all the NPC's financial obligations and stranded contract costs.
- PEMC: Non-stock, nonprofit and non-independent corporation established by the DOE to perform the market operator (MO) functions in the WESM.
- NPC: State-owned and responsible for the missionary electrification function for the government in areas that are not connected to the main grids through the SPUG, manage watershed areas supporting power plants, manage dams nationwide and coordinate with all stakeholders to ensure safety of its communities and environs, and manage undisposed power generating assets.
- TransCo: Provides and maintains the physical infrastructure (transmission network and associated facilities) necessary to transport electricity.
- National Grid Corporation of the Philippines (NGCP): Private sector concessionaire that is operating the national transmission system.

- National Electrification Administration (NEA): Government agency mandated to develop and implement programs to prepare and strengthen electric cooperatives for the deregulated electricity market.

The EPIRA also mandated the WESM to be established within one year of the Act. As a result, rules and regulation with regards to the conduct of the WESM were announced in 2002. In November 2003, the Philippine Electricity Market Corporation (PEMC) was incorporated as a non-stock, non-profit corporation tasked to establish and govern an efficient, competitive, transparent, and reliable market for the wholesale purchase of electricity and ancillary services. In August the following year, PEMC was designated as the MO to undertake the preparations for and the initial operations of the WESM until an independent market operator (IMO) is established and ready to take it over. The system operator (SO), TransCo, is responsible for operating the power system in accordance with the WESM Rules, Grid Code or any instruction from the MO or the ERC to ensure security and reliability of the power system. On 28 February 2008, NGCP was granted the right to take over and operate the whole of TransCo's regulated transmission business while the ownership of all transmission assets and related real properties remained with TransCo. A congressional franchise to operate the transmission network was granted to NGCP on 8 December 2008.

Figure 1 shows the governance of the Philippines' WESM. From the figure, it can be seen that there are four trading participants in WESM; generators (GenCo), distribution utilities (DU), retail electricity suppliers (RES) and directly connected customers (DCC) [20]. GenCos are at the supply side, comprising the privatised GenCos and the IPPs. As at April 2018, there were 113 generation companies in the Luzon-Visayas grid alone and all of them are WESM participants [21]. At the demand side, there are DUs, RESs and DCCs. DUs are responsible to provide open and non-discriminatory access to its system and provide wheeling services to the end users (captive customers (Monthly usage of less than 750 kW)) within its franchise area. Suppliers/aggregators engage in the supply of electricity to the contestable customers (Monthly usage of at least 750 kW) after securing an RES license from the ERC. They are allowed to supply electricity to the contestable customers within a contiguous area. The DCCs are industrial or bulk electricity customers who are connected to the transmission grid and directly supplied with electricity by a GenCo, PSALM or NPC [22]. The respective functions of the MO, SO and WESM participants are summarised in Table 2.

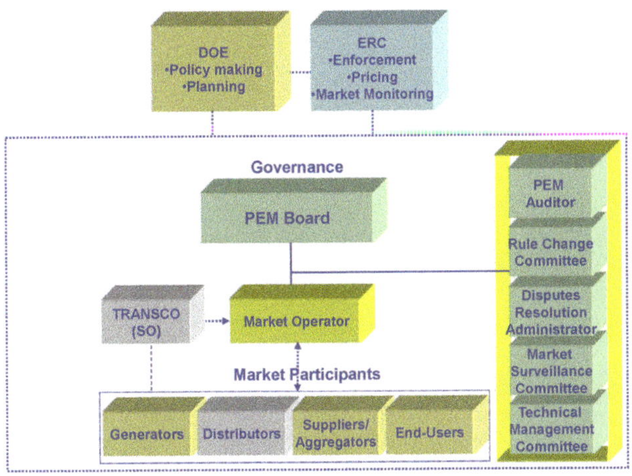

Figure 1. Wholesale Electricity Spot Market (WESM) governance.

Table 2. Key players in WESM.

Player	Role in WESM
MO	Administers the operation of the WESM Determines the dispatch schedule for each hour of all facilities and submits it to the SO Monitor daily the trading activities in the market Clear all settlements after every billing period
SO	In charge of the central dispatch to all generating facilities and loads in accordance with the dispatch schedule submitted by the MO Ensures the reliability and security of the power system at all times
GenCo	Submit offers to sell energy by offering it for sale at a certain price on an hourly basis. Price offers will be ranked from the highest to lowest.
DUs	Purchase electricity directly in bulk from the generating entities or from the spot market Distributors must also procure energy from the Wholesale Electricity Spot Market and secure bilateral contracts to serve the remainder of the customers in their franchise area
Suppliers/ aggregators	Purchase electricity directly in bulk from the generating entities or from the spot market
End-users (Customers)	Purchase electricity directly from the spot market Can opt for demand bidding in the pool or a rather passive participation Can voluntarily engage in bilateral contracting

After several months of trial operations, the WESM commenced commercial operations in the Luzon grid on 26 June 2006. Four years into the commercial operations in Luzon, the Visayas grid was integrated into the WESM and commenced commercial operations on 26 December 2010. In June 2017, WESM in Mindanao was officially launched but until now, it is yet to begin its commercial operation. It was expected to be commercially operational in January 2020, after a number of revisions. However, in a more recent news, further delay is expected [23]. Beginning 26 September 2018, the Independent Electricity Market Operator of the Philippines, Inc. (IEMOP) becomes the operator of the electricity market to manage the registration of market participants, receive generation offers, come out with market prices and dispatch schedules of the generation plants, and handle billing, settlement, and collections, among other functions [24]. This is an important milestone towards more independent and transparent operation of the WESM.

At the retail market level, the EPIRA also provides for the implementation of the retail competition and open access (RCOA) upon the fulfilment of the five pre-conditions, which were achieved in 2011 [25]. Thus, the RCOA was launched in June 2013 for contestable customers. The threshold for customer contestability began with 1 MW, which means that consumers whose power usage reach a monthly average of at least 1 MW are required to choose and buy their electricity from the retail electricity suppliers (RES). The threshold for contestability is gradually decreased. In May 2016, customers with an average monthly peak demand of at least 750 kW are mandated to enter into a retail electricity contract with a RES by June 2017. If it is not due to the temporary restraining order (TRO) by the Supreme Court resulting from the petition submitted by a few contestable customers [26,27], the contestability level would have been lowered further to include customers with an average peak demand between 500 kW and 749 kW [28]. At the time of writing, the high court has yet to lift its order. The number of customers enrolled in the open access scheme of the electricity retail market, as reported by the PEMC, reached 940 in November 2017 [26], nearly four times the number of contestable customers when it was launched in 2013. From the 940 customers, 862 are contestable customers with an average monthly peak demand of 1 MW and 78 with an average monthly peak demand of between 750 kW and 999 kW. This represents 23% of the energy share in the market, while DUs and bulk users have the remaining 77%. As at December 2019, there is a total of 1408 contestable customers from Luzon (1264) and Visayas (144) regions [29]. There are also 33 licensed RES, 25 local RES and 44 Supplier of Last Resort (SOLR) in the RCOA system. Table 3 shows the retail electricity market concentration of the Philippines. SOLR is an entity that provides last resort supply to contestable customers who suddenly find themselves without a RES. In spite of the TRO, more and more contestable customers who were earlier issued certificates of contestability by the ERC are opting to voluntarily migrate. The remaining customers, know as the captive customers, continue to be served by the DUs in respective areas.

Table 3. Retailers' market share in the Philippines based on the number of contestable customers [29].

Player	Market Share (%)
MERALCO Group	35.0
Aboitiz Group	24.2
Ayala Group	17.0
San Miguel Group	12.3
Others	11.5

Concerning RE, the Philippines has set to triple the RE capacity than that achieved in 2017 to 15,304 MW by the end of 2030 [30]. In the Philippines, RE is traded separately in WESM. The Renewable Energy Act (R. A. 9513) or the RE Act provides for the establishment of the RE Market (REM). Pursuant to Section 8 of the RE Act, the DOE shall establish the REM and shall direct the PEMC to incorporate changes to the WESM to accommodate RE trading. The Philippines Renewable Energy Market System (PREMS), a platform to facilitate market competitiveness, efficiency, and transparency in the trading of RE certificates (REC) based on the REM Rules should have been in operation by now [31]. DUs have also been mandated to source a minimum portion of energy from renewable sources to guarantee a market for RE generators. Beginning with 1% in 2020, this portion will be increased on a yearly basis and this will be monitored via PREMS.

3. ESI Liberalisation in Singapore

Beginning 1995, the Singapore's ESI has been progressively liberalised and restructured. A regulatory framework is established, the generation and retail markets are opened up to the commercial players, and a wholesale electricity market is introduced. The aim was to gradually introduce competition in electricity generation and retail such that Singapore would have an electricity market that allows the investment, production and pricing decisions to be driven by market forces rather than central planning. The phases of liberalisation can be divided into four; corporatisation from 1995 to 1998, regulatory infrastructure from 1998 to 2001, privatisation and divestment from 2001 to 2009 and full market contestability from 2009 until now [32]. The first (reform) move began in 1995 when the government decided to corporatise the Public Utilities Board (PUB), who were in charge of the supply of water, electricity and piped gas. Under this first move, the electricity and gas undertakings of the PUB were put under the government's investment arm, Temasek Holdings (TH). Under TH, Singapore Power (SP), which at the time was a vertically integrated monopoly wholly owned by the government, was created as the holding company for several new companies including PowerSenoko (now known as Senoko Energy), PowerSeraya, PowerGrid (now known as SP PowerAssets) and Power Supply (now known as SP Services). SP then took over the electricity and piped gas functions from the PUB, and the electricity generation, transmission, distribution, wholesale and retail became the responsibility of the SP. To undertake the responsibility, the companies under SP above were put in charge of the different segments of the ESI. Generation was put under PowerSenoko, PowerSeraya and Tuas Power (now known as Tuas Power Generation, a generation company that was set up as an independent company directly under TH). Power Grid was in charge of the transmission and distribution, and Power Supply, the electricity supply and utilities support services company, became the sole retailer. PUB was reconstituted to continue to supply water and undertake the new role of regulating the electricity and piped gas industries as a result.

Moving on, the Singapore Electricity Pool (SEP) was launched in 1998. It was a day-ahead market with PowerGrid as the system operator and pool administrator. Singapore thus became the first ASEAN country to have the wholesale electricity trading market. At the time, Power Supply was the only buyer in the SEP where the three power generation companies above and a government-owned waste incineration plant competed to sell electricity in the market. Electricity tariffs were unbundled at the same time, making Singapore's ESI to be the most liberalised in ASEAN by the late 1990s. Comprehensive review done by the government in 1999 resulted in the decision to liberalise the electricity market further being made in March 2000. The Electricity Act (Chap. 89A) was enacted in 2001, with the aim to create a competitive market framework for the electricity industry and provide safety, technical and economic regulation of the generation, transmission, supply and use of electricity. With it, the Energy Market Authority (EMA) of the Singapore Act (Chap. 92B) was also enacted to establish and incorporate the EMA of Singapore, to provide for its functions and powers. Consequently, the Energy Market Authority (EMA) was set up in 2001 as statutory board under the Ministry of Trade and Industry to regulate the energy market including electricity and gas industries. EMA's main goals are to ensure a reliable and secure energy supply, promote effective competition in the energy market and develop a dynamic energy sector in Singapore [33]. With the EMA now regulating the ESI, PUB is further restructured to become the sole water authority under the Ministry of the Environment effective 1 April 2001. A division under the EMA undertakes the responsibility as the power system operator (PSO) to ensure secure operation of the power system. PowerGrid therefore only operates and maintains the transmission and distribution grid. A new company, the Energy Market Company Pte Ltd (EMC), a joint venture between EMA and M-co Pte Ltd from New Zealand, was established to undertake the responsibility for the operation and administration of the wholesale electricity market. The SEP lasted until December 2002. Beginning 1 January 2003, the New Electricity Market of Singapore (NEMS) took over its role. It represents a progression from the pool to fully competitive wholesale and retail electricity markets. Similar to the WESM in the Philippines, there is a total of eight key players in NEMS [33] as shown in Figure 2; EMA, EMC, PSO, transmission licensee (SP Power Assets), market support services licensee (MSSL) (SP Services), generators, retailers and consumers (contestable and non-contestable). Their respective descriptions are shown in Table 4.

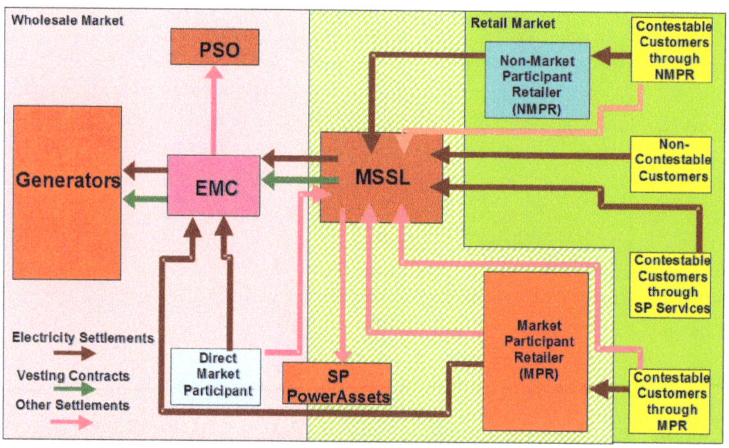

Figure 2. New Electricity Market of Singapore (NEMS) governance [33].

Table 4. Key players in NEMS.

Agency	Role
EMA	The government agency that oversees Singapore's electricity and gas sectors The regulator of the electricity sector and has the ultimate responsibility of ensuring that the NEMS meets the needs of Singapore
EMC	The company that operates and administers the wholesale electricity market
Transmission licensee *	SP PowerAssets, who owns and is responsible for the operation and maintenance of the transmission system comprising high voltage and low voltage networks
PSO	A division of the EMA responsible for ensuring the reliable supply of electricity to consumers and the secure operation of the power system Controls the dispatch of facilities in the wholesale market, coordinates outage and emergency planning and directs the operation of the Singapore high-voltage transmission network under the terms of an "operating agreement" with SP PowerAssets, the transmission licensee
MSSL	Provides market support services such as meter reading, settlement of bills, meter data management, as well as customer transfer services for contestable consumers Facilitates access to the wholesale market for contestable consumers and retailers Responsible for supplying electricity to all non-contestable consumers
Generators *	Dispatchable generators of 10MW or above licensed by the EMA Compete to generate and sell electricity in the wholesale electricity market every half-hour
Retailers *	Can be a market participant retailer (MPR) or non-MPR (NMPR) who compete to sell electricity to consumers MPRs buy electricity in bulk from the wholesale electricity market and NMPRs obtain supply through the MSSL
Consumers *	Contestable consumers may purchase electricity from a retailer, directly from the wholesale market (and become direct market participants) or from the wholesale market indirectly through a MSSL Non-contestable consumers are supplied with electricity by a MSSL

* All market participants need to be licensed by the EMA and registered with the EMC.

Starting July 2001, EMA has gradually opened the retail electricity market to competition to provide business consumers with more options to manage their electricity cost, beginning with those with maximum power requirement of 2 MW. They become known as 'contestable consumers' and at this stage of partial liberalisation, contestable consumers were able to purchase power from either retailers at a price plan that best meet their needs, or from the wholesale electricity market at the half-hourly wholesale electricity prices, or from the MSSL. In June 2003 and December 2003, the thresholds were lowered to customers having average monthly consumption of 20,000 kWh and 10,000 kWh respectively [34], down to 8000 kWh in 2004, to 4000 kWh in 2014 and 2000 kWh in 2015 [35,36]. The non-contestable consumers, mainly households and small businesses, buy electricity at the regulated tariff from the SP Services, the designated public electricity supplier that sells electricity to the non-contestable customers. Since 1 April 2018, EMA commenced the soft launch of open electricity market, where households and businesses in Jurong can choose to buy electricity from a retailer at a price plan that best meets their needs. From 1 November 2018, the open electricity market is extended to all consumers across Singapore by zones. This initiative provides about 1.4 million households and business accounts with more choice and flexibility when buying electricity, while being provided with the same electricity supply through the national power grid. Consumers who wish to remain with the SP Services and buy electricity at the regulated tariff can choose to do so. The switching is not compulsory and there is no deadline for switching. Transmission and distribution of the electricity is still owned and operated by the PowerGrid, and the prices remain regulated. Table 5 and Table 6 shows the concentration in Singapore's generation and retail markets respectively Information obtained from https://www.ema.gov.sg/Statistics.aspx).

Table 5. Key generation players' share in Singapore.

Player	Market Share (%)
Tuas Power Generation	20.6
Senoko Energy	17.3
YTL PowerSeraya	15.6
Keppel Merlimau Cogen	11.7
SembCorp Cogen	10.2
PacificLight Power	9.2
Tuaspring	4.0
Others	11.4

Table 6. Retailers' market share in the Singapore.

Player	Market Share (%)
SP Services Ltd.	24.6
Keppel Electric Pte Ltd.	14.4
Tuas Power Supply Pte Ltd.	13.7
SembCorp Power Pte Ltd.	11.9
Seraya Energy Pte Ltd.	11.9
Senoko Energy Supply Pte Ltd.	11.8
PacificLight Energy Pte Ltd.	6.5
Red Dot Power Pte Ltd.	1.5
Others	3.7

With regards to RE, Singapore aims to achieve up to 350 MW by 2020 and 2 GWp by 2030 of installed solar capacity, the latter amount is sufficient to power 350,000 homes in Singapore [37]. However, RE prospect is limited in Singapore due to the limited land area. Nevertheless, efforts are made to source the RE from the neighbouring countries including Laos using blockchain-enabled market place for RE [38]. Both the Philippines and Singapore are relying largely on fossil fuel to generate electricity namely coal, natural gas and petroleum-based. Singapore has only about 3% of RE share due to the limited renewable energy options other than solar. Hydro resource is not available, wind speeds and mean tidal range are low, and geothermal energy is not economically viable. Figure 3 summarises the transitions towards liberalisation that the Philippines and Singapore have gone through.

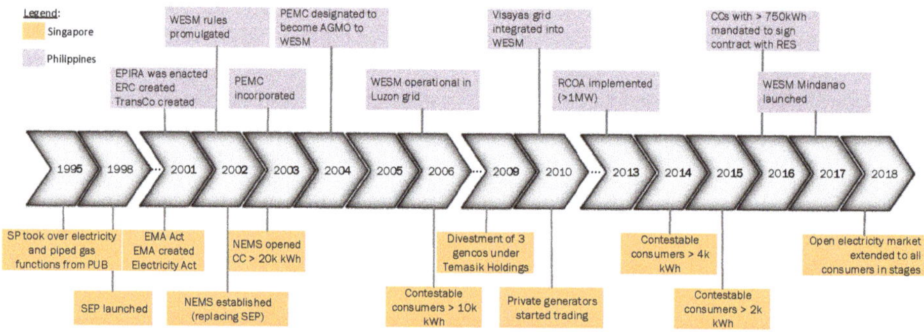

Figure 3. Timeline for the electricity market reform in the Philippines and Singapore.

4. Analysis and Discussion

Taking into consideration the stark contrast between the two countries in many aspects, including geography and demography, the success of the Philippines and Singapore in liberalising their ESIs is worthy of study. Table 1 shows some of the differences. In terms of the country size, vast difference can be seen between the two countries, even though both are islands. The Philippines is an archipelago of more than 7000 islands clustered into three major island groups, Luzon, Visayas and Mindanao. On the other hand, Singapore is a tiny island and the smallest country in ASEAN with a total size of 716 square kilometres (km^2). The country size is in itself posing different kinds of challenge to both countries in reforming their ESIs. It translates into the difference in terms of the population size, with the Philippines having 165 times more than Singapore. However, looking into the population density, Singapore is way more populous with 8358 people per square kilometre compared to just 363 overall in the Philippines. This serves as an indication of the economic growth of the countries, where Singapore is more advanced in terms of the economy in spite of the lower population and smaller size. As can be seen from Table 7, Singapore has the highest GDP growth of 7.1% in 2018 and has the highest GDP of USD 364 billion. Singapore has also been enjoying 100% electrification rate since at least 1990, while the Philippines is continuously working to improve access to electricity, which currently stands at 93% and brownouts are still common. Consequent to these, the electricity consumption in Singapore is more than half of the Philippines', so does its installed capacity. Additionally, in Table 7, information about Malaysia is also included. As can be seen from the table, Malaysia sits between the two countries in most of the attributes included for comparison, which provides a good starting point for the discussion on the prospect of ESI liberalisation in Malaysia. Representing the two extreme conditions, the Philippines' and Singapore's success in attaining full liberalisation of their ESIs can serve as the motivation for Malaysia and other ASEAN countries to follow suit. In the following subsections, analysis and discussion with respect to the four perspectives mentioned in Section 1 are presented.

Table 7. Selected demographic information of the Philippines and Singapore.

Attribute	The Philippines *	Singapore *	Malaysia *
Land size (km^2)	300,000	716	329,847
Population	100,981,437	5,612,000	31,620,000
Population density (people/km^2)	363	8,358	95
Access to electricity (%)	93	100	100
Installed capacity (MW)	20,055	13,649.4	33,023
Electricity consumption (GWh)	99,765	49,643	145
Gross domestic product (GDP) (USD billion)	330.91 (6.2%)	364.14 (7.1%)	358.58 (4.7%)

* All data are captured as at 2018.

4.1. Regulatory Framework and Institutional Bodies

Apparent from the experience of both countries is the defining role of the legislative framework that subsequently entails the setting up of the regulatory and institutional bodies to enforce reform. The legislation must be enacted, and the established regulatory and institutional bodies must be given the authority they need to execute the transformation. The Philippines began to see the progress towards ESI liberalisation after the enactment of the EPIRA. Actions can be seen taken involving the revision and redefinition of the existing agencies' scope of work and the establishment of the new ones to undertake the various tasks to reform. Similarly, Singapore's more serious move towards liberalisation started with the enactment of the Electricity Act (Chapter 89A) that was followed by the EMA Act (Chapter 92B) to define the scope of work to be done and by whom. Furthermore, the authority given to the regulatory and institutional bodies has to cover the entire value chain of the ESI. For instance, the ERB of the Philippines only plays an active role in distribution rate regulation while transmission and generation (wholesale) rates are largely developed by the NPC with only minimal oversight by the ERB [39], which has led to issues in performance accountability, incentive for efficiency and financial burden on the government [39]. The legislation and institutional environment also needs to be designed in such a way that it survives the changes in the government because ESI liberalisation is not a one-off event [40,41]. Continuous development is foreseen and there probably will be no end to it [42]. Thus, it requires strong and sustained political commitment, extensive and detailed preparation, and continuous development to foster long-term investment.

In Malaysia, there is already a handful of laws and acts to govern the ESI, as listed below. Each comes with their respective regulations, orders, rules and other sub-legislation, where applicable [43].

- Electricity Supply Act (ESA) 1990
- Renewable Energy Act 2011
- Gas Supply Act 1993
- Renewable Energy Act 2011
- Environmental Quality Act 1974
- Occupational Safety and Health Act 1994
- Factories and Machinery Act 1967
- Petroleum Development Act 1974
- Petroleum (Safety Measures) Act 1984
- Petroleum and Electricity Control of Supplies Act 1974

For both MESI 1.0 and MESI 2.0 reforms, Malaysia Programme Office for Power Electricity Reform (MyPOWER) has been entrusted to spearhead the initiatives. The agency has been actively engaging the stakeholders through various workshops and discourses in order to gain inputs from them with regards to the subsequent reform plan (MESI 2.0). This is certainly an important move to ensure that the latter's concerns are taken into account in instituting the reform and their support can hence be obtained. The next move to realise MESI 2.0 would be to define or revise the necessary legislation. Based on the timeline presented by the minister during the 10th International Greentech & Eco Products Exhibition & Conference Malaysia (IGEM 2019) in October 2019, a number of work on this has started since the fourth quarter of 2019 and is still on-going with varying expected completion dates. This includes detailing of rules, incentive mechanisms, reviewing and drafting amendment and regulatory controls, detailing of energy and capacity market design and rules, and detailing of retail regulatory framework including SOLR's responsibilities and remuneration [44]. The new Act to cater for the whole reform initiatives is planned to be tabled in the second half of 2022. In formulating the required regulations, there is a number of lessons that can be learnt from the Philippines and Singapore. EPIRA has especially received unfavourable remarks from the critics, stating that it is not achieving its objectives and calling for its amendment [45]. In particular, the electricity price has been and keeps on increasing post EPIRA, which is claimed to be due to the ineffectiveness and inefficiency of its structure and the institutional bodies that constitute the ESI [17,46,47]. Therefore, Malaysia has to carefully identify the agencies and institutional bodies needed to implement the reform. It is also important to clearly delineate the boundary of each, especially in areas where their functions can potentially overlap. In this respect, Reference [48] pointed out the importance of a well designed reform structure as one of the lessons learned from the electricity reform moves around the world. Another reason that causes the benefits of the WESM not be realised is because only a very small amount of about 8.2% (As of 2015) electricity is being traded in the market, while the rest is sold through bilateral contracts [49], which defies the purpose of having the market. On the contrary, Singapore has made it compulsory for the generators with capacity of at least 1MW to register with NEMS [50]. The issue in Singapore on the other hand is concerning the competition at the retail market level. Since the opening of the Singapore's retail electricity market in 2001, the market has been dominated by the "Gentailers", which is the term used for retailers who share parent organisations with generation companies, as opposed to the "independent" retailers, who do not have any connection with the generators. There is also a plan to allow gentailers to operate in the MESI's wholesale market [44]. Therefore, careful analysis and implementation mechanisms are needed to prevent oligopoly and to avoid the independent retailers from shying away when they are not able to develop successful business cases. Finally, it is also recommended for the government of Malaysia to consider not privatising the electricity industry fully and entirely. The government should be allowed to own and operate some assets, either on its own or through the government-linked company like Tenaga Nasional Berhad (TNB). This is especially important during critical and trying times when private companies, which are profit oriented, may decide to stop operation or shut down when business is not good [51]. This is also one of the problems faced in the Philippines because the EPIRA at the moment does not allow the government to own and operate assets.

4.2. Clearly Defined Implementation Phases and Targets

Apart from flawed design, flawed implementation of a reasonable reform design can be the cause of problematic ESI reform [52]. Liberalising the ESI is a long-term endeavour, involving many parties and possibly transcending governments. Singapore did it in about 24 years from corporatisation of the PUB to the full opening of the retail electricity market, which was attained in 2019 [53]. The Philippines, which started five years later than Singapore is in the last few steps of the staggered opening of the retail electricity market, which is now put on hold due to the TRO explained earlier. For such a long-term transformation process, meticulous planning with clearly defined stages and milestones is important. The importance of implementing the reform in 'proper' sequence is also emphasised in Reference [48]. The logical sequence ideally begins with raising the prices to cost-recovering levels, followed by creating regulatory institutions and restructuring the sector, and then privatisation [48]. Such sequence can lead to significant improvements in several dimensions of operating performance and in a variety of country settings. Furthermore, the transformation can be made more systematic and smoother with clearly defined and communicated activities and targets. Singapore for example can clearly divide their electricity industry reform into four distinct phases; corporatisation, regulatory infrastructure, privatisation and divestment, and full market contestability, as explained earlier in Section 3. Although the stages of the ESI liberalisation in the Philippines are not found specifically mentioned in any source, similar approach to Singapore is seen as shown in Figure 3. In general, it can be seen that the first step would be the detachment of the monopolistic agency that governs the ESIs through privatisation or corporatisation. It ends with the staggered opening of the retail markets by gradually lowering the contestability thresholds of the consumers. With respect to privatisation, findings from Reference [54] show that privatisation improves efficiency if accompanied by independent regulation, privatisation and independent regulation have no significant effect on prices and private investment is stimulated by independent regulation. As for Malaysia, privatisation of the NEB, the government agency in charge of the electricity supply industry was already accomplished in 1990, which subsequently known as TNB. However, the private, government-linked company that took over was still a monopoly. The presence of the first generation IPPs has reduced the monopoly at the generation sector a little bit. Until now, the IPPs account up to about 50% of the total electricity generation in the country. In this respect, Malaysia bears similarity with the Philippines where at the onset of the liberalisation is the involvement of IPPs, albeit the number is not as many as the Philippines'. The first reform series that started in 2011 (MESI 1.0) had envisaged the managed market model that will eventually lead to the presence of the franchisers at the retail market with contestable and non-contestable consumers to be achieved by 2020 [55]. However, the progress has been slower than expected. It only started to take a more serious turn recently after the new government took over with the launching of the next wave of reform, MESI 2.0 [12], which indicates strong political influence in determining its course. A MESI reform roadmap comprising four stages as shown in Figure 4 has been presented to the Panel Perundingan Tenaga (energy consultative panel) in March 2018. The first stage focuses on the long run efficiency while the second stage focuses on the short run efficiency. The first three foci of the third stage are to optimise regional resources, to open the market for fuel supplies and to create independent single buyer (SB) and grid system operator (GSO). The first two are already in implementation and the focus of discussion during the meeting was the third focus as the next move forward. In stage 4, choice of qualified distribution customers will be made and the independent SB is expected to become the electricity market operator. However, the timeline is not clearly stated in the roadmap except that beyond 2030, managed competition with customers having the choice of retailers would be the focus. In the revised roadmap [44] presented in October 2019, aggregated timeline has been included with more details. Among others, it is planned that the hybrid wholesale market to be ready by 2029. The hybrid market comprises the capacity market on top of the existing energy market, while still honouring the last batch of the PPAs that will end in 2045 [56]. While capacity market ensures reliability by paying generators to commit generation for delivery in years to come, the energy market pays generators only when they provide power day-to-day [57]. The need for capacity market in a liberalised ESI is also in line with the recommendation made in Reference [58] concerning the risks of energy-only market in supporting intermittent RE. The full price-based retail is also expected to be ready by 2029, with the pilot opening of the retail market expected to happen in the second quarter of 2021 [44]. Thus, it can be seen that the plan forward has been clearly laid out with the activities and respective targets made explicit known. The next step would be the execution of the activities, monitoring and revising them in the course of doing so, taking into account the challenges and delays in the first reform series, as well as lessons from other countries. It is also important to ensure timeliness of the execution, allowing the affected parties to adapt to the changes.

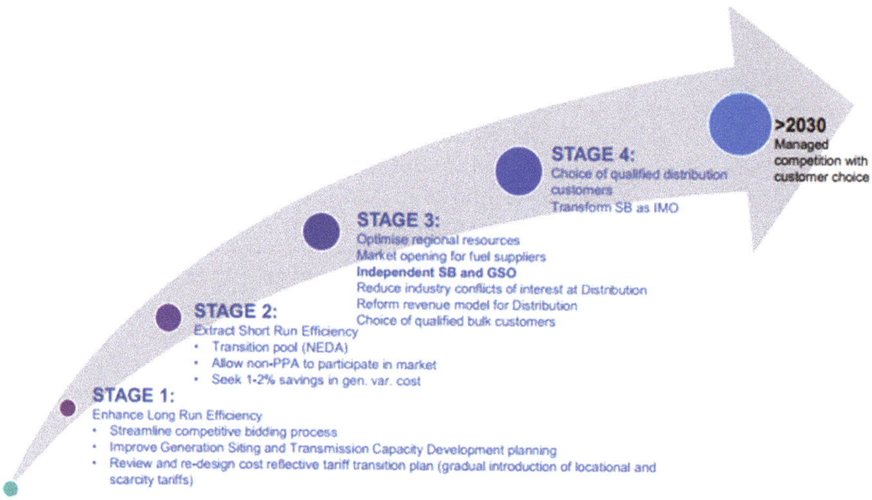

Figure 4. Malaysia electricity supply industry (MESI) roadmap—https://www.singlebuyer.com.my/MESI.php.

4.3. Composition of the Market Components

Next, our analysis looks into the components that constitute the electricity markets. The essence that differentiate between a liberalised ESI and a monopolised one is the presence of the electricity market, both at the wholesale and retail levels. At the wholesale market, generators (the supply side) bid to offer the lowest price to the potential buyers (the demand side). At the retail market, the buyers will subsequently supply electricity to their respective customers, who are the consumers of electricity or the end-users. Despite the different progress achieved, approaches taken and challenges faced by the Philippines and Singapore in liberalising the ESIs and setting up their open electricity markets, similarity is seen in terms of the components that constitute the markets as shown in Table 8. These components can therefore be regarded as the main components of an open ESI. The agencies and/or companies that correspond to the components in the respective markets are also shown in the table.

Table 8. Components in an electricity market.

Component	The Philippines	Singapore	Malaysia (Proposed)
Regulatory oversight	ERC	EMA	EC (rebranded)
Market operator	PEMC	EMA	(Independent SB)
System operator	NGCP	EMA arm	(Independent GSO)
Participant (Wholesale)	Generators, DUs, suppliers, bulk customers and transmission provider	Generators, MSSL, MPR, bulk customers and transmission licensee	Large power consumers, industrial and commercial consumers, TNB Grid
Participant (Retail)	Contestable customers	Contestable customers	TNB Retail and other retailers (yet to have)
Backup supplier (SOLR)	DUs	MSSL	TNB Retail
Central repository	CRB	MSSL	TNB Retail

The current structure of MESI is still largely monopolistic. Competition only exists at the generation sector where the generators sell electricity to the Single Buyer (SB) as shown in Figure 5. Wholesale market in the sense discussed above does not exist and electricity trading is done according to the bilateral agreements (PPA/SLA/NEDA) signed between the generators and the SB. TNB, being the main electricity provider in Malaysia, is still a main player in the generation sector, having more that 50% generation capacity in the peninsular Malaysia alone [59]. It is still monopolising the transmission and distribution sectors, and in almost full control of the retail sector. The Energy Commission (EC) has been regulating the MESI since 2001. It is expected that the EC will continue to assume the role as the regulator once the open electricity market is operationalised, with the necessary redefinition of the scope and possibly renaming or rebranding too. Rebranding and redefining the scope of the regulator are needed to emphasise the independence from the government, to change public perception and to clarify the refreshed scope of the EC. In the Philippines too, the ERB had its name changed to ERC after liberalisation and in Singapore, EMA was established to replace the PUB.

A market operator (MO) also needs to be created to manage and coordinate the activities in the wholesale electricity market. Ideally, an *independent* MO (IMO) is expected. However, transition period is foreseen where government agency, a division of the regulator or a government-owned company will operate during the initial phase of the market. For example, the WESM is run by the government through the PEMC [60] since the market went into operation in 2006 until 2018 [24], although the initial intention was to have the IMO in within the first year of the market launch. The other important component of an electricity market is the System Operator (SO), who is responsible for the dispatch of the electricity traded in the wholesale market. Thus, the SO is not involved in the trading process. TransCo was the first SO in WESM, which was later given to NGCP, while in NEMS, a division of EMA (the regulator) is the SO. Looking at the current MESI structure shown in Figure 5, SB and the Grid System Operator (GSO) would most likely be the MO and SO respectively to undertake the initial operation of the open electricity market to be established, based on the experience they have. At the moment, both are ring-fenced entities under TNB [61]. As mentioned earlier, the plan to free them from TNB is already on the MESI roadmap even prior to the start of MESI 2.0 as can be seen from Figure 4. Commercial operation date for the enhanced SB and GSO is expected to be in the first quarter of 2021 [44]. Transformation of the independent SB into IMO is also on the roadmap. The transmission on the other hand should remain a regulated monopoly as per practiced in the Philippines and Singapore. The same arrangement is also seen in the Nordic countries.

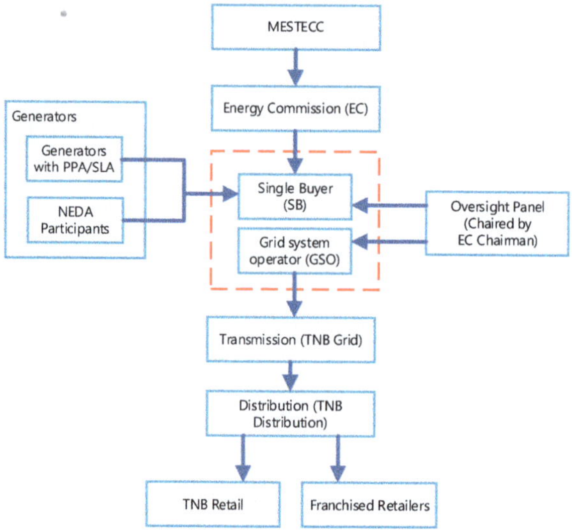

Figure 5. The current MESI structure.

Currently, there are three types of generators in MESI; generators under the TNB who have service level agreements (SLAs) with the SB, IPPs who trade electricity with the SB according to the power purchase agreements (PPAs) and other generators who sell electricity to the SB following the new enhanced dispatch agreements (NEDA) [62]. These generators would be the candidates for the wholesale sellers in MESI wholesale electricity market. Ideally, all electricity trading should be done through the wholesale market. Bilateral contracts between the sellers and buyers may still be allowed on a very small number of specific cases, but with limited percentage. In both WESM and NEMS, three categories of wholesale buyers are identified; suppliers, distribution utilities and directly connected end-users. Depending on the categories, buyers at the wholesale market either consume the electricity themselves or become the sellers at the retail market. Suppliers serve contestable customers, that is, customers whose electricity usage exceed certain imposed limit and as a result, are allowed to choose suppliers whom they deem can offer them the best values. Distribution utilities (MSSL in NEMS) are the default retailers that supply electricity to the rest of the consumers, known as non-contestable or captive customers, at the regulated price. They are not able to choose suppliers until they are eligible to become contestable customers. The last category of the wholesale buyers is the directly connected customers. These are the large power consumers who buy electricity in bulk directly from the wholesale market for their own use. Electricity consumers in Malaysia are categorised into residential, industrial and commercial [63]. Almost all of them are served by TNB retail at the moment. Retail electricity market is currently non-existent in MESI. Thus, in essence, there is no competition at the retail level. Franchised retailers do exist, but with the scope limited to managing the supply of electricity to the tenants of shopping malls or office complexes. The industrial customers are either directly connected to the grid (via switching stations) or to the substations depending on the amount of the maximum demand levels [64]. They are therefore foreseen to be the potential directly connected customers who will buy electricity directly from the wholesale market for their own use. TNB Retail, as the incumbent retailer is expected to play the role as the SOLR, due to the experience and extensive infrastructure that they already have. Thus, it can be seen that current MESI structure is able to prepare itself well into liberalisation. Nevertheless, it is worth noting that the electricity markets have evolved or are also evolving into hybrid forms, liberalised but not fully due to the lack of some of the institutional and other preconditions for the full and effective implementation of the standard reform model, which is beginning to be acceptable new norms of liberalisation [48].

4.4. Influence on the RE Penetration

Finally, the extent to which the electricity market liberalisation has influenced the RE penetration into the grid, if any, is also analysed in this study. This is a topic of interest because Malaysia has set the aim to achieve 20% RE penetration by the year 2030 [65] and 50% by the year 2050 [66], excluding those from large (>100 MW) hydro power stations [59]. Thus, it is important to ascertain in what ways MESI liberalisation can help to achieve the target. Promoting energy efficiency and the use of RE resources are amongst the objectives of ESI liberalisation [67]. One of the means is by having more RE actors into the industry as a result of liberalisation [68]. Findings from Reference [69] suggest that reduction in the monopolistic power of state-owned utilities due to liberalisation has had positive effect on RE policies when various types of actors are ensured access to the grid instead of it being provided to only a few large private firms. It also found that liberalisation increases public support for renewable energy. As explained earlier in the respective sections above, the Philippines and Singapore have also put their own targets with regards to RE. Figure 6 shows the current state with respect to the electricity generation mix in the Philippines, Singapore and Malaysia. At a glance, no significant relationship can be seen between the RE penetration and electricity market liberalisation. It can be seen from the figure that Malaysia is already doing good with respect to the RE penetration, with about 20% RE in its generation mix, a small percentage away behind the Philippines. Looking over a period of time from 2005 to 2018, Malaysia is also showing a good progress with respect to the increased RE penetration as shown in Figure 7. Based on this raw information, for the moment, there is no clear evidence on the influence of the liberalised electricity market per se on the increased RE penetration. In addition, from the experience of Japan, its government does not rely on electricity liberalisation alone to give further impetus to RE development [68]. It is being supported by other initiatives too. Government intervention is also found needed to achieve the objectives of liberalised ESI [58]. For example, as pointed out by Reference [70], generators should have a wide market for the products they are creating, to ensure that they receive the fairest price and the government can help to institute the necessary policies to facilitate this. In this regards, it can be seen that the government of Malaysia has already been introducing a number of schemes to promote RE, including the large scale solar (LSS) (A solar PV plant producing between 1 MWac and 50 MWac connected to either distribution or transmission network [71]), the feed-in tariff (FiT) and the net energy metering (NEM) [72]. Based on the current setting, once the wholesale electricity market is available in MESI, the LSS generators can become the market participants and compete to sell electricity to the wholesale buyers. This is plausible especially with the solar generation cost that is becoming lower and lower than the generation costs from the conventional resources [73]. At the moment, the LSS generators are already selling electricity to the SB together with the other generators through PPAs as shown earlier in Figure 5. For the lower scale solar and other distributed RE generators, FiT and NEM schemes are available. However, as of June 2018, only 10 MW of the allocated 500 MW quota for NEM are taken [72]. Thus, the Malaysian government should continue its effort to promote and support RE penetration through various initiatives, along with the ESI reform initiatives.

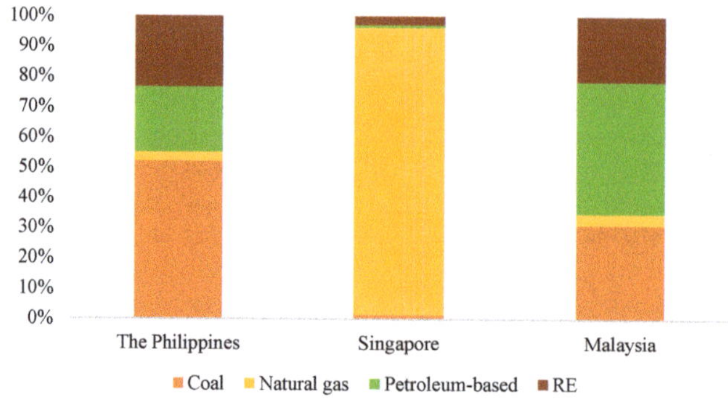

Figure 6. Energy mix in the Philippines, Singapore and Malaysia.

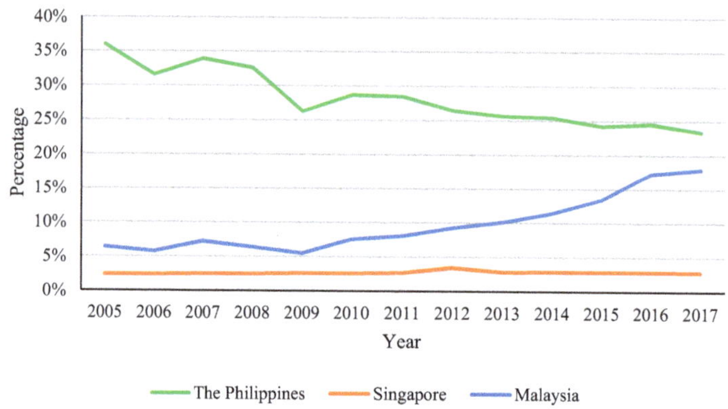

Figure 7. Percentage of renewable energy (RE) the generation mix in the Philippines, Singapore and Malaysia.

5. Conclusions and Outlook

In this paper, the chronological review of the Philippines' and Singapore's experience in liberalising their ESIs is presented. While the Philippines went through an eventful transformation, it was a smoother sailing for Singapore. Nevertheless, both have respective lessons and best practices that Malaysia can learn in the process of liberalising its ESI. The purpose of focusing on the Philippines and Singapore in the review, as mentioned earlier, is because of their current state of ESI liberalisation, which is considered to be in the more advanced stage compared to the other ASEAN countries. The importance of legislative framework and phased implementation are amongst the lessons that Malaysia can learn from their experience. Similarities in terms of components that constitute the electricity markets would serve as the starting point for Malaysia in determining its future electricity market components. Analysis on the impact of the liberalised electricity markets on the RE penetration into the grids provides useful insight for Malaysia in retaining the existing initiatives towards meeting its RE targets. A foreseen challenge for Malaysia in liberalising MESI is in terms of the public acceptance, particularly the households or the residential consumers. Energy liberalisation in general has led to positive and globally widespread but modest efficiency gains across all energy sectors but a lack of clearly visible direct benefits to the households in many countries [10]. Pricing in particular is expected to be one of the main issues. Electricity prices can rise for some or all customer groups (reducing their welfare) while the overall social welfare goes up [10]. This is expected to be the case for Malaysia where the government has been subsidising the electricity tariff for so long. Therefore, there is an urgent need to identify electricity pricing schemes that strike a more satisfactory balance between economic efficiency and social equity as proposed in Reference [48], which will be a subject for a future study. Beyond Malaysia, liberalisation of the ESIs in all ASEAN countries can also be considered in realising the APG. Experience from the other regions, such as the Nordic, shows that liberalised ESI aligns very well in accommodating renewables and attaining efficiency. With the depleting fossil fuel and the energy demand that is predicted to exceed the production of indigenous energy resources of the region based on the recent Southeast Asia energy outlook, increased penetration of renewables into the grid, and more efficient use of energy would be the silver bullets in alleviating the impact of depleting natural indigenous energy resources for the benefits of the generations to come. Thus, it is a potential that future direction of the ASEAN countries with regards to the energy and electricity is towards the liberalisation of the ESIs. Furthermore, extending the ESI liberalisation to the whole of ASEAN in realising the APG requires a more authoritative form of coalition, similar to the European Union where each member state is, by law, obliged to comply to the directions. Having said that, integrated and fully connected APG does not have to wait until all ASEAN countries have liberalised their ESIs in order to materialise. It would be sufficient to have the necessary regulations imposed for the setting up of compatible ESIs that can harmoniously connect and communicate regionally. For this, the ESI in each ASEAN country needs to be studied in detail to identify the extent to which they can be harmonised. However, it is not within the ambit of this research to provide the detailed review on the ESI of each ASEAN country. It can be a separate study on its own, and a future work that we are considering.

Author Contributions: Conceptualization, H.A.; methodology, H.A.; validation, I.S.M.Z., B.N.J.; formal analysis, H.A.; investigation, H.A. and I.S.M.Z.; resources, H.A. and I.S.M.Z.; writing–original draft preparation, H.A.; writing–review and editing, H.A., I.S.M.Z. and B.N.J.; supervision, B.N.J.; project administration, H.A.; funding acquisition, H.A. All authors have read and agree to the published version of the manuscript.

Funding: This research was funded by Universiti Tenaga Nasional grant number RJO10517844/008.

Conflicts of Interest: The authors declare no conflict of interest.

References

1. Zamin, N.Z.M.; Abidin, N.Z.Z.; Ibrahim, J.B. Single buyer - A step forward in Malaysian Electricity Supply Industry reform. In Proceedings of the IEEE 2013 Tencon - Spring, Sydney, Australia, 17–19 April 2013; pp. 391–397. [CrossRef]
2. Aris, H.; Jøgensen, B.N. ASEAN Power Grid 20 years after: An overview of its progress and achievements. *IOP Conf. Ser. Earth Environ. Sci.* **2020**, *463*, 1–8. [CrossRef]
3. Li, Y. ASEAN Sets New Momentum Moving Towards an Integrated Electricity Market in the Region. Available online: http://www.eria.org/news-and-views/asean-sets-new-momentum-moving-towards-an-integrated-electricity-market-in-the-region/ (accessed on 4 April 2019).
4. Amundsen, E.S.; Bergman, L. Why has the Nordic electricity market worked so well? *Util. Policy* **2006**, *14*, 148–157. [CrossRef]
5. Mundaca, L.T.; Dalhammar, C.; Harnesk, D. The Integrated NORDIC Power Market and the Deployment of Renewable Energy Technologies: Key Lessons and Potential Implications for the Future ASEAN Integrated Power Market. In *Energy Market Integration in East Asia: Renewable Energy and its Deployment into the Power System*; Kimura, S., Phoumin, H., Jacobs, B., Eds.; ERIA Research Project Report 2012-26; ERIA: Jakarta, Indonesia, 2013; pp. 25–97.
6. Stridbaek, U. *Lessons from Liberalised Electricity Markets*; Technical report; OECD iLibrary: Paris, France, 2006.
7. Bustos-Salvagno, J.; Fuentes H, F. Electricity Interconnection in Chile: Prices versus Costs. *Energies* **2017**, *10*, 1438. [CrossRef]
8. Department of Trade Industry. *The Social Effects of Energy Liberalisation The UK Experience*; Technical report; Department of Trade Industry: UK, 2000.
9. Owen, A.D.; Finenko, A.; Tao, J. *Power Interconnection in Southeast Asia*; Routledge: London, UK, 2019. [CrossRef]
10. Pollitt, M.G. The role of policy in energy transitions: Lessons from the energy liberalisation era. *Energy Policy* **2012**, *50*, 128–137. [CrossRef]

11. Mohamad, H.F.; Amran, S.N.M.E. Electricity Industry to Undergo Transformation with MESI 2.0. Available online: https://www.nst.com.my/business/2018/09/412549/electricity-industry-undergo-transformation-mesi-20 (accessed on 10 January 2019).
12. Aris, H.; Jørgensen, B.N.; Hussain, I.S. Electricity supply industry reform in Malaysia: Current state and way forward. *Int. J. Recent Technol. Eng.* **2019**, *8*, 6534–6541.
13. Sharma, D. Electricity Reforms in the ASEAN: A Panoramic Discourse. *Econ. Political Wkly.* **2005**, *40*, 5318–5326.
14. Woodhouse, E.J. *The IPP Experience in the Philippines*; Technical report; Stanford University: Stanford, CA, USA, 2005.
15. Hall, D.; Nguyen, T.A. Electricity Liberalisation in Developing Countries. *Prog. Dev. Stud.* **2017**, *17*, 99–115. [CrossRef]
16. Sharma, D.; Madamba, S.E.; Chan, M.L. Electricity industry reforms in the Philippines. *Energy Policy* **2004**, *32*, 1487–1497. [CrossRef]
17. Santiago, A.; Roxas, F. Understanding Electricity Market Reforms and the Case of Philippine Deregulation. *Electr. J.* **2010**, *23*, 48–57. [CrossRef]
18. Afable, P.P.R. *The Players: Philippine Power Industry*; Technical report; KPMG Global Energy Institute: Cebu, Philippines, 2014.
19. Asian Development Bank. *Philippines: Electricity Market and Transmission Development Project*; Technical report; Asian Development Bank: Mandaluyong, Philippines, 2016.
20. Navarro, A.M.; Detros, K.C.; dela Cruz, K.J. *Post-EPIRA Impacts of Electric Power Industry Competition Policies*; Technical report; Philippine Institute for Development Studies: Quezon City, Philippines, 2016.
21. Oplas, B.S., Jr. Electricity Competition, EPIRA, and WESM. Available online: https://www.bworldonline.com/electricity-competition-epira-and-wesm/ (accessed on 17 March 2019).
22. Rudnick, H.; Velasquez, C. *Learning from Developing Country Power Market Experiences The Case of the Philippines*; Technical report; Policy Research Working Paper 8721; The World Bank: Washington, DC, USA, 2019.
23. Rivera, D. DOE Expects WESM Mindanao Opening to be Delayed. Available online: https://www.philstar.com/business/2020/01/11/1983734/doe-expects-wesm-mindanao-opening-be-delayed (accessed on 22 April 2020).
24. Rivera, D. Independent Operator Takes over Wholesale Power Spot Market. Available online: https://www.philstar.com/business/2018/09/27/1854925/independent-operator-takes-over-wholesale-power-spot-market (accessed on 16 January 2019).
25. Department of Energy. DEPARTMENT CIRCULAR NO. DC 2012-05-2005. Available online: https://www.officialgazette.gov.ph/2012/05/09/doe-dc-no-2012-05-005-s-2012/ (accessed on 15 May 2019).
26. Lagare, J.B. 940 Customers Listed in Power Retail Market—WESM. Available online: https://www.manilatimes.net/940-customers-listed-power-retail-market-wesm/370648/ (accessed on 24 June 2019).
27. Dimalanta, M.C.; Balot, S.F.; Zareno, J.G.B. Chapter 27 Philippines. In *The Energy Regulation and Markets Review*, 7th ed.; Schwartz, D.L., Ed.; Law Business Research Ltd.: London, UK, 2018; pp. 337–351.
28. Inquirer.Net. WESM CEO: Fully Implement RCOA to Lower Electricity Prices. Available online: https://business.inquirer.net/261010/wesm-ceo-fully-implement-rcoa-to-lower-electricity-prices (accessed on 5 May 2019).
29. Corporation, P.E.M. *QUARTERLY RETAIL MARKET ASSESSMENT REPORT 26 September–25 December 2019*; Technical report; PEMC: Philippines, 2020.
30. WWF Philippines. Renewables Best Way to Power Philippine Development. Available online: https://wwf.org.ph/what-we-do/climate/renewables/renewables-to-power-ph-dev/ (accessed on 15 April 2020).
31. Lectura, L. DOE Launches First Renewable-Energy On-Line Trading Platform in Philippines. Available online: https://businessmirror.com.ph/2019/12/18/doe-launches-first-renewable-energy-on-line-trading-platform-in-philippines/ (accessed on 30 March 2020).
32. Low, M. Liberalized Electricity Markets? A Case Study of Singapore. Paper Presented at the International Conference of Applied Energy (ICAE) 2012, Suzhou, China, 5–8 July 2012.
33. Energy Market Authority. *Introduction to the National Electricity Market of Singapore*; Technical report; Energy Market Authority: Singapore, 2010.
34. Singapore Power. *Retail Competition for Electricity Supply in Singapore*; Singapore Power: Singapore, 2016.

35. Law, F.M. More Choices for all Electricity Users in 2018. Available online: https://www.todayonline.com/singapore/singapores-electricity-market-be-fully-liberalised-h2-2018 (accessed on 21 October 2019).
36. Energy Market Authority. *Smart Energy Sustainable Future Energy Market Authority Annual Report 2018/19*; Technical report; Energy Market Authority: Singapore, 2019.
37. Min, A.H. Singapore Sets Solar Energy Target for 2030 That Would Provide Enough Power for 350,000 Homes. Available online: https://www.channelnewsasia.com/news/singapore/solar-power-target-energy-350000-homes-2030-hdb-rooftops-12042228 (accessed on 28 February 2020).
38. Goh, M. Blockchain Tech Is Taking on Renewable Energy Trading in Singapore. Available online: https://www.cnbc.com/2018/11/07/blockchain-tech-is-taking-on-renewable-energy-trading-in-one-country.html (accessed on 5 May 2020).
39. Stanley Bowden, R.; Ellis, M. Restructing the Philippine electric power industry. *Electr. J.* **1995**, *8*, 38–44. [CrossRef]
40. Pollitt, M. Electricity reform in Argentina: Lessons for developing countries. *Energy Econ.* **2008**, *30*, 1536–1567. [CrossRef]
41. Hall, D.; Thomas, S.; Corral, V. *Global Experience with Electricity Liberalisation*; Paramadina University: Jakarta, Indonesia, 2009; Seminar, 19 January 2010.
42. Nepal, R.; Jamasb, T. Caught between theory and practice: Government, market, and regulatory failure in electricity sector reforms. *Econ. Anal. Policy* **2015**, *46*, 16–24. [CrossRef]
43. Aziz, F.A.; Khor, K. Chapter 21 Malaysia. In *The Energy Regulation and Markets Review*, 7th ed.; Schwartz, D.L., Ed.; Law Business Research Ltd.: London, UK, 2018; pp. 261–276.
44. Ministry of Energy, Science, Technology, Environment and Climate Change. *Reimagining Malaysian Electricity Supply Industry (MESI 2.0)*; Ministry of Energy, Science, Technology, Environment and Climate Change: Malaysia, 2019.
45. Gamboa, R. Amending Epira. Available online: https://www.philstar.com/business/2013/01/08/894346/amending-epira (accessed on 20 March 2020).
46. Carino, L.V. *Regulatory Governance in the Philippines: Lessons for Policy and Institutional Reform*; Technical report; University of Manchester: Manchester, UK, 2005.
47. Fe Villamejor-Mendoza, M. Bringing Electricity Reform to the Philippines. *Electr. J.* **2008**, *21*, 42–58. [CrossRef]
48. Kessides, I.N. The Impacts of Electricity Sector Reforms in Developing Countries. *Electr. J.* **2012**, *25*, 79–88. [CrossRef]
49. Ho, E. Electricity Derivative Market for the Philippines. Available online: https://asian-power.com/regulation/commentary/electricity-derivative-market-philippines (accessed on 20 May 2020).
50. Energy Market Company. Generation Facility Registration. Available online: https://www.emcsg.com/aboutthemarket/generationfacilityregistration (accessed on 4 April 2020).
51. Gonzales, I. Special Report: What's Wrong with EPIRA? Available online: https://www.philstar.com/headlines/2014/01/14/1278583/special-report-whats-wrong-epira (accessed on 10 April 2020).
52. Roxas, F.; Santiago, A. Broken dreams: Unmet expectations of investors in the Philippine electricity restructuring and privatization. *Energy Policy* **2010**, *38*, 7269–7277. [CrossRef]
53. Liew, E. Open Electricity Market (OEM) Singapore—10 Important Things to Know. Available online: https://sg.finance.yahoo.com/news/open-electricity-market-oem-singapore-160000715.html (accessed on 12 March 2020).
54. Pollitt, M. Evaluating the evidence on electricity reform: Lessons for the South East Europe (SEE) market. *Util. Policy* **2009**, *17*, 13–23. The Political Economy of Electricity Market Reform in South East Europe. [CrossRef]
55. Hasan, A.F. *Malaysian Electricity Supply Industry (MESI) Reform Initiatives*; Energy Commission: Putrajaya, Malaysia, 2014.
56. Tan, C.C. Disruption is Coming to the Local Power Industry. Available online: https://www.theedgemarkets.com/article/disruption-coming-local-power-industry (accessed on 15 June 2019).
57. Ihnen, J. Capacity Market v Energy Market—What's the Diff? Available online: https://michaelsenergy.com/capacity-market-v-energy-market-whats-diff/ (accessed on 30 April 2020).
58. Owen, A.D. Do Liberalized Electricity Markets Discourage Investment in Renewable Energy Technologies? *Electr. J.* **2014**, *27*, 53–59. [CrossRef]

59. Suruhanjaya Tenaga. *Peninsular Malaysia Electricity supply Industry Outlook 2019*; Technical report; Suruhanjaya Tenaga: Putrajaya, Malaysia, 2019.
60. Gatdula, D.L. Independent Operator Expected to Handle WESM Operations. Available online: https://www.philstar.com/business/2010/08/11/601183/independent-operator-expected-handle-wesm-operations (accessed on 20 July 2019).
61. Single Buyer. Ring-Fencing. Available online: https://www.singlebuyer.com.my/ringfencing.php (accessed on 23 March 2020).
62. Zainul, I.F. NEDA to Encourage Healthy Competition. Available online: https://www.thestar.com.my/business/business-news/2015/10/10/neda-to-encourage-healthy-competition (accessed on 20 January 2020).
63. Suruhanjaya Tenaga. *Review on Electricity Tariff in Peninsular Malaysia under the Incentive-Based Regulation Mechanism*; Suruhanjaya Tenaga: Putrajaya, Malaysia, 2013.
64. Tenaga Nasional Berhad. *Electricity Supply Application Handbook*; Technical report; Tenaga Nasional Berhad: Kuala Lumpur, Malaysia, 2011.
65. Eusoff, N.S. Malaysia Sets New Goal of 20% Clean Energy Generation by 2030. Available online: https://www.theedgemarkets.com/article/malaysia-sets-new-goal-18-clean-energy-generation-2030 (accessed on 14 April 2020).
66. Inus, K. Malaysia Will Meet 2050 target of 50% Renewable Energy. Available online: https://www.nst.com.my/news/nation/2018/01/328695/malaysia-will-meet-2050-target-50-renewable-energy (accessed on 14 April 2020).
67. Willems, B.; Ehlers, E. Cross-Subsidies in the Electricity Sector. *Compet. Regul. Netw. Ind.* **2008**, *9*, 201–227. [CrossRef]
68. Gao, A.M.Z.; Fan, C.T.; Liao, C.N. Application of German energy transition in Taiwan: A critical review of unique electricity liberalisation as a core strategy to achieve renewable energy growth. *Energy Policy* **2018**, *120*, 644–654. [CrossRef]
69. Nicolli, F.; Vona, F. Energy market liberalization and renewable energy policies in OECD countries. *Energy Policy* **2019**, *128*, 853–867. [CrossRef]
70. Alder, M. Renewable energy and liberalised electricity markets. *Renew. Energy* **2001**, *24*, 409–413. [CrossRef]
71. Abidin, N.Z.Z.; Jamaluddin, M.; Dawood, A.R. Harnessing the Power of the Sun—Malaysia's Maiden Journey in Large-Scale Solar PV. In *Proceedings of the JCNC—Conference 2016*; CIGRE: Paris, France, 2016; pp. 1–10.
72. Sustainable Energy Development Authority (SEDA) Malaysia. Net Energy Metering (NEM). Available online: http://www.seda.gov.my/reportal/nem/ (accessed on 14 September 2019).
73. Zainuddin, A. Latest Solar Farm Bids Lowest at 17.8 Sen. Available online: https://themalaysianreserve.com/2019/09/04/latest-solar-farm-bids-lowest-at-17-8-sen/ (accessed on 20 January 2020).

energies

Article

Influence of Two-Phase Crossflow for Void Prediction in Bundles Using Thermal-Hydraulic System Codes †

Yunseok Lee and Taewan Kim *

Department of Safety Engineering, Incheon National University, Incheon 22012, Korea; passyun2244@inu.ac.kr
* Correspondence: taewan.kim@inu.ac.kr

† This paper is an extended and revised article presented at the International Conference on Sustainable Energy and Green Technology 2019 (SEGT 2019) on 11–14 December 2019 in Bangkok, Thailand.

Received: 11 June 2020; Accepted: 14 July 2020; Published: 17 July 2020

Abstract: The previous study, where the void fraction predictability of three different thermal-hydraulic system codes was assessed against PSBT (PWR Subchannel and Bundle Test) benchmark data, indicated a general overprediction tendency of all system codes, especially in bundles. Because all codes have been utilized for best-estimate analyses, it is necessary to conduct further assessments in order to find the root cause of the overprediction. A further assessment has been performed using two thermal-hydraulic system codes, TRACE V5.0 patch 5 and MARS-KS 1.4, and the assessment has been carried out for both one- and multi-dimensional components. The results indicate that there is no significant difference in the predictability of the void fraction between one- and multi-dimensional components. In addition, it is found that the turbulent mixing model implemented for the multi-dimensional component of MARS-KS does not play an important role in the prediction of void distribution. Meanwhile, TRACE reveals a significant overprediction due to much less crossflow calculation compared to MARS-KS. By conducting an additional analysis with the modified one-dimensional models, it is clearly confirmed that crossflow significantly affects the void distribution. Therefore, it is concluded that the model for the thermal hydraulic mixing by crossflow in each system code should be improved in order to predict the void distribution in bundles appropriately.

Keywords: TRACE; MARS-KS; PSBT; void fraction; crossflow

1. Introduction

Two-phase flow is a prevailing condition, which applies to a wide range of industrial applications [1]. Especially in the nuclear field, the most relevant application is the boiling water reactor, in which a phase change of flowing liquid directly occurs within the reactor core. In addition, in a pressurized water reactor, the relevance exists to heat exchangers, such as steam generators [2]. However, it is well known that the characteristics of two-phase flow depend on the flow regime, which differs depending on flow conditions and geometry [3–5]. Conventionally, the void fraction is employed to distinguish the transition between the specified regimes. As depicted in Figure 1, it is applied as one of the criteria for flow regime prediction, and this form of flow map is generally utilized in the best-estimate system analysis codes in the nuclear field. Therefore, the precise prediction of the void fraction has great importance in the best-estimate safety analysis methodology since additional conservatism could be induced by the inaccurate prediction of the void fraction, which plays a negative role from a coolability point of view.

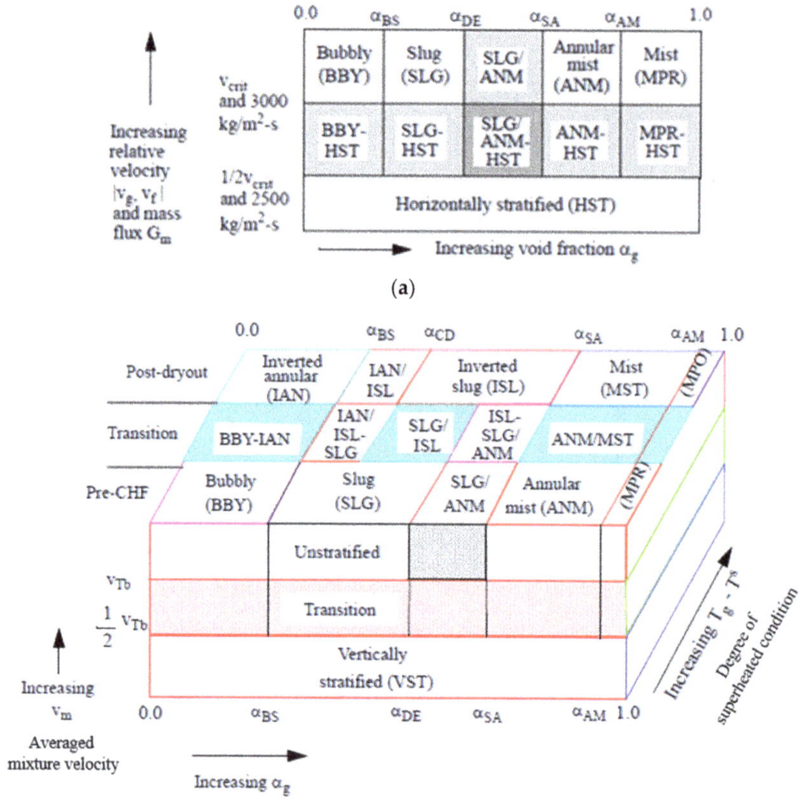

Figure 1. Schematic of flow regime map in MARS-KS [6]: (**a**) horizontal flow regime; (**b**) vertical flow regime.

In the previous study [7], the void fraction predictability of three different best-estimate system analysis codes was assessed against the experimental data from the NUPEC (Nuclear Power Engineering Corporation) test facility used for the OECD/NRC PSBT benchmark [8]. The assessment was performed by comparing the result from one-dimensional components in each system code. The results showed that all codes generally predicted the void fraction greater than the experiment. Especially in the case of TRACE V5.0 Patch 5 [9], a significant overprediction tendency in bundles was revealed, especially compared to MARS-KS 1.4 [10] and RELAP5/MOD3.3 Patch 5 [11]. Both MARS-KS and RELAP5 showed almost identical predictions, but they also showed an overprediction tendency at low void conditions. The general overprediction tendency of the system codes was also illustrated from the benchmark results [12]. Because all codes were utilized for the best-estimate safety analysis for nuclear reactor systems, it is necessary to figure out the root cause of the overprediction and to improve the predictability of the void distribution.

Meanwhile, as the system codes were improved with the multi-dimensional capacity for more realistic analyses of complex components, such as a reactor pressure vessel, it was decided that the follow-up study should be expanded to include an assessment with multi-dimensional components in the codes. Thus, it is of interest in this study to figure out the characteristics of one- and multi-dimensional components in the prediction of void distribution and to find out the root cause of the systematic overprediction in void fraction by the system codes.

In this study, a series of assessments on void distribution predictability of the system codes was performed using one- and multi-dimensional components in order to figure out the root cause of the overprediction in void fractions by the system codes. Because RELAP5 does not have the multi-dimensional component, MARS-KS and TRACE have been employed for the assessment. The results from both codes are analyzed both physically and statistically.

2. Materials and Methods

2.1. *Review of NUPEC Test Facility*

Figure 2 depicts the NUPEC test facility, where the bundle void measurement tests were performed. The test loop was a high-pressure and high-temperature recirculation loop, of which the design pressure and temperature were 19.2 MPa and 635.15 K, respectively, which covered PWR (Pressurized Water Reactor) operating conditions. Once the coolant flowed into the heated section of the bundle, which had an active heated length of 3.658 m, the coolant was electrically heated from the heater rods, and the phase change occurred. At three different locations in the heated section, which consisted of a lower (2.216 m), middle (2.669 m), and upper point (3.177 m), respectively, the void fraction was measured within four central subchannels of the bundle. The steady-state bundle test consisted of three different test series, which were distinguished by test assembly ID, namely B5, B6, and B7, respectively. As listed in Table 1, the test series B5 and B6 utilized an identical 5 × 5 assembly. However, in the case of B7, the assembly contained a thimble rod, which described the guide tube at the center of the test section. The uniform axial power shape was applied to B5, but B6 and B7 used a cosine axial power profile. Each assembly was equipped with three different types of spacer grids, which consisted of a simple spacer grid, a spacer grid with no mixing vane, and with mixing vane, respectively. A more detailed description of the specification of the test facility is available in Reference [8].

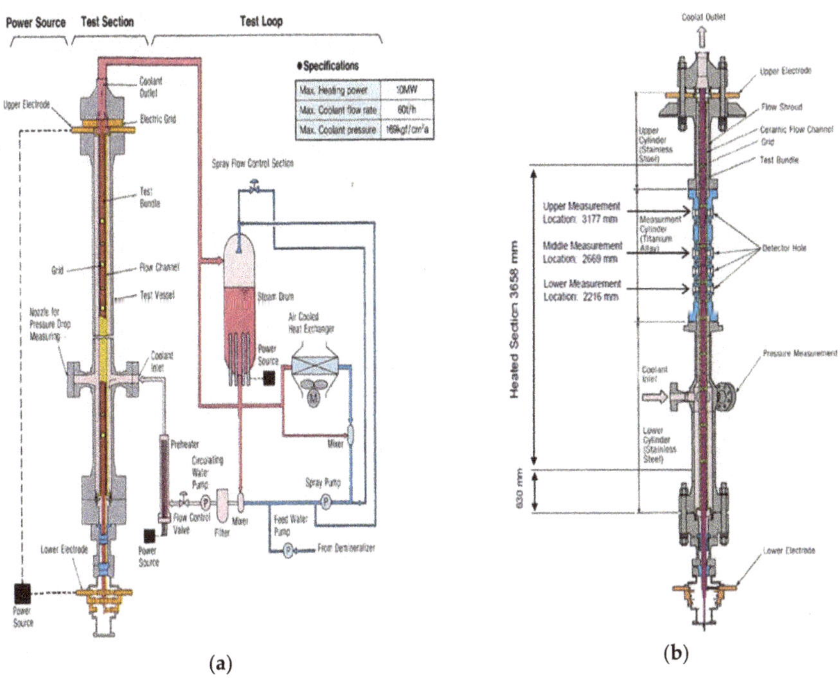

Figure 2. NUPEC test facility [8]: (**a**) whole scheme of the test recirculation loop; (**b**) scheme of the test section where bundle void measurement tests were performed.

Table 1. Specification of the bundle test section [8].

Item	B5	B6	B7
Cross-sectional scheme	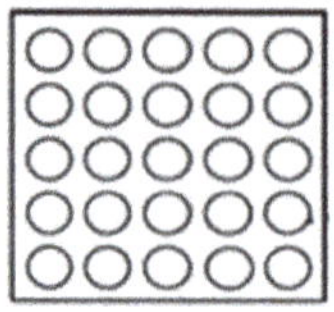	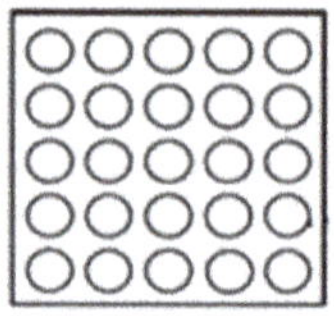	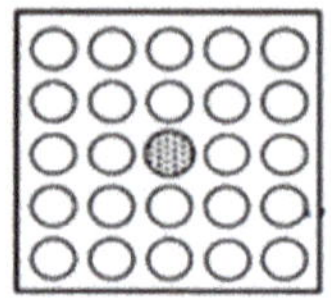
Number of heater rods	25	25	24
Number of thimble rods	–	–	1
Radial power distribution	0.85 0.85 0.85 0.85 0.85 0.85 1.00 1.00 1.00 0.85 0.85 1.00 1.00 1.00 0.85 0.85 1.00 1.00 1.00 0.85 0.85 0.85 0.85 0.85 0.85	0.85 0.85 0.85 0.85 0.85 0.85 1.00 1.00 1.00 0.85 0.85 1.00 1.00 1.00 0.85 0.85 1.00 1.00 1.00 0.85 0.85 0.85 0.85 0.85 0.85	0.85 0.85 0.85 0.85 0.85 0.85 1.00 1.00 1.00 0.85 0.85 1.00 0.00 1.00 0.85 0.85 1.00 1.00 1.00 0.85 0.85 0.85 0.85 0.85 0.85
Axial power distribution	Uniform	Cosine	Cosine
Number of simple spacers	8	8	8
Number of spacers w/o mixing vane	2	2	2
Number of spacers w/ mixing vane	7	7	7

2.2. Multi-Dimensional Modeling

2.2.1. Common Features for Consistent Modeling

The test section of interest was modeled by 36 subchannels as depicted in Figure 3. In total, 864 ($6 \times 6 \times 24$) hydraulic volumes were modeled for the heated section by using the multi-dimensional component of each system code. As with the one-dimensional model, instead of modeling the actual shape of the spacer grid, additional pressure loss was modeled for each specific location by giving the pressure loss coefficient provided by the specification [8]. In addition, the power condition in each subchannel was implemented by the heat structure components, which gave an equivalent heat condition according to the equivalent power fraction within the subchannel. With the connection to additional hydraulic volumes upstream and downstream, which were modeled by the multi-dimensional component, the test section of interest was connected to the common boundary conditions. Whole nodalization for each system code is depicted in Figure 4. The detailed description of the model of each system code is given in the following sections.

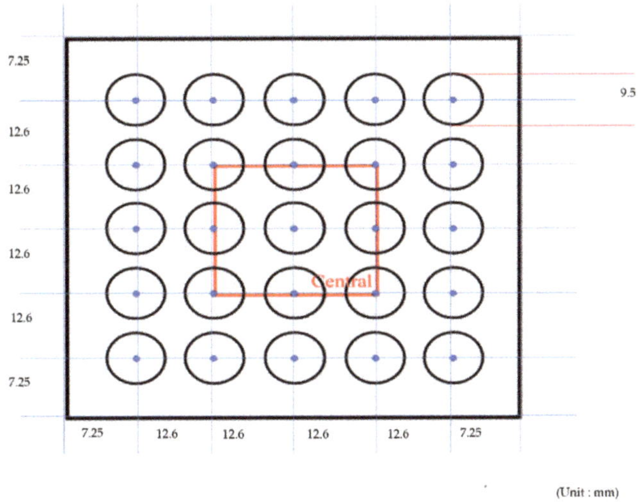

Figure 3. Cross-sectional view of nominal bundle test section.

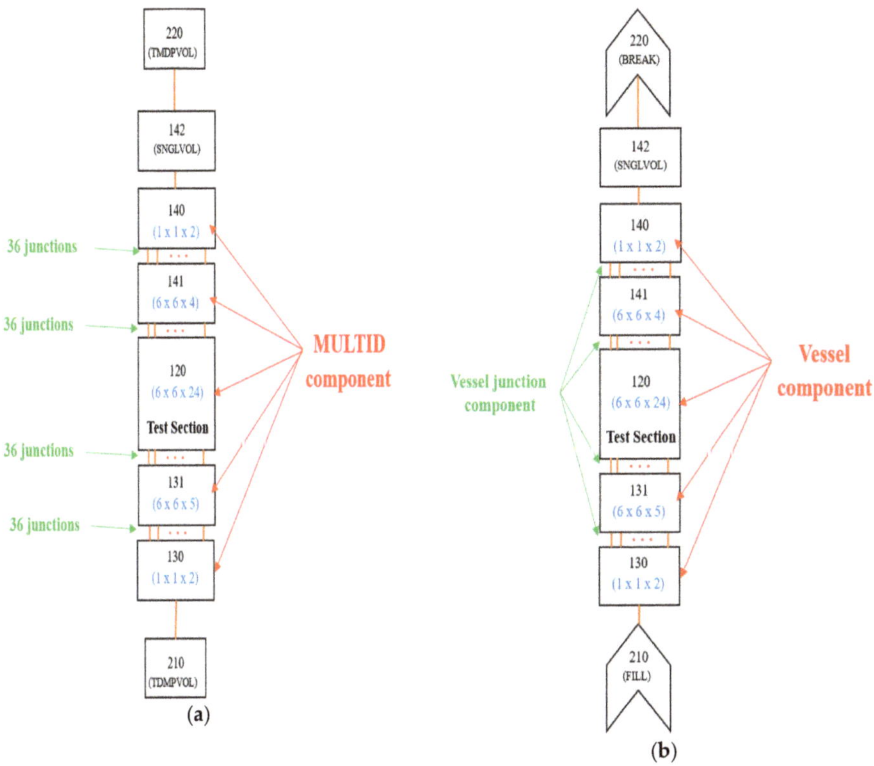

Figure 4. Nodalization scheme of multi-dimensional model of each system code: (**a**) multi-dimensional model of MARS-KS; (**b**) multi-dimensional model of TRACE.

2.2.2. Multi-Dimensional Model of MARS-KS

For modeling the heated section, the multi-dimensional component, namely MULTID, was employed. The connections to the heated sections upstream and downstream were modeled by MULTIPLE JUNCTION components. The system pressure and inlet flow boundary conditions were modeled by TIME DEPENDENT VOLUME and TIME DEPENDENT JUNCTION components, respectively. The MULTID has the capability to model the turbulent stress and energy mixing between channels, and these are proportional to the square of a user-defined Prandtl *mixing length*. The required *mixing length* is given by Equation (1), which is recommended by the user guide [13] as the reference correlation for the channel *mixing length*:

$$Mixing\ length = L\left[0.14 - 0.08(1 - \frac{y}{L})^2 - 0.06(1 - \frac{y}{L})^4\right] \tag{1}$$

where L is channel half-width, and y is given by the range from 0 to L. In this study, the maximum value of *mixing length* was utilized in the reference model. In addition, a sensitivity study without the *mixing length* term was conducted, and the results will be shown in the following section.

2.2.3. Multi-Dimensional Model of TRACE

In the case of TRACE, the VESSEL component was employed to model the heated section. As with the multi-dimensional model of MARS-KS, the additional hydraulic volumes, which were modeled by VESSEL component as well, were connected to each upstream and downstream segment of the heated section, respectively. Inter-connections between the heated section and the additional volumes were given by VESSEL JUNCTION components. The system pressure and inlet flow boundary conditions were given by BREAK and FILL components, respectively. Since TRACE has no additional turbulent model as MARS-KS, only diversion crossflows were implemented between channels.

3. Results

3.1. Model Comparison of Each System Code

In total, 219 steady-state bundle test cases were employed for the assessment. Figure 5 shows the comparison of results from the one- and multi-dimensional models of each system code by plotting the calculated void fraction with respect to the measured value. The results indicated that each system code showed no significant difference in the void prediction from the one-and multi-dimensional models. In fact, the multi-dimensional models generally showed a slightly higher void fraction than the one-dimensional models, but the difference was negligible. Even though the multi-dimensional model of MARS-KS simulated the additional turbulent mixing, no significant difference from the one-dimensional model was captured. Therefore, further assessment was made to evaluate the sensitivity of the *mixing length*. The results with the turbulent *mixing length* model were compared with the additional calculations without the model. Figure 6, where the comparison is depicted, revealed that no significant difference was observed in the void predictions from the two calculations. Furthermore, this was clearly supported by the statistical results by one-sample t-test, as listed in Table 2. For the statistical comparison, absolute error against the experimental data was calculated. In total, 657 error samples for each model were applied to the statistical test in order to assess the prediction quantitatively with respect to the allowed measurement error range, 2σ. In turn, the results were compared with each other. A hypothesis test was performed and divided into two categories based on a void fraction of 30%. This division was made due to the difference in the prediction tendency with respect to the condition, beyond which the general prediction tendency of the code showed underprediction, whereas an overprediction tendency was captured at low void regions less than 30%. The results indicated that there was no significant difference in the prediction capability, which was supported by the almost identical statistical values between the models. This means that the turbulent mixing model in MARS-KS does not play an important role in predicting the void fraction in bundles. This is because the turbulent mixing model of MARS-KS applied the same equal mass method [14] used for single-phase flow. Thus, the code cannot model direct mass and energy exchange between channels under two-phase conditions. Therefore, the model is not appropriate to give effective influence to the void prediction.

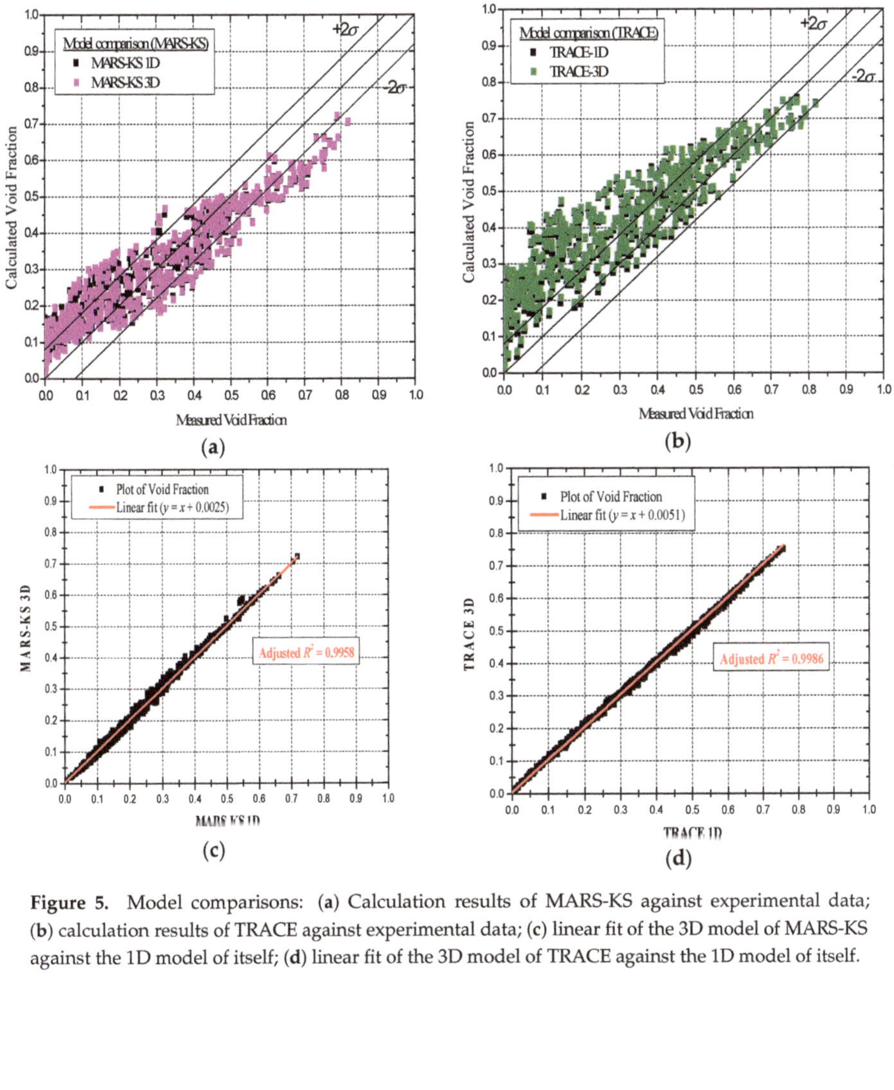

(a) (b) (c) (d)

Figure 5. Model comparisons: (**a**) Calculation results of MARS-KS against experimental data; (**b**) calculation results of TRACE against experimental data; (**c**) linear fit of the 3D model of MARS-KS against the 1D model of itself; (**d**) linear fit of the 3D model of TRACE against the 1D model of itself.

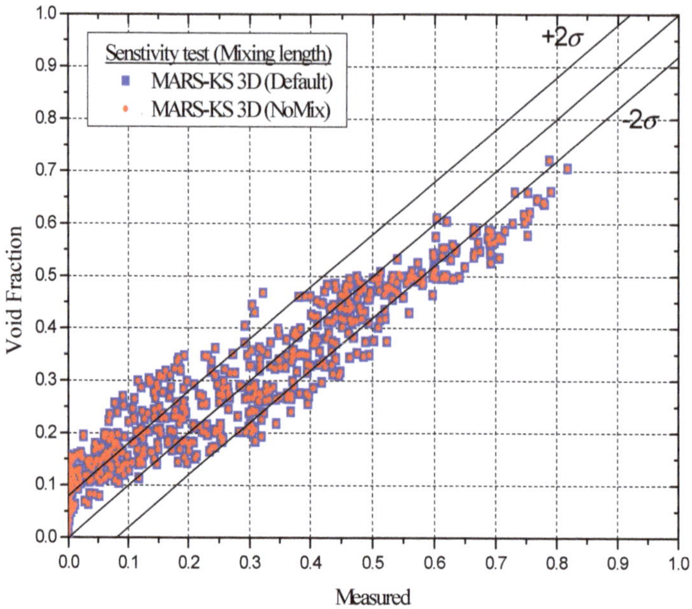

Figure 6. Model comparison of MARS-KS for sensitivity check to *mixing length*.

Table 2. Results of one-sample *t*-test for assessing model predictions of MARS-KS.

Void Fraction (Number of Samples)	Value	MARS-KS 3D (Default)	MARS-KS 3D (No Mixing)
Void .LT. 30 (382 samples)	Mean	0.0773	0.0773
	Standard deviation	0.0418	0.0418
	Hypothesis test	Is the test mean significantly greater than 2σ (=0.08)?	
	Results	No	No
	t-Value	−1.245	−1.255
	Probability > *t*	0.893	0.895
Void .GT. 30 (275 samples)	Mean	0.0662	0.0661
	Standard deviation	0.0433	0.0434
	Hypothesis test	Is the test mean significantly greater than 2σ (=0.08)?	
	Results	No	No
	t-Value	−5.281	−5.302
	Probability > *t*	1.000	1.000

3.2. Code-to-Code Comparison

Figure 7 depicts the code-to-code comparison of one- and multi-dimensional results by both codes, respectively. The results clearly showed that TRACE predicted high void fractions compared to MARS-KS in both one- and multi-dimensional cases. As listed in Table 3, the results of the one-sample t-test based on absolute error samples of each code against the measured data also supported the significant difference in the void prediction between the codes. The negative t-values of MARS-KS for all categories proved that its prediction did not exceed the allowed measurement error, 2σ, generally. However, as TRACE showed positive *t*-values significantly, this supported that its prediction generally exceeded the allowable range. In other words, this meant that TRACE showed poor prediction capability compared to MARS-KS. It was found that such a significant overprediction tendency of TRACE came from the difference in crossflow calculation. As depicted in Figure 8, it was clearly confirmed that TRACE generally calculated much less vapor crossflow compared to MARS-KS for all test cases. Moreover, in the multi-dimensional cases, TRACE calculated the net vapor inflow to the measuring sections, whereas MARS-KS generally calculated significant vapor outflow from the measuring sections. As the accompanying liquid crossflow showed no significant difference, the difference in the void fraction prediction was totally attributable to the vapor crossflow. From these, it was clearly identified that TRACE had the tendency to keep more vapor in the central channel with much less vapor crossflow. Thus, the higher void fraction in the measuring section was the result.

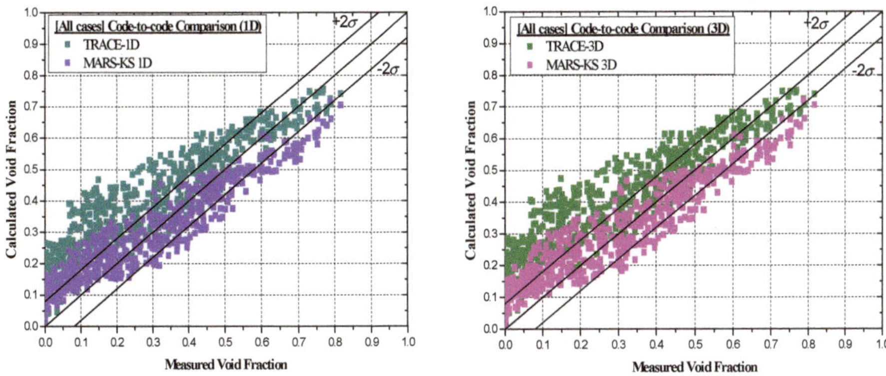

(**a**) (**b**)

Figure 7. Code comparisons: (**a**) 1D void predictions; (**b**) 3D void predictions.

Table 3. Results of one-sample t-test for assessing code predictions.

Void (Samples)	Value	TRACE 1D	TRACE 3D	MARS-KS 1D	MARS-KS 3D
All (657)	Mean	0.1114	0.1158	0.0726	0.0727
	Standard deviation	0.0762	0.0779	0.0415	0.0428
	Hypothesis test	Is the test mean significantly greater than 2σ (=0.08)?			
	Results	Yes	Yes	No	No
	t Value	10.571	11.792	−4.589	−4.391
	Probability > t	1.586×10^{-24}	1.522×10^{-29}	1.000	1.000
LT.30 (382)	Mean	0.1451	0.1525	0.0759	0.0773
	Standard deviation	0.0743	0.0738	0.0397	0.0418
	Hypothesis test	Is the test mean significantly greater than 2σ (=0.08)?			
	Results	Yes	Yes	No	No
	t Value	17.136	19.200	−2.018	−1.245
	Probability > t	1.646×10^{-49}	2.945×10^{-58}	0.978	0.893
GT.30 (275)	Mean	0.0646	0.0648	0.0679	0.0662
	Standard deviation	0.0493	0.0495	0.0436	0.0433
	Hypothesis test	Is the test mean significantly greater than 2σ (= 0.08)?			
	Results	No	No	No	No
	t Value	−5.188	−5.080	−4.590	−5.281
	Probability > t	1.000	1.000	1.000	1.000

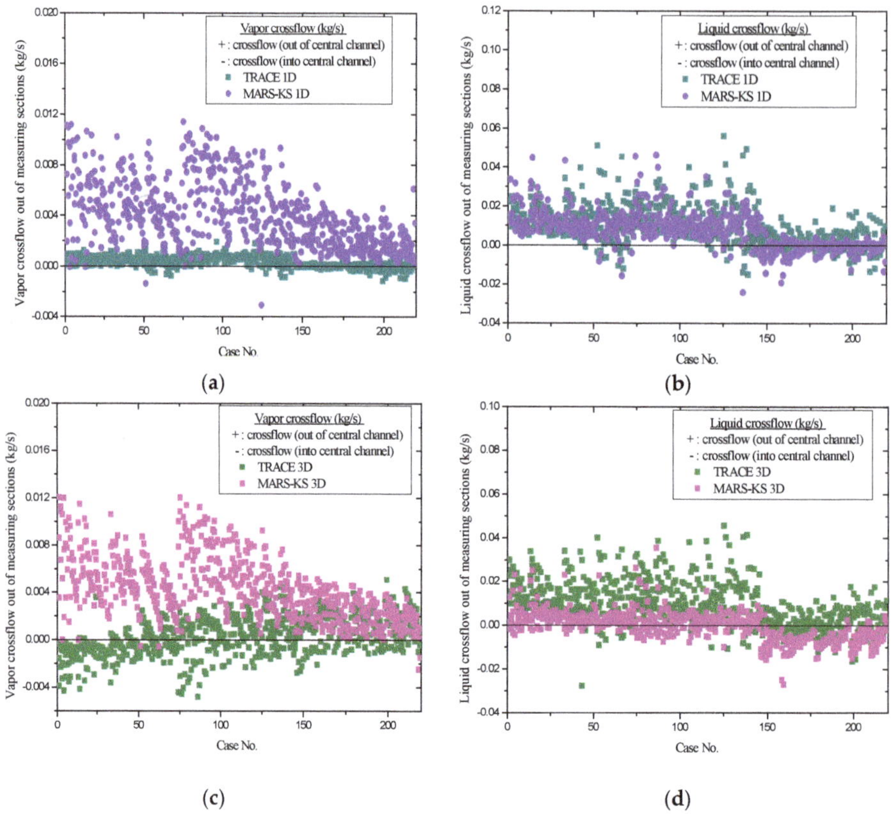

Figure 8. Code comparisons: (**a**) 1D vapor crossflow; (**b**) 1D liquid crossflow; (**c**) 3D vapor crossflow; (**d**) 3D liquid crossflow.

3.3. *Influence of Crossflow-to-Code Predictions*

As it was found that there was a significant difference in crossflow calculations between both codes, and the difference made TRACE predict higher void fractions compared to MARS-KS, crossflow was considered as the root cause of the overprediction tendency. Therefore, an additional evaluation of the sensitivity against crossflow was conducted in order to clearly identify the influence of crossflow on each code prediction. For this, a simple modification of the one-dimensional model of each code, which was convenient to modify, was made as depicted in Figure 9. The modification was made by deleting crossflow junction connections and giving averaged mass flow conditions to the central measuring channel and peripheral channel, respectively. Except for test series B7, of which the averaged central mass flow fraction was 13%, the nominal mass flow condition for the central channel was given as 15% of total mass flow. The calculated result of each modified model was compared with the previous one-dimensional results of each system code, as depicted in Figure 10. Especially in the case of MARS-KS, the influence of crossflow was clearly captured, as the results showed that the modified model predicted higher void fractions compared to the default model when the crossflow was disabled. Moreover, the comparison of the modified model showed higher void fractions from MARS-KS compared to TRACE. In addition, in the case of TRACE, there was no drastic change by the crossflow because of negligible crossflow even with the default model. From these, it was clearly concluded that crossflow was the root cause of the overpredictions. Thus, further improvement to the crossflow model is necessary in order to reduce the significant overpredictions of TRACE and to improve the predictions of MARS-KS as well. For the improvement, additional constitutive models should be employed to improve the prediction of two-phase crossflow. This could be done by employing a turbulent mixing model based on the Equal Volume and Void Drift (EVVD) method, which is adopted by state-of-the-art subchannel analysis codes, in order to enhance two-phase crossflow [15]. The EVVD method enables direct mass and energy mixing between channels, by which void drift from a higher void channel to a lower void channel is induced by not only inter-channel void difference but also net liquid flow from low void to high void. Therefore, it is expected that this model will enable the enhancement of two-phase crossflow for both codes by modeling additional net mass and energy exchanges directly in the field equation of each code.

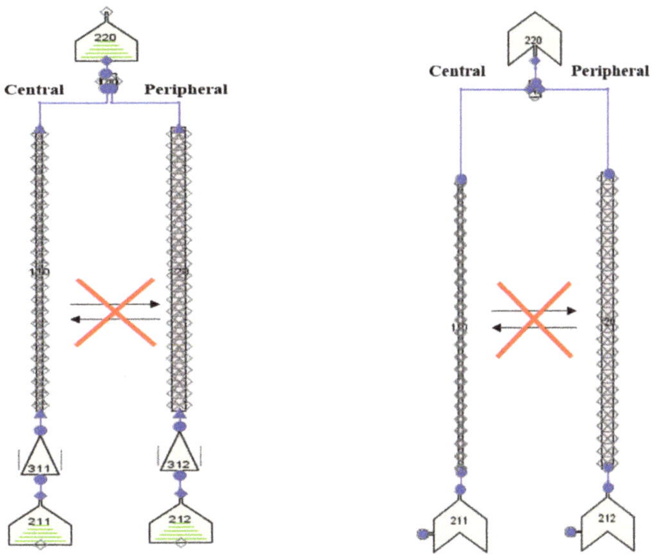

(**a**) (**b**)

Figure 9. Modified one-dimensional model for crossflow influence test: (**a**) nodalization of MARS-KS; (**b**) nodalization of TRACE.

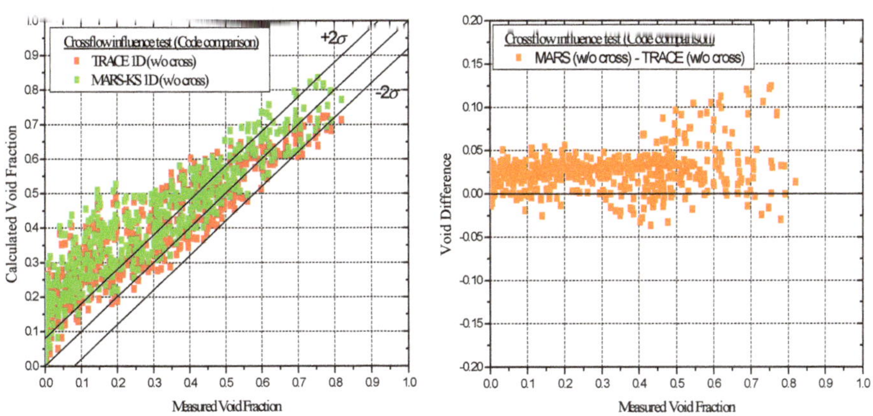

(**a**) (**b**)

Figure 10. *Cont.*

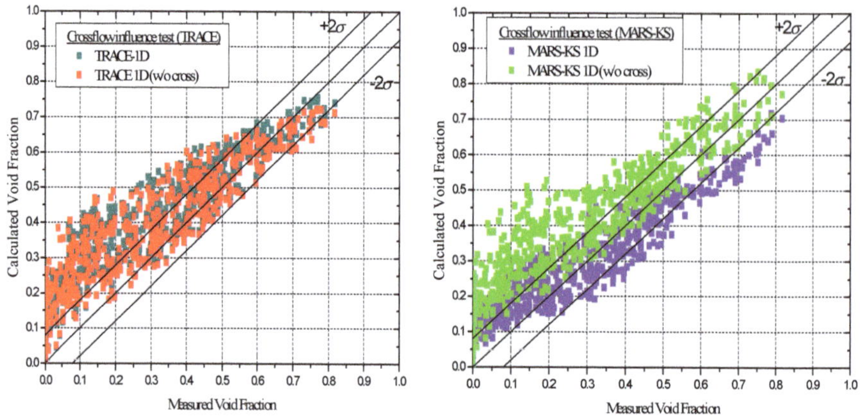

(c) (d)

Figure 10. Comparisons for checking influence of crossflow-to-code predictions: (**a**) code comparison of modified models, in which the crossflows are disabled; (**b**) void difference between modified models of MARS-KS and TRACE; (**c**) model comparison of TRACE; (**d**) model comparison of MARS-KS.

4. Conclusions

As a follow-up study, a further assessment of void fraction predictability of the system codes, MARS-KS 1.4 and TRACE V5.0, was performed with one- and multi-dimensional models, mainly to find the root cause of overprediction tendencies in bundle cases identified by the previous study. In total, 219 steady-state bundle test cases from the OECD/NRC PSBT benchmark were utilized for the assessment. From the model comparison of each system code, it was found that each code showed no significant difference in void prediction between one- and multi-dimensional models. In the case of MARS-KS, because the multi-dimensional model cannot implement direct mass and energy exchange due to turbulence, no clear difference emerged when compared to the one-dimensional model of itself. Therefore, it was concluded that the turbulent mixing model of MARS-KS is not appropriate to cover two-phase mixing flow within the bundle. Meanwhile, as TRACE showed significant void predictions due to fewer crossflow calculations compared to MARS-KS, crossflow was considered as the root cause of the overprediction tendency. From the additional assessment with the modified one-dimensional models, it was clearly confirmed that crossflow significantly affects the code predictions, and thus crossflow is the root cause of the overpredictions in the bundle cases. From these, it was concluded that further improvement of the crossflow model is necessary in order to predict the void fraction more realistically. As state-of-the-art subchannel codes adopt the Equal Volume and Void Drift (EVVD) method as the turbulent mixing model to improve the inter-channel crossflow, this study concludes that an improvement should be made with the application of the EVVD method to model the direct net mass and energy interchanges between channels under two-phase flow conditions. Therefore, as a future work, modification of the crossflow model will be made by applying the EVVD method to both one- and multi-dimensional solutions of each system code, respectively.

Author Contributions: Conceptualization, funding acquisition, project administration, resources, and supervision, T.K.; methodology and investigation, Y.L. and T.K.; formal analysis, validation, visualization, writing—original draft preparation, and writing—review and editing, Y.L. All authors have read and agreed to the published version of the manuscript.

Funding: This work was supported by Incheon National University (International Cooperative) Research Grant in 2018.

Conflicts of Interest: The authors declare no conflicts of interest.

References

1. Kalaga, D.V.; Yadav, A.; Goswami, S.; Bhusare, V.; Pant, H.J.; Dalvi, S.V.; Joshi, J.B.; Roy, S. Comparative analysis of liquid hydrodynamics in a co-current flow-through bubble column with densely packed internals via radiotracing and Radioactive Particle Tracking (RPT). *Chem. Eng. Sci.* **2017**, *170*, 332–346. [CrossRef]
2. Breeze, P. Nuclear Power. In Nuclear Power Generation Technologies. 2019; pp. 399–429. Available online: https://www.sciencedirect.com/science/article/pii/B9780081026311000171?via%3Dihub (accessed on 16 July 2020). [CrossRef]
3. Wu, B.; Firouzi, M.; Mitchell, T.; Rufford, T.E.; Leonardi, C.; Towler, B. A critical review of flow maps for gas-liquid flows in vertical pipes and annuli. *Chem. Eng. J.* **2017**, *326*, 350–377. [CrossRef]
4. Schlegel, J.P.; Sawant, P.; Paranjape, S.; Ozar, B.; Hibiki, T.; Ishii, M. Void fraction and flow regime in adiabatic upward two-phase flow in large diameter vertical pipes. *Nucl. Eng. Des.* **2009**, *239*, 2864–2874. [CrossRef]
5. Liu, H.; Hibiki, T. Flow regime transition criteria for upward two-phase flow in vertical rod bundles. *Int. J. Heat Mass Transf.* **2017**, *108*, 423–433. [CrossRef]
6. Korea Atomic Energy Research Institute. *MARS Code Manual*; Volume V: Models and Correlations; KAERI/TR-3872/2009; Korea Atomic Energy Research Institute: Daejeon, Korea, 2009.
7. Lee, Y.S.; Kim, T.W. Assessment of void fraction predictability of system codes in subchannels. *Kerntechnik* **2018**, *83*, 414–425. [CrossRef]
8. Rubin, A.; Schoedel, A.; Avramova, M.; Utsuno, H.; Bajorek, S.; Velazquez-Lozada, A. OECD/NRC Benchmark Based on NUPEC PWR Subchannel and Bundle Tests (PSBT); Volume 1: Experimental Database and Final Problem Specifications. OECD/NEA Nuclear Science Committee. 2012. Available online: http://www.oecd-nea.org/globalsearch/download.php?doc=77653 (accessed on 16 July 2020).
9. United States Nuclear Regulatory Commission. *TRACE V5.0 Theory Manual: Field Equations, Solution Methods, and Physical Models*; ML071000097; United States Nuclear Regulatory Commission: Washington, DC, USA, 2012.
10. Korea Atomic Energy Research Institute. *MARS Code Manual*; Volume I: Code Structure, System Models, and Solution Methods; KAERI/TR-2812/2004; Korea Atomic Energy Research Institute: Daejeon, Korea, 2009.
11. United States Nuclear Regulatory Commission. *RELAP5/MOD3.3 Code Manual*; Volume I: Field Equations, Solution Methods, and Physical Models; NUREG/CR-5535/Rev P5; United States Nuclear Regulatory Commission: Washington, DC, USA, 2016.
12. Nuclear Energy Agency. International Benchmark on Pressurised Water Reactor Sub-Channel and Bundle Tests; Volume II: Benchmark Results of Phase I–Void Distribution. NEA/NSC/R(2015)4. 2016. Available online: https://www.oecd-nea.org/science/docs/2015/nsc-r2015-4.pdf (accessed on 16 July 2020).
13. Korea Institute of Nuclear Safety. *MARS-KS Code Manual*; Volume II: Input Requirements; KINS/RR-1282 Rev.1; Korea Institute of Nuclear Safety: Daejeon, Korea, 2016.
14. Todreas, N.E.; Kazimi, M.S. Subchannel Analysis. In Nuclear Systems II: Elements of Thermal Hydraulic Design. Massachusetts Institute of Technology, 1990; pp. 246–251. Available online: http://www.gammaexplorer.com/wp-content/uploads/2014/03/Nuclear-Systems-II-Elements-of-Thermal-Hydraulic-Design-Todreas.pdf (accessed on 16 July 2020).
15. Mao, H.; Yang, B.W.; Wang, S. A Review of Void Drift Models in Subchannel Analysis. *Nucl. Sci. Eng.* **2018**, *193*, 33–45. [CrossRef]

energies

MDPI

Article

An Estimation of Hydraulic Power Take-off Unit Parameters for Wave Energy Converter Device Using Non-Evolutionary NLPQL and Evolutionary GA Approaches †

Mohd Afifi Jusoh, Mohd Zamri Ibrahim *, Muhamad Zalani Daud, Zulkifli Mohd Yusop and Aliashim Albani

Renewable Energy & Power Research Interest Group (REPRIG), Eastern Corridor Renewable Energy Special Interest Group, Faculty of Ocean Engineering Technology and Informatics, Universiti Malaysia Terengganu, Kuala Nerus 21030, Terengganu, Malaysia; mohd.afifi.jusoh@gmail.com (M.A.J.); zalani@umt.edu.my (M.Z.D.); zul_12521@yahoo.com (Z.M.Y.); a.albani@umt.edu.my (A.A.)

* Correspondence: zam@umt.edu.my; Tel.: +60-96683328

† This paper is an extended and revised article presented at the International Conference on Sustainable Energy and Green Technology 2019 (SEGT 2019) on 11–14 December 2019 in Bangkok, Thailand.

Abstract: This study is concerned with the application of two major kinds of optimisation algorithms on the hydraulic power take-off (HPTO) model for the wave energy converters (WECs). In general, the HPTO unit's performance depends on the configuration of its parameters such as hydraulic cylinder size, hydraulic accumulator capacity and pre-charge pressure and hydraulic motor displacement. Conventionally, the optimal parameters of the HPTO unit need to be manually estimated by repeating setting the parameters' values during the simulation process. However, such an estimation method can easily be exposed to human error and would subsequently result in an inaccurate selection of HPTO parameters for WECs. Therefore, an effective approach of using the non-evolutionary Non-Linear Programming by Quadratic Lagrangian (NLPQL) and evolutionary Genetic Algorithm (GA) algorithms for determining the optimal HPTO parameters was explored in the present study. A simulation–optimisation of the HPTO model was performed in the MATLAB/Simulink environment. A complete WECs model was built using Simscape Fluids toolbox in MATLAB/Simulink. The actual specifications of hydraulic components from the manufacturer were used during the simulation study. The simulation results showed that the performance of optimal HPTO units optimised by NLPQL and GA approaches have significantly improved up to 96% and 97%, respectively, in regular wave conditions. The results also showed that both optimal HPTO units were capable of generating electricity up to 62% and 77%, respectively, of their rated capacity in irregular wave circumstances.

Keywords: wave energy converter; hydraulic power take-off unit; parameter estimation; genetic algorithm; non-linear programming by quadratic Lagrangian

Citation: Jusoh, M.A.; Ibrahim, M.Z.; Daud, M.Z.; Yusop, Z.M.; Albani, A. An Estimation of Hydraulic Power Take-off Unit Parameters for Wave Energy Converter Device Using Non-Evolutionary NLPQL and Evolutionary GA Approaches †. *Energies* **2021**, *14*, 79. https://dx.doi.org/10.3390/en14010079

Received: 2 December 2020
Accepted: 20 December 2020
Published: 25 December 2020

1. Introduction

Ocean waves are one of the renewable energy resources that potentially can be exploited to produce usable electricity due to their excellent features of predictability, high energy density and high source availability [1,2]. Currently, numerous wave energy converters (WECs) have been designed, developed, tested and patented through a variety of harnessing techniques that are subjected to the characteristics of the target location such as shoreline, nearshore and offshore, as reported in [3–6]. In general, WECs are a combination of three main parts, such as a wave energy converter (WEC) device, power take-off (PTO) unit and control system unit. Recently, various types of PTO units have been developed for WEC devices based on different working principles, such as the air and water turbine-based, direct-electrical drive-based, direct-mechanical drive-based and hydraulic-based, as reported in [7,8]. A hydraulic PTO (HPTO) is considered to be the most effective PTO for wave-activated-body (WAB) or point-absorber based WECs due to the outstanding features, including high-efficiency, high-controllability, well-adapted to the large power density ocean waves and low-frequency [9]. It has been reported in the literature that this type of PTO system's efficiency could be achieved up to 90% [10]. Furthermore, the HPTO unit is also easily constructed using standard hydraulic components, which are commonly used in other applications. Due to such a promising characteristic, the HPTO system finds its application in the majority of the WAB-WECs field.

Recently, many HPTO unit applications in various WECs have been published [11–17]. From the preliminary survey, most of the studies concentrated on the HPTO unit's efficiency without taking into account the optimal parameters of the HPTO model, such as hydraulic cylinder size, hydraulic accumulator capacity and pre-charge pressure and hydraulic motor displacement. The optimal configuration parameters of the HPTO unit is a crucial issue as it can affect the system's efficiency and the amount of output to be produced [13]. Only a few reports, for example, in [13–15], have considered this critical problem. However, the optimal parameters setting has been obtained by manually tuning these configuration parameters [13]. This method is usually prone to human error and easy to cause a non-optimal selection of the HPTO system's parameters. In addition, this approach also frequently involves a long-time process in order to obtain the optimal configuration parameters.

Presently, the optimisation of design parameters using a mathematical algorithm is an attractive method to estimate the accurate parameters during the design phase. It is due to the advancement of fast computing technologies that can be reliably used. A variety of studies were performed using different types of mathematical algorithms, such as non-linear programming by quadratic Lagrangian (NLPQL) and genetic algorithm (GA), Particle Swarm Optimization (PSO), Ant Colony Optimization (ACO), Tabu Search (TS), et cetera, in order to obtain the best parameters for the design model [18–23]. For example, GA has been used to optimise the parameters of state-of-charge (SOC) controllers for battery energy storage in photovoltaic device applications [21]. The authors emphasized that the GA-based optimisation method has accelerated the optimisation process of the considered design parameters and effectively improves the design model's performance. Similarly, in [22], different types of heuristic optimisation approaches were applied, including Gravitational Search Algorithm (GSA) and PSO, to conduct model optimization-based studies for improving the efficiency of developed power converter units. The authors had concluded that GSA-based optimisation provides the highest convergence speed and best fitness value compared to the other algorithms. Motivated by the studies presented in [21,22], an optimized new design of WEC with an HPTO unit is presented in this study.

From the WECs point of view, a similar optimisation approach has been implemented in optimising the performance of the WECs. From the preliminary survey, several studies of the GA applications for WECs optimisation have been done [24–28]. For example, in [24], GA has been used to obtain some WECs parameters, such as buoy radius, draft, generator damping and the optimal spatial layout of a WECs park. In [25], GA has been adopted to optimally design the shape and dimensions of a WEC and also the PTO and other subsystems parameters. The techno-economic aspects of energy productivity and WECs device cost have also been considered in the study. In [28], GA has been used to obtain the optimal HPTO parameters of WECs, such as hydraulic cylinder size, hydraulic accumulator capacity and pre-charge pressure, and hydraulic motor displacement without considering the hydrodynamic effects of the floater. Since the hydrodynamic effects of the floater are the vital factors in WECs design, the optimal HPTO obtained in the study is inapplicable for real wave application. Therefore, the HPTO optimisation with the consideration of the floater's hydrodynamic effects using two different types of algorithm, such as NLPQL and GA, were investigated in this present study. The optimisation approaches presented in this study can be a useful reference to other researchers and engineers of WECs in order to design an accurate and reliable HPTO unit for the future WECs application.

The paper is organised as follows. The technical descriptions of the considered WEC with HPTO unit and its important configuration parameters are given in Section 2.

The simulation studies of the HPTO unit, which includes the simulation set-up process, optimisation process and evaluation of HPTO unit performance, are described in Section 3. Results and discussion are provided in Section 4, and finally, the Conclusions are given in Section 5.

2. Mathematical Modelling of WEC with HPTO Unit

The design of the WECs depends on the characteristics of the installed location. In the present study, the rotation-based WEC attached to the fixed body concept was considered, as illustrated in Figure 1. This WECs concept has been implemented in numerous studies for various investigation goals, for example, in [29–32]. This concept is suitable to be installed at shoreline, nearshore and offshore locations. In this concept, the WEC device consists of a single or multiple floating buoy or floater attached to the rotatable arm and connected to the fixed body directed towards the dominant wave direction, as depicted in Figure 1. The multi-design of floater can be used, which is dependent on the direction of the ocean wave, either single or multi-direction. Usually, the semisphere-shaped and boat-shaped floater have been considered for the offshore and nearshore location, as investigated in [29,30,33–35]. In this concept, the HPTO unit is utilised to convert the absorbed energy by the WEC device from the ocean wave to become usable electricity. A hydraulic actuator module of the HPTO unit is attached to the rotatable floater arm in order to absorb the mechanical energy produced by the WEC device, as presented in Figure 1. Meanwhile, the rest of the HPTO unit components are placed in the PTO house located on the top of the fixed-structure. In the present study, the model of WECs with a capacity of 0.1 kW was considered.

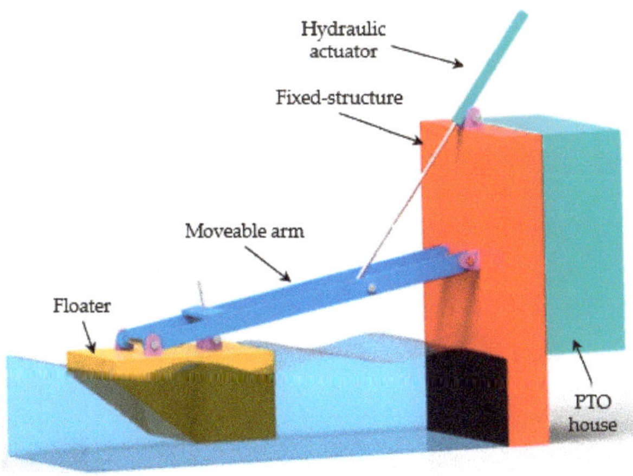

Figure 1. Conceptual design of future wave energy converter (WEC) with the hydraulic power take-off (HPTO) unit.

2.1. Hydrodynamic Motion of the Floater

In general, the hydrodynamic motion of the WEC device in real waves can be formulated in the time domain using the linear wave theory, as described in Equation (1). M_A is the D'Alembert moment of inertia, M_{ex} is the moment due to the diffracted waves, M_{rad} is the moment due to radiated waves, M_{res} is the hydrostatic restoring moment and M_{PTO} is the moment due to the HPTO unit.

$$M_A = M_{ex} - M_{rad} - M_{res} - M_{PTO} \tag{1}$$

The equation of the hydrodynamic motion in Equation (1) can be expended as given in Equation (2). Here, J_{WEC} is the inertia moment of the floater and arm. Whereas, $J_{add,\infty}$

is the added mass at the infinite frequency and $\ddot{\theta}_{arm}$ is the angular acceleration of a WEC device during the pitch motion. Then, $k_{rad}(t)$ is the radiation impulse response function, τ is the time delay and $\dot{\theta}_{arm}$ is the angular velocity of the floater's arm. Other variables such as k_{res} is the hydrostatic restoring coefficient and θ_{arm} is the angular of the floater's arm during the pitch motion. Finally, $h_{ex}(t-\tau)$ is the impulse response function of the excitation moment and η_W is the undisturbed wave elevation at the floater center point.

$$(J_{WEC}+J_{add,\infty})\ddot{\theta}_{arm}(t)+\int_0^t k_{rad}(t-\tau)\,\dot{\theta}_{arm}(t)+k_{res}\,\theta_{arm}(t)+M_{PTO}(t) = \int_{-\infty}^{\infty} h_{ex}(t-\tau)\eta_W(\tau)d\tau \tag{2}$$

The impulse response function in Equation (2) can be obtained from the hydrodynamic diffraction analysis using Computational Fluid Dynamics (CFD) software. In the present study, the hydrodynamic diffraction analysis of the WEC model was performed using ANSYS/AQWA software.

In addition, the moment due to the HPTO unit, M_{PTO} can be defined using Equations (3)–(5), where F_{PTO} is the feedback force from the HPTO unit applied to the WEC device. The variables L_1, L_2, L_3 and L_4 are the lengths between point a, b, c and d, as illustrated in Figure 2 [30,36–38]. x_p is the displacement of hydraulic cylinder piston and $L_{3,0}$ is the initial stroke of the hydraulic cylinder.

$$M_{PTO} = F_{PTO}L_4 \tag{3}$$

$$L_4 = \frac{L_1L_2sin(\theta_{arm,0}-\theta_{arm})}{L_{3,0}+x_c} \tag{4}$$

$$x_p = L_{3,0}-\sqrt{L_1{}^2+L_2{}^2-2L_1L_2cos(\theta_{arm,0}-\theta_{arm})} \tag{5}$$

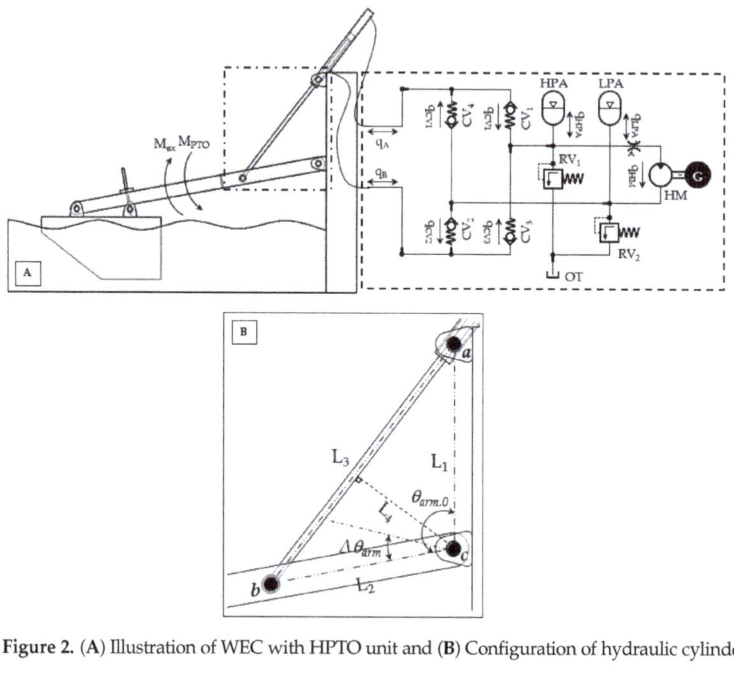

Figure 2. (**A**) Illustration of WEC with HPTO unit and (**B**) Configuration of hydraulic cylinder motion.

2.2. Hydraulic Power Take-off (HPTO) Mechanism

Figure 2A illustrates the considered HPTO unit, which includes a hydraulic actuator, set of control check valves (CV), high-pressure and low-pressure accumulator (HPA and LPA), hydraulic motor (HM) and electrical generator (G). In the HPTO unit, the large chamber of the hydraulic cylinder (chamber A) is connected to the CV_4 (outlet) and CV_1 (inlet), while the small chamber of DAC (chamber B) is connected to the CV_2 (outlet) and CV_3 (inlet), respectively. Meanwhile, HPA and LPA are placed at the inlet and outlet of the hydraulic motor. During the operation, the wave force generated from the passing ocean wave causes a floater to swing upward and downward repeatedly, as illustrated in Figure 2B. The mechanical force produced by the WEC device forces the rod and piston of the hydraulic cylinder at the specified velocity ($\dot{x}_p$) relatively subjected to the PTO load force. During the upward motion, the high-pressurised fluid in chamber A flows to the chamber B through CV_1, HPA, HM, LPA and CV_2. On the other hand, the process is vice-versa during downward motion, where high-pressure fluid in chamber B flows to chamber A through CV_3, HPA, HM, LPA and CV_4. The high-pressure fluid flows through HM lead to the HM, and G rotates simultaneously at the specified rotation speed (ω_G) subjected to the load torque of the G (τ_G). As a result, the usable electricity can be generated by the continuous motion of this mechanism.

In general, the behaviour of the HPTO unit is strongly nonlinear. Equations (6)–(21) theoretically explain the operation of the considered HPTO unit illustrated in Figure 2. According to Equation (6), the F_{PTO} from the HPTO unit applied to the WEC device depends on the pressure in both hydraulic chambers (p_A and p_B) and the effective piston area (A_P). Further, the effects of piston friction (F_{fric}) and initial force of rod, piston and oil (F_{in}) are also considered. These effects can be expressed using Equations (7) and (8), where η_{fric} is a friction coefficient, $\ddot{x}_p$ is the piston acceleration, g is a gravitational acceleration, M_p, M_r and M_{oil} are the mass of the piston, rod, and oil, respectively [12,36].

$$F_{PTO} = A_p(|p_A - p_B|) + F_{fric} + F_{in} \tag{6}$$

$$F_{fric} = |A_p(p_B - p_A)|\left(1 - \eta_{fric}\right) \tag{7}$$

$$F_{in} = \ddot{x}_p\left(M_p + M_r + M_{oil}\right) + \left(M_p + M_r\right)g \tag{8}$$

Since a double-acting-cylinder with a single rod piston is considered a hydraulic actuator, the F_{PTO} is unbalanced during the upward and downward motion of the WEC device due to the unbalanced pressure in both chambers of the hydraulic cylinder. Based on the configuration of the HPTO unit in Figure 2, the F_{PTO} during the upward movement is greater than the F_{PTO} during the downward movement. The dynamics of p_A and p_B can be described using a fluid continuity equation as in Equations (9) and (10) [12,37]. β_{eff}, q_A and q_B are the effective bulk modulus and the in/out volumetric flows in the hydraulic cylinder actuator. x_p, $\dot{x}_p$ and L are position, velocity and stroke length of the piston. $A_{p,A}$ and $A_{p,B}$ are the effective piston area in the hydraulic chamber A and B, that can be expressed by Equations (11) and (12), where the d_p and d_r are the diameter of the piston and rod, respectively.

$$\frac{d}{dt}p_A = \frac{\beta_{eff}}{A_{p,A}(L - x_p)}\left(q_A - \dot{x}_p A_{p,A}\right) \tag{9}$$

$$\frac{d}{dt}p_B = \frac{\beta_{eff}}{A_{p,B}(L - x_p)}\left(\dot{x}_p A_{p,B} - q_B\right) \tag{10}$$

$$A_{p,A} = \pi {d_p}^2/4 \tag{11}$$

$$A_{p,B} = \pi\left({d_p}^2 - {d_r}^2\right)/4 \tag{12}$$

For the check valve, the spring-loaded non-return valves are used in this HPTO model. The flow across the valve (q_{CV}) can be described by the orifice equation, as expressed in Equation (13), where C_d is the discharge coefficient, A_{CV} is the check valve opening area and ρ_{oil} is the fluid density. The p_{CVin} and p_{CVout} are the pressure at the inlet and outlet of the check valve [12,38].

$$q_{CV} = \begin{cases} C_d A_{CV}\sqrt{2|p_{CV_{in}} - p_{CV_{out}}|/\rho_{oil}},\ if\ p_{CV_{in}} > p_{CV_{out}} \\ 0,\ else \end{cases} \tag{13}$$

Besides that, the gas compression and expansion in the HPA and LPA, which are based on the isentropic process, can be described according to Equations (14) and (15), respectively. Where p_{HPA}, p_{LPA}, $p_{0,HPA}$ and $p_{0,LPA}$ are the pressure and pre-charge pressure in the HPA and LPA. γ is the adiabatic index of the compressed gas in the HPA and LPA, while, $V_{HPA.}$, V_{LPA}, $V_{0,HPA}$ and $V_{0,LPA}$ are the initial and the instantaneous volume of gas in the HPA and LPA, respectively. The instantaneous volume of gas can be expressed by Equations (16) and (17), where q_{HPA} and q_{LPA} are the volumetric flow in the HPA and LPA.

$$p_{HPA} \cdot V_{HPA}{}^{\gamma} = p_{0,HPA} \cdot V_{0,HPA}{}^{\gamma} \tag{14}$$

$$p_{LPA} \cdot V_{LPA}{}^{\gamma} = p_{0,LPA} \cdot V_{0,LPA}{}^{\gamma} \tag{15}$$

$$V_{HPA}(t) = V_{0,HPA} - \int_0^t q_{HPA} dt \tag{16}$$

$$V_{LPA}(t) = V_{0,LPA} - \int_0^t q_{LPA} dt \tag{17}$$

Meanwhile, the fluid continuity in the HPTO model should satisfy the following equations:

$$q_{HPA} = q_{CV_1} + q_{CV_2} - q_{HM} \tag{18}$$

$$q_{LPA} = q_{CV_3} + q_{CV_4} - q_{HM} \tag{19}$$

where q_{HM} is the volumetric flow through the hydraulic motor. Here, q_{HM} is given by Equation (20), where D_{HM}, ω_{HM}, and $q_{HM,loss}$ are displacement, speed and volumetric flow losses of the hydraulic motor, respectively. The output torque of the hydraulic motor, τ_{HM} can be expressed by Equation (21), where Δp_{HM} is the pressure difference in the hydraulic motor.

$$q_{HM} = D_{HM}\omega_{HM} - q_{HM,loss} \tag{20}$$

$$\tau_{HM} = D_{HM}\Delta p_{HM} \tag{21}$$

Based on the theoretical descriptions provided in Equations (6)–(21), the most important component parameters, which influence the operation of the HPTO model can be defined as summarised in Table 1. The inaccuracy of the selected component parameters will reduce the HPTO unit's capability in converting the absorbed wave energy to electrical energy. Thus, the optimisation of these important component parameters using mathematical algorithms is considered in this study.

Table 1. Important component parameters of the HPTO system.

No.	Parameter Setting	Unit
1	Diameter of piston, d_p	m
2	Diameter of rod, d_r	m
3	Volume capacity of HPA, $V_{cap,HPA}$	L
4	Volume capacity of LPA, $V_{cap,LPA}$	L
5	Pre-charge gas pressure of HPA, $p_{0,HPA}$	Bar
6	Pre-charge gas pressure of LPA, $p_{0,LPA}$	Bar
7	Displacement of HM, D_{HM}.	cc/rev

3. Simulation Studies of WECs

3.1. Ocean Wave Input Data

A previous study reported that the ranges of the wave height (H_W) and wave period (T_W) at several locations in Terengganu, Malaysia, were equal to the range of 0.2–1.2 m and 2–8 s, respectively [39]. In addition, a further forecast analysis found that the most annual occurrences sea-state at the considered installed location were equal to 0.8 m and 2.5 s. From these statistical results, the regular and irregular wave inputs data were generated based on Airy's wave theory and Joint North Sea Wave Observation Project (JONSWAP) spectrum, as illustrated in Figure 3A,B, respectively. For the irregular wave data profile generation, the peak enhancement factor (γ) of JONSWAP was set to 2. Regular wave input profile data were used in determining the optimal parameters of the HPTO unit process. While the irregular wave input profile data were used to evaluate the effectiveness of the optimal HPTO unit in generating the electricity in inconsistent wave circumstances.

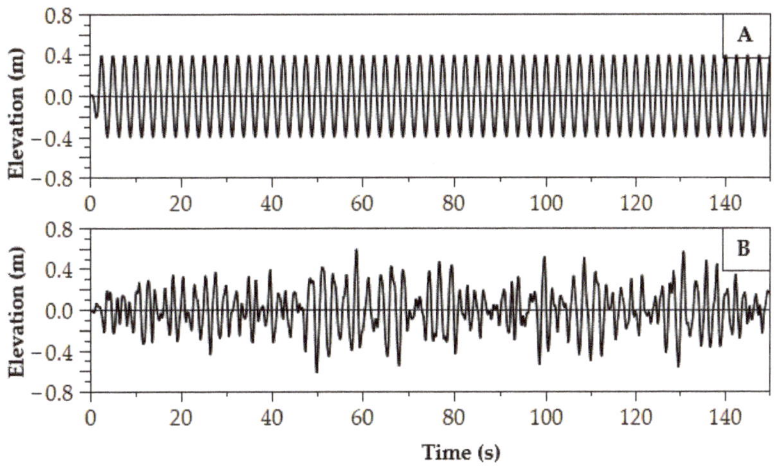

Figure 3. Ocean wave elevation inputs, (**A**) regular wave and (**B**) irregular wave.

3.2. Simulation Set-up of WEC with HPTO Unit Model

In the present study, the main specifications of the computer device that was used for the simulation studies are given in Table 2. As can be seen from Equations (1) and (2), the frequency domain analysis was required to determine the hydrodynamic parameters of the WEC device. Thus, hydrodynamic simulation of the WEC model was preliminarily performed using ANSYS/AQWA software. The hydrodynamic simulation method presented in [40,41] was considered. The results from the preliminary hydrodynamic simulation are presented in Figure 4. The parameters obtained from the hydrodynamic simulation were used to build the complete simulation model of WEC with the HPTO unit in *MATLAB*®® /Simulink software, as illustrated in Figure 5. A WEC model based on the linear wave motion, as mentioned in Equations (1)–(5), was developed using the function blocks.

Table 2. Main specifications of the computer device.

Item	Details
Type	Desktop
Windows	Windows 10 Pro
Memory (RAM)	12 GB
CPU	Intel (R) Core (TM) i7-9750H 2.60 GHz
MATLAB Version	R2019b

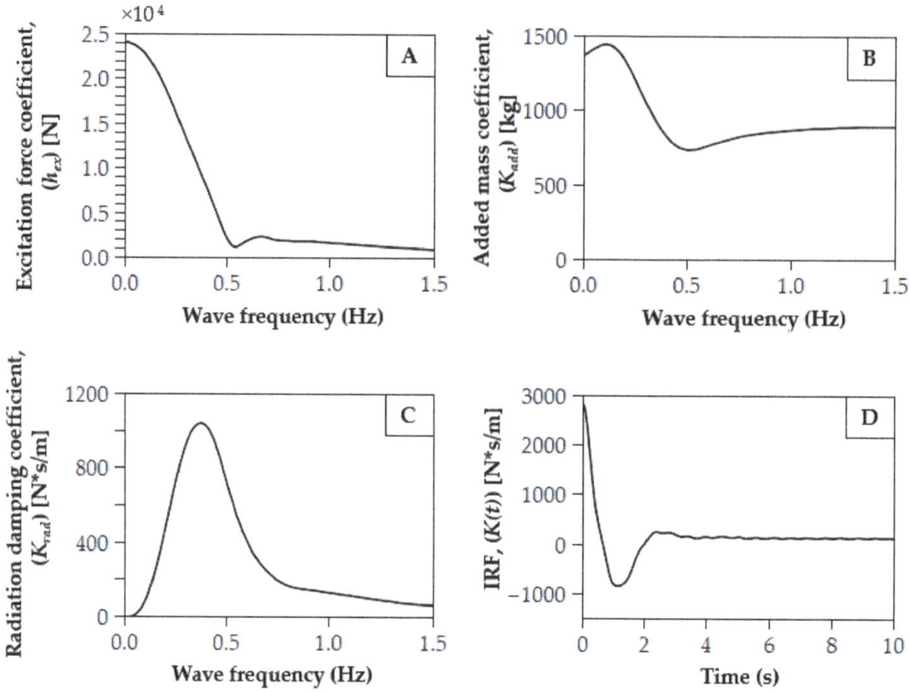

Figure 4. Hydrodynamic analysis parameters. (**A**) Excitation force coefficient, (**B**) added mass coefficient, (**C**) radiation damping coefficient, and (**D**) impulse response function.

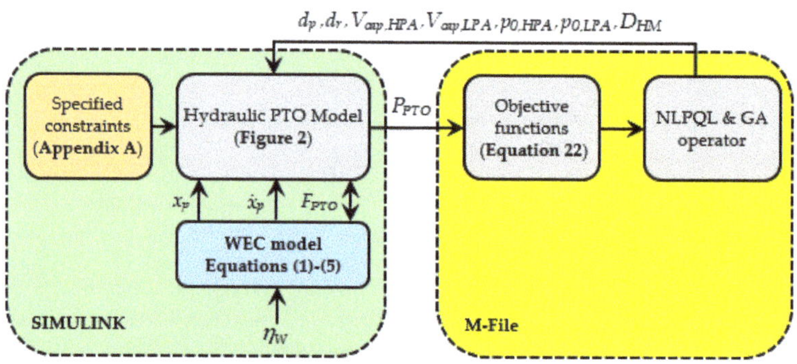

Figure 5. Illustration of simulation model set-up in MATLAB software.

Meanwhile, the HPTO model was developed using the hydraulic components in the Simscape Fluid toolbox, such as double hydraulic chamber single rod jack, hydraulic motor, hydraulic accumulator, hydraulic check valve with saturation, et cetera. The actual parameters of the hydraulic components from manufacturers were used to configure the HPTO model. Since the selection of the HPTO components was incredibly complex due to the variety of hydraulic products from the manufacturers and suppliers, the specification data of hydraulic components from a well-known manufacturer such as Parker Hannifin was considered, as summarised in Appendix A. The data in Appendix A were used as a guideline in determining the optimal configuration parameters of each element in the HPTO model simulation.

Furthermore, a simple dynamic sub-model of a rotary load was utilised to represent the permanent magnet synchronous generator (PMSG) unit. The generated electrical power output from the HPTO model was calculated based on the speed-power curve of PMSG, which was obtained from the manufacturer. In addition, the PTO force, hydraulic motor torque, hydraulic motor speed and electrical power were the acquired outputs from the HPTO model. The detailed specifications of each component that were used in the HPTO model are presented in Table 3.

Table 3. Technical specifications of the developed HPTO model.

Descriptions (Unit)	Value
Generator	
Rated power, P_{rated} (kW)	0.1
Rated speed, $\omega_{G,rated}$ (rpm)	200
Rated torque $\tau_{G,rated}$ (Nm)	6.0
Viscous friction coefficient, (Nm/rpm)	0.024
Moment of inertia, (kgm^2)	0.0036
Hydraulic cylinder	
Diameter of the piston, d_p (m)	0.035 *
Diameter of the piston rod, d_r (m)	0.025 *
Length of stroke, l_{stroke} (m)	0.3
HP accumulator	
Pre-charge gas pressure, $p_{0,HPA}$ (bar)	40 *
Volume capacity, $V_{cap,HPA}$ (L)	8 *
Adiabatic index, γ	1.4
LP accumulator	
Pre-charge gas pressure, $p_{0,LPA}$ (bar)	5 *
Volume capacity, $V_{cap,HPA}$ (L)	2 *
Adiabatic index, γ	1.4
Hydraulic motor	
Displacement, D_{HM} (cc/rev)	8 *
Oil properties	
Viscosity, Vis_{oil} (cSt)	50
Density, D_{oil} (kg/m^3)	850

* Initial value by manual estimation.

Simulation results of the WEC with non-optimal HPTO unit using the regular waves input profile data are shown in Figure 6. Figure 6A shows that the displacement of WEC was relatively lower than the wave displacement due to the PTO force applied to the WEC device. The PTO force profile applied to the WEC device is shown in Figure 6B. The figure shows that the PTO forces applied to the WEC device during the upward and downward motion were equal to 1.5 kN and 0.7 kN, respectively. Meanwhile, Figure 6C shows the electrical power generated from the non-optimal HPTO unit only can be reached up to an average of 71 W, which was 71% of its rated capacity.

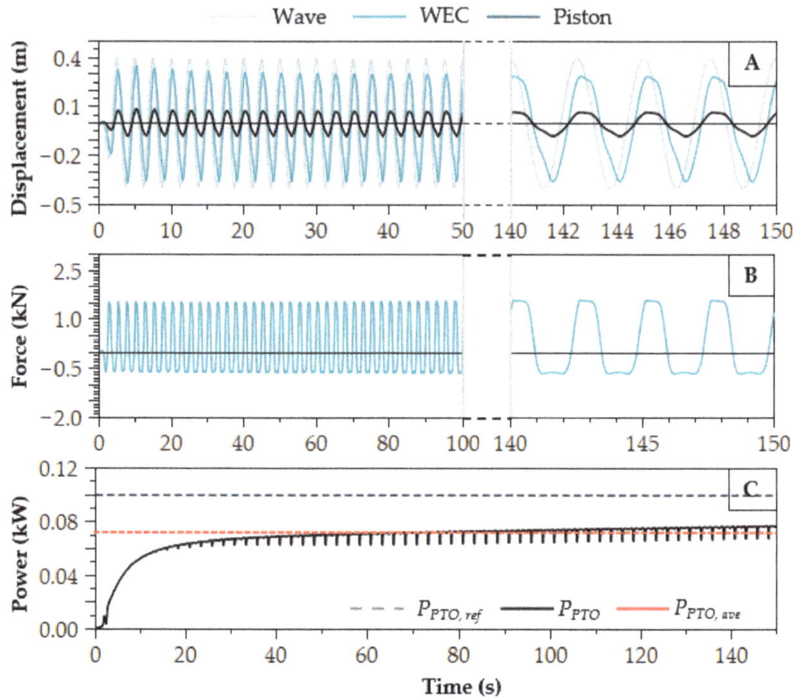

Figure 6. Preliminary simulation results of WEC with non-optimal power take-off (PTO) unit (H_W = 0.8, T_W = 2.5 s). (**A**) Wave, WEC and piston displacements, (**B**) PTO force, and (**C**) PTO power.

3.3. Optimisation of Configuration Parameter

As the sea state was relatively unstable throughout the year, a suitable HPTO unit was compulsory for a WEC device to ensure that the ocean wave energy can be maximally absorbed and converted to electrical energy. Conventionally, the optimal parameters of the HPTO unit were obtained by iteratively simulating the HPTO model using any sophisticated analysis software. In this process, the designer was required to manually specify a set of configuration parameters value, evaluate the HPTO unit model and analyse the PTO model output. Normally, this process may be repeated many times due to unsatisfactory results from the HPTO performance. Consequently, the designer again proposed a new set of HPTO parameters value based on experience and intuition, which probably will result in a better output of the HPTO model. This optimisation process will end when the time runs out. Unfortunately, sophisticated analysis software and high-speed computer technology were unable to help the designer in determining the optimal parameters of the HPTO unit using this technique.

Alternatively, the optimisation technique HPTO unit parameters using a computer algorithm was presented in this study. By using this technique, the designer was taken out from the trial-and-error loop process. The sophisticated computer was now utilised to conduct a complete determination process of the optimal configuration parameters. Through this technique, the designer workload can be reduced, in which the designer only focused on the interpretation of the optimisation results. Moreover, the determination of the optimal configuration parameters can be found in a shorter and more accurate time compared to the case using a conventional technique. In the algorithm-based optimisation technique, many kinds of algorithms can be applied to solve the optimisation problem.

In the present study, the simulation–optimisation using two major types of optimisation algorithms was explored in this present study. A specific objective function (OF) was designed to maximise the electrical energy generation of the HPTO unit, as described in Equation (22). Here, $P_{PTO,ref}$ and P_{PTO} represented the desired and the actual electrical power output of the HPTO unit. The optimisation problem in Equation (22) was solved by two kinds of optimisation algorithms, i.e., NLPQL and GA. In order to provide a fair-ground for comparison between two optimisation algorithms, the same constraints, design parameters, and objective function were considered for both cases under study. The details of the considered algorithms are described in the following subsections.

$$OF(x) = min\left[\frac{\int_0^T \left(\left|P_{PTO}(t) - P_{ref}(t)\right|\right) dt}{\int_0^T P_{PTO,ref}(t)\, dt}\right] \tag{22}$$

3.3.1. Non-Evolutionary NLPQL-Based Optimisation

The NLPQL algorithm was a local optimiser and has the advantages of fast convergence and high-stability [42]. In several studies, the NLPQL-based optimisation was applied to solve and optimise various non-linear problems during the design stage [42–45]. Figure 7A shows the flowchart of the NLPQL-based optimisation technique. Initially, the NLPQL-based optimisation process was started by randomly generating the guest point of each study parameter (d_p, d_r, $p_{0,HPA}$, $V_{cap,HPA}$, $p_{0,LPA}$, $V_{cap,LPA}$, D_{HM}). Then, in the first iteration, the generated random guest point was chosen for each study parameter, and the HPTO model was then evaluated based on the objective function in Equation (22). The linear search calculation method was then implemented in order to determine the convergence satisfaction of the objective function. As presented in Figure 7A, the new iteration will be started if the objective function does not meet the convergence criterion. A new iteration was initially started to determine the new search direction and step size using the sequential quadratic programming (SQP) method. Then, the variables for each study parameters were determined based on the new search direction and step size. Finally, the Hessian approximation was updated by the modified BFGS-formula, as described in [42]. The parameters setting of the NLPQL-based optimisation is listed in Table 4. This process was repeated until the NLPQL algorithm met the termination accuracy.

Table 4. Parameters setting of NLPQL.

Setting	Value
Maximum number of function evaluations	7
Maximum number of iterations	100
Step size for finite difference step	0.001
Final accuracy	0.0001

3.3.2. Evolutionary GA-Based Optimisation

In contrast to the NLPQL, GA was an evolutionary algorithm that was inspired by the natural evolution process. GA has been effectively applied to a wide range of real-world problems. In this algorithm, the variables of the optimisation problem were coded in chromosomes. Figure 7B presents the flowchart of the GA pseudo-code. The GA-based optimisation process was initially started by randomly generating a population of chromosomes (study parameters: d_p, d_r, $p_{0,HPA}$, $V_{cap,HPA}$, $p_{0,LPA}$, $V_{cap,LPA}$, D_{HM}), as presented in Figure 7B. Thereafter, for the first iteration, the random values from the generated population were chosen for each study parameter. The HPTO model was then evaluated based on the objective function in Equation (22). The chromosomes of the population were then sorted according to the least cost or highest fitness. Some percentages of the best chromosomes were transferred directly to the next generation based on their merit. Then, three GA operators named as selection, crossover and mutation were implemented to manipulate the rest of the chromosomes for the next generation. During the selection rule, the parent's chromosome that contributed to the current population was selected for the next generation process. Then, pairs of selected parents were recombined by a crossover operator to produce new chromosomes. A mutation rule was then applied to the new chromosomes to avoid the GA converging to the local optimum. Finally, this process was iterated until the satisfactory fitness level was reached. The parameters setting of GA was gathered in Table 5.

Table 5. Parameters setting of GA.

Setting	Value
Population size	50
Reproduction ratio (%)	80
Maximum number of generations	100
Mutation probability (%)	10
Mutation amplitude	0.1
Seed	1
Final accuracy	0.0001

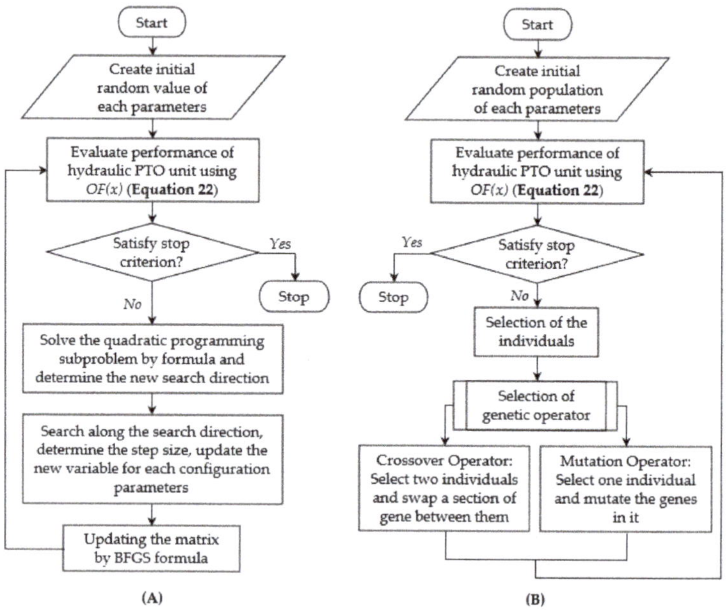

Figure 7. Optimisation procedures using (**A**) Non-Linear Programming by Quadratic Lagrangian (NLPQL) and (**B**) Genetic Algorithm (GA).

4. Results and Discussion

4.1. Comparisons between NLPQL and GA Optimisation of HPTO Unit

In order to evaluate the best optimisation approaches for the HPTO unit, a critical comparison analysis was performed. The comparison in terms of the final objective function, the best-estimated parameter values and the HPTO unit's performance were considered.

4.1.1. Chronological Variation of the Objective Function and Parameters Variables

Figures 8 and 9 depicted the chronological variation of the objective function and parameters variables with respect to the number of generations of the optimisation processes done by NLPQL and GA operators. The red vertical line in both figures indicated the optimisation process's termination at the lowest objective function value. The lowest objective function value was of interest for optimisation purposes in a feasible solution framework. Both of the optimisation processes were terminated after the algorithms met the optimum point, which was determined based on the termination criterion (final accuracy), as previously mentioned in Tables 4 and 5.

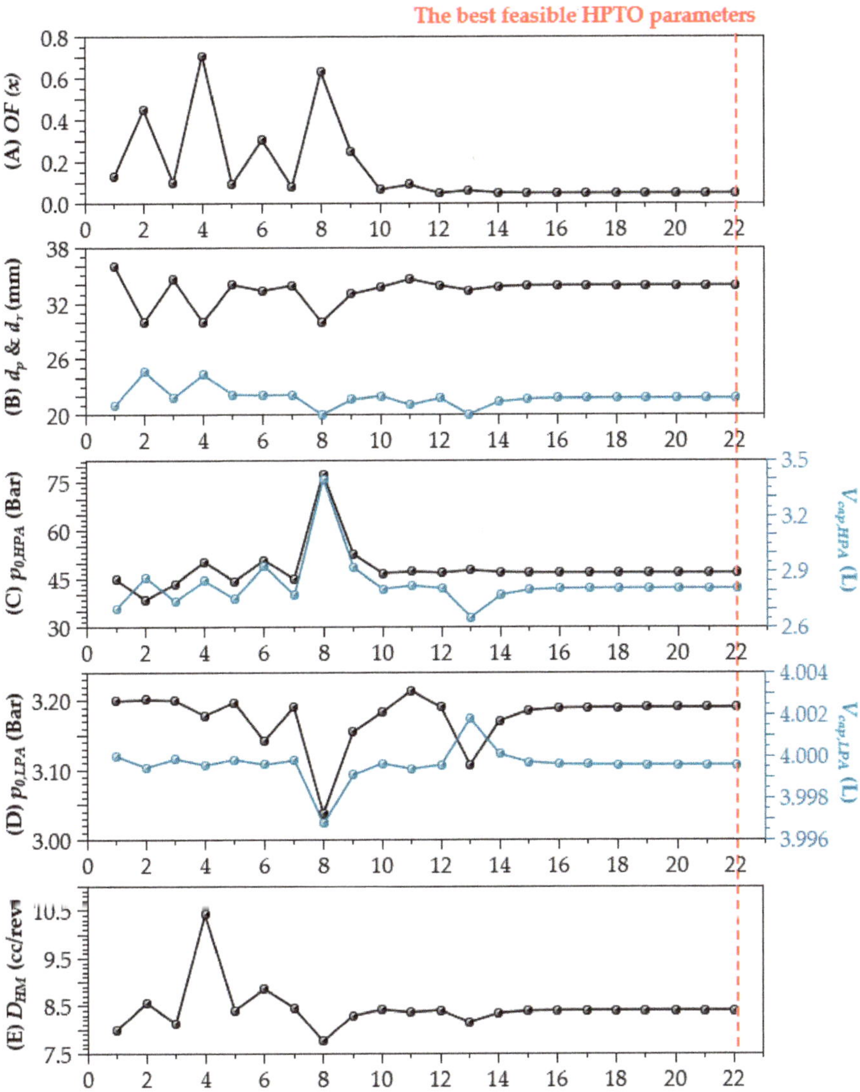

Figure 8. Chronological variation of (**A**) objective function, (**B**) diameter of piston and rod, (**C**) pre-charge gas pressure and capacity of HPA, (**D**) pre-charge gas pressure and capacity of LPA, and (**E**) displacement of HM for NLPQL algorithm.

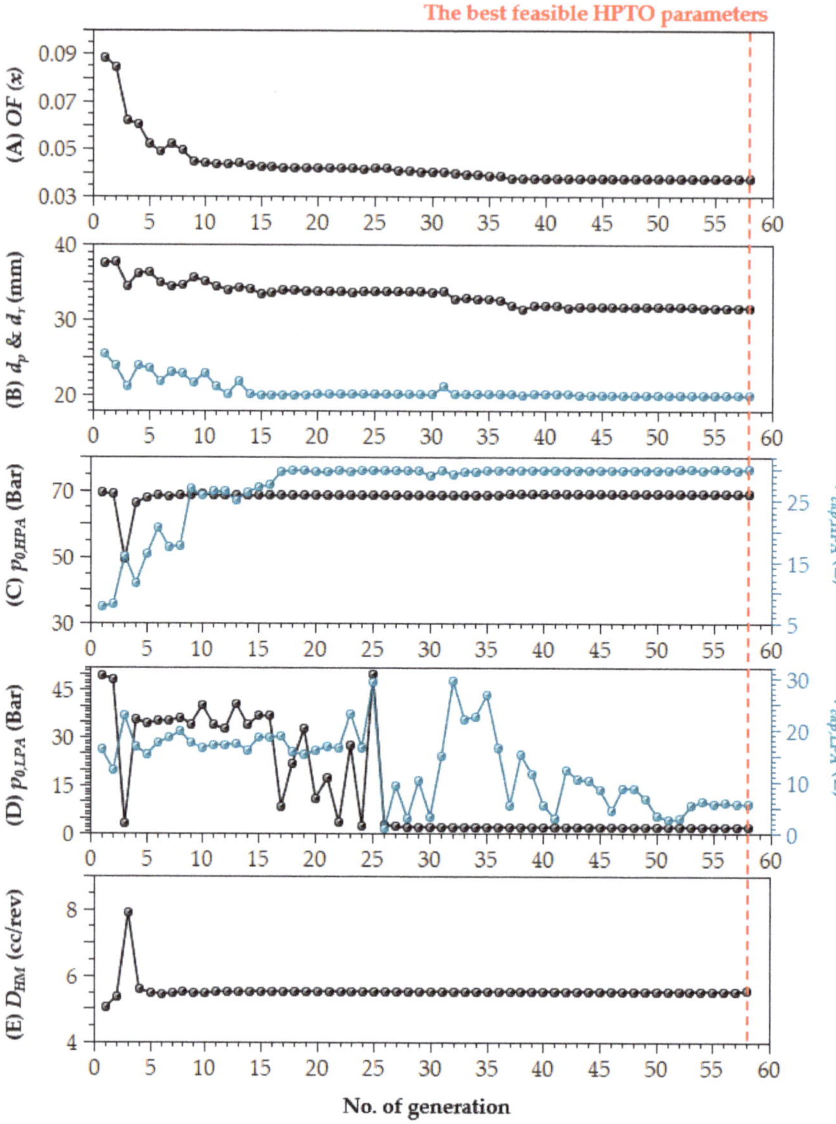

Figure 9. Chronological variation of (**A**) objective function, (**B**) diameter of piston and rod, (**C**) pre-charge gas pressure and capacity of HPA, (**D**) pre-charge gas pressure and capacity of LPA, and (**E**) displacement of HM for GA algorithm.

Figure 8A showed that the estimation process of the best configuration parameters was completed after the 22 number of iterations since the NLPQL operator had satisfied its accuracy requirement. The overall simulation–optimisation process using the NLPQL algorithm was carried out for 3237 s (approximately 53 m 57 s). As shown in the figure, the lowest objective function value at 22 iterations was obtained at 0.0492. Meanwhile, 56 numbers of iterations were needed to find the optimum case by the GA operator, as exhibited in Figure 9A. A complete simulation–optimisation process by the GA operator was performed for 7 h and 32 min. The lowest objective function was obtained equal to 0.0375, as illustrated in Figure 9A. Besides that, Figure 8B–E showed the chronological variation of the HPTO parameters throughout the optimisation process by the NLPQL algorithm. From these figures, the HPTO parameters seemed to approach the optimum conditions starting from 14 number of iterations. On the other hand, for the GA optimisation, Figure 9B–E showed that some of the HPTO parameters reached the best condition after 5 iterations.

In summary, the comparison of the chronological results of both optimisation approaches in Figure 8, and Figure 9 found that the optimisation using the NLPQL algorithm was much faster than the GA optimisation case. The reason was that since the NLPQL was the local optimisation approach, this algorithm depended on the initial point of each HPTO parameter. As reported in [42], the numerical test showed that different initial points required different time consumed and would give different optimal results. In contrast to the NLPQL algorithm, since the GA is a global algorithm, it takes more time in its exploration and exploitation processes that consider more points in search space in order to find the optimum condition. Thus, it returns more accurate and reliable results as depicted in Figure 9. In order to improve the performance of the NLPQL algorithm, the hybridisation of the NLPQL algorithm with the other global optimisation operators can be considered, as presented in [46]. In [46], the optimal starting points of the NLPQL algorithm were set by GA, and better optimum results have turned up.

4.1.2. Best Estimated Parameters

Table 6 presents the best configuration parameters sets of the HPTO unit that were successfully estimated using NLPQL and GA optimisation approaches. As shown in the table, the d_p and d_r parameters of the hydraulic cylinder were estimated at 3% and 12.8% smaller than their initial values for the NLPQL case, which equaled to 34.9 mm and 21.8 mm, respectively. For the GA case, the d_p and d_r were estimated closely to their minimum constraints, which equal to 37.6 mm and 10 mm. Apart from that, the data in Table 6 reported that the best-estimated values of the $p_{0,HPA,}$ and $V_{cap,HPA}$ from the NLPQL, and GA optimisation were significantly different from their initial condition. The optimal value of $p_{0,HPA}$ was estimated larger than its initial value for both cases. While, for the $V_{cap,HPA}$, Table 6 clearly shows that the best values of $V_{cap,HPA}$ were estimated 65% lower and 275% larger than its initial value for the NLPQL and GA cases. For the $p_{0,LPA}$ and $V_{cap,LPA}$, the best-estimated values were not too significantly different from their initial values for both cases. Furthermore, it can be found in Table 6 that the best values of D_{HM} were significantly different between both optimisation cases. The result from the table shows that the GA operator successfully estimated a smaller value of D_{HM} compared to the NLPQL case.

In summary, based on the comparison of best configuration parameters estimated from both optimisation approaches, a few preliminary conclusions in terms of physical size, cost of the HPTO unit and others can be drawn. Practically, the physical size, weight and cost of the HPTO unit depend on its configuration parameters. Based on the results in Table 6, it can be preliminarily concluded that the physical size and weight of the HPTO unit for the NLPQL approach were much smaller than the GA approach case. This was due to the larger capacity of HPA as estimated by the GA approach. The physical size and weight of the HPTO unit were vital to being reduced since they can influence the complete design of the WECs, as reported in [13]. Moreover, the configuration parameters also influenced the total cost of the HPTO unit. As reported in [10], the hydraulic accumulator and the hydraulic motor were the most expensive HPTO unit components. Thus, from Table 6, it can be concluded that the overall cost of the WECs from the NLPQL optimisation approach was much lower than the GA case.

Table 6. Best configuration parameters from NLPQL and GA parameter estimation approaches.

Parameter (Unit)	Non-Optimal Case	Optimal Case by	
		NLPQL	GA
Hydraulic cylinder			
Diameter of piston, d_p (mm)	36	34.9	31.6
Diameter of piston rod, d_r (mm)	25	21.8	20.0
HP accumulator			
Pre-charge gas pressure, $p_{0,HPA}$ (bar)	40	46.9	68.9
Volume capacity, $V_{cap,HPA}$ (L)	8	2.8	30.0
LP accumulator			
Pre-charge gas pressure, $p_{0,LPA}$ (bar)	5	3.2	2.2
Volume capacity, $V_{cap,LPA}$ (L)	2	4.0	5.8
Hydraulic motor			
Displacement, D_{HM} (cc/rev)	8	8.4	5.5

4.1.3. Operational Behaviour of the HPTO Unit

Table 7 compares the operational behaviour of the HPTO unit for the non-optimal, NLPQL-optimal, and GA-optimal cases. By comparing the data in Table 7 and Appendix A, the HPTO unit's operations were satisfied with its operational constraints for all cases. As reported in Table 7, the overall operating pressure of the HPTO unit increased for both optimal cases. For example, the operating pressures of the hydraulic cylinder chambers for the NLPQL-optimal case increased by 8.6% (side A) and 8.8% (side B). While, for the GA-optimal case, the operating pressures of the hydraulic cylinder increased by 56.6% (side A) and 56.8% (side B) from the non-optimal case, respectively. Besides that, the pressures in the HPA for both cases also increased up to 10.1% and 60.8%, respectively. The hydraulic motor pressure also significantly increased by 9.4% and 60% for both cases, up to 47.5 bar and 69.8 bar.

Table 7. Comparison of the operational behaviour of the HPTO unit for non-optimal, NLPQL-optimal, and GA-optimal cases.

Descriptions (Unit)		Non-Optimal Case	Optimal Case by	
			NLPQL	GA
Hydraulic cylinder				
Max. operating pressure, (bar)	Side A	46.48	50.5	72.8
	Side B	46.40	50.5	72.8
Max. operating flow rate, (L/min)	Side A (In)	2.75	3.49	2.71
	Side A (Out)	4.73	4.29	4.05
	Side B (In)	2.10	2.08	2.33
	Side B (Out)	2.07	2.23	2.25
HP accumulator				
Max. operating pressure, (bar)		43.4	47.8	69.8
Max. operating flow rate, (L/min)	In	3.78	3.23	3.06
	Out	1.36	1.66	1.10
LP accumulator				
Max. operating pressure, (bar)		5.05	3.25	2.61
Max. operating flow rate, (L/min)	In	1.28	1.66	1.09
	Out	1.30	1.82	0
Hydraulic motor				
Max. operating pressure, (bar)	Inlet	43.4	47.5	69.8
	Outlet	5.03	3.25	2.63
Max. operating flow rate, (L/min)		5.03	1.67	1.10
Max. operating speed, (rpm)		174	202	204
Max. operating torque, (Nm)		5.24	6.04	6.05

The increasing pressure in the HPTO unit significantly increased the speed and torque of the hydraulic motor. As depicted in Table 7, the hydraulic motor speed and torque increased to its rated (200 rpm, 6 Nm) for both cases. In short, the results in Table 7 clearly show that the operational speed and torque of the hydraulic were influenced by the pressure of the other components in the HPTO unit. Thus, the presented results in Table 7 proved that the optimisation process by NLPQL and GA were highly effective in estimating the best component parameters of the HPTO unit.

4.1.4. Performance of the WECs

Technically, the force of the HPTO unit was directly proportional to its operational pressure [15,37]. Since the HPTO unit's pressure significantly increased, the HPTO force applied to the WEC device also increased, as depicted in Figure 10. Comparing Figure 10 with Figure 6B showed that the HPTO force applied to the WEC obviously increased for both cases. As depicted in Figure 10A, the HPTO forces applied to the WEC for the NLPQL case can be reached up to 1.65 kN (upward) and 0.78 kN (downward). While, for the GA-optimal case, the HPTO forces applied to the WEC can be reached up to 2.3 kN (upward) and 1.2 kN (downward), respectively. The results clearly showed that the overall HPTO force applied to the WEC device for the GA-optimal case was significantly larger than the HPTO force in the NLPQL case. This significant difference was attributed due to the larger pre-charge gas pressure and volume capacity of the HPA in the HPTO unit for the GA-optimal case, as depicted in Table 6. A larger pre-charge gas pressure required a more massive flow of high-pressure fluid [10]. In addition, from both figures, it can be seen that the HPTO forces applied to the WEC device during the upward movement were larger compared to the downward movement for both cases. This was due to the unsymmetrical double-acting hydraulic cylinder used in the HPTO unit. Since the hydraulic cylinder chambers were unsymmetrical, the fluid pressure in the chamber, which comprises a large effective piston area, was higher than the fluid pressure in the small effective area chamber, as clearly described in [10].

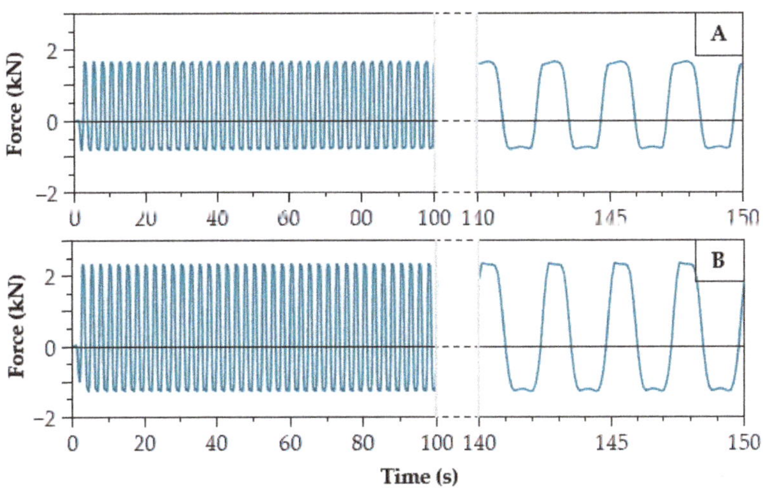

Figure 10. HPTO force applied to the WEC device ($H_W = 0.8$, $T_W = 2.5$ s), (**A**) NLPQL and (**B**) GA cases.

Furthermore, the HPTO force's effect on the displacement of the WEC device and hydraulic cylinder piston can be seen in Figure 11. Figure 11A,B illustrated the displacement of the wave, WEC, and hydraulic cylinder piston during the HPTO unit operation for both optimal cases. In Figure 11A, it was depicted that the average displacements of the WEC device and hydraulic cylinder piston for the NLPQL-optimal were 77.5% and 19.3% of the average wave elevation. Meanwhile, for the GA-optimal case, the average displacement of the WEC device and hydraulic cylinder piston was 65% and 16% of average wave elevation. The comparison of the results in Figures 6A and 11 showed that the displacement of the WEC device and hydraulic cylinder piston was slightly reduced for the NLPQL-optimal and GA-optimal cases. The reduction was due to the larger HPTO force applied to the WEC device in both cases. In addition, the comparison of Figure 11A,B showed that the average displacement of the WEC device and hydraulic cylinder piston for GA-optimal was less than the NLPQL-optimal case.

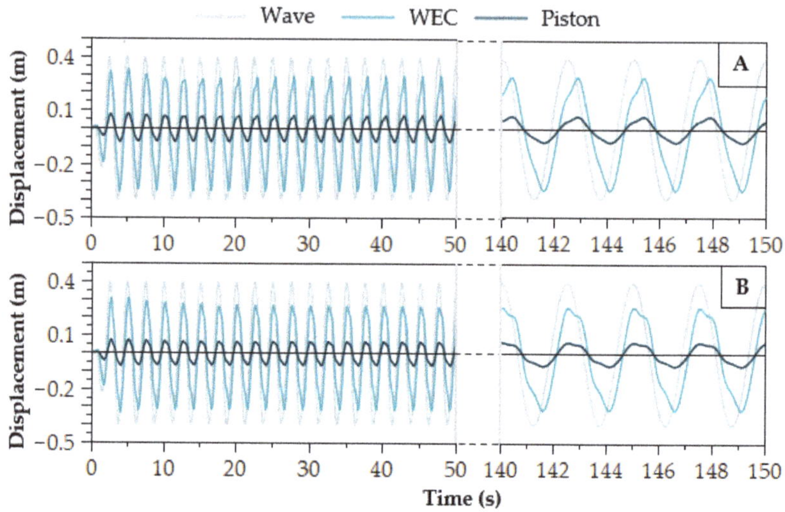

Figure 11. Displacement of the wave, WEC, and piston for three different cases ($H_W = 0.8$, $T_W = 2.5$ s), (**A**) NLPQL and (**B**) GA cases.

Apart from that, Figure 12 illustrated the comparison of the electrical power generation profiles of WECs for both optimal cases. Comparing the results in Figure 12 with Figure 6C, the overall electrical power generated from the HPTO unit optimised by NLPQL and GA approaches was successfully enhanced. For the non-optimal case, the electrical power profile in Figure 6C clearly indicated that the electrical power generated from the HPTO unit was lower than its rated capacity. Figure 6C showed the electrical power generated from the non-optimal HPTO was up to 71% (71 W) of its rated capacity. In contrast to both optimal cases. From Figure 12, the result showed the electrical generated output of HPTO was close to its rated capacity. The average electrical power generated from the HPTO unit for the NLPQL-optimal and GA-optimal cases was calculated up to 96% (96 W) and 97% (97 W) rated capacity, respectively. The comparison results in Figure 12 also showed that the electrical power generated from the HPTO unit of the GA-optimal case fluctuated less compared to the NLPQL-optimal case. This was due to the larger HPA used in the HPTO unit of the GA-optimal case. In addition, both of the optimal HPTO units reached their steady-state condition around 80 s.

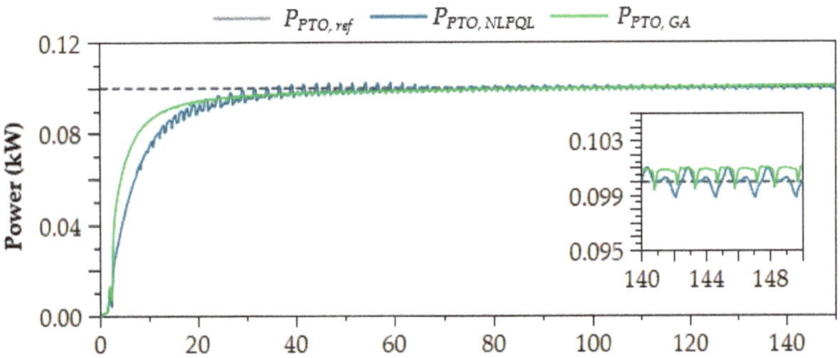

Figure 12. Comparison of electrical power generated from the best HPTO unit estimated by NLPQL and GA optimisations (H_W = 0.8, T_W = 2.5 s).

4.2. Evaluations of Optimal WECs Using Irregular Wave Data

The optimal HPTO units obtained from the optimisation processes were evaluated using irregular wave elevation input in order to evaluate their performance in generating the electricity in irregular wave circumstances. The results in Figure 13 provided the hydraulic cylinder piston responses for both cases. The figure showed that the displacement of the hydraulic cylinder piston for GA-optimal was smaller than the NLPQL-optimal case. This was due to the different pressures in the hydraulic cylinder chambers, as shown in Figure 14. Figure 14A showed that the reciprocating motions of the piston for the NLQPL case produced high-pressure liquid in both hydraulic cylinder chambers that reached up to 54 Bar. At the same time, the pressure of the hydraulic cylinder chamber for the GA case reached up to 75 Bar. The pressure difference for both cases was due to the difference in the HPTO unit's parameter design.

The high-pressure liquid produced in the hydraulic cylinder chamber then flowed to HPA and hydraulic motor. The HPA was used as liquid energy storage to smooth out pressure fluctuation in the HPTO unit. Thus, the liquid's pressure, which exceeded the HPA pre-charge pressure setting, was accumulated in the HPA ballast. In contrast, the HPA released the high-pressure liquid stored in its ballast when the HPTO system's pressure was lower than its pre-charge pressure setting. Figure 15 showed the pressure of the HPA for both optimal cases. For both cases, the pre-charge pressures of HPA were set to 46.9 Bar and 68.9 Bar, as previously given in Table 6. In Figure 15A, the pressure of the HPA reached up to 49 Bar, which was 4.5% higher than its pre-charge pressure setting several times. For the GA case, the highest pressure of the HPA can be reached up to 69.01 Bar, which was 0.16% higher than its pre-charge pressure setting, as depicted in Figure 15B. The difference in the pressure variation rate of HPA in both cases was due to the different HPA volume capacity. As given in Table 6, the volume capacity of HPA for the GA-optimal case was larger than the NLPQL-optimal case. In addition, the comparison of results in Figure 15A,B showed that the high-pressure liquid accumulation was more often for the GA-optimal case. This was due to a larger volume capacity of HPA used in the HPTO unit.

The smoothing effects of the HPA unit on the pressure in the HPTO unit for both optimal cases can be clearly seen in Figure 16. Figure 16A,B showed the smoothed pressure of the hydraulic motor for both cases. The comparison results in Figure 16A,B showed that the smoothing effect of the hydraulic motor pressure for the GA-optimal case was higher than the NLPQL-optimal case. It can be seen from the figures, the pressure of the hydraulic motor fluctuated less for the GA-optimal case compared to the NLPQL-optimal case. However, at the initial state of both cases, the hydraulic motor's pressure was more fluctuating due to insufficient energy stored in the HPA, as depicted in Figure 15. In addition, Figure 17 illustrated the comparison of the hydraulic motor speed for both cases. From the figure, the average speed of the hydraulic motor for the GA-optimal case was higher than the NLQPL-optimal case, which was 163 rpm instead of 137 rpm.

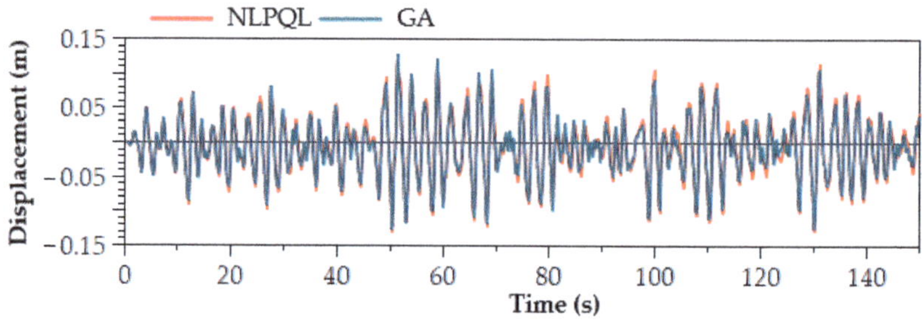

Figure 13. Displacement of hydraulic cylinder piston of HPTO unit (H_W = 0.8, T_W = 2.5 s).

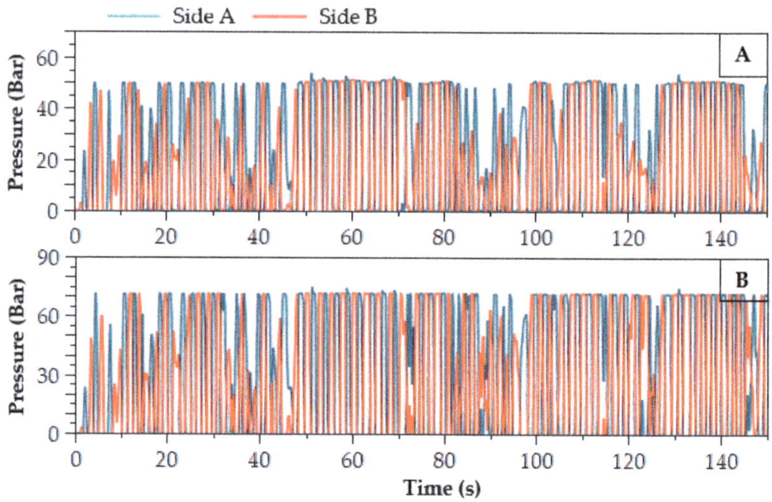

Figure 14. The pressure of the hydraulic cylinder chamber of HPTO unit (H_W = 0.8, T_W = 2.5 s), (**A**) NLPQL and (**B**) GA cases.

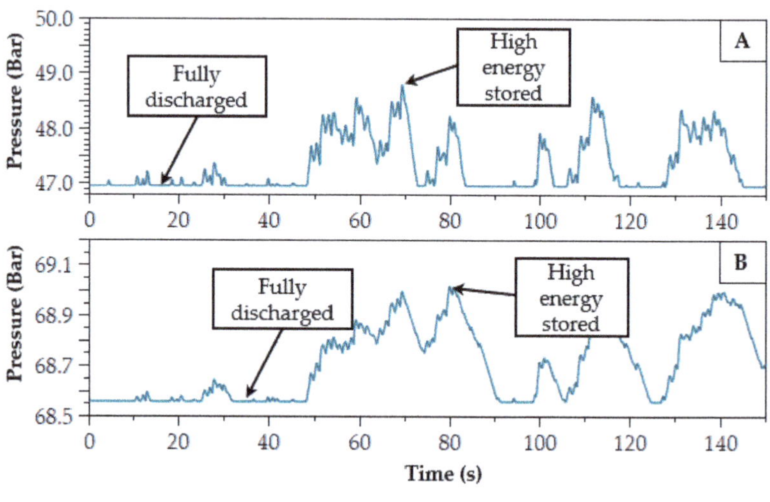

Figure 15. Pressure of high-pressure accumulator of HPTO unit (H_W = 0.8, T_W = 2.5 s), (**A**) NLPQL and (**B**) GA cases.

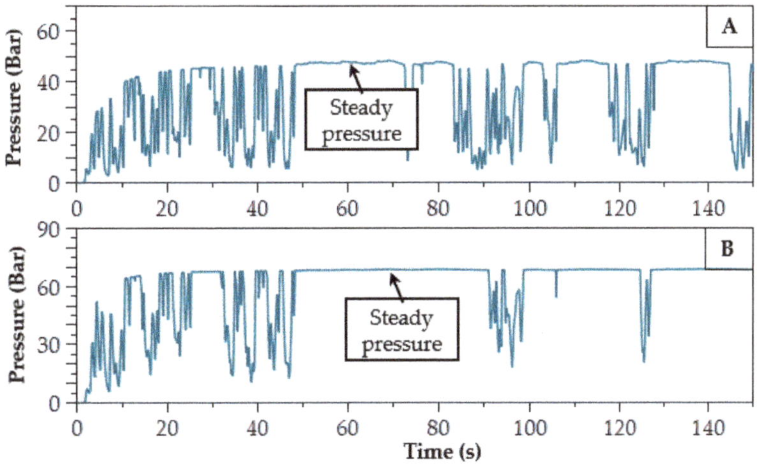

Figure 16. The pressure of hydraulic motor of HPTO unit (H_W = 0.8, T_W = 2.5 s), (**A**) NLPQL and (**B**) GA cases.

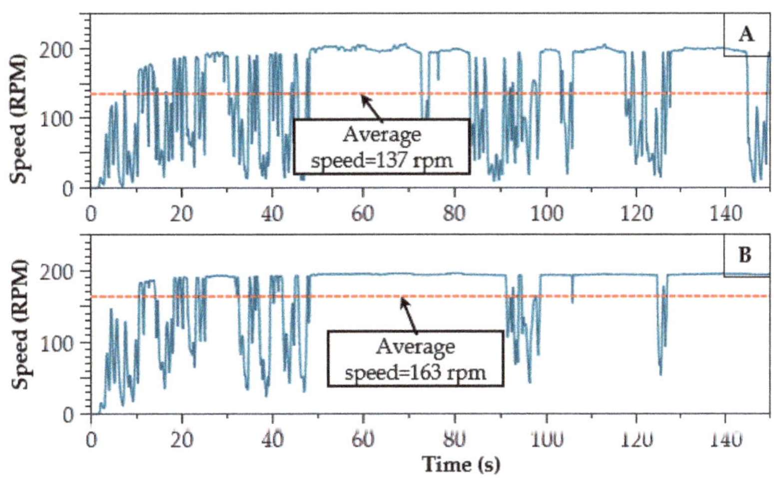

Figure 17. Speed of hydraulic motor of HPTO unit (H_W = 0.8, T_W = 2.5 s), (**A**) NLPQL and (**B**) GA cases.

Figure 18A–C presented the electrical power profiles of the HPTO unit for the non-optimal, NLPQL-optimal and GA-optimal cases. For the non-optimal case, the average electrical power generated from the PMSG generator in the HPTO unit was equal to 55 W, which was only 55% of its rated capacity, as shown in Figure 18A. For this case, the highest electrical power that was generated only reached up to 71 W. This was significantly different for the cases of the optimal HPTO unit optimised by NLPQL and GA approaches. It can be seen in Figure 18B,C, both optimal HPTO units capable of generating electricity of up to an average of 62 W and 77 W, which was 62% and 77% of its rated capacity, respectively.

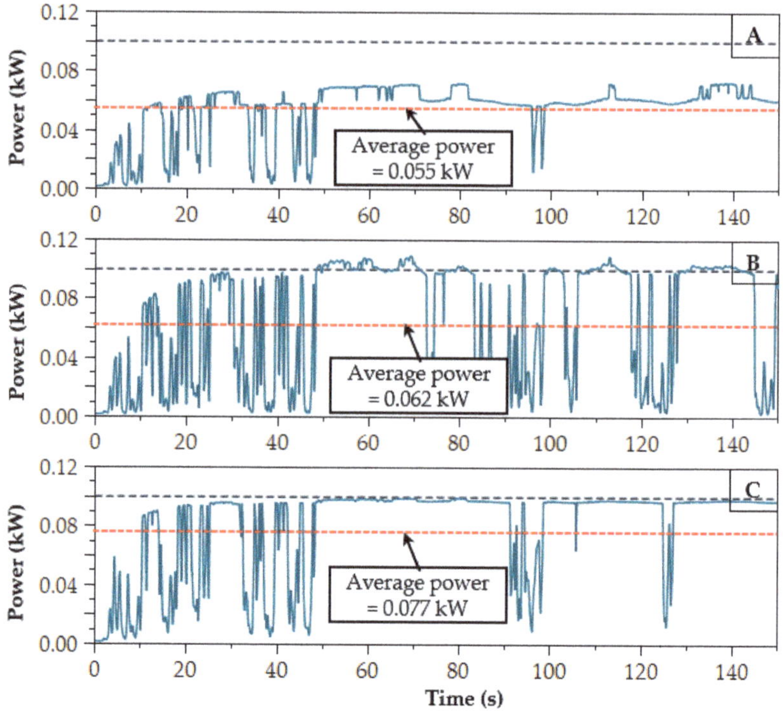

Figure 18. Comparison of electrical power generated from the best HPTO unit in irregular wave condition (H_W = 0.8, T_W = 2.5 s), (**A**) Non-optimal, (**B**) NLPQL-optimal, and (**C**) GA-optimal cases.

5. Conclusions

A comprehensive study was conducted to estimate the configuration parameters of the HPTO unit for a wave energy conversion device using a non-evolutionary NLPQL and evolutionary genetic algorithm. Seven important configuration parameters of the HPTO unit were considered in this optimisation study. The simulation–optimisation of HPTO model parameters was performed using MATLAB®®/Simulink software. The optimisation function problem was designed to maximise the output power generated from the HPTO unit. The optimal HPTO unit was then evaluated using irregular wave input to evaluate its performance in irregular circumstances. From the simulation studies, the key results can be listed as follows:

- The simulation–optimisation using the NLPQL algorithm was completed after the 22 number of iterations with the duration of 3237 s (approximately 53 m 57 s) after the NLPQL operator had satisfied its accuracy requirement. Importantly, the overall performance of HPTO has significantly improved up to 96% in regular wave conditions.
- The simulation–optimisation duration using the GA technique is longer than the NLPQL approach, which was completed after 7 h and 32 min. However, the overall performance of HPTO has significantly improved up to 97% in regular wave conditions.
- The HPTO unit estimated by the NLPQL approach is much smaller in terms of size, weight and cost compared to the GA approach. Thus, the HPTO unit's cost estimated by NLPQL is cheaper than the HPTO unit cost estimated by the GA approach.
- The results show that both optimal HPTO units can generate electricity up to 62% and 77% of rated capacity in irregular wave circumstances.
- In conclusion, both of the optimisation approaches were effective in determining the optimal parameters of the HPTO unit. For the sake of quickness, the NLPQL

approach is more relevant. While, for the sake of effectiveness, the GA approach is more recommended.

The simulation–optimisation framework presented may help the engineers and researchers of WECs to design a reliable and high-efficiency HPTO unit for wave energy converter devices. It is suggested that further researches should be conducted in the following areas:

- Further experimental validation of the best estimated HPTO unit is needed to verify the accuracy of the developed model simulation.
- The simulation–optimisation of the HPTO unit using other types of the optimisation algorithm, such as Particle Swarm Optimization, Gravitational Search Algorithm, et cetera, needs to be explored to achieve a good trade-off between cost and performance.
- The simulation–optimisation using other software types, such as Simcenter Amesim software invented by Siemens, is highly recommended.

Author Contributions: M.A.J., conceptualisation, methodology, software, data curation, analysis, writing—original draft; M.Z.I., writing—review and editing and supervision, project administration, funding acquisition; M.Z.D., conceptualisation, methodology, writing—review and editing and supervision; Z.M.Y., software, data curation, analysis and writing—original draft; A.A. data curation, analysis and writing—review and editing. All authors have read and agreed to the published version of the manuscript.

Funding: This project was funded by the Ministry of Higher Education (MOHE) under Fundamental Research Grant Scheme (FRGS/1/2019/TK07/UMT/01/1).

Acknowledgments: The authors would like to thank the Ministry of Higher Education (MOHE) and Universiti Malaysia Terengganu (UMT) for financial support for this research.

Conflicts of Interest: The authors declare no conflict of interest.

Abbreviations

ACO	Ant Colony Optimization
CFD	Computational Fluid Dynamics
CV	Check Valve
GA	Genetic Algorithm
GSA	Gravitational Search Algorithm
HM	Hydraulic Motor
HPA	High-Pressure Accumulator
HPTO	Hydraulic Power Take-Off
JONSWAP	Joint North Sea Wave Observation Project
LPA	Low-Pressure Accumulator
NLPQL	Non-Linear Programming by Quadratic Lagrangian
OF	Objective Function
PMSG	Permanent Magnet Synchronous Generator
PSO	Particle Swarm Optimization
PTO	Power Take-Off
SOC	State-Of-Charge
SQP	Sequential Quadratic Programming
TS	Tabu Search
WAB	Wave-Activated-Body
WEC	Wave Energy Converter
WECs	Wave Energy Converters

Appendix A

Table A1. Specifications of hydraulic components from Parker Hannifin Manufacturer.

HPTO Component (Unit)	Ranges		Ref.
	Minimum	Maximum	
Hydraulic cylinder [a]			[47]
Available piston diameter, (mm)	30	203	
Available rod diameter, (mm)	10	140	
Operating pressure, (bar)	0	207	
Operating flow rate, (L/min)	0	900	
HP accumulator [b]			[48]
Available nominal volume, (L)	0.2	57	
Operating pressure, (bar)	0	690	
Operating flow rate, (L/min)	0	900	
LP accumulator [c]			[49]
Available nominal volume, (L)	0	565	
Operating pressure, (bar)	0	80	
Operating flow rate, (L/min)	0	3000	
Hydraulic motor [d]			[50]
Available motor displacement, (cc/rev)	20	23,034	
Operating pressure, (bar)	0	420	
Operating speed, (rpm)	0	1000	
Operating flow rate, (L/min)	0	200	
Operating torque, (Nm)	0	1428	

[a] Heavy Duty Roundline Welded Series, [b] High-Pressure Bladder Accumulator Series, [c] Low-Pressure Bladder Accumulator Series, [d] High Torque Radial Piston Motors Series.

References

1. Sang, Y.; Karayaka, H.B.; Yan, Y.; Yilmaz, N.; Souders, D. Ocean (Marine) Energy. In *Comprehensive Energy Systems*; Elsevier: Amsterdam, The Netherlands, 2018; Volumes 1–5, pp. 733–769, ISBN 9780128095973.
2. Mustapa, M.A.; Yaakob, O.B.; Ahmed, Y.M.; Rheem, C.K.; Koh, K.K.; Adnan, F.A. Wave energy device and breakwater integration: A review. *Renew. Sustain. Energy Rev.* **2017**, *77*, 43–58. [CrossRef]
3. de Falcão, A.F.O. Wave energy utilization: A review of the technologies. *Renew. Sustain. Energy Rev.* **2010**, *14*, 899–918. [CrossRef]
4. Titah-Benbouzid, H.; Benbouzid, M. An up-to-date technologies review and evaluation of wave energy converters. *Int. Rev. Electr. Eng.* **2015**, *10*, 52–61. [CrossRef]
5. Rusu, E.; Onea, F. A review of the technologies for wave energy extraction. *Clean Energy* **2018**, *2*, 10–19. [CrossRef]
6. Al Shami, E.; Zhang, R.; Wang, X. Point absorber wave energy harvesters: A review of recent developments. *Energies* **2019**, *12*, 47. [CrossRef]
7. Têtu, A. *Power Take-Off Systems for WECs*; Springer: Cham, Switzerland, 2017; pp. 203–220.
8. Kukner, A.; Erselcan, İ.Ö. A review of power take-off systems employed in wave energy. *J. Nav. Sci. Eng.* **2014**, *10*, 32–44.
9. Gaspar, J.F.; Calvário, M.; Kamarlouei, M.; Guedes Soares, C. Power take-off concept for wave energy converters based on oil-hydraulic transformer units. *Renew. Energy* **2016**, *86*, 1232–1246. [CrossRef]
10. Jusoh, M.A.; Ibrahim, M.Z.; Daud, M.Z.; Albani, A.; Yusop, Z.M. Hydraulic power take-off concepts for wave energy conversion system: A review. *Energies* **2019**, *12*, 4510. [CrossRef]
11. Galván-Pozos, D.E.; Ocampo-Torres, F.J. Dynamic analysis of a six-degree of freedom wave energy converter based on the concept of the Stewart-Gough platform. *Renew. Energy* **2020**, *146*, 1051–1061. [CrossRef]
12. Penalba, M.; Davidson, J.; Windt, C.; Ringwood, J.V. A high-fidelity wave-to-wire simulation platform for wave energy converters: Coupled numerical wave tank and power take-off models. *Appl. Energy* **2018**, *226*, 655–669. [CrossRef]
13. Liu, C.; Yang, Q.; Bao, G. Influence of hydraulic power take-off unit parameters on power capture ability of a two-raft-type wave energy converter. *Ocean Eng.* **2018**, *150*, 69–80. [CrossRef]
14. Sheng, W.; Lewis, A. Power takeoff optimization for maximizing energy conversion of wave-activated bodies. *IEEE J. Ocean. Eng.* **2016**, *41*, 529–540. [CrossRef]
15. Cargo, C.J.; Hillis, A.J.; Plummer, A.R. Optimisation and control of a hydraulic power take-off unit for a wave energy converter in irregular waves. *Proc. Inst. Mech. Eng. Part A J. Power Energy* **2014**, *228*, 462–479. [CrossRef]
16. Brito, M.; Teixeira, L.; Canelas, R.B.; Ferreira, R.M.L.; Neves, M.G. Experimental and numerical studies of dynamic behaviors of a hydraulic power take-off cylinder using spectral representation method. *J. Tribol.* **2018**, *140*. [CrossRef]
17. Brito, M.; Ferreira, R.M.L.; Teixeira, L.; Neves, M.G.; Canelas, R.B. Experimental investigation on the power capture of an oscillating wave surge converter in unidirectional waves. *Renew. Energy* **2020**, *151*, 975–992. [CrossRef]

18. Amaran, S.; Sahinidis, N.V.; Sharda, B.; Bury, S.J. Simulation optimization: A review of algorithms and applications. *Ann. Oper. Res.* **2016**, *240*, 351–380. [CrossRef]
19. Jusoh, M.A.; Daud, M.Z. Particle swarm optimisation-based optimal photovoltaic system of hourly output power dispatch using Lithium-ion batteries. *J. Mech. Eng. Sci.* **2017**, *11*, 2780–2793. [CrossRef]
20. Jusoh, M.A.; Daud, M.Z. Control strategy of a grid-connected photovoltaic with battery energy storage system for hourly power dispatch. *Int. J. Power Electron. Drive Syst.* **2017**, *8*, 1830–1840. [CrossRef]
21. Daud, M.Z.; Mohamed, A.; Hannan, M.A. An improved control method of battery energy storage system for hourly dispatch of photovoltaic power sources. *Energy Convers. Manag.* **2013**, *73*, 256–270. [CrossRef]
22. Daud, M.Z.; Mohamed, A.; Ibrahim, A.A.; Hannan, M.A. Heuristic optimization of state-of-charge feedback controller parameters for output power dispatch of hybrid photovoltaic/battery energy storage system. *Meas. J. Int. Meas. Confed.* **2014**, *49*, 15–25. [CrossRef]
23. Jusoh, M.A.; Daud, M.Z. Accurate battery model parameter identification using heuristic optimization. *Int. J. Power Electron. Drive Syst.* **2020**, *11*, 333–341. [CrossRef]
24. Giassi, M.; Göteman, M. Layout design of wave energy parks by a genetic algorithm. *Ocean Eng.* **2018**, *154*, 252–261. [CrossRef]
25. Sirigu, S.A.; Foglietta, L.; Giorgi, G.; Bonfanti, M.; Cervelli, G.; Bracco, G.; Mattiazzo, G. Techno-Economic optimisation for a wave energy converter via genetic algorithm. *J. Mar. Sci. Eng.* **2020**, *8*, 482. [CrossRef]
26. McCabe, A.P.; Aggidis, G.A.; Widden, M.B. Optimizing the shape of a surge-and-pitch wave energy collector using a genetic algorithm. *Renew. Energy* **2010**, *35*, 2767–2775. [CrossRef]
27. Calvário, M.; Gaspar, J.F.; Kamarlouei, M.; Hallak, T.S.; Guedes Soares, C. Oil-hydraulic power take-off concept for an oscillating wave surge converter. *Renew. Energy* **2020**, *159*, 1297–1309. [CrossRef]
28. Jusoh, M.A.; Ibrahim, M.Z.; Daud, M.Z.; Yusop, Z.M.; Albani, A.; Rahman, S.J.; Mohad, S. Parameters estimation of hydraulic power take-off system for wave energy conversion system using genetic algorithm. In *Proceedings of the IOP Conference Series: Earth and Environmental Science*; Institute of Physics Publishing: Bristol, UK, 2020; Volume 463, p. 12129.
29. Hansen, R.H.; Kramer, M.M.; Vidal, E.; Hansen, R.H.; Kramer, M.M.; Vidal, E. Discrete displacement hydraulic power take-off system for the wavestar wave energy converter. *Energies* **2013**, *6*, 4001–4044. [CrossRef]
30. Hansen, A.H.; Asmussen, M.F.; Bech, M.M. Model predictive control of a wave energy converter with discrete fluid power power take-off system. *Energies* **2018**, *11*, 635. [CrossRef]
31. Garcia-Rosa, P.B.; Cunha, J.P.V.S.; Lizarralde, F.; Estefen, S.F.; Costa, P.R. Efficiency optimization in a wave energy hyperbaric converter. In Proceedings of the 2009 International Conference on Clean Electrical Power, ICCEP 2009, Capri, Italy, 9–11 June 2009; pp. 68–75.
32. Estefen, S.F.; Esperança, P.D.T.; Ricarte, E.; Da Costa, P.R.; Pinheiro, M.M.; Clemente, C.H.P.; Franco, D.; Melo, E.; De Souza, J.A. Experimental and numerical studies of the wave energy hyperbaric device for electricity production. In *Proceedings of the International Conference on Offshore Mechanics and Arctic Engineering—OMAE*; American Society of Mechanical Engineers Digital Collection: New York, NY, USA, 2008; Volume 6, pp. 811–818.
33. Windt, C.; Davidson, J.; Ransley, E.J.; Greaves, D.; Jakobsen, M.; Kramer, M.; Ringwood, J.V. Validation of a CFD-based numerical wave tank model for the power production assessment of the wavestar ocean wave energy converter. *Renew. Energy* **2020**, *146*, 2499–2516. [CrossRef]
34. Ransley, E.J.; Greaves, D.M.; Raby, A.; Simmonds, D.; Jakobsen, M.M.; Kramer, M. RANS-VOF modelling of the Wavestar point absorber. *Renew. Energy* **2017**, *109*, 49–65. [CrossRef]
35. Gibraltar Project—Eco Wave Power. Available online: https://www.ecowavepower.com/gibraltar-project/ (accessed on 4 May 2019).
36. Penalba, M.; Sell, N.P.; Hillis, A.J.; Ringwood, J.V. Validating a wave-to-wire model for a wave energy converter—Part I: The hydraulic transmission system. *Energies* **2017**, *10*, 977. [CrossRef]
37. Cargo, C.J.; Plummer, A.R.; Hillis, A.J.; Schlotter, M. Determination of optimal parameters for a hydraulic power take-off unit of a wave energy converter in regular waves. *Proc. Inst. Mech. Eng. Part A J. Power Energy* **2012**, *226*, 98–111. [CrossRef]
38. Do, H.T.; Dang, T.D.; Ahn, K.K. A multi-point-absorber wave-energy converter for the stabilization of output power. *Ocean Eng.* **2018**, *161*, 337–349. [CrossRef]
39. Muzathik, A.M.; Wan Nik, W.B.; Samo, K.B.; Ibrahim, M.Z. Ocean wave measurement and wave climate prediction of Peninsular Malaysia. *J. Phys. Sci.* **2011**, *22*, 77–92.
40. Chen, Q.; Yue, X.; Geng, D.; Yan, D.; Jiang, W. Integrated characteristic curves of the constant-pressure hydraulic power take-off in wave energy conversion. *Int. J. Electr. Power Energy Syst.* **2020**, *117*, 105730. [CrossRef]
41. Jianan, X.; Tao, X. MPPT Control of Hydraulic Power Take-Off for Wave Energy Converter on Artificial Breakwater. *J. Mar. Sci. Eng.* **2019**, *8*, 304. [CrossRef]
42. Zhang, S.; Tezdogan, T.; Zhang, B.; Xu, L.; Lai, Y. Hull form optimisation in waves based on CFD technique. *Ships Offshore Struct.* **2018**, *13*, 149–164. [CrossRef]
43. Navid, A.; Khalilarya, S. Evaluation of a diesel engine optimized by non-evolutionary NLPQL and evolutionary genetic algorithms and assessing second law efficiency: Analysis in exergy loss and chemical exergy. *Appl. Therm. Eng.* **2019**, *159*. [CrossRef]

44. Navid, A.; Khalilarya, S.; Taghavifar, H. Comparing multi-objective non-evolutionary NLPQL and evolutionary genetic algorithm optimization of a DI diesel engine: DoE estimation and creating surrogate model. *Energy Convers. Manag.* **2016**, *126*, 385–399. [CrossRef]
45. Chen, Y.; Lv, L. The multi-objective optimization of combustion chamber of DI diesel engine by NLPQL algorithm. *Appl. Therm. Eng.* **2014**, *73*, 1332–1339. [CrossRef]
46. Hu, N.; Zhou, P.; Yang, J. Comparison and combination of NLPQL and MOGA algorithms for a marine medium-speed diesel engine optimisation. *Energy Convers. Manag.* **2017**, *133*, 138–152. [CrossRef]
47. Hydraulic Cylinders—Heavy Duty Roundline Welded—Series RDH | Malaysia. Available online: https://ph.parker.com/my/en/heavy-duty-hydraulic-roundline-cylinders-series-rdh (accessed on 19 January 2020).
48. Bladder Accumulator—High Pressure (EHV) | Malaysia. Available online: https://ph.parker.com/my/en/bladder-accumulator-high-pressure-ehv (accessed on 19 January 2020).
49. Bladder Accumulator—Low Pressure (EBV Series) | Malaysia. Available online: https://ph.parker.com/my/en/low-pressure-bladder-accumulator-ebv (accessed on 19 January 2020).
50. High Torque Radial Piston Motors—Series MR* | Malaysia. Available online: https://ph.parker.com/my/en/high-torque-radial-piston-motors-series-mr (accessed on 19 January 2020).

energies

MDPI

Article

Water Conservation Potential of Self-Funded Foam-Based Flexible Surface-Mounted Floatovoltaics

Koami Soulemane Hayibo [1], Pierce Mayville [2], Ravneet Kaur Kailey [2] and Joshua M. Pearce [1,2,3,*]

1 Department of Electrical & Computer Engineering, Michigan Technological University, Houghton, MI 49931, USA; khayibo@mtu.edu
2 Department of Material Science & Engineering, Michigan Technological University, Houghton, MI 49931, USA; pjmayvil@mtu.edu (P.M.); rkailey@mtu.edu (R.K.K.)
3 School of Electrical Engineering, Aalto University, FI-00076 Esburg, Finland
* Correspondence: pearce@mtu.edu

Received: 20 October 2020; Accepted: 25 November 2020; Published: 28 November 2020

Abstract: A potential solution to the coupled water–energy–food challenges in land use is the concept of floating photovoltaics or floatovoltaics (FPV). In this study, a new approach to FPV is investigated using a flexible crystalline silicon-based photovoltaic (PV) module backed with foam, which is less expensive than conventional pontoon-based FPV. This novel form of FPV is tested experimentally for operating temperature and performance and is analyzed for water-savings using an evaporation calculation adapted from the Penman–Monteith model. The results show that the foam-backed FPV had a lower operating temperature than conventional pontoon-based FPV, and thus a 3.5% higher energy output per unit power. Therefore, foam-based FPV provides a potentially profitable means of reducing water evaporation in the world's at-risk bodies of fresh water. The case study of Lake Mead found that if 10% of the lake was covered with foam-backed FPV, there would be enough water conserved and electricity generated to service Las Vegas and Reno combined. At 50% coverage, the foam-backed FPV would provide over 127 TWh of clean solar electricity and 633.22 million m^3 of water savings, which would provide enough electricity to retire 11% of the polluting coal-fired plants in the U.S. and provide water for over five million Americans, annually.

Keywords: water; floatovoltaic; photovoltaic; energy water nexus; dual use; water conservation; FPV; floating photovoltaic; solar energy

1. Introduction

Water scarcity [1,2], the energy crisis [3], and food scarcity [4,5] are the largest currently coupled challenges [6] facing the global community, where they most severely affect the arid and semiarid regions of the world [7]. There is a wide scientific consensus that combustion of fossil fuels for energy is increasing atmospheric carbon dioxide (CO_2) concentrations and driving climate change [8]. This anthropogenic climate change is increasing globally averaged mean annual air temperatures and driving changes in precipitation, which are expected to continue and increase [9,10]. The IPCC (Intergovernmental Panel on Climate Change) warns that the climate change over the next century will affect rainfall, river flows and sea levels all over the world [11], which will negatively impact agricultural yield [12]; particularly in already-malnourished sub-Saharan Africa. de Wit and Stankiewicz [13] predict rainfall in sub-Saharan Africa could drop by 10% causing surface drainage to drop 30–50% by midcentury, which would cause major water shortages. It is widely agreed that to prevent the worst of climate change, humanity needs to rapidly convert fossil fuel-based energy systems to renewable energy systems [14]. Solar photovoltaic (PV) technology is the most widely accessible, sustainable, and clean renewable source of energy that can be scaled to meet humanity's energy needs [15,16]. To meet these needs, however, a substantial amount of land is still needed for PV to replace fossil fuels and this creates competition for limited land resources between food and energy [17]. A utility-scale PV plant land occupation varies between 20 km^2/GWh and 40 km^2/GWh depending on the type of solar panels used [18]. Despite life cycle carbon emissions [19], PV is more land efficient than all carbon capture and sequestration plans for coal [20], but with nearly a billion people already living undernourished, further reductions in agricultural land are not acceptable during a world food crisis [21].

A potential solution to these coupled water–energy–food challenges is the concept of floating photovoltaics or floatovoltaics (FPV), which has been rapidly gaining a base in scientific literature [22–28]. FPV is growing fast and is expected to have an average growth rate of above 20% in the next five years due to extremely low costs (with an FPV bid recently coming in for a system in Thailand at under USD 0.50/Wp) [29]. FPV are easier to install and simpler to decommission than conventional PV systems and the racking costs are less, which lead to these overall cost savings [29]. As FPV are located near or immersed in water, the operational temperature is reduced, which raises the solar energy conversion efficiency [23,26,30–34]. In regions where water scarcity is an issue and particularly when this issue is likely to be aggravated by climate change, FPV can also be used to reduce water loss because it can reduce water evaporation by more than 70% [32,35–37]. The Penman–Monteith daily evaporation method indicates that FPV could even cut evaporation by as much as 90% [38]. Studies in China [39] and India [40,41] have all indicated massive potential water savings for both small and large FPV coverage areas. This is particularly important for preservation of water sources in arid and semi-arid regions, especially with water shortages in the region [42]. FPV, therefore, also holds substantial promise when coupling with existing hydro power to make dual use of the electrical infrastructure while improving the water resource itself [39,43]. Similar advantages are to be expected for hybrid systems with pumped storage [44]. Finally, there is also evidence that FPV deployment reduces the PV degradation rate below 0.5% per year [45], which improves the levelized cost of solar electricity further.

FPV research has focused on several system design strategies [46]:

(1) Tilted arrays of solid modules (normally on top of pontoon structures) [36,47–49];
(2) Submerged PV modules (with and without a pontoon) [24,30,33,34,50];
(3) Micro-encapsulated phase change material (MEPCM)-based pontoon modules [51–53];
(4) Thin-film PV (no ridged pontoon supporting structure) [24,26,54].

The thin-film FPV design has the advantage of reducing racking costs even more so than pontoon style FPV, as it clearly stops more evaporation and gains an advantage by the operational temperature being lower. However, the temperature coefficients are better for amorphous silicon (a-Si:H) thin film materials than those of crystalline silicon (c-Si) so the benefits of the water cooling are muted for a-Si:H-based FPV.

In this study, a new approach is used with a flexible crystalline silicon module on a similar foam system to that described by Pierce et al. [54] for a-Si:H FPV. This approach enables a larger solar electric output gain (or FPV boost) and as solar is largely already profitable, there is an opportunity for the electricity production value of c-Si flexible foam-backed FPV to subsidize a means of water conservation by cutting water evaporation losses. To build on past FPV work and investigate the potential of FPV coupled to hydro power in the U.S., the water saving potential at Lake Mead using FPV is investigated in this study. Lake Mead is an artificial reservoir created by the United States government to run the Hoover Dam, which was built in 1935 [55,56]. This novel form of FPV is analyzed for water-saving using an evaporation calculation adapted from the Penman–Monteith daily evaporation model [57] that is approved by the Food and Agriculture Organization of the United Nations (FAO) [58]. An energy production analysis is performed and an open source spreadsheet was developed to simulate the evaporation and the energy yield of the flexible FPV [59], as well as to investigate the impact of using passive water-cooled FPV, where the cooling potential was measured experimentally for a foam-based FPV. The potential is determined for a case study based on the coverage of FPV ranging from 10% to

50% [60] of Lake Mead. The results are compared to "conventional" tilted pontoon-style FPV and are discussed in the context of the energy–water–food nexus.

2. Materials and Methods

2.1. Data Collection

2.1.1. Lake Evaporation Data

Most of the weather data used in this study were collected on Lake Mead through buoys installed by the United States National Oceanic and Atmospheric Administration's National Data Buoy Center (NOAA-NDBC) [61]. The rest of the data were obtained from open-access weather data made available by the McCarran International Airport's weather station in Las Vegas [62], and from SOLCAST, a solar data provider [63].

The main characteristics of the lake differ slightly from one study to another and depend on the year the study was conducted. In this study, the lake characteristics' values used for the evaporation calculation are taken from the National Park Service (NPS) website [64]. According to the NPS, as of 2010, the lake has a maximum surface area of 159,866 acres (647 million m^2), and a maximum capacity of 29,686,054 acre-feet (36,617 million m^3). The mean depth of the lake is estimated to be 55.5 m by the National Park Service [56]. The elevation of the lake is 328.574 m above sea water level. The weather buoy used to collect the data is located in the North Boulder Basin of the lake at a geographical position of latitude 36.087 N and longitude 114.728 W. The temperature sensor for air temperature collection is located at a height of 2 m above the lake surface while the anemometer is at 3 m above. Additionally, the atmospheric pressure sensor is located at 330.574 m above sea water level or 2 m above the lake surface, and the water temperature is measured at 0.5 m below the lake surface [65].

The buoy installed in Lake Mead's North Boulder basin by the NOAA-NDBC has been capturing different types of variables since 2016, which are stored in a historical database on the agency's website. Among the data required to conduct an evaporation calculation using the Penman–Monteith model, the wind speed (w_s), the atmospheric pressure (P), the maximum ($T_{w,max}$), minimum ($T_{w,min}$), and daily mean (T_w) water temperature; and the air temperature were obtained from the NOAA-NDBC historical database. The rest of the data were not captured by the buoy; therefore, alternative methods have been used to gather the required data. According to Moreo and Swancar, when data are not available for the study location, nearby airport weather data can be used instead [55]. In this study, the nearest airport close to Lake Mead is the Las Vegas Airport. The relative humidity (Rh) data have thus been obtained from the Weather Underground website that has made data from the Las Vegas Airport available. The remaining variable is the daily incoming solar irradiation or global horizontal irradiation (R_S) that has been obtained from SOLCAST's historical database [63]. This variable is also used for the solar energy production modeling.

The raw data from the NOAA-NDBC database were collected with an interval of 10 min starting at 00 h 00 min each day while the data from the Las Vegas Airport were measured with an 1 h interval starting at 00 h 56 min each day. Since daily data were required for the calculation, a mean daily value has been calculated for each variable. First, the data obtained from the NOAA-NDBC were cleaned by keeping only hourly data at the beginning of the hour (00 min) in order to match the data from Las Vegas Airport. A MATLAB code [66] was developed to perform this operation. Then, the same code was used to strip the missing data from the data table. A line of data was considered missing from the data table if one or more of the variables were not recorded by either the NOAA-NDBC station sensors or the Las Vegas Airport station sensors. After that, the data were reported in a spreadsheet that was used to calculate the mean daily value of the wind speed (w_s), the atmospheric pressure (P), the water temperature (T_w) and the air temperature (T_a) by averaging the hourly data for each day. Another method used in the literature to calculate daily mean weather data is to calculate the average of the maximum and minimum value of the day [67]. However, studies have shown that if data are available, it is best to calculate the mean daily temperature by averaging the hourly values [68,69]. The spreadsheet was also used to retrieve the maximum ($T_{w,max}$), and minimum ($T_{w,min}$) daily temperatures as well as the maximum (Rh_{max}), and minimum (Rh_{min}) daily relative humidity. The number of missing data points was 246 hourly data. Instead of having total hourly data of 8760 points, 8514 data points were used for this study after the data cleaning process. There was no more than 3 missing data points for a single day except for 5 specific days that are the 4th, 60th, 97th, 318th, and 347th day of the year 2018. These 5 days were, respectively, missing 4, 4, 10, 5, and 16 data points. The days with the highest number of missing data were the 97th and 347th day of the year. Since there are only two such days among the 365 that populated the year 2018, it has been considered that it will not have a significant impact on the results. Therefore, the available data were representative in estimating the mean daily values of the variables for each day.

2.1.2. FPV Panel Data Collection

In a previous study, it was found that polyethylene (PE) foam was the most cost-effective way to add buoyancy to flexible solar modules [54]. This study uses this after-market conversion method to convert SunPower SPR-E-Flex PVs into FPVs [70]. The density of the green polyethylene 1.2 lb $\frac{1}{2}$″ (12.7 mm) was used to determine the area of foam needed to make the panel rise by approximately 10 mm above the water's surface [71] using the calculations detailed in [54]. The foam was cut into about 50 mm by 240 mm sections that were placed evenly on the backside of module. The sections were then adhered using 3 M 5200 fast-set waterproof adhesive. Each foam piece had a line of adhesive caulked onto its perimeter and through the center. Then, the foam piece was pressed on the surface of the panel to adhere it, see Figure 1. The FPV with PV control was deployed in Chassell Bay, MI during the summer of 2020 to determine operational temperature and performance. This resulted in the FPV floating directly above the water surface, but still enabling wave action to clear the modules as shown in Figure 2.

(a) (b)

Figure 1. Cut away view showing adhesive underneath foam attached to c-Si-based flexible photovoltaic (PV) module: (**a**) top view and (**b**) orthogonal view.

The NanoDAQ monitoring board used in [54] was used in this study to measure module power and temperature of both the control (flat land-based mounted dry PV set at zero degree tilt angle) and wet FPV (floating on lake surface). The thermistors used for measuring temperature were held in place on the panels using 3M VHB tape. The air and water temperature were also measured with thermistors. The SunPower panels came with MC4 connectors installed on 12 AWG (2 mm^2) wires. MC4 connectors were added to the 14 AWG (1.6 mm^2) wires coming from the NanoDAQ, including the load wires. An additional hole was made in the NanoDAQ waterproof case and sealed using 3M 5200 to use the battery's USB port to power it. An AC load with a timer was used to drain the battery during mid-day to ensure there was a load to produce the power measurement. The schematic of the wiring diagram for the experimental set up is shown in Figure 3.

Figure 2. Closeup of floating photovoltaic/floatovoltaic (FPV) corner after deployment, showing water coverage from a modest wave (top left).

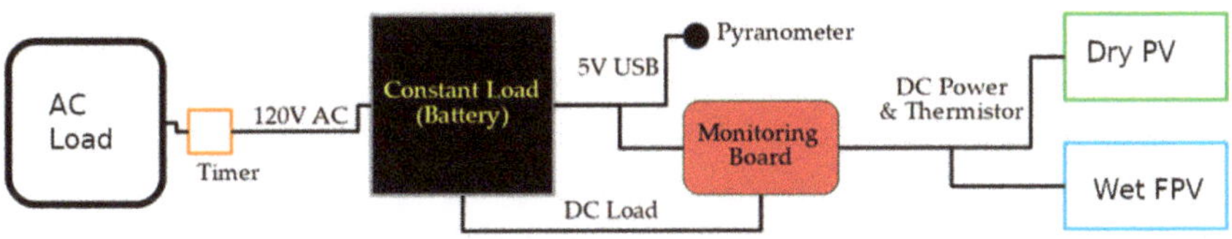

Figure 3. Wiring diagram for NanoDAQ monitoring board.

The FPV utilized mooring similar to that used in [54] except for the inclusion of a buoy. The wet FPV was moored by using an anchor and a towing ring on land. A rope was looped through the grommets in the solar PV and overhand loop knots were tied to secure the FPV in place. Energy generation of dry PV and wet FPV, temperature of air, water, PV, and FPV were recorded in 15 min increments.

2.2. Water Evaporation Modeling

The Penman–Monteith model used in this study is a datum intensive water evaporation model because it requires the measurement of several weather data. Some of the data can be calculated, but the accuracy of the model is increased if they are measured. The Penman–Monteith model was originally designed to calculate the evapotranspiration losses from leaves' and canopies' surfaces [57]. However, the method has been adapted in several studies to estimate the evaporation of surface water [72–75]. One important thing to note regarding the use of the Penman–Monteith evapotranspiration model for lake evaporation is the use of water temperature instead of air temperature in some of the parameters' calculations: the outgoing longwave radiation, the partial vapor pressure at the water surface and slope of the temperature saturation water vapor curve. The original Penman–Monteith model estimates the evapotranspiration of crops; therefore, the model only uses the air temperature in its implementation. The use of water temperature instead of air temperature has been validated in several lake evaporation studies [72,74,75].

The Penman–Monteith [57] equation adapted to open water surfaces is [74,75]:

$$E_L = \frac{1}{\lambda} \times \frac{\left(\Delta \times (R_N - H_S) + 86400 \times \rho_a \times Cp_a \times \frac{(P_w - P_a)}{r_a}\right)}{\Delta + \gamma} \quad \left(\text{mm}\cdot\text{day}^{-1}\right) \tag{1}$$

where E_L (mm/day) is the daily evaporation rate and Cp_a (kJ/kg/°C) and ρ_a (kg/m^3) are, respectively, the heat capacity, and the density of air. The other parameters in the Penman–Monteith equation need to be calculated and depend on several weather data. The weather data needed to calculate these parameters are comprised of: the daily maximum ($T_{a,max}$) (°C) and daily minimum ($T_{a,min}$) (°C) air temperature; the daily maximum ($T_{w,max}$) (°C), daily minimum ($T_{w,min}$) (°C), and daily mean water temperature (T_w) (°C); the daily maximum (Rh_{max}) (%), and daily minimum relative humidity (Rh_{min}) (%); the daily mean dew temperature (T_d) (°C), the daily mean atmospheric pressure (P) (kPa); the daily mean wind speed (w_s) (m/s) at a height of 2 m above the water surface; and the daily incoming solar radiation (R_S) (MJ/m^2/day). The other parameters that are needed to calculate the components in the evaporation model of Penman–Monteith include: the altitude of the lake's location (h) (m); the surface area (A) (m^2), and the effective depth (d_w) (m) of the lake reservoir; and the latitude of the location of the water surface (ϕ) (rad).

When all the listed parameters are available, the computation of the lake water evaporation using the Penman–Monteith model starts with the calculation of the mean saturation vapor pressure (P_w) (kPa), and the actual vapor pressure of the air (P_a) (kPa) [58,67,73]:

$$P_w = \frac{1}{2} \times 0.6108 \times \left(\exp\left(\frac{17.27 \times T_{w,max}}{T_{w,max} + 237.3} \right) + \exp\left(\frac{17.27 \times T_{w,\min}}{T_{w,\min} + 237.3} \right) \right) \quad \text{(kPa)} \tag{2}$$

$$P_a = \tfrac{1}{2} \times 0.6108 \quad \times \left(\tfrac{Rh_{min}}{100} \times \exp\left(\tfrac{17.27 \times T_{w,max}}{T_{w,max}+237.3} \right) + \tfrac{Rh_{max}}{100} \times \exp\left(\tfrac{17.27 \times T_{w,min}}{T_{w,min}+237.3} \right) \right) \quad \text{(kPa)} \tag{3}$$

After the calculation of the two vapor pressures, the slope of the saturation vapor pressure curve (Δ) (kPa/°C) is calculated [58,67,73]:

$$\Delta = \frac{4096 \times P_w}{(T_w + 237.3)^2} \quad \left(\text{kPa} \cdot {}^\circ\text{C}^{-1} \right) \tag{4}$$

Then, the latent heat of vaporization (λ) (MJ/kg), which depends on the water temperature, is calculated [58,73]:

$$\lambda = 2.501 - 0.002361 \times T_w \quad \left(\text{kPa} \cdot {}^\circ\text{C}^{-1} \right) \tag{5}$$

From the latent heat of vaporization, the psychrometric constant (γ) (kPa/°C) can be deduced [58,67],

$$\gamma = \frac{Cp_a \times P}{R_{MW} \times \lambda} \quad \left(\text{kPa} \cdot {}^\circ\text{C}^{-1} \right) \tag{6}$$

In Equation (6), $R_{MW} = 0.622$ and is equal to the molecular weight of water vapor over the molecular weight of dry air.

After that, the wind function f_w (MJ/m^2/kPa/day) is needed to estimate the aerodynamic resistance of the water surface. The formula used to calculate the wind function is proposed by McJannet et al. [76]. The formula was found to work well with the Penman–Monteith evaporation model. The wind function calculation by McJannet's formula depends on the wind speed as well as on the surface area of the lake.

$$f_w = (2.36 + 1.67 \times w_s) \times A^{-0.05} \quad \left(\text{MJ} \cdot \text{m}^{-2} \cdot \text{kPa}^{-1} \cdot \text{day}^{-1} \right) \tag{7}$$

Once the wind function is known, a combination of the Penman–Monteith model equations presented in the works of Zotarelli et al. and Finch et al. gives the value of the aerodynamic resistance (s/m) [67,75]:

$$r_a = \frac{\rho_a \times Cp_a \times 86400}{1000 \times \gamma \times f_w} \quad \left(\text{s} \cdot \text{m}^{-1} \right) \tag{8}$$

The two remaining terms are the net solar radiation (R_N) (MJ/m^2/day) and the change in water heat storage flux (H_S) (MJ/m^2/day). The net solar radiation's calculation depends on the net longwave radiation (R_{NL}) (MJ/m^2/day) and the net shortwave radiation (R_{NS}) (MJ/m^2/day) [58,67,73].

$$R_N = R_{NS} - R_{NL} \quad \left(\text{MJ}\cdot\text{m}^{-2}\cdot\text{day}^{-1}\right) \tag{9}$$

The net shortwave radiation is calculated using the albedo (a) and the measured incoming solar radiation (R_S) (MJ/m^2/day) [58,67,73–75]:

$$R_{NS} = (1-a)\times R_S \quad \left(\text{MJ}\cdot\text{m}^{-2}\cdot\text{day}^{-1}\right) \tag{10}$$

The net longwave radiation is calculated by taking the difference between the outgoing longwave radiation (R_{OL}) (MJ/m^2/day) and the incoming longwave radiation (R_{IL}) (MJ/m^2/day). The incoming longwave radiation is given by the Equation (11) [77,78]

$$R_{IL} = \sigma\left(C_f + \left(1-C_f\right)\left(1-\left(0.261\times\exp\left(-7.77\times10^{-4}T_a^2\right)\right)\right)\right)(T_a+273.15)^4 \quad \left(\text{MJ}\cdot\text{m}^{-2}\cdot\text{day}^{-1}\right) \tag{11}$$

In Equation (11), σ [MJ/m^2/T^4/day] is the Stefan–Boltzmann's constant, T_a is the daily mean air temperature and C_f is the cloud coverage fraction that has been estimated as follows [79]:

$$\begin{array}{l} C_f = 1.1 - R_{Ratio}\ ;\ R_{Ratio} \le 0.9 \\ C_f = 2(1-R_{Ratio})\ ;\ R_{Ratio} > 0.9 \end{array} \tag{12}$$

The parameter R_{Ratio} is the ratio of the incoming solar radiation (R_S) to the clear sky radiation R_{CS} (MJ/m^2/day). The clear sky radiation is calculated using Equation (13) [75,78,79]:

$$R_{CS} = \left(0.75 + 2\cdot10^{-5}\times h\right)\times R_{EX} \quad \left(\text{MJ}\cdot\text{m}^{-2}\cdot\text{day}^{-1}\right) \tag{13}$$

The extraterrestrial radiation R_{EX} (MJ/m^2/day) depends on the latitude of the lake, the sunset hour angle, the solar declination angle, the solar constant, and the inverse relative distance from the sun to earth. This calculation is a well-known procedure that has been detailed in the guide for crop evapotranspiration calculations by the FAO [58]. The outgoing longwave radiation depends on the water surface temperature and is calculated as:

$$R_{OL} = \varepsilon\times\sigma\times(T_w+273.15)^4 \quad \left(\text{MJ}\cdot\text{m}^{-2}\cdot\text{day}^{-1}\right) \tag{14}$$

T_w (°C) is the mean daily water temperature and ε is the emissivity of the water surface. The emissivity of water surface varies between 0.95 and 0.99 for water temperatures below 55 °C [80]. An average value of ε = 0.97 has been used in this study. The net longwave radiation is therefore:

$$R_{NL} = R_{IL} - R_{OL} \quad \left(\text{MJ}\cdot\text{m}^{-2}\cdot\text{day}^{-1}\right) \tag{15}$$

The water heat storage flux (H_S) (MJ/m^2/day) expresses the change in the heat stored in the water from one day to another. The heat storage flux calculation methods used in two different studies by Abtew et al., and Finch et al. are suitable for shallow water bodies evaporation [73,75]. Since Lake Mead is a deep lake, the equilibrium temperature approach proposed by de Bruin has been used instead. In this approach, an equilibrium temperature is used to estimate a mean daily uniform temperature of the water body for each day [81]. The heat storage flux's formula using de Bruin's method is [78,81–83]:

$$H_S = \rho_w C p_w d_w \times \frac{\left(T_{uw,j} - T_{uw,j-1}\right)}{\Delta t} \quad \left(\text{MJ}\cdot\text{m}^{-2}\cdot\text{day}^{-1}\right) \tag{16}$$

The constants' values ρ_w (kg/m^3), Cp_w (MJ/kg/°C), d_w (m) are, respectively, the density of water, the heat capacity of water, and the depth of the lake. $T_{uw,j}$ and $T_{uw,j-1}$ are, respectively, the mean uniform water temperature for day (j), and day ($j-1$). Δt is the time step for the temperature estimation. The mean uniform water temperature ($T_{uw,j}$) depends on the equilibrium temperature (T_e) (°C) and the time constant (τ) (day):

$$T_{uw,j} = T_e + \left(T_{uw,j-1} - T_e\right) \times \exp\left(\frac{-1}{\tau}\right) \quad (°C) \tag{17}$$

$$T_e = T_{wb} + \frac{R_{N,wb}}{4 \times \sigma \times (T_{wb} + 273.15)^3 + f_w \times (\Delta_{wb} + \gamma)} \quad (°C) \tag{18}$$

$$\tau = \frac{\rho_w \times Cp_w \times d_w}{4 \times \sigma \times (T_{wb} + 273.15)^3 + f_w \times (\Delta_{wb} + \gamma)} \quad (\text{day}) \tag{19}$$

$R_{N,wb}$ (MJ/m^2/day), T_{wb} (°C), and Δ_{wb} (kPa/K) are, respectively, the net radiation at the wet-bulb temperature, the wet-bulb temperature, and the slope of the saturation vapor pressure curve at the wet-bulb temperature. The wet-bulb temperature (T_{wb}) is calculated using the following equation [78,83]:

$$T_{wb} = \frac{0.00066 \times 100 \times T_a + \frac{(4098 \times P_a \times T_d)}{(T_d + 237.3)^2}}{0.00066 \times 100 + \frac{(4098 \times P_a \times T_d)}{(T_d + 237.3)^2}} \quad (°C) \tag{20}$$

The saturation vapor pressure curve at the wet-bulb temperature Δ_{wb} (kPa/K) is calculated by:

$$\Delta_{wb} = \frac{4096 \times 0.6108 \times \exp\left(\frac{17.27 \times T_{wb}}{T_{wb} + 237.3}\right)}{(T_{wb} + 237.3)^2} \quad \left(\text{kPa}\cdot\text{K}^{-1}\right) \tag{21}$$

The net radiation ($R_{N,wb}$) at the wet-bulb temperature is:

$$R_{N,wb} = (1 - a) \times R_S + \left(R_{IL} - R_{OL,wb}\right) \quad \left(\text{MJ}\cdot\text{m}^{-2}\cdot\text{day}^{-1}\right) \tag{22}$$

In Equation (22), $R_{OL,wb}$ (MJ/m^2/day) is the outgoing longwave radiation at the wet-bulb temperature and is calculated by:

$$R_{OL,wb} = C_f \times \sigma \times \left((T_a + 273.15)^4 + 4 \times (T_a + 273.15)^3 \times (T_{wb} - T_a)\right) \quad \left(\text{MJ}\cdot\text{m}^{-2}\cdot\text{day}^{-1}\right) \tag{23}$$

After the calculation of all parameters, the lake evaporation's value (E_L) can be calculated using Equation (1).

2.3. Energy Production Modeling

The power output of a PV module (P_{out}) (W) is calculated by applying different losses to the incoming solar irradiance and is given by:

$$P_{out} = I_S \times A_P \times \eta_P \quad (\text{W}) \tag{24}$$

where I_S (W/m^2) is the incoming solar irradiance, A_P (m^2) is the effective area of the solar panel, and η_P (%) is the efficiency of the PV system. In this study, the efficiency of the system includes the electrical efficiency of the module, which is dependent on the operating temperature, the shading losses, the soiling and hotspot losses, and the mismatch losses. Additionally, the solar irradiation component used is the global horizontal irradiation because the inclination of the panels is 0°. The power output is calculated hourly and summed up to determine the energy production of the system over a year.

2.3.1. FPV Operating Temperature

The energy produced by a photovoltaic system depends on the electrical efficiency of the modules. The electrical efficiency of the modules (η_e) changes with the operating temperature of the cell and is calculated using Equation (25) [45,84]:

$$\eta_e = \eta_{ref} \times \left[1 - \beta \times \left(T_{eo} - T_{ref}\right)\right] \qquad (\%) \tag{25}$$

where η_{ref} (%), β_{ref} (%/°C), T_{eo} (°C), and T_{ref} (°C) are, respectively, the reference efficiency of the panel, the temperature coefficient of the panel, the effective operating temperature of the panel, and the reference temperature.

The data collected from the FPV test bed are used to determine the effective operating temperature (T_{eo}) of the FPV. The model describing the temperature dependence on the ambient temperature and the solar power in nominal operating cell temperature (NOCT) conditions is a linear model [84–86]:

$$T_{cell} = T_{me} + k \times I_S \qquad (°C) \tag{26}$$

T_{Cell} (°C) is the operating temperature of the solar cells, k (°C. m^2/W) is the coefficient of the relationship, I_S (W/m^2) is the solar irradiance, and T_{me} (°C) is the ambient temperature of the location of the solar module. This model is well-adapted for offshore, roof or ground mounted, PV systems but needs to be modified to accurately describe FPV systems. A study conducted by Kamuyu et al. [45] has proposed a solar cell temperature calculation in FPV using the air temperature, the water temperature, the solar irradiance, and the wind speed. Kamuyu et al.'s study focused on FPV mounted at a tilt angle relative to the water's surface where the air temperature and wind speed played a larger role in determining the module temperature than the water temperature. In this study, because the modules are on/under the water surface, wind speed is neglected, and the water temperature plays the dominant role in module temperature. Thus, the Kamuyu approach for pontoon-based FPV was adapted and used here with experimental data for solar flux, air temperature, water temperature, and module temperature. The approach used was a multilinear variable regression. The regression has three independent variables that are the solar irradiance (I_S), the water temperature (T_w), and the air temperature (T_a). The last variable of the regression, the FPV module's effective operating temperature (T_{eo}), depends on the previous three. The goal of the regression is to find a linear relationship between the module's effective operating temperature (T_{eo}), and the three independent variables in the form of:

$$T_{eo} = \alpha_0 + \alpha_1 T_w + \alpha_2 T_a + \alpha_3 I_S \qquad (°C), \tag{27}$$

where α_0 is a constant term; α_1, α_2, and α_3 are the regression coefficients relative to the water temperature, air temperature, and solar irradiance, respectively.

The solar module temperature dataset from the test bed has been stored in a MATLAB column vector, and the independent variables have been stored within a MATLAB numeric matrix to which an additional unit column has been added at the beginning to account for the coefficient α_0. Then, the regression is performed using a dedicated MATLAB function called "regress" [87]. The "regress" function performs a multivariable regression on experimental data and outputs the coefficients of the regression as well as other values such as the R-squared value of the regression and the residuals. The experimentally obtained coefficients are used in the case study of Lake Mead to estimate the effective operating module temperatures that are, in turn, used in the energy yield simulation.

2.3.2. Other Loss Factors

This study focuses on the FPV system; therefore, the other factors considered are only related to the panels. In the case of a complete system design, losses from other equipment such as the inverter or transformer need to be considered. Other factors that could impact the efficiency of the floating solar

PV modules are the same as conventional land-based PV systems. These factors are solar irradiance losses, shading losses, soiling, mismatch losses, and DC cabling losses [60,88,89].

The foam-based support as well as the PV are mounted flat on the water surface (e.g., tilt angle = 0 degrees); therefore, they are not exposed to the optimum amount of solar irradiation for any location other than those on the equator. A study conducted by Jacobson et al. has provided an estimate of the optimal tilt for fixed tilt solar PV systems for different locations throughout the world [90]. The loss due to the tilt angle has been taken into account in this study and only the global horizontal irradiation for the energy yield calculation is used.

The impact of shading losses on FPV is low because water surfaces are flat and there are no nearby obstacles that could cause a direct shade to the modules. In the case of foam-based FPV, there is no mutual shade between the modules either because the mounting systems are flat on the water surface. Lake Mead is located in a mountainous region; therefore, far horizon shading may occur during certain times of the day or the year but is expected to be minimal. A detailed shading losses analysis has not been conducted during this study and an estimated value of zero percent has been used.

Soiling can be significant on FPV panels. Soiling in the case of FPV systems is mostly due to bird dropping or algae growth [54]. According to a report on FPV systems by the World Bank Group, nesting birds have been found to prefer the use of FPV modules as a nesting place [60]. In the report of the World Bank Group, however, the floating systems used were inclined; thus, allowing the presence of sheltered places where the birds were nesting. In the case of foam-based FPV, it has been assumed that the effect of birds will be lower because the modules are mounted directly on the water surface and the mounting system offers no sheltered space for nesting. A detailed study of the impact of birds nesting on foam-based FPV panels would be interesting for future studies. In addition, by ensuring the FPV are above the water surface, the growth of algae on the front surface of the FPV can be minimized.

Mismatch losses and DC cable losses can be higher in FPV systems due to the relative movement of the modules on the water surface, but an optimum design can minimize these losses [60].

2.3.3. Parameters Used for Energy Yield Simulation

The energy production model simulates a floating solar PV system on the surface of Lake Mead. The area covered by the solar PV installation is described in Section 2.4. The values used for the energy production simulation as well as the sources of the values are given in Table 1.

Table 1. Energy modeling simulation parameters.

Parameters	Value	Source
Solar PV temperature model	(Equation (27))	This study
Reference efficiency of the module	23%	[70]
Module inclination	0°	This study
Shading losses	0%	This study
Soiling	3%	[60]
Mismatch losses	6%	[89]
DC cable losses	3%	[89]

2.4. Water Savings Capability and Efficiency of the System

The water savings capability of the FPV system investigated in this study has been estimated to be 90% of the volume of water corresponding to the surface covered by the FPV. This assumption is supported by previous studies that found that covering water surfaces with pontoon-based FPV could reduce the evaporation losses by more than 90% [38,91]. Thus, the resulting values are extremely conservative as here the FPV covers the entire water surface and is not a tilted FPV mount as in [38,91]. When planning an FPV installation on a water surface, the percentage coverage of the water by the solar systems depends on the type of activities that are being performed on the body of water. According to the World Bank Group, the FPV system should not cover more than 50% of the water surface if used for fishing and not more than 60% if the water body is not used for fishing [60]. Therefore, a sensitivity analysis will be run on the coverage percentage to investigate the energy production and water saving capability of the foam-based FPV system in this study from 10% to 50% in 10% increments because Lake Mead is used for fishing. Then, the water saving capability is estimated by multiplying the water evaporation rate and the surface coverage. The result is adjusted by 90%. The cost of water saved annually is estimated using the average water cost in Nevada where Lake Mead serves as a clean water source. The cost of water according to Las Vegas Valley Water District ranges from USD 0.35/m^3 to 1.37/m^3 for a family size residential home, according to the size of the installed water meter [92]. On the other hand, the wholesale electricity rate of the power produced at the Hoover Dam, located in Nevada, is USD 0.02/kWh [93,94]. These values are used to estimate the lowest and highest potential energy revenues of the foam-based FPV system.

3. Results

3.1. Water Evaporation

The results from the water evaporation model simulation at Lake Mead show an evaporation rate estimate of 1957 mm in 2018. This result is in agreement with the results of the study conducted by Moreo and Swancar [55] on Lake Mead during the period of March 2010 through February 2012 using the eddy covariance evaporation method. The study estimated the lake evaporation from March 2010 to February 2011, and from March 2011 to February 2012. According to the two authors, the evaporation rate for the first study period had a minimum value of 1958 mm and a maximum value of 2190 mm; while the minimum value was 1787 mm and the maximum value was 1975 mm for the second study period. The result obtained in this present study is located within the result interval of Moreo and Swancar's study. Another early study by Westenburg et al. provided the evaporation data for Lake Mead from 1997 to 1999 [95]. The average evaporation rate for that period was 2281 mm.

Figure 4a shows the monthly results of the evaporation rates' simulation using 2018 data. The evaporation rate is low in the winter and increases in summer. The evaporation rate at the peak of the summer, in June, is approximately five times more important than the lowest evaporation rate of the winter, in December. Figure 4b shows the daily evaporation estimates throughout the year.

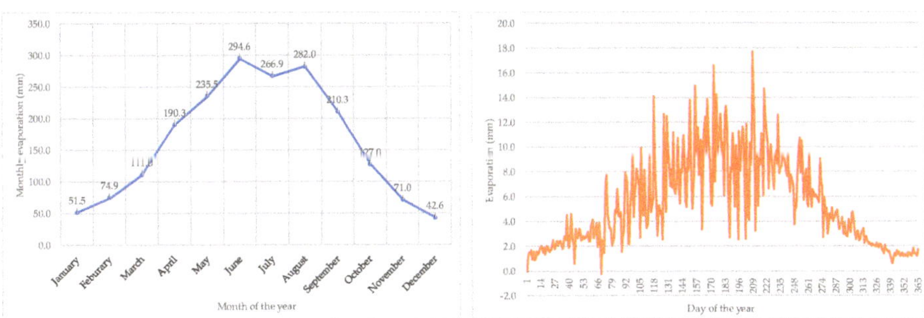

(a) (b)

Figure 4. Water evaporation simulation results for Lake Mead: (**a**) simulated evaporation values (mm) for each month of the year 2018; (**b**) simulated evaporation values (mm) for each day of the year 2018.

3.2. Energy Production

3.2.1. FPV Operating Temperature Model

The multilinear regression on the collected data yielded the coefficients α_0, α_1, α_2, and α_3, which describe the relationship between the FPV effective operating temperature (T_{eo}) and the independent variables: the water temperature (T_w), the air temperature (T_a), and the solar irradiance (I_S). The regression coefficients have been obtained with an R-squared value of 0.8276. Figure 5 shows the statistical results of the regression. The R-squared value combined with the random distribution of the residuals' plot on Figure 5b show that there is a linear relationship between T_{eo} and the independent variables.

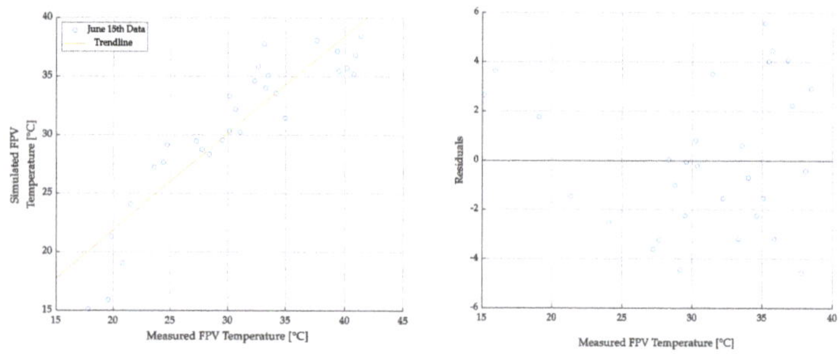

(a) (b)

Figure 5. Multilinear regression results of the FPV panels' effective operating temperature (T_{eo}): (**a**) simulated FPV temperature plotted against the measured temperature for 15 June 2020; (**b**) residuals' distribution plotted against the simulated FPV temperature for 15 June 2020.

Equation (28) is proposed as a model that represents the effective operation temperature of FPV mounted on a foam-based support.

$$T_{eo} = -13.2554 - 0.0875 \times T_w + 1.2645 \times T_a + 0.0128 \times I_S \qquad (°C) \tag{28}$$

Figure 6 shows the simulated operating temperature using the proposed model, the operating temperature of a titled aluminum pontoon-based mount FPV model based on the original Kamuyu et al.'s model (for pontoon-based tilted FPV), and the measured operating temperature for 15 June 2020. The simulated temperature is at times higher or lower than the measured temperature, but the overall trend of the two temperature profiles matches. The model proposed in this study is compared to the unadapted tilted FPV model and the current model (which is an adaptation of Kamuyu et al.'s model for foam-backed flat-surface FPV) and provides a better description of a foam-based FPV panel's operating temperature. The proposed model in this study has a similar profile to Kamuyu's model, and the proposed model provides a better description of the foam-based solar module's behavior.

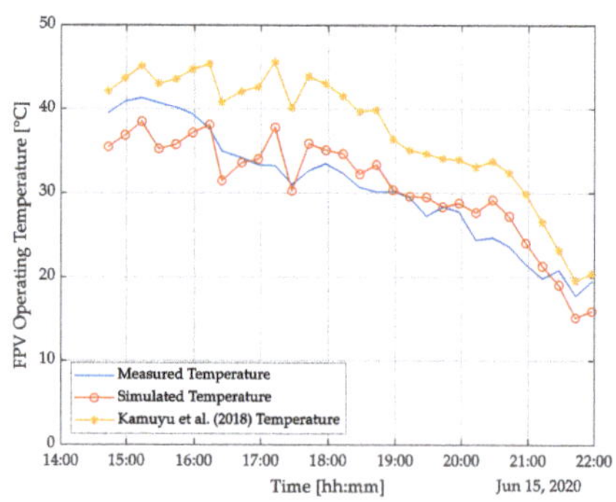

Figure 6. Measured FPV operating temperature compared to simulated FPV operating temperature for 15 June 2020.

The temperature profile of a foam-based FPV panel installed on Lake Mead has been simulated using the proposed FPV operating temperature model and compared to a pontoon-based FPV as described by Kamuyu's model in Figure 7.

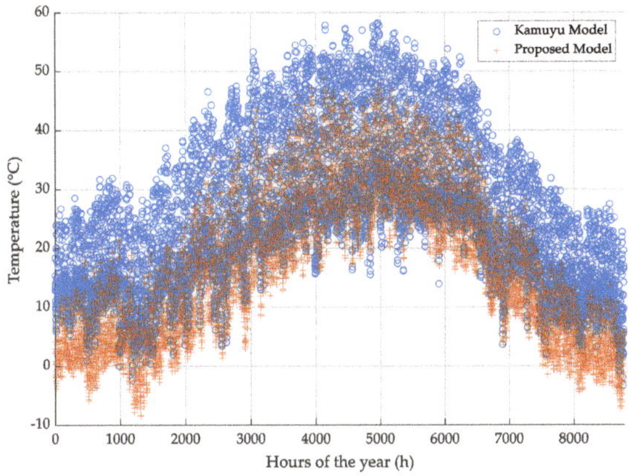

Figure 7. Operation temperature of an FPV installed on the surface of Lake Mead. (**+**) Operating temperature using the proposed model in this study for foam-based FPV. (**o**) Operating temperature using a ponton-based tilted FPV described by Kamuyu's model.

The maximum temperature obtained with the model proposed in this study is 48.7 °C and the minimum temperature is −8.5 °C. On the other hand, the maximum temperature and the minimum temperature obtained if the FPV system was tilted are, respectively, 58.2 °C and −3.4 °C. Overall, the temperature model used here based on experimental data during the summer months predicts a lower temperature when the panels are in direct contact with the water surface.

3.2.2. Energy Yield and Water Savings of an FPV System Installed on Lake Mead

The temperature profile is used to estimate the electrical efficiency of the solar panel, which is in turn used to simulate the energy yield of an FPV system installed on Lake Mead with historical weather data. The energy yield has been simulated by assuming a coverage of the lake surface between 10 and 50% in 10% increments. The results are shown in detail for the 10% coverage case and the total energy production is shown for the other cases.

Figure 8 shows the comparison between the monthly energy production obtained using the proposed model and the energy production of a tilted FPV for 10% coverage of the lake's surface. As can be seen in Figure 8 and expected from Figure 7, the proposed model predicts a slightly higher energy production, about 3.5%, which is correlated with the lower operating temperature of the modules. The maximum energy per month production is 3.2 TWh and occurs in the month of June, while the minimum energy per month production is 1.1 TWh and occurs in December.

Figure 9 shows the result for the daily energy simulation when 10% of Lake Mead's surface is covered with a foam-based solar FPV system. The maximum daily energy production is 570 MWh on 6 January while the minimum daily production is 21 MWh on 18 June.

Figure 8. Monthly energy yield of a simulated foam-based FPV system installed on 10% of Lake Mead's surface using historical data from 2018. Comparison between the proposed model (c-Si flexible foam-backed FPV) and a tilted FPV based on Kamuyu's model (c-Si aluminum mount FPV).

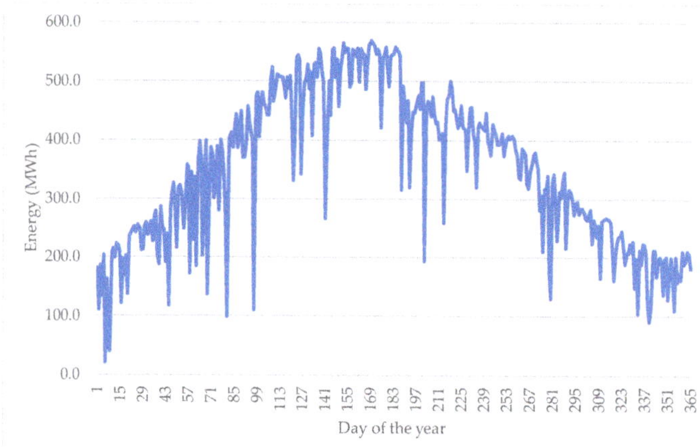

Figure 9. Daily energy production results using the temperature model proposed in this study for 10% coverage of Lake Mead's surface.

Figure 10 shows the simulated annual energy production, and the water saving capabilities of a foam-based solar FPV system installed on the surface of Lake Mead as a function of coverage area from 10–50%. For a coverage of 10%, the annual production using collected temperature data is 25.59 TWh, corresponding to a saved water volume of 126.64 million m^3. When the percentage coverage is increased, the energy production is increased linearly. For a coverage of 50% of the lake's surface, it is possible to harvest 127.93 TWh of electrical energy and save 633.22 million of m^3 of water using foam-based FPV panels.

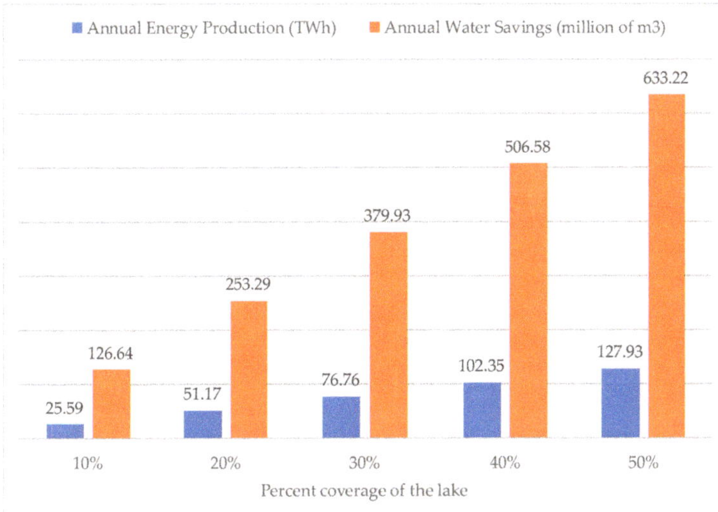

Figure 10. Simulated annual energy production (TWh) and water saving capability (millions of m^3) of a foam-based solar FPV system installed on Lake Mead's surface using historical temperature data and the proposed model depending on the percentage coverage of the lake's surface.

Table 2 shows the annual water and energy savings estimates related to the water savings and energy production from the FPV. With houses with the least water consumption, the cost of the water saved is estimated to be USD 44 million when 10% of the lake surface is covered, and USD 220 million when 50% of the lake surface is covered. On the other hand, when the consumers' water consumption is on the high side, these costs increase, amounting to USD 172 million when 10% of the lake is covered and USD 861 million when 50% is covered. Furthermore, the results for the energy production show that USD 0.5 billion of energy can be generated when 10% of the lake surface is covered. The value of the energy generated when 50% of the lake surface is covered is estimated as USD 2.6 billion.

Table 2. Estimation of the yearly cost of water saved and energy produced using water and energy cost range from Nevada for an FPV system covering 10–50% of Lake Mead's surface.

Lake Surface Percent Coverage	Water Savings at $0.35/m^3 (Millions of $)	Water Savings at $1.37/m^3 (Millions of $)	Energy Revenues at 2¢/kWh (Billions of $)
10%	43.99	172.19	0.51
20%	87.98	344.37	1.02
30%	131.97	516.56	1.54
40%	175.96	688.75	2.05
50%	219.95	860.94	2.56

The relative values of the water and energy provided by the foam-based FPV indicate that the electricity production from the FPV could be used to subsidize water conservation in arid and semi-arid areas. Thus, FPV could be a self-funded water conservation approach.

4. Discussion

The water evaporation calculation performed in this study predicts a significant water saving potential for foam-based FPV systems on Lake Mead. The evaporation calculation using historical data has shown an annual evaporation estimate of 1957.6 mm for the lake. The result of the calculation performed in this study is in agreement with previous evaporation studies on Lake Mead [55,95].

The simulation results show annual water savings ranging from 126.64 to 633.22 million m^3 depending on the percentage of the lake surface covered by the FPV system. According to the United States Environmental Protection Agency (US EPA), each American uses, on average, 88 gallons of water per day, resulting in an annual water consumption of 32,120 gallons or 121.59 m^3 per capita [96]. The amount of water saved using foam-based FPV on Lake Mead will therefore be enough to supply water to more than five million Americans in the case that 50% of the lake surface is covered. This would make a significant impact on the cities near Lake Mead. The value is more than the four million population of the second largest city in the country, Los Angeles [97] or the entire population of Nevada of 3.1 million [98]. When 10% of the lake is covered by FPV panels, the amount of water saved is enough to supply water to the populations of both Henderson (320,189) and Las Vegas (651,319) or Las Vegas and Reno (255,601) in Nevada [99]. According to an analysis performed by Barsugli et al., Lake Mead has a 50% percent chance of going dry between 2035 and 2047 if nothing is done to stop the current draw down and evaporation rate of the lake [100]. Other studies on the management of the lake have resulted in the same conclusion [101,102]. These studies have shown the need for new ways to mitigate lake evaporation not only on Lake Mead, but on other lakes in the world, especially those located in arid environments. Floating solar photovoltaic technology provides a solution to limit evaporation of water surfaces and provide electrical energy for the surrounding populations.

The energy production analysis has yielded an annual energy production ranging from 25.9 TWh to 127.93 TWh for a coverage of the lake of 10%, and 50%, respectively. The energy production profile is in accordance with a previous FPV study conducted by Kamuyu et al. [45], showing an improvement of 10% from a ground-mount system. This is confirmed by the study of Pierce et al., who determined that the energy production improvement of an FPV systems is 5–10% compared to a ground-mount system for mono and polycrystalline silicon [54]. This is due to the cooling effect of the water on the FPV modules. According to the United States Energy Information Agency (US EIA), the average American household electricity consumption is 10,649 kWh per year. This means that the energy production of an FPV system covering 10% of Lake Mead has the capacity to power more than two million American homes [103]. This is more than the electricity needed to power the homes in Las Vegas, Henderson and Reno combined. On the other hand, the total electricity consumption in the U.S. according to 2018 statistics is 4178 TWh. This implies that the electricity production from a solar FPV system covering 50% of Lake Mead can supply 3% of the total electricity consumed in the U.S. and can replace more than 11% of the coal-fired power plants operating in the country; thus, contributing in a significant way to the reduction in the national carbon dioxide emissions [104] and the concomitant air pollution-related mortality [105–108]. This study is in agreement with past work showing enormous potential for FPV on water bodies in the U.S. [109].

The results of this study show that there are several benefits to implementing a foam-based FPV solar plant. Foam-based FPV avoids the issues related to land use in ground-mounted solar PV [110] and since the floating device is made of low-cost material, the racking cost is lower than other raft racking FPV technologies [54]. In addition, FPV systems in general have the potential to form agrivoltaic type systems [111] by merging with aquaculture to form aquavoltaics [112,113]. The flexible foam-backed FPV approach used here even makes mobile FPV possible. The FPV approach demonstrated here is less expensive than conventional pontoon-based FPV and has a slightly higher energy output per W because of the modules' close proximity to the water. FPV racking in general is less costly than conventional ground mounted PV. Thus, as PV is already often the least costly method for new electricity production, it provides a potentially profitable means of reducing water evaporation in the world's dwindling bodies of fresh water. Overall, the results of this study appear extremely promising. Solar FPV is a fairly new technology that is growing at a tremendous rate, but for it to reach its full potential, future work is needed to explore policies that sustain the development of this technology while also minimizing negative externalities. To accomplish this, a full life cycle analysis (LCA) study is needed on this technology.

Future work is also needed to experimentally verify the results of this study in different locations throughout the world. In addition, future work is needed to investigate fouling (and means to prevent it) in different bodies of water. More data should also be collected to further refine the temperature model and improve the energy production accuracy of the results shown here. Foam-based technology used as a floating device needs to be investigated more in order to have a commercially viable mass-produced FPV foam racking. The work shown here and completed previously was accomplished using after-market alterations of flexible PV modules. It should be pointed out that economic calculations used here assumed a 25 year lifetime for the PV modules. Although they are rated for extreme environments, guaranteed to resist corrosion and waterproof, the flexible SunPower modules only carry a 5 year warranty rather than the industry standard 25–30 year warranty. Future work to test the long-term performance of such systems is needed to ensure the reliability and safety of a foam-backed FPV as described in this article. In addition, future technical work is needed to investigate the potential for making flexible modules rated for high voltages that would be more appropriate for utility scale systems such as described in this study. The cost of the FPV racking would be further reduced by integrating bulk purchased foam into the PV manufacturing process. In addition, closed loop, circular economy [114–116] and industrial symbiosis [117] could be applied to the FPV manufacturing process. This would be expected to further reduce the cost of the FPV as well but may also necessitate policy intervention to ensure end of life recycling [118]. The polyethylene foam used here could be fabricated from recycled plastic waste [119–121], thereby further improving the environmental balance sheet for foam-backed FPV. Future studies can potentially look into the long-term stability of foam in water by analyzing the effect of different qualities of water on this material. Another aspect of foam-based rack where future work is needed is the mooring technology used to secure the FPV. Finally, the environmental impacts of the floating solar systems on marine life have not been fully established [60] and will be an interesting subject for future studies.

5. Conclusions

This study introduced a new approach to FPV using a flexible crystalline silicon module backed with foam, which is less expensive than conventional pontoon-based FPV racking and land-based PV racking. The results show that the foam-backed FPV had a lower operating temperature than conventional pontoon-based fixed tilt out-of-water FPV and thus a higher energy output per unit power because of the modules' close proximity to the water. Thus, because PV costs are now normally the least costly method of new electricity production, the hypothetical large-scale foam-based FPV provides a potentially profitable means of reducing water evaporation in the world's at-risk bodies of fresh water.

The case study of Lake Mead found that if 10% of the lake was covered with foam-backed FPV, there would be more than enough solar electricity generated to power the homes in Las Vegas, Henderson and Reno combined and enough water savings for Las Vegas and Reno. At 50% lake coverage, the foam-backed FPV would provide over 127 TWh of clean solar electricity and 633.2 million m^3 of water savings, which would provide enough electricity to retire 11% of the polluting coal-fired plants in the U.S. and water for over five million Americans, annually.

Author Contributions: Conceptualization, J.M.P.; methodology, K.S.H., P.M. and R.K.K.; software, K.S.H.; validation, K.S.H. and R.K.K.; formal analysis, K.S.H., P.M., R.K.K. and J.M.P.; investigation, K.S.H. and R.K.K.; resources, J.M.P.; data curation, K.S.H. and P.M.; writing—original draft preparation, K.S.H., P.M., R.K.K. and J.M.P.; writing—review and editing, K.S.H., P.M., R.K.K. and J.M.P.; visualization, K.S.H. and P.M.; supervision, J.M.P.; project administration, J.M.P.; funding acquisition, J.M.P. All authors have read and agreed to the published version of the manuscript.

Funding: This research was supported by the Witte endowment.

Acknowledgments: The authors would like to acknowledge technical support from Shane Oberloier, helpful conversations with Nelson Sommerfeldt, as well as the support of SOLCAST who provided historical solar data for the simulations performed in the study.

Conflicts of Interest: The authors declare no conflict of interest.

Glossary

Symbol	Name	Unit
P_a	actual saturation vapor pressure	(kPa)
r_a	aerodynamic resistance	(s/m)
ρ_a	air density	(kg/m^3)
a	albedo	-
h	altitude	(m)
T_a	average daily air temperature	(°C)
T_a	average daily air temperature	(°C)
P	average daily atmospheric pressure	(kPa)
T_d	average daily dew temperature	(°C)
T_w	average daily water temperature	(°C)
w_s	average daily wind speed	(m/s)
R_{CS}	clear sky radiation	(MJ/m^2/day)
C_f	cloud coverage fraction	-
d_w	effective depth of the lake	(m)
T_{eo}	effective operating temperature	(°C)
η_{ref}	efficiency at reference temperature	(%)
η_e	electrical efficiency	(%)
ε	emissivity of water	-
T_e	equilibrium temperature	(°C)
R_{EX}	extraterrestrial radiation	(MJ/m^2/day)
I_S	global horizontal irradiation	(W/m^2)
R_S	global horizontal irradiation	(MJ/m^2/day)
Cp_a	heat capacity of air	(kJ/kg/°C)
Cp_w	heat capacity of water	(MJ/kg/°C)
H_S	heat storage flux	(MJ/m^2/day)
R_{IL}	incoming longwave radiation	(MJ/m^2/day)
E_L	lake evaporation	(mm)
λ	latent heat of vaporization of water	(MJ/kg)
ϕ	latitude	(rad)
$T_{a,max}$	maximum daily air temperature	(°C)
Rh_{max}	maximum daily relative humidity	(%)
$T_{w,max}$	maximum daily water temperature	(°C)
P_w	mean saturation vapor pressure	(kPa)
T_{uw}	mean uniform temperature of water	(°C)
$T_{a,min}$	minimum daily air temperature	(°C)
Rh_{min}	minimum daily relative humidity	(%)
$T_{w,min}$	minimum daily water temperature	(°C)
R_{NL}	net longwave radiation	(MJ/m^2/day)
$R_{N,wb}$	net radiation at wet-bulb temperature	(MJ/m^2/day)
R_{NS}	net shortwave radiation	(MJ/m^2/day)
R_N	net solar radiation	(MJ/m^2/day)
R_{OL}	outgoing longwave radiation	(MJ/m^2/day)
$R_{OL,wb}$	outgoing longwave radiation at wet-bulb temperature	(MJ/m^2/day)
P_{out}	output power	(W)
A_P	photovoltaic surface	(m^2)
η_P	photovoltaic system efficiency	(%)
γ	psychrometric constant	(kPa/°C)
T_{ref}	reference temperature	(°C)
Δ_{wb}	saturation vapor pressure curve at wet-bulb temperature	(kPa/K)
Δ	slope of saturation vapor pressure curve	(kPa/°C)
σ	Stephan–Boltzmann constant	(MJ/m^2/K^4/day)

A	surface of the lake	(m^2)
β	temperature coefficient of the PV panel	(%/°C)
τ	time constant	(day)
Δt	time step	(h/day)
ρ_w	water density	(kg/m^3)
T_{wb}	wet-bulb temperature	(°C)
f_w	wind function	($MJ/m^2/kPa/day$)

References

1. Arnell, N. Climate change and global water resources. *Glob. Environ. Chang.* **1999**, *9*, S31–S49. [CrossRef]
2. Kummu, M.; Ward, P.J.; de Moel, H.; Varis, O. Is physical water scarcity a new phenomenon? Global assessment of water shortage over the last two millennia. *Environ. Res. Lett.* **2010**, *5*, 034006. [CrossRef]
3. Coyle, E.D. (Ed.) *Understanding the Global Energy Crisis*; Purdue Studies in Public Policy; Purdue University Press: West Lafayette, IN, USA, 2014; ISBN 978-1-55753-661-7.
4. Brown, L.R. *Full Planet, Empty Plates: The New Geopolitics of Food Scarcity*, 1st ed.; W.W. Norton & Company: New York, NY, USA, 2012; ISBN 978-0-393-08891-5.
5. Baum, S.D.; Denkenberger, D.C.; Pearce, J. Alternative Foods as a Solution to Global Food Supply Catastrophes. *Solutions* **2016**, *7*, 31–35.
6. Rockström, J.; Falkenmark, M.; Karlberg, L.; Hoff, H.; Rost, S.; Gerten, D. Future water availability for global food production: The potential of green water for increasing resilience to global change: Water availability for food production. *Water Resour. Res.* **2009**, 45. [CrossRef]
7. Misra, A.K. Climate change and challenges of water and food security. *Int. J. Sustain. Built Environ.* **2014**, *3*, 153–165. [CrossRef]
8. Cook, J.; Oreskes, N.; Doran, P.T.; Anderegg, W.R.L.; Verheggen, B.; Maibach, E.W.; Carlton, J.S.; Lewandowsky, S.; Skuce, A.G.; Green, S.A.; et al. Consensus on consensus: A synthesis of consensus estimates on human-caused global warming. *Environ. Res. Lett.* **2016**, *11*, 048002. [CrossRef]
9. Solomon, S.; Intergovernmental Panel on Climate Change (Eds.) *Climate Change 2007: The Physical Science Basis: Contribution of Working Group I to the Fourth Assessment Report of the Intergovernmental Panel on Climate Change*; Cambridge University Press: New York, NY, USA, 2007; ISBN 978-0-521-88009-1.
10. Kellogg, W.W.; Schware, R. *Climate Change and Society: Consequences of Increasing Atmospheric Carbon Dioxide*, 1st ed.; Routledge: New York, NY, USA, 2019; ISBN 978-0-429-04873-9.
11. Bates, B.; Kundzewicz, Z.W.; Wu, S.; Palutikof, J.P. (Eds.) *Climate Change and Water. Technical Paper of the Intergovernmental Panel on Climate Change*; IPCC Technical Paper 6; IPCC Secretariat: Geneva, Switzerland, 2008; ISBN 978-92-9169-123-4.
12. Baker, E.; Fowlie, M.; Lemoine, D.; Reynolds, S.S. The Economics of Solar Electricity. *Annu. Rev. Resour. Econ.* **2013**, *5*, 387–426. [CrossRef]
13. de Wit, M.; Stankiewicz, J. Changes in Surface Water Supply Across Africa with Predicted Climate Change. *Science* **2006**, *311*, 1917–1921. [CrossRef]
14. Edenhofer, O.; Pichs-Madruga, R.; Sokohana, Y.; Farahani, E.; Kadner, S.; Seyboth, K.; Adler, A.; Baum, I.; Brunner, S.; Eickemeier, P.; et al. (Eds.) *IPCC Climate Change 2014: Mitigation of Climate Change: Working Group III Contribution to the Fifth Assessment Report of the Intergovernmental Panel on Climate Change*; Cambridge University Press: New York, NY, USA, 2014; ISBN 978-1-107-05821-7.
15. Pearce, J.M. Photovoltaics—A path to sustainable futures. *Futures* **2002**, *34*, 663–674. [CrossRef]
16. Creutzig, F.; Agoston, P.; Goldschmidt, J.C.; Luderer, G.; Nemet, G.; Pietzcker, R.C. The underestimated potential of solar energy to mitigate climate change. *Nat. Energy* **2017**, *2*, 17140. [CrossRef]
17. Calvert, K.; Pearce, J.M.; Mabee, W.E. Toward renewable energy geo-information infrastructures: Applications of GIScience and remote sensing that build institutional capacity. *Renew. Sustain. Energy Rev.* **2013**, *18*, 416–429. [CrossRef]
18. SEIA Siting, Permitting & Land Use for Utility-Scale Solar. Available online: https://www.seia.org/initiatives/siting-permitting-land-use-utility-scale-solar (accessed on 13 November 2020).
19. Kenny, R.; Law, C.; Pearce, J.M. Towards real energy economics: Energy policy driven by life-cycle carbon emission. *Energy Policy* **2010**, *38*, 1969–1978. [CrossRef]

20. Groesbeck, J.G.; Pearce, J.M. Coal with Carbon Capture and Sequestration is not as Land Use Efficient as Solar Photovoltaic Technology for Climate Neutral Electricity Production. *Sci. Rep.* **2018**, *8*, 13476. [CrossRef] [PubMed]
21. FAO (Ed.) *How does International Price Volatility Affect Domestic Economies and Food Security*; The state of food insecurity in the world; FAO: Rome, Italy, 2011; ISBN 978-92-5-106927-1.
22. Yasmeena, S.; Das, G.T.R. A Review on New Era of Solar Power Systems: Floatovoltaic Systems or Floating Solar Power Plants. *JIC* **2015**, *3*, 1–8. [CrossRef]
23. Majid, Z.A.A.; Ruslan, M.H.; Sopian, K.; Othman, M.Y.; Azmi, M.S.M. Study on Performance of 80 Watt Floating Photovoltaic Panel. *J. Mech. Eng. Sci.* **2014**, *7*, 1150–1156. [CrossRef]
24. Trapani, K.; Millar, D.L. The thin film flexible floating PV (T3F-PV) array: The concept and development of the prototype. *Renew. Energy* **2014**, *71*, 43–50. [CrossRef]
25. do Sacramento, E.M.; Carvalho, P.C.; de Araújo, J.C.; Riffel, D.B.; da Cruz Corrêa, R.M.; Neto, J.S. Scenarios for use of floating photovoltaic plants in Brazilian reservoirs. *Iet Renew. Power Gener.* **2015**, *9*, 1019–1024. [CrossRef]
26. Trapani, K.; Santafé, M.R. A review of floating photovoltaic installations: 2007–2013. *Prog. Photovolt. Res. Appl.* **2015**, *23*, 524–532. [CrossRef]
27. Patil, S.S.; Wagh, M.M.; Shinde, N.N. A review on floating solar photovoltaic power plants. *Int. J. Sci. Eng. Res.* **2017**, *8*, 789–794.
28. Kumar, N.M.; Kanchikere, J.; Mallikarjun, P. Floatovoltaics: Towards improved energy efficiency, land and water management. *Int. J. Civ. Eng. Technol.* **2018**, *9*, 1089–1096.
29. Haugwitz, F. Floating Solar PV Gains Global Momentum. Available online: https://www.pv-magazine.com/2020/09/22/floating-solar-pv-gains-global-momentum/ (accessed on 3 October 2020).
30. Rosa-Clot, M.; Rosa-Clot, P.; Tina, G.M.; Scandura, P.F. Submerged photovoltaic solar panel: SP2. *Renew. Energy* **2010**, *35*, 1862–1865. [CrossRef]
31. Tina, G.M.; Rosa-Clot, M.; Rosa-Clot, P.; Scandura, P.F. Optical and thermal behavior of submerged photovoltaic solar panel: SP2. *Energy* **2012**, *39*, 17–26. [CrossRef]
32. Ferrer-Gisbert, C.; Ferrán-Gozálvez, J.J.; Redón-Santafé, M.; Ferrer-Gisbert, P.; Sánchez-Romero, F.J.; Torregrosa-Soler, J.B. A new photovoltaic floating cover system for water reservoirs. *Renew Energy* **2013**, *60*, 63–70. [CrossRef]
33. Abdulgafar, S.A.; Omar, O.S.; Yousif, K.M. Improving the efficiency of polycrystalline solar panel via water immersion method. *IJIRSET* **2014**, *3*, 96–101.
34. Mehrotra, S.; Rawat, P.; Debbarma, M.; Sudhakar, K. Performance of a solar panel with water immersion cooling technique. *Int. J. Sci. Environ. Technol.* **2014**, *3*, 1161–1172.
35. McKay, A. *Floatovoltaics: Quantifying the Benefits of a Hydro-Solar Power Fusion*; Pomona College: Claremont, CA, USA, 2013.
36. Santafé, M.R.; Gisbert, P.S.F.; Romero, F.J.S.; Soler, J.B.T.; Gozálvez, J.J.F.; Gisbert, C.M.F. Implementation of a photovoltaic floating cover for irrigation reservoirs. *J. Clean. Prod.* **2014**, *66*, 568–570. [CrossRef]
37. Sharma, P.; Muni, B.; Sen, D. Design parameters of 10 KW floating solar power plant. In Proceedings of the International Advanced Research Journal in Science, Engineering and Technology (IARJSET), National Conference on Renewable Energy and Environment (NCREE-2015), Ghaziabad, India, 1 May 2015; Volume 2.
38. Rosa-Clot, M.; Tina, G.M.; Nizetic, S. Floating photovoltaic plants and wastewater basins: An Australian project. *Energy Procedia* **2017**, *134*, 664–674. [CrossRef]
39. Liu, L.; Wang, Q.; Lin, H.; Li, H.; Sun, Q.; Wennersten, R. Power Generation Efficiency and Prospects of Floating Photovoltaic Systems. *Energy Procedia* **2017**, *105*, 1136–1142. [CrossRef]
40. Mittal, D.; Saxena, B.K.; Rao, K.V.S. Floating solar photovoltaic systems: An overview and their feasibility at Kota in Rajasthan. In Proceedings of the 2017 International Conference on Circuit, Power and Computing Technologies (ICCPCT), Kollam, India, 20–21 April 2017; IEEE: New York, NY, USA, 2017; pp. 1–7.
41. Mittal, D.; Saxena, B.K.; Rao, K.V.S. Potential of floating photovoltaic system for energy generation and reduction of water evaporation at four different lakes in Rajasthan. In Proceedings of the 2017 International Conference on Smart Technologies for Smart Nation (SmartTechCon), Bengaluru, India, 17–19 August 2017; pp. 238–243.

42. Abid, M.; Abid, Z.; Sagin, J.; Murtaza, R.; Sarbassov, D.; Shabbir, M. Prospects of floating photovoltaic technology and its implementation in Central and South Asian Countries. *Int. J. Environ. Sci. Technol.* **2019**, *16*, 1755–1762. [CrossRef]
43. Teixeira, L.E.; Caux, J.; Beluco, A.; Bertoldo, I.; Louzada, J.A.S.; Eifler, R.C. Feasibility Study of a Hydro PV Hybrid System Operating at a Dam for Water Supply in Southern Brazil. *JPEE* **2015**, *3*, 70–83. [CrossRef]
44. Vasco, G.; Silva, J.S.; Beluco, A. Feasibility Study of a PV Hydro Hybrid System, With Photovoltaic Panels on Floating Structures. *Iop Conf. Ser. Mater. Sci. Eng.* **2018**, *366*, 012011. [CrossRef]
45. Kamuyu, W.C.L.; Lim, J.; Won, C.; Ahn, H. Prediction Model of Photovoltaic Module Temperature for Power Performance of Floating PVs. *Energies* **2018**, *11*, 447. [CrossRef]
46. Ranjbaran, P.; Yousefi, H.; Gharehpetian, G.B.; Astaraei, F.R. A review on floating photovoltaic (FPV) power generation units. *Renew. Sustain. Energy Rev.* **2019**, *110*, 332–347. [CrossRef]
47. Lee, A.-K.; Shin, G.-W.; Hong, S.-T.; Choi, Y.-K. A study on development of ICT convergence technology for tracking-type floating photovoltaic systems. *SGCE* **2014**, *3*, 80–87. [CrossRef]
48. Song, J.; Choi, Y. Analysis of the Potential for Use of Floating Photovoltaic Systems on Mine Pit Lakes: Case Study at the Ssangyong Open-Pit Limestone Mine in Korea. *Energies* **2016**, *9*, 102. [CrossRef]
49. Choi, Y.K.; Choi, W.S.; Lee, J.H. Empirical Research on the Efficiency of Floating PV Systems. *Sci. Adv. Mater.* **2016**, *8*, 681–685. [CrossRef]
50. Stachiw, J.D. Performance of Photovoltaic Cells in Undersea Environment. *J. Manuf. Sci. Eng.* **1980**, *102*, 51–59. [CrossRef]
51. Rathod, M.K.; Banerjee, J. Thermal stability of phase change materials used in latent heat energy storage systems: A review. *Renew. Sustain. Energy Rev.* **2013**, *18*, 246–258. [CrossRef]
52. Ho, C.J.; Chou, W.-L.; Lai, C.-M. Thermal and electrical performance of a water-surface floating PV integrated with a water-saturated MEPCM layer. *Energy Convers. Manag.* **2015**, *89*, 862–872. [CrossRef]
53. Ho, C.J.; Chou, W.-L.; Lai, C.-M. Thermal and electrical performances of a water-surface floating PV integrated with double water-saturated MEPCM layers. *Appl. Therm. Eng.* **2016**, *94*, 122–132. [CrossRef]
54. Mayville, P.; Patil, N.V.; Pearce, J.M. Distributed Manufacturing of After Market Flexible Floating Photovoltaic Modules. *J. Clean. Prod.* **2020**, in press. [CrossRef]
55. Moreo, M.T.; Swancar, A. *Evaporation from Lake Mead, Nevada and Arizona, March 2010 through February 2012*; Scientific Investigations Report; U.S. Geological Survey: Reston, VA, USA, 2013; p. 52.
56. NPS Overview of Lake Mead-Lake Mead National Recreation Area (U.S. National Park Service). Available online: https://www.nps.gov/lake/learn/nature/overview-of-lake-mead.htm (accessed on 29 June 2020).
57. Monteith, J.L. Evaporation and environment. *Symp. Soc. Exp. Biol.* **1965**, *19*, 205–234. [PubMed]
58. Allen, R.G.; FAO. *Crop Evapotranspiration: Guidelines for Computing Crop Water Requirements*; FAO Irrigation and Drainage Paper; Food and Agriculture Organization of the United Nations: Rome, Italy, 1998; ISBN 978-92-5-104219-9.
59. Hayibo, K.S.; Pearce, J.M. Calculations for Water Conservation Potential of Self-funded Foam-Based Flexible Surface-Mounted Floatovoltaics. *OSF* **2020**. Available online: https://osf.io/tweuy/ (accessed on 16 October 2020)
60. World Bank Group; ESMAP; SERIS. *Where Sun Meets Water: Floating Solar Handbook for Practitioners*; World Bank Group: Washington, DC, USA, 2019.
61. US Department of Commerce, N.O. and A.A. NDBC Station Page. Available online: http://www.ndbc.noaa.gov/station_page.php?station=nbba3 (accessed on 22 June 2020).
62. Weather Underground Las Vegas, NV Weather History | Weather Underground. Available online: https://www.wunderground.com/history/daily/us/nv/las-vegas/KLAS/date/2018-2-28 (accessed on 22 June 2020).
63. Solcast Solar Irradiance Data 2020. Available online: https://solcast.com (accessed on 19 September 2020).
64. NPS Storage Capacity of Lake Mead-Lake Mead National Recreation Area (U.S. National Park Service). Available online: https://www.nps.gov/lake/learn/nature/storage-capacity-of-lake-mead.htm (accessed on 22 June 2020).
65. US Department of Commerce, N.O. and A.A. NDBC Station History Page. Available online: http://www.ndbc.noaa.gov/station_history.php?station=nbba3 (accessed on 23 June 2020).
66. Hayibo, K.S. *Soul-Ash/floating-pv: Lake Mead Data Cleaning Code*; Zenodo, 2020; Available online: https://zenodo.org/record/3960777 (accessed on 26 July 2020).
67. Zotarelli, L.; Dukes, M.D.; Romero, C.C.; Migliaccio, K.W. *Step by Step Calculation of the Penman-Monteith Evapotranspiration (FAO-56 Method)*; University of Florida: Gainesville, FL, USA, 2018.

68. Weiss, A.; Hays, C.J. Calculating daily mean air temperatures by different methods: Implications from a non-linear algorithm. *Agric. For. Meteorol.* **2005**, *128*, 57–65. [CrossRef]
69. Shi, T.-T.; Guan, D.-X.; Wu, J.-B.; Wang, A.-Z.; Jin, C.-J.; Han, S.-J. Comparison of methods for estimating evapotranspiration rate of dry forest canopy: Eddy covariance, Bowen ratio energy balance, and Penman-Monteith equation. *J. Geophys. Res. Atmos.* **2008**, 113. [CrossRef]
70. Sunpower®SunPower Flexible Solar Panels | SPR-E-Flex-110. Available online: https://us.sunpower.com/sites/default/files/110w-flexible-panel-spec-sheet.pdf (accessed on 13 October 2020).
71. Foam Factory Foam Factory. Data Sheets | Foam Factory, Inc. Available online: https://www.foambymail.com/datasheets.html (accessed on 13 October 2020).
72. Domany, M.A.; Touchart, L.; Bartout, P.; Nedjai, R. The Evaporation From Ponds In The French Midwest. *Lakes Reserv. Ponds* **2013**, *7*, 75–88.
73. Abtew, W.; Melesse, A. *Evaporation and Evapotranspiration*; Springer: Dordrecht, The Netherlands, 2013; ISBN 978-94-007-4736-4.
74. Jensen, M.E.; Dotan, A.; Sanford, R. Penman-Monteith Estimates of Reservoir Evaporation. In *Proceedings of the Impacts of Global Climate Change*; American Society of Civil Engineers: Anchorage, AK, USA, 2005; pp. 1–24.
75. Finch, J.W.; Hall, R.L.; Great Britain; Environment Agency. *Estimation of Open Water Evaporation: A Review of Methods*; Environment Agency: Bristol, UK, 2005; ISBN 978-1-85705-604-4.
76. McJannet, D.L.; Webster, I.T.; Cook, F.J. An area-dependent wind function for estimating open water evaporation using land-based meteorological data. *Environ. Model. Softw.* **2012**, *31*, 76–83. [CrossRef]
77. Idso, S.B.; Jackson, R.D. Thermal radiation from the atmosphere. *J. Geophys. Res.* **1969**, *74*, 5397–5403. [CrossRef]
78. Mekonnen, M.M.; Hoekstra, A.Y. The blue water footprint of electricity from hydropower. *Hydrol. Earth Syst. Sci.* **2012**, *16*, 179–187. [CrossRef]
79. Jegede, O.O.; Ogolo, E.O.; Aregbesola, T.O. Estimating net radiation using routine meteorological data at a tropical location in Nigeria. *Int. J. Sustain. Energy* **2006**, *25*, 107–115. [CrossRef]
80. Niclòs, R.; Valor, E.; Caselles, V.; Coll, C.; Sánchez, J.M. In situ angular measurements of thermal infrared sea surface emissivity—Validation of models. *Remote Sens. Environ.* **2005**, *94*, 83–93. [CrossRef]
81. de Bruin, H.A.R. Temperature and energy balance of a water reservoir determined from standard weather data of a land station. *J. Hydrol.* **1982**, *59*, 261–274. [CrossRef]
82. Finch, J.W. A comparison between measured and modelled open water evaporation from a reservoir in south-east England. *Hydrol. Process.* **2001**, *15*, 2771–2778. [CrossRef]
83. McJannet, D. *Estimating Open Water Evaporation for the Murray-Darling Basin*; CSIRO: Water for a Healthy Country National Research Flagship: Canberra, Australia, 2008. [CrossRef]
84. Duffie, J.A.; Beckman, W.A. Chapter 23-Design of Photovoltaic Systems. In *Solar Engineering of Thermal Processes*; John, A.D., William, A.B., Eds.; John Wiley: Hoboken, NJ, USA, 2013; ISBN 978-0-470-87366-3.
85. Shaari, S.; Sopian, K.; Amin, N.; Kassim, M.N. The Temperature Dependence Coefficients of Amorphous Silicon and Crystalline Photovoltaic Modules Using Malaysian Field Test Investigation. *Am. J. Appl. Sci.* **2009**, *6*, 586–593. [CrossRef]
86. Kamkird, P.; Ketjoy, N.; Rakwichian, W.; Sukchai, S. Investigation on Temperature Coefficients of Three Types Photovoltaic Module Technologies under Thailand Operating Condition. *Procedia Eng.* **2012**, *32*, 376–383. [CrossRef]
87. MathWorks Multiple Linear Regression-MATLAB Regress. Available online: https://www.mathworks.com/help/stats/regress.html (accessed on 30 July 2020).
88. Maghami, M.R.; Hizam, H.; Gomes, C.; Radzi, M.A.; Rezadad, M.I.; Hajighorbani, S. Power loss due to soiling on solar panel: A review. *Renew. Sustain. Energy Rev.* **2016**, *59*, 1307–1316. [CrossRef]
89. Fouad, M.M.; Shihata, L.A.; Morgan, E.I. An integrated review of factors influencing the perfomance of photovoltaic panels. *Renew. Sustain. Energy Rev.* **2017**, *80*, 1499–1511. [CrossRef]
90. Jacobson, M.Z.; Jadhav, V. World estimates of PV optimal tilt angles and ratios of sunlight incident upon tilted and tracked PV panels relative to horizontal panels. *Sol. Energy* **2018**, *169*, 55–66. [CrossRef]
91. Taboada, M.E.; Cáceres, L.; Graber, T.A.; Galleguillos, H.R.; Cabeza, L.F.; Rojas, R. Solar water heating system and photovoltaic floating cover to reduce evaporation: Experimental results and modeling. *Renew. Energy* **2017**, *105*, 601–615. [CrossRef]

92. Las Vegas Valley Water District Water Rates. Available online: https://www.lvvwd.com/customer-service/pay-bill/water-rates.html (accessed on 16 October 2020).
93. Karambelkar, S. *Hydropower Operations in the Colorado River Basin: Institutional Analysis of Opportunities and Constraints*; Hydropower Foundation: Littleton, CP, USA, 2018; p. 91.
94. Trabish, H.K. Hoover Dam, the Drought, and a Looming Energy Crisis. Available online: https://www.utilitydive.com/news/hoover-dam-the-drought-and-a-looming-energy-crisis/281133/ (accessed on 13 November 2020).
95. Westenburg, C.L.; De Meo, G.A.; Tanko, D.J. *Evaporation from Lake Mead, Arizona and Nevada, 1997–1999*; U.S. Geological Survey: Reston, VA, USA, 2006; p. 34.
96. US EPA, O. Statistics and Facts. Available online: https://www.epa.gov/watersense/statistics-and-facts (accessed on 10 October 2020).
97. U.S. Census Bureau. U.S. Census Bureau QuickFacts: Los Angeles City, California. Available online: https://www.census.gov/quickfacts/losangelescitycalifornia (accessed on 10 October 2020).
98. U.S. Census Bureau. U.S. Census Bureau QuickFacts: Nevada. Available online: https://www.census.gov/quickfacts/NV (accessed on 13 October 2020).
99. U.S. Census Bureau. U.S. Census Bureau QuickFacts: Reno City, Nevada; Las Vegas City, Nevada; Henderson City, Nevada. Available online: https://www.census.gov/quickfacts/fact/table/renocitynevada,lasvegascitynevada,hendersoncitynevada/PST045219 (accessed on 13 October 2020).
100. Barsugli, J.J.; Nowak, K.; Rajagopalan, B.; Prairie, J.R.; Harding, B. Comment on "When will Lake Mead go dry?" by T. P. Barnett and D. W. Pierce: Commentary. *Water Resour. Res.* **2009**, *45*. [CrossRef]
101. Rajagopalan, B.; Nowak, K.; Prairie, J.; Hoerling, M.; Harding, B.; Barsugli, J.; Ray, A.; Udall, B. Water supply risk on the Colorado River: Can management mitigate? *Water Resour. Res.* **2009**, 45. [CrossRef]
102. Barnett, T.P.; Pierce, D.W. When will Lake Mead go dry? *Water Resour. Res.* **2008**, 44. [CrossRef]
103. U.S. EIA. *Frequently Asked Questions (FAQs)-U.S. Energy Information Administration (EIA)*. Available online: https://www.eia.gov/tools/faqs/faq.php (accessed on 10 October 2020).
104. U.S. EIA. *Electric Power Annual 2018*; U.S. Energy Information Administration: Washington, DC, USA, 2019; p. 239.
105. Prehoda, E.W.; Pearce, J.M. Potential lives saved by replacing coal with solar photovoltaic electricity production in the U.S. *Renew. Sustain. Energy Rev.* **2017**, *80*, 710–715. [CrossRef]
106. Burney, J.A. The downstream air pollution impacts of the transition from coal to natural gas in the United States. *Nat. Sustain.* **2020**, *3*, 152–160. [CrossRef]
107. Thurston, G.D.; Burnett, R.T.; Turner, M.C.; Shi, Y.; Krewski, D.; Lall, R.; Ito, K.; Jerrett, M.; Gapstur, S.M.; Diver, W.R.; et al. Ischemic Heart Disease Mortality and Long-Term Exposure to Source-Related Components of U.S. Fine Particle Air Pollution. *Environ. Health Perspect.* **2016**, *124*, 785–794. [CrossRef]
108. Krewski, D.; Jerrett, M.; Burnett, R.T.; Ma, R.; Hughes, E.; Shi, Y.; Turner, M.C.; Pope, C.A., III; Thurston, G.; Calle, E.E.; et al. *Extended Follow-Up and Spatial Analysis of the American Cancer Society Study Linking Particulate Air Pollution and Mortality*; Health Effects Institute: Boston, MA, USA, 2009; p. 154.
109. Spencer, R.S.; Macknick, J.; Aznar, A.; Warren, A.; Reese, M.O. Floating Photovoltaic Systems: Assessing the Technical Potential of Photovoltaic Systems on Man-Made Water Bodies in the Continental United States. *Environ. Sci. Technol.* **2019**, *53*, 1680–1689. [CrossRef]
110. Gorjian, S.; Sharon, H.; Ebadi, H.; Kant, K.; Scavo, F.B.; Tina, G.M. Recent technical advancements, economics and environmental impacts of floating photovoltaic solar energy conversion systems. *J. Clean. Prod.* **2021**, *278*, 124285. [CrossRef]
111. Dinesh, H.; Pearce, J.M. The potential of agrivoltaic systems. *Renew. Sustain. Energy Rev.* **2016**, *54*, 299–308. [CrossRef]
112. Pringle, A.M.; Handler, R.M.; Pearce, J.M. Aquavoltaics: Synergies for dual use of water area for solar photovoltaic electricity generation and aquaculture. *Renew. Sustain. Energy Rev.* **2017**, *80*, 572–584. [CrossRef]
113. Moustafa, K. Toward Future Photovoltaic-Based Agriculture in Sea. *Trends Biotechnol.* **2016**, *34*, 257–259. [CrossRef]
114. Sica, D.; Malandrino, O.; Supino, S.; Testa, M.; Lucchetti, M.C. Management of end-of-life photovoltaic panels as a step towards a circular economy. *Renew. Sustain. Energy Rev.* **2018**, *82*, 2934–2945. [CrossRef]
115. Farrell, C.C.; Osman, A.I.; Doherty, R.; Saad, M.; Zhang, X.; Murphy, A.; Harrison, J.; Vennard, A.S.M.; Kumaravel, V.; Al-Muhtaseb, A.H.; et al. Technical challenges and opportunities in realising a circular economy for waste photovoltaic modules. *Renew. Sustain. Energy Rev.* **2020**, *128*, 109911. [CrossRef]

116. Lisperguer, R.C.; Cerón, E.M.; de la Casa Higueras, J.; Martín, R.D. Environmental Impact Assessment of crystalline solar photovoltaic panels' End-of-Life phase: Open and Closed-Loop Material Flow scenarios. *Sustain. Prod. Consum.* **2020**, *23*, 157–173. [CrossRef]
117. Pearce, J.M. Industrial symbiosis of very large-scale photovoltaic manufacturing. *Renew. Energy* **2008**, *33*, 1101–1108. [CrossRef]
118. McDonald, N.C.; Pearce, J.M. Producer responsibility and recycling solar photovoltaic modules. *Energy Policy* **2010**, *38*, 7041–7047. [CrossRef]
119. Kabamba, E.T.; Rodrigue, D. The effect of recycling on LDPE foamability: Elongational rheology. *Polym. Eng. Sci.* **2008**, *48*, 11–18. [CrossRef]
120. Bedell, M.; Brown, M.; Kiziltas, A.; Mielewski, D.; Mukerjee, S.; Tabor, R. A case for closed-loop recycling of post-consumer PET for automotive foams. *Waste Manag.* **2018**, *71*, 97–108. [CrossRef]
121. Al-Sabagh, A.M.; Yehia, F.Z.; Eshaq, G.H.; Rabie, A.M.; ElMetwally, A.E. Greener routes for recycling of polyethylene terephthalate. *Egypt. J. Pet.* **2016**, *25*, 53–64. [CrossRef]

Publisher's Note: MDPI stays neutral with regard to jurisdictional claims in published maps and institutional affiliations.

energies

MDPI

Article

Performance Analysis of a Printed Circuit Heat Exchanger with a Novel Mirror-Symmetric Channel Design

Cheng-Yen Chang [1], Wei-Hsin Chen [1,2,3,*], Lip Huat Saw [4], Arjay Avilla Arpia [1] and Manuel Carrera Uribe [1,5]

1 Department of Aeronautics and Astronautics, National Cheng Kung University, Tainan 701, Taiwan; 10810016@gs.ncku.edu.tw (C.-Y.C.); arjay.arpia@upd.edu.ph (A.A.A.); p06107015@gs.ncku.edu.tw (M.C.U.)
2 Research Center for Smart Sustainable Circular Economy, Tunghai University, Taichung 407, Taiwan
3 Department of Mechanical Engineering, National Chin-Yi University of Technology, Taichung 411, Taiwan
4 Lee Kong Chian Faculty of Engineering and Science, Tunku Abdul Rahman University, Kajang 43000, Malaysia; sawlh@utar.edu.my
5 International Master Degree Program on Energy Engineering, National Cheng Kung University, Tainan 701, Taiwan
* Correspondence: chenwh@mail.ncku.edu.tw

Abstract: The printed circuit heat exchanger (PCHE) is a promising waste heat recovery technology to improve energy efficiency. The current investigation presents the experimental results on the thermal performance of a novel PCHE for low-temperature waste heat recovery. The novel PCHE was manufactured using precision machining and diffusion bonding. The thermal performances, such as effectiveness and NTU values at different temperatures, are evaluated, and water is used as a working fluid. The experimental results indicate that the PCHE's effectiveness is around 0.979 for an inlet flow temperature of 95 °C. The predominant factors affecting the thermal performance of the PCHE are the inlet flow temperature and the flow rate of the working fluid. In addition, a comparison of the experimental results and the literature shows that the effectiveness of the PCHE is better than the others, which have fewer layers of PCHE fins.

Keywords: print circuit heat exchanger; PCHE; efficiency; nusselt number; heat transfer coefficient; NTU value; thermal performance

Citation: Chang, C.-Y.; Chen, W.-H.; Saw, L.H.; Arpia, A.A.; Carrera Uribe, M. Performance Analysis of a Printed Circuit Heat Exchanger with a Novel Mirror-Symmetric Channel Design. *Energies* **2021**, *14*, 4252. https://doi.org/10.3390/en14144252

Academic Editors: Ron Zevenhoven and Mahmoud Bourouis

Received: 31 May 2021
Accepted: 8 July 2021
Published: 14 July 2021

1. Introduction

Waste heat generation is inevitable during energy utilization in an industrial process and grows evidently alongside global industrialization. Industrial waste heat is categorized into three temperature ranges, namely: low temperature (<230 °C, e.g., paper, textile, food processing industry, etc.), medium-temperature (230–650 °C, e.g., ceramic and cement industry, etc.), and high-temperature (>650 °C, e.g., steel and metal processing industry, etc.) [1]. Half of input energy is lost in different forms of waste heat into its surroundings [2], wherein the low-temperature range accounted for about 66% of the total waste heat generated [3,4]. Globally, low-temperature waste heat from industrial activities was extensively observed; for example, roughly 34% was generated in Europe, 50% in China, and 60% in the United States [4]. The main application of the print circuit heat exchanger (PCHE) is in a supercritical CO_2 (S-CO_2) power cycle, which is promising electricity generation using waste heat recovery. This is due to the high efficiency and compact configuration of the PCHE, which reduces the system footprint area [5]. In addition, the heat exchanger can be designed using various fin geometries and working fluids to fulfill the system requirements [6]. Taiwan's rich geothermal resources have been well-developed as hot springs and are an integral part of the tourism industry. Sodium bicarbonate hot spring (pH value: 6.2~8.6; temp. 60–99 °C) accounts for 70% of hot springs throughout Taiwan. The heat generated from these hot springs could be recovered and shows great potential for low-temperature waste heat recovery from geothermal heat. Taking advantage of this waste heat from low-temperature sources, not only improves energy conversion efficiency, but also reduces harmful emissions. There are many technical applications for waste heat recovery, such as passive waste heat recovery systems for heat pipes [7] and static power generation technology by automotive exhaust with a thermoelectric generation (TEG) system [8]; however, these options are not suitable for the hot springs scenario.

Heat exchangers, such as a linchpin unit, are widely used in waste heat recovery systems, working independently or combined with other systems. Heat exchangers commonly used in the industry still have many issues that need to be addressed. For example, a shell and tube heat exchanger (STHE) and a finned-tube heat exchanger (FTE) require a large space for installation and operation and have a lower distribution of channels within a small space. In addition, for plate heat exchangers (PHE), the plates connected by welding [9] are weaker than the body. In addition, high channel distribution density has contributed to high-pressure drop and high power of the pump is needed. Increasing the area of heat transfer of the heat exchanger flow channel is a good method to improve heat recovery, but it will incur a high-pressure drop, thus, energy consumption during operation is also increased [10,11]. For low-temperature waste heat recovery, the temperature difference between the heat source and the heat collector is too small to recover, thus, low-temperature waste heat recovery is always harder than middle and high temperatures. Hence, there is less low-temperature waste heat recovery technology in industrial applications using a heat exchanger and the technology is immature compared to other types of waste heat recovery technologies.

A diffusion bonding application is viewed to be a suitable technology for the fabrication of a PCHE. It is a bonding method wherein the gap between the two materials is within an atomic level. This makes the device approximately one body without adding materials. Printed circuit heat exchangers (PCHEs) are high-integrity and density-compact heat exchangers [12]. Due to diffusion bonding technology, the strength and properties of the entire PCHE are unified, and the endurance to stress is excellent. The heat exchanger is well compacted and the channels are densely distributed. Particularly, the design using microchannels offers a better heat transfer performance [13]; however, it is accompanied by a large pressure drop. On the other hand, diffusion bonding technology can offer various advantages [14], such as the reduction of unpredictable risks, such as leakage and malfunctions in service. In addition, the lifetime of heat exchangers fabricated through diffusion bonding is longer than others. Most recent studies focused on carbon dioxide and helium as the working supercritical fluid as well as geometry designs of the channel [15–18], or applying PCHE technology as the heat exchanger for a Brayton cycle and using a nanofluid as a working fluid [19,20]. Most of the temperature conditions are medium or high temperatures and only a few examples are in low-temperature regions, as shown in Table 1 [5,21–28].

Table 1. Literature review of printed circuit heat exchangers through diffusion bonding with low Reynolds number.

Working Fluid	Shape of Channel	Correlations/Outcome		Reference
Water	Straight	$Nu = (0.01352 \pm 0.0094)\ Re^{(0.80058 \pm 0.0921)}$ $Nu = (3.6361 \pm 0.0094)\ Re^{(1.2804 \pm 0.0273)}$	$1200 < Re \leq 1850$ $1850 < Re \leq 2900$	[1]
He-water He-He & He-water	Zigzag	$f_P \cdot Re = 15.78 + 0.0557\ Re^{0.82}$ $Nu = 4.089 + 0.00497\ Re^{0.95} \cdot Pr^{0.55}$	$0 < Re < 3000$ $0 < Re < 3000$ $0.66 < Pr < 13.41$	[2]
He	Zigzag	$Nu = 4.089 + 0.00365 \cdot Re \cdot Pr^{0.58}$; $f_p \cdot Re = 15.78 + 0.004868\ Re^{0.8416}$	$0 < Re < 2500$	[2]
CO_2 & Water	Zigzag	$Nu = 0.8405 \cdot Re^{0.5704}$; $f = 0.0758 \cdot Re^{-0.19}$ $Nu = 0.2829 \cdot Re^{0.6686}$; $f = 6.9982 \cdot Re^{-0.766}$	$15{,}000 < Re < 85{,}000$ $50 < Re < 200$	[3]
Water	Straight	$Nu_h = 0.7203\ Re^{0.1775} Pr^{1/3} (\mu/\mu)^{0.14}$; $f = 1.3383Re - 0.5003$	$100 < Re < 850$	[4]
s-CO_2	Straight	Hot fluid ($470K < T_b < 630K$): $Nu = 87.56(\frac{Lp}{12})^{-0.178}(\frac{\alpha}{116})^{-0.9306}$; $f = 0.0375(\frac{Lp}{12})^{-0.9639}(\frac{\alpha}{116})^{-2.409}$ Cold fluid ($400K < T_b < 520K$): $Nu = 85.95(\frac{Lp}{12})^{-0.171}(\frac{\alpha}{116})^{-0.8912}$; $f = 0.0395(\frac{Lp}{12})^{-0.9479}(\frac{\alpha}{116})^{-2.239}$	$5000 < Re < 32{,}000$	[5]

Table 1. *Cont.*

Working Fluid	Shape of Channel	Correlations/Outcome		Reference
Helium	Zigzag	$Nu = (0.05516 \pm 0.00160) \cdot Re^{(0.69195 \pm 0.00559)}$ $Nu = (0.09221 \pm 0.01397) \cdot Re^{(0.62507 \pm 0.01949)}$ $Nu = 5.05 + (0.02 \cdot \alpha + 0.003) \cdot Re \cdot Pr^{0.6}$	$1400 \leq Re \leq 2200$ $2200 \leq Re \leq 3558$	[6]
Helium	Zigzag	$Nu = (0.71\alpha + 0.289) \left(\frac{l_R}{D_h}\right)^{-0.087} \cdot Re^{(-0.11(\alpha-0.55)^2 - 0.004(l_R/D_h)\alpha + 0.54)} Pr^{0.56}$ $Nu = (0.18\alpha + 0.457) \left(\frac{l_R}{D_h}\right)^{-0.038} \cdot Re^{(-0.23(\alpha-0.74)^2 - 0.004(l_R/D_h)\alpha + 0.56)} Pr^{0.58}$	$100 \leq Re \leq 2000$ $Pr \leq 1.0$	[7]
s-CO_2 & Al_2O_3-water nanofluid	Straight	$Nu = \frac{\frac{f}{8}(Re - 100) \cdot Pr}{1 + 12.7 \cdot \sqrt{\frac{f}{8}} \cdot (P_r{}^{2/3} - 1)}$	$2300 < Re < 10^6$ $0.5 < Pr < 2000$	[8]
s-CO_2	Straight	$Nu = \frac{\frac{f}{8}(Re_b - 1000) \cdot Pr_b}{1 + 12.7 \cdot \sqrt{\frac{f}{8}} \cdot (P_{r_b}{}^{2/3} - 1)}$	$2300 < Re < 5 \times 10^6$ $0.5 < Pr < 2000$	[9]
Water	S-type	$Nu = 0.03428 \cdot Re^{0.6135}$	$50 \leq Re \leq 310$	This paper

The above literature shows that few studies focused on heat transfer performance and efficiency of PCHEs with water as a working fluid for waste heat recovery at low-temperatures. In view of this, this work aims to design a novel PCHE for low-temperature application fields, especially for hot springs. In this study, bezels are designed on the flow plates to increase heat exchange efficiency. Each bezel is arranged at an interval and with a mirror design for the cold and hot fluid flow plates. The current arrangement of the flow channel in the flow plates will keep the fluid flow in the special flow direction through a bezel. On the other hand, the mirror-symmetric design of the channel in the fluid flow plates is to improve heat exchange performance. Furthermore, the microchannel design is applied to the runner plates to increase the area of heat transfer and thus improve the overall efficiency of the PCHE. Although increasing the fin density will lead to an increase in pressure drop, this effect can be ignored as the current PCHE is targeted at hot-spring waste heat recovery, which does not require any pump for operation. This study investigated the relationship between the heat transfer and flow rate for the PCHE, with a particular focus on different inlet temperatures, flow rate ratio, and heat exchanger effectiveness. In addition, the performance of using precision machining manufacturing runner microchannels under various flow rates and temperature profiles is also investigated. Lastly, a comparison of the current PCHE with other heat exchangers is made, where an etching manufacturing method is used to fabricate the flow channel. This will enhance our understanding of the PCHE performance fabricated using a low-cost manufacturing method and, in return, contribute to the future development of PCHEs for special applications.

2. Materials and Methods

2.1. Design and Manufacture of PCHE

The printed circuit heat exchanger (PCHE) used in this research consisted of four flat stainless steel plates: two hot plates and two cold plates with lengths, widths, and heights of 0.11 m, 0.145 m, and 0.012 m, respectively, as shown in Figure 1a,b. The rectangular channels had a hydraulic diameter of 0.0021 m and were manufactured using precision machining. The flow area of the plate was divided into three zones. The first sone was the three-channel areas in the plate composed of 13 channels. The second zone was the four pool areas at the side of the channel areas. The channel was designed to guide the fluid flow path turn at 180 degrees to the next channel. The last zone had two baffles with a 0.003-m width on the plate, which guided the flow and separated the channel areas and the aforementioned configuration of the channel. Therefore, the working fluid flowed on the plate in an S-shape. This design was to ensure that each channel was fully filled with working fluid and extended the staying time for the fluid in the flow plate. Moreover, hot and cold fluid plates showed symmetric geometry to increase the area of heat transfer [29].

According to the literature, a high compactness and channel distribution density in the PCHE will improve the overall efficiency [12]. The flow directions of cold and hot fluids and the configuration of the PCHE are shown in Figure 2. During the manufacturing process, the stacked plates, including four runner mirror-symmetrically plates and a top plate with a thickness of 0.004 m, were put in a vacuum (10−3 tor) with a working temperature of 1045 °C. The function of the top plate was to protect the PCHE channels from damage during the pressurization process. Next, the plates were subjected to a pressure of 200 bar in the vertical direction to ensure that the distance between the two plates achieved the atomic distance, while cooling down to the surrounding temperature. The fluid flow direction is illustrated in Figure 2c, in which the blue and red arrows represent the cold and hot fluid flows, respectively. After the fluids enter the PCHE, they go into each channel of the plate and then go out. In order to ease the testing of the PCHE, the screw holes at the edge of the plate were designed with a diameter of 0.012 m to facilitate the installation of a 0.375 inch copper joint. Then, heat-resistant silicone hoses were installed on the inlet and outlet of the PCHE.

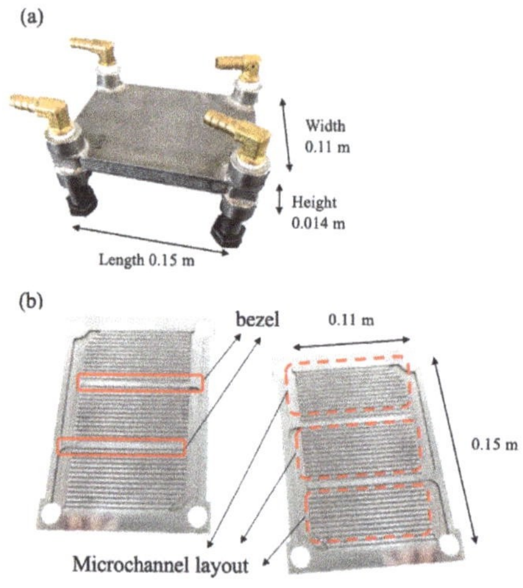

Figure 1. (**a**) The design of the PCHE; (**b**) hot fluid flow plate and cold fluid flow plate in the PCHE.

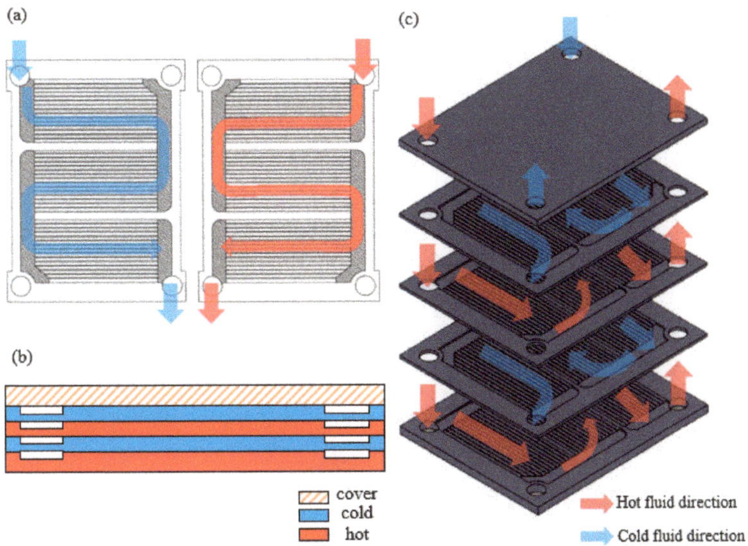

Figure 2. Flow plate geometric of PCHE and flow direction: (**a**) cold and hot fluid flow plate; (**b**) side view and three-dimensional explosion map; and (**c**) schematic diagram of the PCHE.

2.2. Experiment Setup

The experimental apparatus and setup to measure and calculate the thermal performance of the PCHE is described in Figure 3, which consists of water as the hot and cold working fluids, two flow meters (Shuang Huan, Taiwan, DK800-6) with adjustable valves installed on the hot fluid and cold fluid circulating loop, refrigerated circulating bath (Yih Der, Tainan, Taiwan, BL710-D), pump (HITON, Tainan, Taiwan, HF-8006) and a 40-L water heating system powered by liquified petroleum gas (Ta-Han, Tainan, Taiwan, BDF-23C). The temperatures were measured using eight K-type thermocouples (Chuan Yu, Kaohsiung, Taiwan, K type) connected to an industrial computer, equipped with a thermocouple slot module. Filters were installed at the outlet of the sink and the water heating system to remove impurities in the working fluid and prevent damage to the flow meter. Since the flow layout was set to be in a counterflow, the counterflow pattern was adopted in the flow configuration [30], which resulted in a high effectiveness, where the cold inlet was on the opposite side of the hot inlet. It is also shown in Figure 3 that the flow rate passing through the flowmeters was controlled by a valve; the maximum mass flow rate was 100 L/h. The hot working fluid was driven by a pump and the temperature of the fluid was controlled by a heater. A constant-temperature cold working fluid was supplied by a refrigerated circulating bath to the test rig. The thermocouples were installed at the inlet and outlet of both fluids and inside the tube of the copper joint. They were about 0.001 m away from the entrance of the PCHE. Before performing experiments, the thermocouples were calibrated to an accuracy of ±0.01 °C. In addition, the water heating system had thermocouples to accurately monitor and control the temperature. Finally, all piping system and PCHE were insulated with thermal insulation wool to minimize heat loss.

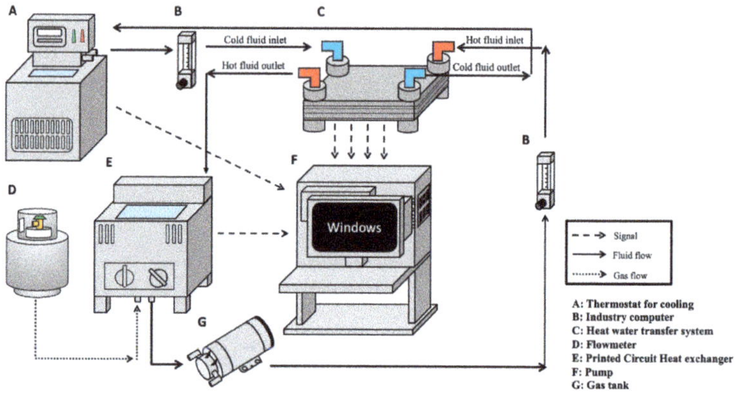

Figure 3. Schematic diagram of the experiment setup.

2.3. *Operations and Data Analysis of Experimentations*

Three different hot fluid inlet temperatures, 75 °C, 85 °C, and 95 °C, were used in this study. The inlet temperature of the cold fluid (water) was 20 °C. The range of the cold and hot fluids' flow rates were 0.1667–1.667 L/min (i.e., 10–100 L/h), they were adjusted by the valve on the flowmeter. This corresponds to the Reynolds numbers of 50–514. The Reynolds number (Re) was calculated in Equation (1) in light of the method suggested by Cowell [31]. The average convective heat transfer coefficient (h) and Nusselt number (Nu) were obtained using Equations (2) and (3) as [9]:

$$\text{Re} = \frac{\rho V D_h}{\mu} \tag{1}$$

$$h = \frac{\left(\dot{m} c_p \Delta T\right)_{hot}}{A\,(T_s - T_\infty)} \tag{2}$$

$$\text{Nu} = \frac{h D_h}{k} \tag{3}$$

In Equation (1), D_h (0.0021 m) is the hydraulic diameter, V is the velocity of the fluid, μ is the dynamic viscosity of the fluid, and ρ is the density of the fluid. In Equation (2), the product in the numerator is the heat flux of hot fluid, including c_p (specific heat), $\dot{m}$ (flow rate), and ΔT (the temperature difference between the inlet and outlet), respectively. In addition, T_s and T_∞ are the temperature of channel surface and fluid temperature, respectively. In Equation (3), k represents the thermal conductivity.

The effectiveness (ε) of the PCHE [9] is assessed as the ratio of the actual heat transfer rate (Q_{actual}) to the theoretical maximum rate of the heat transfer (Q_{max}) and is expressed using the equation below:

$$\varepsilon = \frac{Q_{actual}}{Q_{max}} = \frac{C_{hot}\,(T_{h,in} - T_{h,out})}{C_{min}\,(T_{h,in} - T_{c,in})} = \frac{(\dot{m}\,c_p)_{hot}\,(T_{h,in} - T_{h,out})}{(\dot{m}\,c_p)_{min}\,(T_{h,in} - T_{c,in})} \tag{4}$$

In Equation (4), $T_{c,in}$, $T_{h,in}$, and $T_{h,out}$ represent the temperature of the cold inlet, hot inlet, and hot outlet, respectively. C_{hot} denotes the product of the mass flow and specific heat of the hot and fluid, and C_{min} is the smaller one between the cold and hot fluid.

The value of the overall heat transfer coefficient (U) can be calculated from the heat transfer area and the logarithmic mean temperature difference (LMTD) between the hot and cold fluid flows and average heat transfer rate ($\dot{Q}_{average}$) by Thulukkanam [9]. They are expressed as follows:

$$\dot{Q}_{average} = \frac{\dot{Q}_{hot} + \dot{Q}_{cold}}{2} \tag{5}$$

$$U = \frac{\dot{Q}_{average}}{A_{total}\ LMTD} \tag{6}$$

$$LMTD = \frac{\Delta T_1 - \Delta T_2}{LnN(\frac{\Delta T_1}{\Delta T_2})} \tag{7}$$

where ΔT_1 and ΔT_2 are the temperature difference of the inlet and outlet between the hot and cold working fluid, respectively, as shown below:

$$\Delta T_1 = T_{hot,inlet} - T_{cold,\ outlet} \tag{8}$$

$$\Delta T_2 = T_{hot,outlet} - T_{cold,\ inlet} \tag{9}$$

The number of transfer units (NTU) value can be calculated by the overall heat transfer coefficient [32], as shown in Equation (10). In the equation, A_{total} is the total area of heat transfer of the PCHE, C_{min} is the smaller one between heat capacity of cold and hot fluid, $c_{p,c}$ is the heat capacity of cold fluid. Similar to other literature analysis, the usual heat exchanger effectiveness ε is defined as the relation function between C_r and NTU, as shown in Equation (11) and can be simplified to Equation (12) [33]:

$$NTU = \frac{UA_{total}}{C_{min}} = \frac{UA_{total}}{(\dot{m}c_{p,c})_C} \tag{10}$$

$$\varepsilon = \frac{1 - e^{[-NTU(1-C_r)]}}{1 - C_r e^{[-NTU(1-C_r)]}} \tag{11}$$

$$\varepsilon = \frac{NTU}{1 + NTU} \tag{12}$$

$$C_r = \frac{C_{min}}{C_{max}} \tag{13}$$

2.4. Uncertainties Analysis

The results of the experiment and its measurements were affected by many factors. Uncertainty analysis was used to make sure that the precision of the measurement device was set before the experiments, and to calibrate and ascertain their accuracies. This included electronic load, flowmeters, and thermocouples. The ranges of measuring or operating, resolution, and uncertainty in measurement is tabulated in Table 2, where uncertainty in the measurement of devices is defined as in Equation (14) [34]:

$$Relative\ uncertainty = \frac{0.5 \times resolution}{value\ of\ measuring\ or\ operating} \tag{14}$$

Table 2. Uncertainty analysis of the equipment used in this study.

Equipment Uncertainty			
Equipment	**Operating or Measuring Range**	**Resolution**	**Relative Uncertainly**
Heating system	70–100 °C	0.01 °C	1.8%
Cooling system	20 °C	0.01 °C	1.7%
Flowmeter	10–100 L h^{-1}	1 L h^{-1}	2.4%

Table 2 lists the data for supplying and the deviation analysis of experiments, including temperature, which was controlled between 75 °C and 95 °C and the flow rate was between 10 L/h and 100 L/h. The analysis results of the uncertainty of the measurement due to the equipment was less than 2.5%.

3. Results and Discussion

3.1. *Effectiveness*

Figure 4 illustrates the effectiveness of the PCHE for three different inlet temperatures of the hot fluid flow versus the cold-to-hot fluid flow rate ratio. The effectiveness (ε) can be observed to easily decrease with an increase in flow rate ratio. The effectiveness of the heat exchanger drops by at least 50% when the ratio of flow rate increases from 0.1 to 1, showing the pronounced influence of the ratio of the flow rate on the effectiveness of the PCHE. On the other hand, the differences between the three curves are not significant, revealing that the effect of increasing the inlet temperatures is not as obvious as the flow rate ratio.

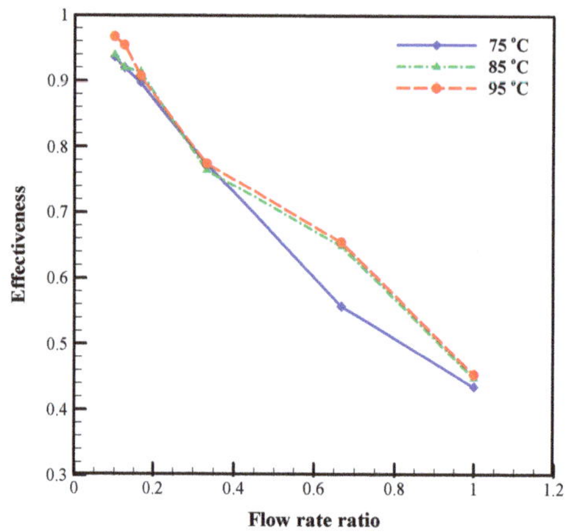

Figure 4. The effectiveness of the PCHE on different hot inlet temperatures.

In Figure 4, the highest effectiveness of 0.979 was achieved at a flow rate ratio of 0.1 (i.e., the minimum ratio of flow rate), whereas the lowest effectiveness of 0.428 is exhibited at a flow rate ratio of 1, rendering a 53% difference in the effectiveness between the two ratios under the same hot inlet temperature. On account of the higher effective value at a ratio of 0.1, this ratio is suitable for the operation of PCHEs. On the contrary, a flow rate ratio of 1 will lead to poor performance of the PCHE. Overall, the effectiveness of the PCHE under the hot inlet temperature of 95 °C is better than the other two temperatures at a low flow rate ratio. In summary, operating the heat exchanger with high inlet temperature and low flow rate ratio is conducive to intensifying the effectiveness and thereby the heat exchange. Attala et al.'s [35] experimental results on the plate heat exchanger had a similar behavior at a low Reynolds number.

3.2. *Temperature Distribution*

The temperature distribution of the hot and cold fluid inlet and outlet at three different hot inlet fluid temperatures (75, 85, and 95 °C), along with a fixed cold inlet temperature (20 °C) for different flow rate ratios, are shown in Figure 5. Altering the flow rate ratio causes variations in the temperature of the distribution. At low flow ratios, such as 0.1, 0.125, and 0.167, the temperature slopes of the hot fluid flow are relatively insignificant, whereas the temperature slopes of the cold fluid flow are steeper. This is ascribed to more heat being contained in the hot fluid and relatively less heat being transferred to the cold fluid, stemming from a lower cold fluid flow rate. As a consequence, the temperature variation of the hot fluid flow is small, whereas it is pronounced in the cold fluid flow. On the contrary, at higher flow rate ratios, such as 0.333, 0.667, and 1, the variation in the temperature of the hot fluid flow tends to become obvious, whereas the rising tendency in the temperature of the cold fluid flow becomes less obvious [36]. For the cases of a flow rate ratio of <0.667, it is noteworthy that, after heat exchange, the outlet temperature of the cold fluid flow is always higher than that of the hot fluid flow, whereas an opposite result is observed at flow rate ratio =0.667 and 1. For the two factors of the hot fluid inlet temperature and the flow rate ratio investigated in this study, Figure 5 indicates that the flow rate ratio is more influential on the temperature profile when compared with the hot fluid inlet temperature. Figley et al. [32] explored the correlation between flow rate ratio and hot fluid inlet temperature, which had a similar temperature changing trend, and established the thermal-hydraulic performance in their PCHE numerical model.

3.3. *Temperature Difference and Effectiveness*

Figure 6a examines the temperature difference of the cold and hot fluids' between the inlet and outlet. The temperature difference of the hot fluid between the outlet and inlet decreases with the rising flow rate ratio, but it shows an opposite trend for the cold fluid. Physically, the higher the temperature difference, the better the heat transfer. At low flow rate ratios, the temperature difference of the cold fluid flow is higher which is contributed by a high flow rate of the cold fluid. Thus, the temperature of the cold fluid flow can be easily raised. Meanwhile, the temperature difference in the hot fluid flow is small, which is ascribed to the high flow rate of the hot fluid. It is not surprising that an increase in the hot fluid's inlet temperature increases the temperature differences of the hot and cold fluids. A past study [37] provided several cross-flow configurations and explained the correlation between the effectiveness and the temperature difference which is in line with the obtained results in the present study.

To further investigate the heat transfer performance, the temperature difference between the cold and hot fluid temperature versus flow rate ratio is shown in Figure 6b. Meanwhile, the profiles of the effectiveness from Equation (12) are also shown in Figure 6b. As a whole, both the D value and effectiveness (ε) decrease monotonically with the increase of the flow rate ratio. Physically, the effectiveness is a ratio between the actual heat transfer rate and the maximum (ideal) heat transfer rate. Accordingly, when the effectiveness is larger, the performance of heat transfer is also better. The maximum value of the effectiveness in Figure 6b is 0.979, occurring at a hot fluid flow inlet of 95 °C and the flow rate ratio of 0.1. In contrast, the minimum value of the effectiveness is 0.428 occurring at a hot fluid flow inlet of 75 °C and a flow rate ratio of 1. This indicates that the low flow rate ratio with low cold fluid flow rate and high hot fluid flow rate is a better combination for optimum thermal performance for the PCHE. Overall, the distributions of D and ε showed high correlation.

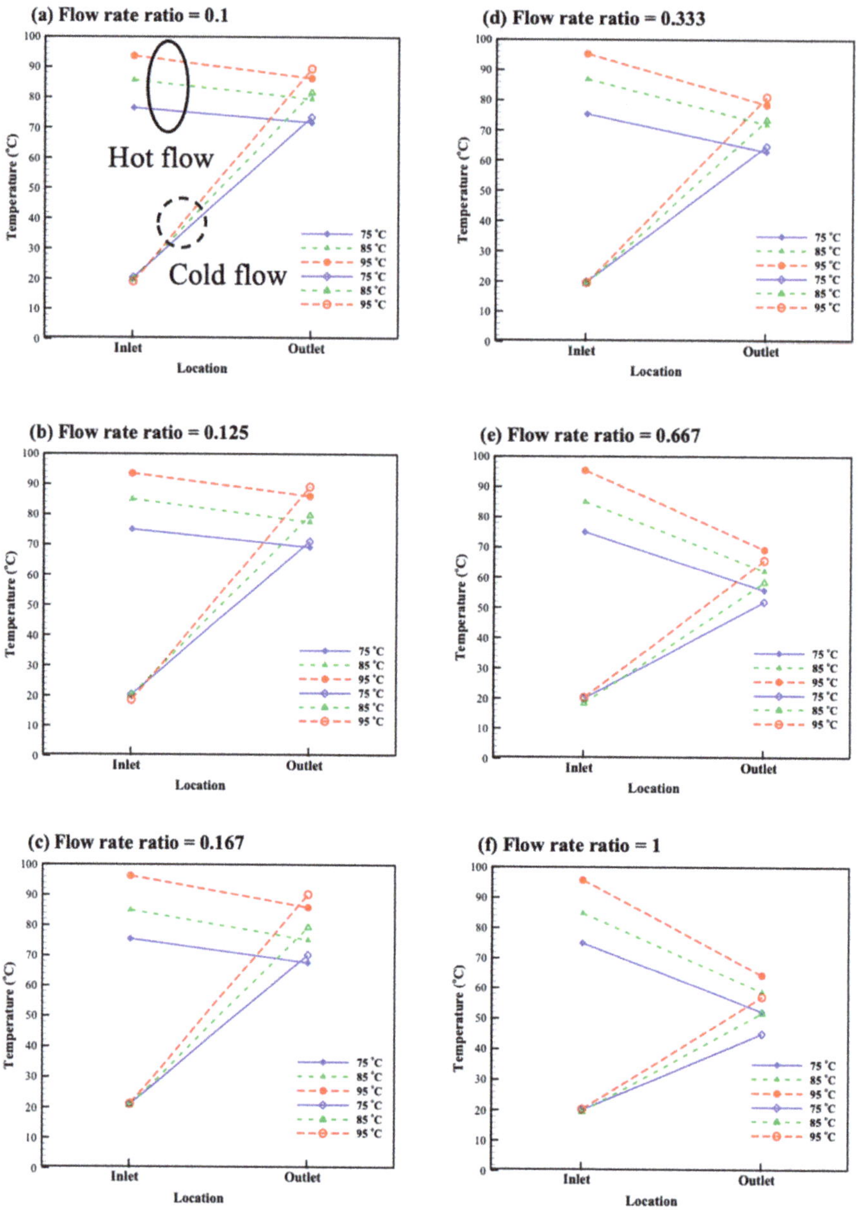

Figure 5. Temperature distributions of hot and cold fluid flow after heat exchange at flow rate ratios of (**a**) 0.1, (**b**) 0.125, (**c**) 0.167, (**d**) 0.333, (**e**) 0.667, and (**f**) 1.

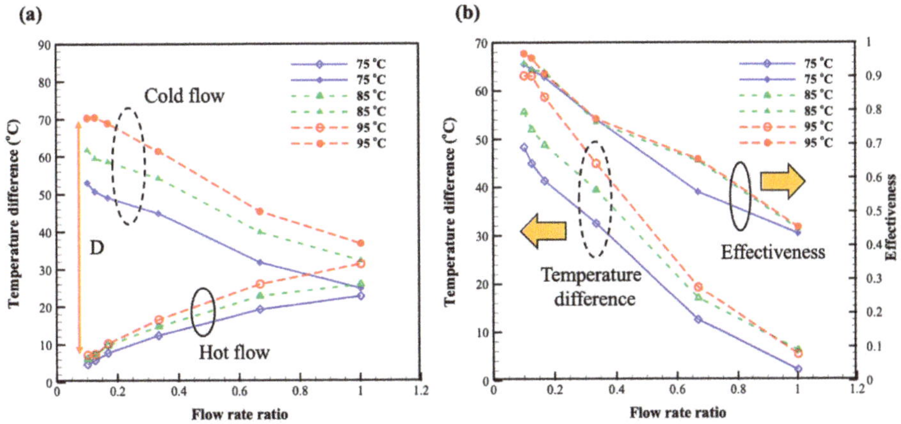

Figure 6. The temperature difference plot for different flow rate ratios and effectiveness (**a**) the temperature difference between inlet and outlet of hot (solid line) and cold (dash line) fluid. (**b**) the temperature difference (solid line) between hot and cold fluid and effectiveness (dash line).

3.4. Characteristics of Heat Transfer Performance

Figure 7 shows the convective heat transfer coefficient of the PCHE under various Reynolds numbers of the cold fluid flows. A higher Reynolds number is conducive to convective heat transfer. A similar behavior was also observed in a previously reported study [38]. This is why there is an increment of convective heat transfer coefficient with a rising Reynolds number. For the hot fluid flow inlet temperature of 95 °C, the heat transfer coefficient value is increased from Re = 50 to Re = 300, rendering an increment of 67.8%, and other conditions of hot inlet temperatures are increasing by at least 64%. At low Re values, such as 50 and 100, the sensitivity of the heat transfer coefficient to the hot fluid flow inlet temperature Re is low and the variation is small. However, at Re = 300, the convective heat transfer coefficient at the hot fluid flow inlet temperature at 95 °C is about 5% higher than that of the convective heat transfer coefficient at the hot fluid flow inlet temperature at 75 °C. The relationship between the effectiveness and the Reynolds number is shown in Figure 7. Unlike the convective heat transfer coefficient, as the Reynolds number increases, the effectiveness shows a downward trend. The convective heat transfer coefficient shows a positive correlation with the Reynolds number of the cold fluid flow where the amount of heat transfer increases. However, the fluid velocity is relatively high and the residence time of the cold fluid in the channel is shorter and reduces the effectiveness of the heat exchanger. The flow plates of PCHE are designed as an S-shape (Figures 1 and 2), which can prolong the residence time of the working fluid. It is obvious that the influence of the residence time prevails over the flow rate. Thus, the effectiveness declines with an increasing Reynolds number. Yan et al. [39] studied the effectiveness values of the different flow rate configurations in which they also observed a similar decreasing trend of the heat exchanger effectiveness with the increasing of the flow rate.

Figure 8 further explores the relationship between the Nusselt number (Nu) and the Reynolds number in the channel. The Nusselt number is proportional to the convective heat transfer coefficient which is a function of the flow rate or Reynolds number. As a consequence, the Nusselt number goes up when the Reynolds number increases, and the entire trend of the Nusselt number curves resembles that of the convective heat transfer coefficient. Yang et al. [38] studied the flow and performance of heat transfer in mini channels configured with hexagonal fins at laminar flow, and obtained a similar heat transfer performance. For the hot fluid flow inlet temperature of 95 °C, the Nusselt number increases by 68% when the Reynolds number increases from 50 to 300. Similar to the convective heat transfer coefficient, the Nusselt number is fairly insensitive to the variation of the hot fluid flow inlet temperature at Re = 50 and 100, while its variation at Re = 300 is also insignificant. This reflects that both the Nusselt number and the convective heat transfer coefficient are mainly governed by the Reynolds number, whereas the hot fluid flow inlet temperature is not affecting the Nusselt number and the convective heat transfer coefficient. To develop the correlation for Nusselt number and Reynold number, the basic logarithm form was used, as follows:

$$Nu = 0.03428Re^{0.6135} \tag{15}$$

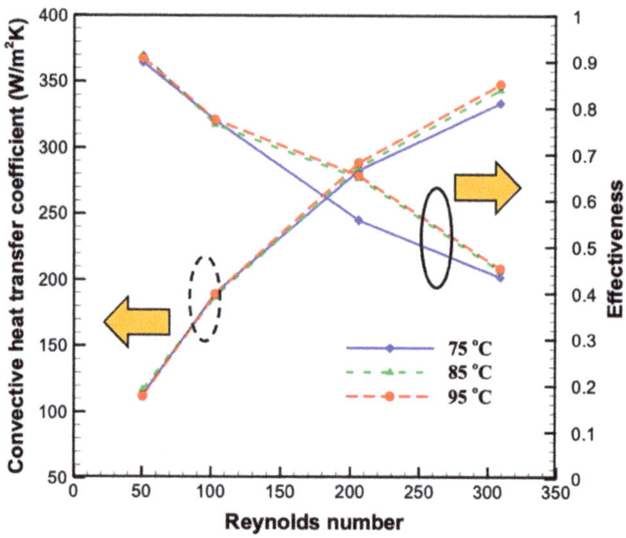

Figure 7. Convective heat transfer coefficient and effectiveness versus Reynolds number for different hot inlet temperatures.

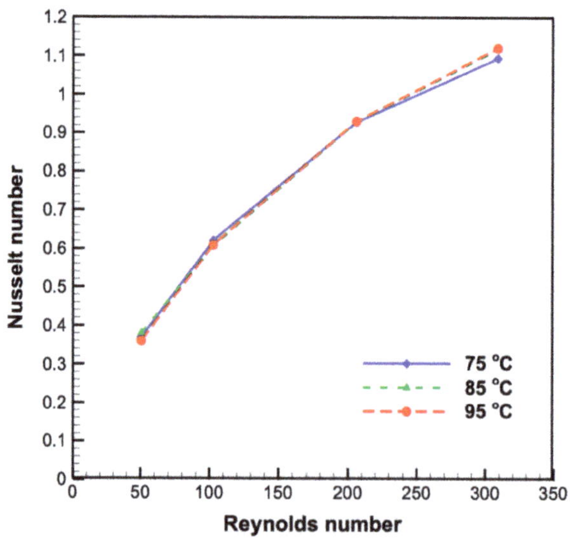

Figure 8. Nusselt number with different inlet temperatures for different Reynolds numbers.

3.5. The Number of Transferred Unit Value (NTU)

The value of the number of transferred unit (NTU) signifies the rate of net heat exchange between two fluids in the PCHE. The profiles of NTU and effectiveness are plotted in Figure 9 to show the relationship between NTU, effectiveness, and flow rate ratio. Both the NTU and effectiveness values decrease with rising flow rate ratios, revealing that the trend of NTU follows the trend of effectiveness. A past simulation study [40] explored the relationship between NTU and effectiveness, and observed that the heat exchanger effectiveness ε and NTU have the same trend in counterflow. When the ratio of flow rate decreases from 1 to 0.1, the NTU value is increased by 70.3% at a low flow rate. These explain that a longer residence time in the PCHE is beneficial to heat transfer. When the C_r value is 1, the values of the effectiveness and NTU with the hot inlet temperature of 95 °C are 0.428 and 0.83, respectively. The calculated effectiveness using Equation (12), in terms of NTU, is 0.453, which is close to the calculated effectiveness using Equation (3), which is 0.428. Sheldon et al. [41] compared the most common estimated efficiency methods for heat exchangers—effectiveness and NTU methods—and mentioned that the two methods were equivalent.

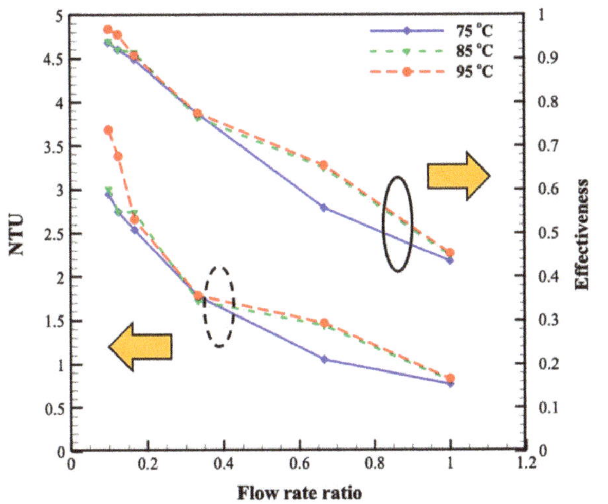

Figure 9. The profiles of NTU value versus the flow rate ratio.

3.6. Comparison to Other Research

This study provides the effectiveness of PCHEs using water as a working fluid at low Reynolds numbers. The experimental results showed that the obtained effectiveness values in this study were between the values of past reported studies, as shown in Figure 10. In past studies, heat exchangers with compact microchannel were constructed and water was used as a working fluid in counterflow at low Reynolds numbers. Hasan et al. [42] studied the influence of channel geometry on the performance of microchannel heat using simulations, where the effectiveness values of the circular and square channels were higher and lower, respectively. The hydraulic diameter was about 24 times smaller than our PCHE. In general, the attained effectiveness in this study is better than that of the Hasan et al. design [42]. On the other hand, the effectiveness of this study is lower than the results of Seyf et al. [43] and Mohammadian et al. [44]. Seyf et al. [43] studied microchannel heat exchangers using simulations. The hydraulic diameter in the study of Seyf et al. [43] was smaller than our PCHE by a factor of around 23 times. This is the reason why their results are better. In the study of Mohammadian et al. [44], they numerically studied nanofluid (Al_2O_3-water) in a counterflow heat exchanger, which had a better heat transfer performance. The nanofluid could easily increase the cold outlet temperature and make the temperature difference increase, so their results are better than those of the present study.

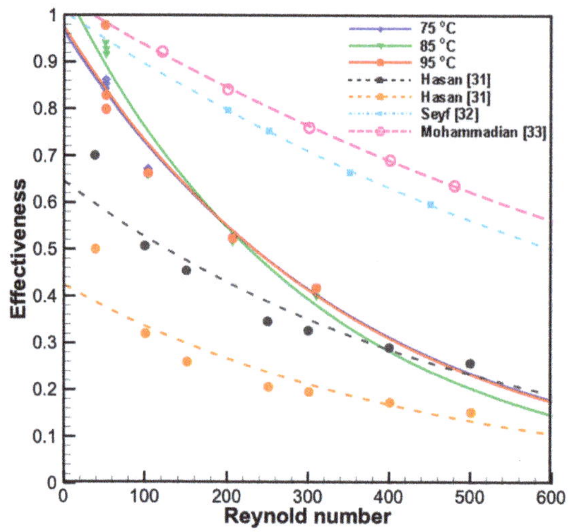

Figure 10. Comparison with the studies from the literature.

4. Conclusions

In the present study, a printed circuit heat exchanger (PCHE) is successfully fabricated using precision manufacturing and diffusion bonding. The printed circuit heat exchanger with an S-shaped meandering design for a flow path was tested at laminar flow (50 < Re < 300). The experimental results showed that the printed circuit heat exchanger provides high effectiveness and thermal performance at low flow rate ratios. The highest effectiveness of the PCHE is about 0.979 for an inlet flow temperature of 95 °C. The highest heat transfer coefficient obtained from the experiment is about 347.8 for Re = 300. The waste heat at a hot spring can be effectively harvested through the S-type meandering design of the flow path on the heat exchanger plate. This design will provide a larger heat transfer area and maximize the heat exchange between the fluids. A combination of the multiple flow rate ratios shows the best operating conditions. Regardless of the inlet temperature, the effectiveness of the PCHE is always higher. When the flow rate ratio is 1, the NTU value and effectiveness are about 0.83 and 0.428, respectively. In future work, the printed circuit heat exchanger will be tested under turbulent flow conditions to further characterize its thermal performance.

Author Contributions: Conceptualization, C.-Y.C. and W.-H.C.; methodology, C.-Y.C.; formal analysis, C.-Y.C. and M.C.U.; investigation, C.-Y.C. and M.C.U.; resources, W.-H.C.; data curation, C.-Y.C., W.-H.C. and M.C.U.; validation, C.-Y.C. and M.C.U.; writing—original draft preparation, C.-Y.C. and A.A.A.; writing—review and editing, W.-H.C. and L.H.S.; visualization, C.-Y.C.; supervision, W.-H.C.; funding acquisition, W.-H.C.; Project administration, C.-Y.C. and W.-H.C. All authors have read and agreed to the published version of the manuscript.

Funding: This research was supported in part by Higher Education Sprout Project, Ministry of Education to the Headquarters of University Advancement at National Cheng Kung University(NCKU). The authors also acknowledge the financial support of the Ministry of Science and Technology, Taiwan, R.O.C., under the contracts The authors acknowledge the financial support of the Ministry of Science and Technology, Taiwan, R.O.C., under the contract MOST 109-2622-E-006-006-CC1.

Institutional Review Board Statement: Not applicable.

Informed Consent Statement: Not applicable.

Data Availability Statement: Not applicable.

Acknowledgments: This research was supported in part by Higher Education Sprout Project, Ministry of Education to the Headquarters of University Advancement at National Cheng Kung University(NCKU). The authors also acknowledge the financial support of the Ministry of Science and Technology, Taiwan, R.O.C., under the contracts. The authors acknowledge the financial support of the Ministry of Science and Technology, Taiwan, R.O.C., under the contract MOST 109-2622-E-006-006-CC1.

Conflicts of Interest: The authors declare no conflict of interest.

Nomenclature

A	Area of heat transfer (m^2)
$c_{p,c}$	Cold work fluid's heat capacity (J/kg °C)
$c_{p,h}$	Hot work fluid's heat capacity (J/kg °C)
C_{min}	Smaller of the two heat capacity rates (J/kg °C)
C_{hot}	Hot fluid's heat capacity rates (J/kg °C)
D_h	Hydraulic diameter (m)
h	Convective heat transfer coefficient ($W/m^2 \cdot °C$)
k	Thermal conductivity (W/m °C)
Nu	Nusselt number
Q	Heat transfer rate (W)
q	Heat flux (W/m^2)
q_{max}	Large of heat flux (W/m^2)
Re	Reynold number (W/m^2)
T	Temperature (°C)
T_m	Average temperature of hot fluid inlet (°C)
$T_{h,in}$	Hot fluid inlet temperature (°C)
$T_{h,out}$	Hot fluid outlet temperature (°C)
$T_{c,in}$	Cold fluid inlet temperature (°C)
$T_{c,out}$	Cold fluid outlet temperature (°C)
T_s	channel surface temperature (°C)
T_∞	fluid temperature in channel (°C)
ΔT_m	Temperature difference (°C)
V	Velocity (m/s)
$\dot{m}$	Mass flow (kg/s)
Greek letters	
ε	Effectiveness
μ	Viscosity (Pa·s)
ρ	Density (kg/m^3)
Subscripts	
C	Cold fluid
H	Hot fluid
in	Inlet
out	Outlet

References

1. Papapetrou, M.; Kosmadakis, G.; Cipollina, A.; La Commare, U.; Micale, G. Industrial waste heat: Estimation of the technically available resource in the EU per industrial sector, temperature level and country. *Appl. Therm. Eng.* **2018**, *138*, 207–216. [CrossRef]
2. Hung, T.C.; Shai, T.Y.; Wang, S.K. A review of organic Rankine cycles (ORCs) for the recovery of low-grade waste heat. *Energy* **1997**, *22*, 661–667. [CrossRef]
3. Zhou, N.J.; Wang, X.Y.; Chen, Z.; Wang, Z.Q. Experimental study on Organic Rankine Cycle for waste heat recovery from low-temperature flue gas. *Energy* **2013**, *55*, 216–225. [CrossRef]
4. Xu, Z.Y.; Wang, R.Z.; Yang, C. Perspectives for low-temperature waste heat recovery. *Energy* **2019**, *176*, 1037–1043. [CrossRef]
5. Baik, S.; Kim, S.G.; Lee, J.; Lee, J.I. Study on CO_2–Water printed circuit heat exchanger performance operating under various CO_2 phases for S-CO_2 power cycle application. *Appl. Therm. Eng.* **2017**, *113*, 1536–1546. [CrossRef]
6. Shi, H.-Y.; Li, M.-J.; Wang, W.-Q.; Qiu, Y.; Tao, W.-Q. Heat transfer and friction of molten salt and supercritical CO_2 flowing in an airfoil channel of a printed circuit heat exchanger. *Int. J. Heat Mass Transf.* **2020**, *150*, 119006. [CrossRef]

7. Orr, B.; Akbarzadeh, A.; Mochizuki, M.; Singh, R. A review of car waste heat recovery systems utilising thermoelectric generators and heat pipes. *Appl. Therm. Eng.* **2016**, *101*, 490–495. [CrossRef]
8. Shen, Z.G.; Tian, L.L.; Liu, X. Automotive exhaust thermoelectric generators: Current status, challenges and future prospects. *Energy Convers. Manag.* **2019**, *195*, 1138–1173. [CrossRef]
9. Thulukkanam, K. *Heat Exchanger Design Handbook*; CRC Press: Boca Raton, FL, USA, 2013.
10. Han, D.H.; Lee, K.-J. Single-phase heat transfer and flow characteristics of micro-fin tubes. *Appl. Therm. Eng.* **2005**, *25*, 1657–1669. [CrossRef]
11. Hosseini, R.; Hosseini-Ghaffar, A.; Soltani, M. Experimental determination of shell side heat transfer coefficient and pressure drop for an oil cooler shell-and-tube heat exchanger with three different tube bundles. *Appl. Therm. Eng.* **2007**, *27*, 1001–1008. [CrossRef]
12. Southall, D.; Dewson, S.J. Innovative compact heat exchangers. *Group* **2010**. 226, (212.6), 583.0. Available online: https://www.heatric.com/app/uploads/2018/04/Innovative-compact-heat-exchangers.pdf (accessed on 13 July 2021).
13. Khan, M.G.; Fartaj, A. A review on microchannel heat exchangers and potential applications. *Int. J. Energy Res.* **2011**, *35*, 553–582. [CrossRef]
14. Son, S.; Lee, Y.; Lee, J.I. Development of an advanced printed circuit heat exchanger analysis code for realistic flow path configurations near header regions. *Int. J. Heat Mass Transf.* **2015**, *89*, 242–250. [CrossRef]
15. Gkountas, A.A.; Stamatelos, A.M.; Kalfas, A.I. Recuperators investigation for high temperature supercritical carbon dioxide power generation cycles. *Appl. Therm. Eng.* **2017**, *125*, 1094–1102. [CrossRef]
16. Zhang, H.; Guo, J.; Cui, X.; Zhou, J.; Huai, X.; Zhang, H.; Cheng, K.; Han, Z. Experimental and numerical investigations of thermal-hydraulic characteristics in a novel airfoil fin heat exchanger. *Int. J. Heat Mass Transf.* **2021**, *175*, 121333. [CrossRef]
17. Zhu, C.-Y.; Guo, Y.; Yang, H.-Q.; Ding, B.; Duan, X.-Y. Investigation of the flow and heat transfer characteristics of helium gas in printed circuit heat exchangers with asymmetrical airfoil fins. *Appl. Therm. Eng.* **2021**, *186*, 116478. [CrossRef]
18. Hu, H.; Li, J.; Xie, Y.; Chen, Y. Experimental investigation on heat transfer characteristics of flow boiling in zigzag channels of printed circuit heat exchangers. *Int. J. Heat Mass Transf.* **2021**, *165*, 120712. [CrossRef]
19. Gkountas, A.A.; Benos, T.L.; Nikas, K.-S.; Sarris, I.E. Heat transfer improvement by an Al_2O_3-water nanofluid coolant in printed-circuit heat exchangers of supercritical CO_2 Brayton cycle. *Therm. Sci. Eng. Prog.* **2020**, *20*, 100694. [CrossRef]
20. Gkountas, A.A.; Stamatelos, A.M.; Kalfas, A.I. Thermodynamic Modeling and Comparative Analysis of Supercritical Carbon Dioxide Brayton Cycle. ASME Turbo Expo 2017: Turbomachinery Technical Conference and Exposition, Charlotte, NC, USA, 26–30 June 2017; Available online: https://asmedigitalcollection.asme.org/GT/proceedings-abstract/GT2017/V003T06A017/242325 (accessed on 13 July 2021).
21. Chen, M.; Sun, X.; Christensen, R.N.; Shi, S.; Skavdahl, I.; Utgikar, V.; Sabharwall, P. Experimental and numerical study of a printed circuit heat exchanger. *Ann. Nucl. Energy* **2016**, *97*, 221–231. [CrossRef]
22. Kim, I.H.; No, H.C. Thermal–hydraulic physical models for a Printed Circuit Heat Exchanger covering He, He–CO_2 mixture, and water fluids using experimental data and CFD. *Exp. Therm. Fluid Sci.* **2013**, *48*, 213–221. [CrossRef]
23. Seo, J.-W.; Kim, Y.-H.; Kim, D.; Choi, Y.-D.; Lee, K.-J. Heat Transfer and Pressure Drop Characteristics in Straight Microchannel of Printed Circuit Heat Exchangers. *Entropy* **2015**, *17*, 3438–3457. [CrossRef]
24. Meshram, A.; Jaiswal, A.K.; Khivsara, S.D.; Ortega, J.D.; Ho, C.; Bapat, R.; Dutta, P. Modeling and analysis of a printed circuit heat exchanger for supercritical CO_2 power cycle applications. *Appl. Therm. Eng.* **2016**, *109*, 861–870. [CrossRef]
25. Chen, M.; Sun, X.; Christensen, R.N.; Skavdahl, I.; Utgikar, V.; Sabharwall, P. Dynamic behavior of a high-temperature printed circuit heat exchanger: Numerical modeling and experimental investigation. *Appl. Therm. Eng.* **2018**, *135*, 246–256. [CrossRef]
26. Yoon, S.-J.; O'Brien, J.; Chen, M.; Sabharwall, P.; Sun, X. Development and validation of Nusselt number and friction factor correlations for laminar flow in semi-circular zigzag channel of printed circuit heat exchanger. *Appl. Therm. Eng.* **2017**, *123*, 1327–1344. [CrossRef]
27. Gkountas, A.A.; Benos, L.T.; Sofiadis, G.N.; Sarris, I.E. A printed-circuit heat exchanger consideration by exploiting an Al_2O_3-water nanofluid: Effect of the nanoparticles interfacial layer on heat transfer. *Therm. Sci. Eng. Prog.* **2021**, *22*, 100818. [CrossRef]
28. Xu, H.; Duan, C.; Ding, H.; Li, W.; Zhang, Y.; Hong, G.; Gong, H. The optimization for the straight-channel PCHE size for supercritical CO_2 Brayton cycle. *Nucl. Eng. Technol.* **2021**, *53*, 1786–1795. [CrossRef]
29. Allen, B.; Gosselin, L. Optimal geometry and flow arrangement for minimizing the cost of shell-and-tube condensers. *Int. J. Energy Res.* **2008**, *32*, 958–969. [CrossRef]
30. Kim, W.; Baik, Y.-J.; Jeon, S.; Jeon, D.; Byon, C. A mathematical correlation for predicting the thermal performance of cross, parallel, and counterflow PCHEs. *Int. J. Heat Mass Transf.* **2017**, *106*, 1294–1302. [CrossRef]
31. Cowell, T. A general method for the comparison of compact heat transfer surfaces. *J. Heat Transf.* **1990**, *112*, 288–294. [CrossRef]
32. Figley, J.; Sun, X.; Mylavarapu, S.K.; Hajek, B. Numerical study on thermal hydraulic performance of a Printed Circuit Heat Exchanger. *Prog. Nucl. Energy* **2013**, *68*, 89–96. [CrossRef]
33. San, J.-Y.; Lin, G.-S.; Pai, K.-L. Performance of a serpentine heat exchanger: Part I—Effectiveness and heat transfer characteristics. *Appl. Therm. Eng.* **2009**, *29*, 3081–3087. [CrossRef]
34. Chen, W.-H.; Liao, C.-Y.; Hung, C.-I.; Huang, W.-L. Experimental study on thermoelectric modules for power generation at various operating conditions. *Energy* **2012**, *45*, 874–881. [CrossRef]

35. Attalla, M.; Maghrabie, H.M. Investigation of effectiveness and pumping power of plate heat exchanger with rough surface. *Chem. Eng. Sci.* **2020**, *211*, 115277. [CrossRef]
36. Pourahmad, S.; Pesteei, S.M. Effectiveness-NTU analyses in a double tube heat exchanger equipped with wavy strip considering various angles. *Energy Conv. Manag.* **2016**, *123*, 462–469. [CrossRef]
37. Magazoni, F.C.; Cabezas-Gómez, L.; Alvariño, P.F.; Sáiz-Jabardo, J.M. Closed form relationships of temperature effectiveness of cross-flow heat exchangers. *Therm. Sci. Eng. Prog.* **2019**, *9*, 110–120. [CrossRef]
38. Yang, X.; Wei, L.; Cao, F.; Zhang, L.; Lu, Z.; Meng, X.; Jin, L. A parametric study of laminar convective heat transfer in fractal minichannels with hexagonal fins. *Int. J. Energy Res.* **2020**, *44*, 9382–9398. [CrossRef]
39. Yan, S.-R.; Moria, H.; Pourhedayat, S.; Hashemian, M.; Asaadi, S.; Sadighi Dizaji, H.; Jermsittiparsert, K. A critique of effectiveness concept for heat exchangers; theoretical-experimental study. *Int. J. Heat Mass Transf.* **2020**, *159*, 120160. [CrossRef]
40. Sammeta, H.; Ponnusamy, K.; Majid, M.A.; Dheenathayalan, K. Effectiveness charts for counter flow corrugated plate heat exchanger. *Simul. Model. Pract. Theory* **2011**, *19*, 777–784. [CrossRef]
41. Jeter, S.M. Effectiveness and LMTD correction factor of the cross flow exchanger: A simplified and unified treatment. In Proceedings of the ASEE Southeast Section Conference, Tuscaloosa, AL, USA, 2–4 April 2006; pp. 1–10.
42. Hasan, M.I.; Rageb, A.A.; Yaghoubi, M.; Homayoni, H. Influence of channel geometry on the performance of a counter flow microchannel heat exchanger. *Int. J. Therm. Sci.* **2009**, *48*, 1607–1618. [CrossRef]
43. Seyf, H.R.; Keshavarz Mohammadian, S. Thermal and hydraulic performance of counterflow microchannel heat exchangers with and without nanofluids. *J. Heat Transf.* **2011**, *133*, 081801. [CrossRef]
44. Mohammadian, S.K.; Reza Seyf, H.; Zhang, Y. Performance augmentation and optimization of aluminum oxide-water nanofluid flow in a two-fluid microchannel heat exchanger. *J. Heat Transf.* **2014**, *136*, 021701. [CrossRef]

energies

MDPI

Article

Value Co-Creation Behavior in Green Supply Chains: An Empirical Study

Xianliang Shi [1], Genzhu Li [1], Ciwei Dong [2] and Yefei Yang [1,*]

1 School of Economics and Management, Beijing Jiaotong University, Beijing 100044, China; xlshi@bjtu.edu.cn (X.S.); ligenzh@163.com (G.L.)

2 School of Business Administration, Zhongnan University of Economic and Law, Wuhan 430073, China; dongciwei@zuel.edu.cn

* Correspondence: yangyefei@bjtu.edu.cn

Received: 10 June 2020; Accepted: 24 July 2020; Published: 30 July 2020

Abstract: As an important supply chain development strategy, green investment and sustainability are concerns of the government and enterprises. However, due to the high cost and low profit of green investment, a large number of small and medium-sized firms can be deterred from their implementation. Value co-creation has become a key measure to solve this problem. This article explores the relationship between the green supply chain (GSC) strategy, value co-creation, and corporate performance in the manufacturing environment, and considers the regulatory effects of internal environmental factors and external environmental pressures on this relationship. Based on data from 115 manufacturers in China, we tested the hypotheses, explained the statistical results, and identified key concerns for implementing GSC through value co-creation. The findings reveal that the GSC strategy can promote a high level of firms' value co-creation with their supply chain partners, and different value co-creation modes have different effects on firm performance (i.e., operational performance, innovation performance, and financial performance). In addition, the findings indicate that macro-level external pressure and micro-level internal support could enhance such effects. This study enriches the literature with value co-creation modes and GSC management by integrating GSC strategies and value co-creation strategies, providing confidence to the firms and their supply chain partners in value co-creation, thus helping them to better implement a GSC strategy.

Keywords: green supply chain; value co-creation; firm performance

---

1. Introduction

Green sustainable development is a very important supply chain strategy, which has received widespread attention worldwide [1,2]. Increasing public awareness, increasingly stringent government regulatory requirements, and market pressures have forced many companies to incorporate green and sustainable development into their supply chain [3]. In China, with its rapid economic development, resource and environmental problems are becoming more and more serious such as severe smog, lack of water, soil pollution, etc. Additionally, these resource waste and pollutant emissions mainly come from manufacturing industries. Therefore, the Chinese government and the Ministry of Environmental Protection (MEP) have promulgated and implemented stricter government regulatory policies, forcing many manufacturers to focus on sustainable supply chain. In fact, many manufacturers have started green sustainable supply chain management, which is more important than some supply chain management issues (such as integration and cooperation), raising environmental awareness such as green procurement, green transportation, green packaging, etc., and strives to promote suppliers implement environmental management.

Effective management of a green supply chain (GSC) requires an expanded perspective beyond a focus company including supply chain partners [4]. Value co-creation is considered as a business cooperation strategy for supply chain partners. As the cost of green innovation is relatively high, some manufacturers have started to cooperate with other enterprises in the supply chain (such as suppliers and retailers) to jointly develop green products and technologies [1,5]. For example, companies such as BMW, Patagonia, and the Body Shop have taken a sustainable approach by developing innovative environmentally friendly products with their consumers and have realized that the development of green products/services has become a new competitive advantage for their firms [1].

The current literature relevant to co-creation indicates that it is conducive to the deep cooperation between supply chain enterprises [4]. For the value co-creation of GSC, it is beneficial for upstream and downstream firms to share green technology investment and improve the efficiency of the whole supply chain [6]. However, there are contradicting views on the impact of co-creation on supply chain performance. For example, some literature suggests that value co-creation could cause organizational conflicts such as relationship conflict (i.e., the incompatibility in buyer-supplier manufacturer relationship) and task conflict (i.e., the differences in positions and ideas of tasks performed by suppliers and manufacturers), thereby causing a negative impact on performance [7–9]. Thus, the connotation of value co-creation and its impact on performance in the context of green supply chain are still unclear and need to be explored.

As a complex system, manufacturing is directly or indirectly related to the creation of economic wealth and impacts on the product life cycle of the natural environment and social systems [6]. Due to different sectors, the internal and external contextual factors of the manufacturer have an important impact on the implementation of the manufacturing supply chain strategy. The internal environment support mainly refers to some internal firm factors that influence the activities of the enterprise (e.g., the staff, culture, and operations process [8,10]). For example, some of the literature indicates that the attitude and ability of managers is highly relevant with the implementation of a green supply chain and value co-creation [8,10]. The external environmental pressures refer to the macro-factors such as marketing pressure and institutional pressures. Indeed, some literature suggests that internal contextual factors could have some influence on the implementation of supply chain strategies [10]. Thus, we need to further explore the actual impact of these contextual factors on the implementation of a green supply chain.

This paper mainly focused on two problems: (1) How does the firm's green supply chain strategy impact the performance by value co-creation behaviors? and (2) How does the internal environment support and external contextual factors impact on the relationship between green strategy adoption and value co-creation? First, this paper proposes a theoretical model of value co-creation under the green supply chain strategy and further posited the hypotheses in Figure 1. We examined 115 manufacturers through a series of statistical methods. Our findings indicate that adoption of the green supply chain strategy by firms can indeed improve operational performance, financial performance, and even innovation performance by a value co-creation strategy. In addition, external and internal contextual factors have a different moderating effect on the relationship between GSC strategy, value co-creation, and performance. Our findings enrich the literature relevant to co-creation by exploring new insights into green supply chain strategies and considering the roles of some contextual factors. In addition, our findings enhance the internal power of manufacturers and their upstream and downstream firms to implement a green supply chain management and provide some suggestions for practitioners.

The main contributions of this paper are as follows: This paper innovatively studies the green supply chain management from the perspective of value co-creation. Regarding the selection of value co-creation behavior factors, this paper considers three different value co-creation behaviors between manufacturers, suppliers, competitors, and retailers. This paper evaluates corporate performance from three aspects: operational performance, innovative performance, and financial performance. At the same time, this paper analyzed the coordination effect of the internal environment and external environment on the value co-creation of a green supply chain. In the internal environment, this paper particularly studied the influence of the digital level of enterprises on green value co-creation, which is innovative to some extent.

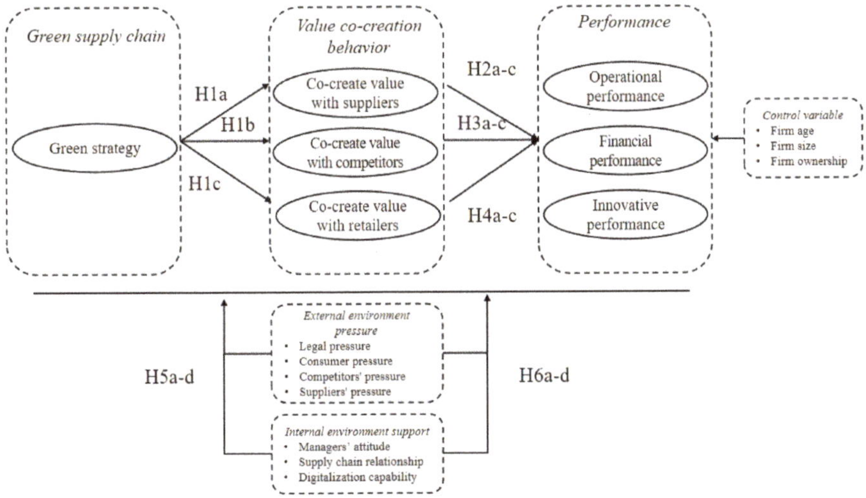

Figure 1. Conceptual model.

The rest of this paper is arranged as follows. Section 2 introduces the literature review of related topics. Section 3 explains the conceptual model of the hypothesis. Section 4 introduces the research methods, and Section 5 introduces the results of the empirical analysis. Section 6 discusses the main findings and Section 7 introduces the conclusions, limitations, and future research directions.

2. Literature Review

2.1. Cooperation in Green Supply Chain

Green supply chain cooperation has an important impact on sustainable product development and operations mode [11–15] and has become a positive change in corporate strategy for various companies [14,15]. Cooperation can be divided into vertical and horizontal cooperation and among the two types of cooperation and their cooperation performance is impacted by different ways [16–22]. The literature shows that vertical cooperation in the supply chain can improve supply chain decision-making and performance through information sharing, thereby achieving a win–win situation [16–18]. In the horizontal cooperation, two competitive retailers can cooperate to purchase and obtain a quantity discount, thereby influencing their profits [19].

In addition, for the purpose of realizing the green and economic performance of products, supply chain members mainly adopt two typical ways in the green cooperation of a supply chain: alliance and cost sharing contracts [23]. For example, Ge et al. showed that both cooperation and competition will tend toward a green technology alliance, and the company's decision to cooperate will change with the change in the endogenous knowledge sharing rate [24–26].The second model is to stimulate the green investment of members through supply chain contracts and deal with the coordination of green channels. For example, Yenipazarli et al. analyzed the effects of retailer supplier cooperation on profit/cost and the environmental effects on the supply chain under two upstream and downstream contracts [27]. When members of the supply chain share the cost, on one hand, it can make products greener. On the other hand, it can make the overall profit of the supply chain increase [21,28–30]. Therefore, it is necessary to discuss how to cooperate with supply chain partners to improve corporate performance.

2.2. Value Co-Creation Behavior in the Supply Chain

Michael (1980) suggests that the value creation of an enterprise is composed of internal production and operations routines that create value. However, with the continuous change in the market competition environment, the creation process of enterprise value cannot be limited to the inside of the enterprise [31], but is co-created by the cooperative network composed of enterprises [32]. Enterprises can improve the efficiency, quality, cost, and other aspects of the whole supply chain through cooperative implementation of strategic management [33].

Prahalad et al. proposed value co-creation as a new method of value creation [34] and believe that value originates not only from producers, but from the joint creation of consumers and enterprises or other relevant stakeholders. Ultimately, the value is determined by the consumer. From the perspective of stakeholders, the meaning of a "company" is a relationship that exists between individuals or groups that affect the company's business or is affected by it, and the purpose for which a business is established forms the basis for whether a stakeholder establishes relationships and cooperates with them [35]. In this relationship, stakeholders are the unity of recipients of value and the value creators/co-creators [36]. Multiple stakeholders have an impact on the business environment and company efficiency by providing resources and benefit from it [37]. Therefore, the joint efforts of stakeholders are the most important part of value co-creation [38]. In addition, stakeholders must first have a common purpose to collaborate [39] and then also encourage the positive contributions of members [40].

Value creation usually occurs in the process of using products or services [41]. The fierce market competition requires the enterprises in the supply chain to strengthen cooperation and establish a value network. Many scholars have studied the vertical cooperation among enterprises, that is, the cooperation between enterprises and suppliers and retailers. Enterprises can create more value for the supply chain through information sharing, resource integration, and process integration with upstream and downstream partners [29,42–47]. Some scholars have also paid attention to the horizontal cooperation between supply chain enterprises and competitors, but there have been few studies. D'Aspremon et al. and Kamien et al. introduced the concept of horizontal Research & Development. cooperation in theory early on [48,49]; Ge et al. and Dai et al. compared different horizontal cooperation modes of green supply chain [22,23]; and Luo et al. studied the green technology investment of two manufacturers in a competitive and cooperative environment [50]. Gnyawali et al. applied practical case analysis, and concluded that competitive cooperation is beneficial to the development of advanced technology [51]. Therefore, our research focused on three different types of value co-creation: co-creation with suppliers, co-creation with retailers, and co-creation with competitors.

2.3. Value Co-Creation and Performance

Research shows that value co-creation behavior benefits both suppliers and customers [52]. For customers, value co-creation mainly focuses on the customer dimension, which is a relatively new research field [53]. Most of the relevant studies have focused on the development of measurement tools to study customer co-creation value across different research backgrounds [54–56]. Co-creation practices enable interaction and communication between customers and enterprises, so as to benefit customers who participate. Active participation in co-creating value makes them more satisfied than those who are not [57], encourages customers to be more innovative [58], and increases the customers' enthusiasm and willingness to take risks [59]. At the same time, the perceived value gained from value co creation will increase customer commitment [60], and psychological attachment, which increases the willingness of customers to make more voluntary efforts in the cooperative relationship. The increase of customer value in value co-creation improves the customer experience in supplier relationships [61], and establishes customer loyalty [62].

For enterprises, value co-creation is realized through the establishment of a co-operation relationship between two or more parties in the supply chain, so as to jointly obtain strategic benefits and improve the operation efficiency, economic performance, and service performance of enterprises [63–65], which is also the key success factor of enterprises in delivering products and services [66]. Through value co-creation, companies have a positive impact on service performance and environmental performance [67,68]. Through the green internal innovation process and green technology investment research and development, it can promote enterprises to develop green energy-saving products and realize energy-saving and emissions reduction of the whole supply chain [69,70]. The optimal strategy of the enterprise and the performance of the supply chain will change due to different cooperation models [17,18,21,71]. Thus, this study will examine different performances (i.e., operational performance, innovation performance and financial performance) of value co-creation in the supply chain and the changes in performance under different contextual factors.

3. Hypotheses Development

3.1. The Relationship between Green Strategy and Value Co-Creation Behavior

A green strategy should not be a burden to firms. On the contrary, it will promote value co-creation among firms and increase the value of firms. A single enterprise may not achieve the goal of a green strategy completely or to a high degree, depending on its limited ability. Therefore, it is necessary to establish a strategic alliance with green environmental protection as the core between firms [72] to achieve the green strategic goals of ecological, economic, and social sustainable development. As a product of the times and social development, the green strategy has been recognized by all participants in the supply chain, and has formed a green strategic alliance with the upstream and downstream enterprises of the supply chain. The green supply chain strategy can promote the alliance participants to take a positive attitude to cooperate, jointly design products and systems that meet the production requirements, and establish a fair and reasonable cost sharing, risk sharing, and benefit distribution mechanism.

The implementation of green strategy promotes green procurement, which is regarded as the starting point to reduce environmental problems [73]. Cooperation between manufacturers and suppliers is the key to green procurement. Therefore, the deeper the implementation of the GSC strategy, the higher the manufacturer's requirements for raw materials from suppliers, and the more likely the suppliers' raw materials are to be customized, resulting in stronger interdependence between manufacturers and suppliers, more frequent value co-creation, and closer alliances between manufacturers and suppliers. Consequently, we propose (see Figure 1):

Hypothesis 1a (H1a). *Green Strategy of Manufacturers is Positively Related to Value Co-creation with their Suppliers.*

With the growing popularity of green products, the consumers' demand for green products has gradually increased. A single manufacturer may not be able to meet the consumers' demand due to the size of the enterprise, so there may be multiple manufacturers producing the same type of green products. As the demand for green products increases, the revenue of the dominant manufacturers will increase significantly. However, the profit of following manufacturers may decrease, even lower than the level before cooperation [74]. Therefore, in order to promote common development, achieve win–win and mutual benefit, a contract may be drawn up between the two manufacturers for revenue distribution. The deeper the implementation of the green strategy, the larger the market size of manufacturers will be, and the more necessary it is for manufacturers to adopt co-creative behavior to secure cooperation. Thus, we propose:

Hypothesis 1b (H1b). *Green Strategy of Manufacturers is Positively Related to Value Co-creation with their Competitors.*

As a downstream company in the supply chain, retailers directly contact consumers to understand their consumer preferences. Under the green strategy, retailers can share sales data with manufacturers to reduce the bullwhip effect [75] and assist manufacturers in designing green products that better meet consumer needs. When the green concept is accepted by more consumers, the value-creation behavior of retailers and manufacturers will become more frequent. Thus, we propose:

Hypothesis 1c (H1c). *Green Strategy of Manufacturers is Positively Related to Value Co-creation with their Retailers.*

3.2. The Impact of Value Co-Creation Behavior on Firm Performance

By closely cooperating with each other to complete value creation (that is, value co-creation), companies can achieve performance that a single company cannot achieve by itself. Manufacturers work with suppliers, manufacturers in the same industry, and retailers to create value together. Through cooperation, they can design different business models to improve the operational efficiency of the enterprise, thereby improving the performance level [65]. The value co-creation behavior between firms can improve the quality of service and response speed. In the supply chain, companies share knowledge and information resources with partners through value co-creation, and use complementary capabilities to jointly deliver products and services to different customers [76], thereby improving the company's operating performance. Consequently, we propose:

Hypothesis 2a (H2a). *Value Co-creation with their Suppliers is Positively Related to the Manufacturers' Operational Performance in a Green Supply Chain.*

Hypothesis 2b (H2b). *Value Co-creation with their Competitors is Positively Related to the Manufacturers' Operational Performance in a Green Supply Chain.*

Hypothesis 2c (H2c). *Value Co-creation with their Retailers is Positively Related to the Manufacturers' Operational Performance in a Green Supply Chain.*

The value co-creation behavior promotes mutual trust between participating companies and through the use of overall synergies. It can promote the improvement of corporate financial performance [77]. In the context of the Green Strategic Alliance, manufacturers sign a certain contract with the supplier to give the supplier a certain preferential price [72] to ensure the possibility of achieving the common goal of green development and establish a brand image of a green strategic alliance, in order to form competitiveness in their respective fields and give products a green value. By complementing resources, manufacturers can solve problems that cannot be solved by a single enterprise, while weakening unilateral opportunistic tendencies and diversifying potential financial risks [78]. Through value co-creation behavior, manufacturers and retailers can enable firms to make flexible adjustments in response to changes in the environment, which can help companies seize market opportunities and improve dynamic capabilities [79]. Firms reach a consensus on cooperation, share resources, and conduct profitability to a certain degree. Such value co-creation behavior will form a strong and mutually beneficial relationship between firms, give products a higher value, reduce operating risks and costs, and increase product competition as well as efforts to expand market share and improve corporate returns. Consequently, we propose:

Hypothesis 3a (H3a). *Value Co-creation with their Suppliers is Positively Related to the Manufacturers' Financial Performance in GSC.*

Hypothesis 3b (H3b). *Value Co-creation with their Competitors is Positively Related to the Manufacturers' Financial Performance in GSC.*

Hypothesis 3c (H3c). *Value Co-creation with their Retailers is Positively Related to the Manufacturers' Financial Performance in GSC.*

The value co-creation activity has changed the relationship between companies. During the exchange of corporate culture and interests, the two sides also improved the company's openness, and the increase in openness in the organization is conducive to organizational innovation. In value co-creation activities, the sharing of technical resources can increase the breadth of employee thinking and promote the innovation of individual employees [80]. Value co-creation behavior can bring about knowledge transfer, which has a positive effect on product and process innovation [81]. Knowledge originally belonging to different organizations is passed between partners, and new products or services are easily born through permutations and combinations. Therefore, companies that adopt value co-creation can be the first to apply new technologies in the industry, obtain corresponding intellectual property rights, increase sales revenue of innovative products, and make products that have good market response and irreplaceability within a certain time. Thus, we propose:

Hypothesis 4a (H4a). *Value Co-creation with their Suppliers is Positively Related to the Manufacturers' Innovative Performance in GSC.*

Hypothesis 4b (H4b). *Value Co-creation with their Competitors is Positively Related to the Manufacturers' Innovative Performance in GSC.*

Hypothesis 4c (H4c). *Value Co-creation with their Retailers is Positively Related to the Manufacturers' Innovative Performance in GSC.*

3.3. The Moderating Effect of External Environmental Pressure

The essence of firms adopting green management is to respond to the natural environment, and the motivation of response is affected by various factors such as stakeholder pressure. Regulatory, competition, marketing pressure, and drivers can improve the environmental awareness of enterprises [82]. We perceive that the external environmental pressure (legal pressure, consumer pressure, competitor pressure, supplier pressure) of the company will have an impact on the above changes. Firms with these external pressures could be inclined to adopt co-creation with supply chain members. Additionally, the relatively mature legal environment, consumption environment, and industry environment will make the advantages of green strategies more obvious, thereby encouraging firms to conduct value co-creation. Suppliers can provide complete green products and services, which will also cause a certain degree of pressure and incentives on the supply chain and promote value creation between firms. Thus, we propose:

Hypothesis 5a (H5a). *The External Environmental Pressures have a Positive Moderating Effect on the Relationship between Green Strategy and Value Co-creation Behavior.*

The change in the external market environment is an important means for enterprises to deal with various complex environments, transfer risks, gain competitive advantages, and improve operation performance [83]. It may play a positive role in the relationship between enterprise value co-creation and operational performance. Thus, we propose:

Hypothesis 5b (H5b). *The External Environment Pressure has a Positive Moderating Effect on the Relationship between Value Co-creation Behavior and Operational Performance.*

A better market environment and higher customer satisfaction can improve the financial performance of enterprises [84]. The improvement of external factors can strengthen the market competitiveness of enterprises, so as to increase the role of value co-creation in promoting the financial performance of enterprises. Thus, we propose:

Hypothesis 5c (H5c). *The External Environment Pressure has a Positive Moderating Effect on the Relationship between Value Co-creation Behavior and Financial Performance.*

The government's actions to improve the external environment such as improving the market mechanism, optimizing the innovative business environment of small and medium-sized enterprises, and formulating supporting policies, etc. can improve the level of innovation of enterprises [85]. In the foundation of value co-creation among enterprises, favorable external factors can improve the innovation performance of enterprises. Thus, we propose:

Hypothesis 5d (H5d). *The External Environmental Pressure has a Positive Moderating Effect on the Relationship between Value Co-creation Behavior and Innovative Performance.*

3.4. The Moderating Effect of Internal Environmental Support

Internal environment support such as the stability of the relationship between enterprises and the reputation of suppliers has an important impact on the results of the supply chain alliance [86]. We perceive that the internal support environment of the enterprise (three categories) will have an impact on the above changes. The recognition of the green strategy by enterprise managers, the relative stability of the green alliance, and the relatively high technology level of enterprises may promote co-creation between enterprises. Thus, we propose:

Hypothesis 6a (H6a). *The Internal Support Environment has a Positive Moderating Effect on the Relationship between Green Strategy and Value Co-creation Behavior.*

By strengthening coordination with suppliers and downstream members of the supply chain including retailers, enterprises will promote their own operation activities [87]. The middle-level supply managers of internal stakeholders also play an important role in the operation activities of enterprises [88]. We believe that the internal support environment of an enterprise will have a positive impact on the relationship between co-creation and operation performance. Thus, we propose:

Hypothesis 6b (H6b). *The Internal Support Environment has a Positive Moderating Effect on the Relationship between Value Co-creation Behavior and Operational Performance.*

The idea of managers determines the operation of an enterprise. At the same time, the relationship between an enterprise and its supply chain partners as well as its digitalization level, also become the key to improving the financial level and building sustainable competitiveness of an enterprise [6]. These internal factors will strengthen the promotion of co-creation on the financial performance of an enterprise. Thus, we propose:

Hypothesis 6c (H6c). *The Internal Support Environment has a Positive Moderating Effect on the Relationship between Value Co-creation Behavior and Financial Performance.*

Enterprise alliance can improve the internal support environment of enterprises, realize knowledge sharing among partners, and promote the innovation performance of enterprises [6]. The enterprise managers' emphasis on innovation and the enterprise's own digital ability also determine the enterprise's innovation ability. Thus, we propose:

Hypothesis 6d (H6d). *The Internal Support Environment has a Positive Moderating Effect on the Relationship between Value Co-creation Behavior and Innovative Performance.*

4. Methodology

4.1. Research Process

This paper conducted an empirical analysis on a sample survey of Chinese firms in the form of a questionnaire. The analysis of this paper was divided into four steps: (1) A reliability analysis was carried out to illustrate the effectiveness of sample extraction; (2) As the questionnaire involved many contents, principal component analysis (PCA) was carried out to screen out several important influential factors; (3) a validity analysis was used to show that there was a good distinction between various factors; and (4) hierarchical regression analysis was used to study the influence of each factor on co-creation and performance.

4.2. Survey Development

In this paper, each factor in the conceptual model was measured in multiple ways (see Figure 1). The questionnaire was divided into descriptive items (see Table 1) and measurement items (see Appendix A). Descriptive questions mainly included the basic information of the respondents and their companies. The measurement items were further divided into three parts: the first part describes the green strategy of the company and the cooperation between the company and each member of the supply chain on green technology; the second part analyzes the legal environment and international environment of the company; and the third part examines the company from the performance of cooperation, financial situation, and innovation ability. This paper adopts the Likert scale commonly used in other articles, in which "1" means "strongly disagree" and "5" means "strongly agree".

To develop a structured questionnaire, a comprehensive review of the related literature was undertaken. In this study, value co-creation with supplier (VCS) was measured from four dimensions including whether enterprises and suppliers could adopt a positive attitude and cooperative behavior; jointly design products and systems to meet production demand; standardize the relationship of responsibilities, rights, benefits and some behaviors in the process of value co-creation; and establish a benefit distribution mechanism [89]. Value co-creation with competitor (VCC) was measured from four dimensions including whether enterprises and competitors can adopt a positive attitude and cooperative behavior; share information, technology and resources to improve competitive advantage; standardize the relationship of responsibilities, rights, benefits and some behaviors in the process of value co-creation; and establish a benefit distribution mechanism [90]. Value co-creation with retailer (VCR) was also measured from four dimensions including whether enterprises and retailers could adopt a positive attitude and cooperative behavior; receive retailers' feedback in time and let retailers participate in product design and development; standardize the relationship of responsibilities, rights, benefits and some behaviors in the process of value co-creation; and establish a benefit distribution mechanism [91].

4.3. Sample Selection

We chose automobile manufacturers in Mainland China as the main research objects. Before the formal issuance of the questionnaire, we sent the questionnaire to the three senior managers of the company. After filling in the questionnaire, we communicated with them, inquired about the rationality of the questionnaire, and modified the questionnaire according to the opinions. We selected 150 manufacturing companies to conduct anonymous surveys on senior executives in the form of electronic questionnaires. Each question was required to be answered to ensure the authenticity of the information obtained. Each respondent was divided into four groups and it took one month to collect the questionnaire. After each group had been completed, it sent a filling request for the next group. If the previous group failed to fill in the questionnaire on time, it sent a reminder again. Finally, we received a total of 115 available questionnaires with a response rate of 76.7%. The profile of the interviewees and their companies can be seen in Table 1.

Among the surveyed enterprises, 54% of them sent their Chief Officer (CEO) or middle manager to participate in the interviews, 24% had been established for more than 20 years, one third of them had an annual income of more than 2 billion yuan, and 90% were local private enterprises and foreign-funded enterprises. Over 50% of the enterprises had environmental management system certification, social responsibility management system certification, and quality management system certification. A total of 84% of the enterprises provided staff training, and three quarters of enterprises carried out total quality management.

Table 1. Descriptive items.

Characteristics	Percentage (%)
Respondent's Position	
CEO	12
Middle manager	42
General staff	46
Age of the Firm (Number of Years)	
Less than 5	25
5–10	34
11–15	13
16–20	4
More than 20	24
Annual Revenue (Million RMB Yuan)	
Less than 50	23
50–100	27
100–200	10
200–2000	7
More than 2000	33
Ownership of the Firm	
State-owned	9
Local private	58
Foreign	32
Joint venture	1
Environmental Management System Certification	
Yes	66
No	34
Social Responsibility Management System Certification	
Yes	54
No	46
Quality Management System Certification	
Yes	75
No	26
Train Employees on Environmental Management	
Never	16
Sometimes	58
Always	26
Total Quality Management	
Yes	75
No	25

5. Results

5.1. Preliminary Study

In order to ensure that the collected data were suitable for all factor analysis, the Kaiser-Meyer-Olkin (KMO) test was carried out. The KMO test result was 0.867, greater than 0.8, and the Bartlett's test showed a satisfactory result when $p < 0.000$, so the samples passed the reliability test, and factor analysis could be carried out.

5.2. Exploratory Study

In this paper, SPSS 21.0 was used for exploratory factor analysis (EFA) to extract principal component factors. As shown in Table 2, nine factors were obtained through principal component analysis, and the cumulative explanatory variance of these factors was 82.7%. In addition, Cronbach's α coefficient was greater than 0.8, and the sample data were good, which shows that these nine factors can be used to explain all measured items.

Table 2. Results of the exploratory factor analysis.

Item	Factor 1	Factor 2	Factor 3	Factor 4	Factor 5	Factor 6	Factor 7	Factor 8	Factor 9
GS1	0.812	-	-	-	-	-	-	-	-
GS2	0.764	-	-	-	-	-	-	-	-
GS3	0.695	-	-	-	-	-	-	-	-
GS4	0.703	-	-	-	-	-	-	-	-
GS5	0.710	-	-	-	-	-	-	-	-
GS6	0.693	-	-	-	-	-	-	-	-
VCS1	-	0.785	-	-	-	-	-	-	-
VCS2	-	0.778	-	-	-	-	-	-	-
VCS3	-	0.790	-	-	-	-	-	-	-
VCS4	-	0.811	-	-	-	-	-	-	-
VCC1	-	-	0.828	-	-	-	-	-	-
VCC2	-	-	0.813	-	-	-	-	-	-
VCC3	-	-	0.830	-	-	-	-	-	-
VCC4	-	-	0.802	-	-	-	-	-	-
VCR1	-	-	-	0.793	-	-	-	-	-
VCR2	-	-	-	0.727	-	-	-	-	-
VCR3	-	-	-	0.847	-	-	-	-	-
VCR4	-	-	-	0.835	-	-	-	-	-
LP1	-	-	-	-	0.837	-	-	-	-
LP2	-	-	-	-	0.809	-	-	-	-
LP3	-	-	-	-	0.812	-	-	-	-
LP4	-	-	-	-	0.811	-	-	-	-
LP5	-	-	-	-	0.849	-	-	-	-
LP6	-	-	-	-	0.811	-	-	-	-
LP7	-	-	-	-	0.823	-	-	-	-
CP1	-	-	-	-	0.806	-	-	-	-
CP2	-	-	-	-	0.792	-	-	-	-
CP3	-	-	-	-	0.763	-	-	-	-
CP4	-	-	-	-	0.806	-	-	-	-
CP5	-	-	-	-	0.769	-	-	-	-
CmP1	-	-	-	-	0.857	-	-	-	-
CmP2	-	-	-	-	0.878	-	-	-	-
CmP3	-	-	-	-	0.860	-	-	-	-
SP1	-	-	-	-	0.762	-	-	-	-
SP2	-	-	-	-	0.789	-	-	-	-
SP3	-	-	-	-	0.805	-	-	-	-
SP4	-	-	-	-	0.803	-	-	-	
SP5	-	-	-	-	0.820	-	-	-	-
MA1	-	-	-	-	-	0.821	-	-	-
MA2	-	-	-	-	-	0.787	-	-	-
MA3	-	-	-	-	-	0.814	-	-	-
MA4	-	-	-	-	-	0.773	-	-	-
SCR1	-	-	-	-	-	0.874	-	-	-
SCR2	-	-	-	-	-	0.860	-	-	-
SCR3	-	-	-	-	-	0.894	-	-	-
SCR4	-	-	-	-	-	0.845	-	-	-
SCR5	-	-	-	-	-	0.876	-	-	-
SCR6	-	-	-	-	-	0.856	-	-	-
DC1	-	-	-	-	-	0.863	-	-	-
DC2	-	-	-	-	-	0.815	-	-	-
DC3	-	-	-	-	-	0.830	-	-	-

Table 2. *Cont.*

Item	Factor 1	Factor 2	Factor 3	Factor 4	Factor 5	Factor 6	Factor 7	Factor 8	Factor 9
DC4	-	-	-	-	-	0.843	-	-	-
DC5	-	-	-	-	-	0.850	-	-	-
DC6	-	-	-	-	-	0.849	-	-	-
DC7	-	-	-	-	-	0.855	-	-	-
DC8	-	-	-	-	-	0.836	-	-	-
OP1	-	-	-	-	-	-	0.881	-	-
OP2	-	-	-	-	-	-	0.849	-	-
OP3	-	-	-	-	-	-	0.838	-	-
FP1	-	-	-	-	-	-	-	0.896	-
FP2	-	-	-	-	-	-	-	0.888	-
FP3	-	-	-	-	-	-	-	0.841	-
FP4	-	-	-	-	-	-	-	0.855	-
FP5	-	-	-	-	-	-	-	0.867	-
IP1	-	-	-	-	-	-	-	-	0.862
IP2	-	-	-	-	-	-	-	-	0.825
IP3	-	-	-	-	-	-	-	-	0.823
IP4	-	-	-	-	-	-	-	-	0.844
IP5	-	-	-	-	-	-	-	-	0.915
IP6	-	-	-	-	-	-	-	-	0.869
IP7	-	-	-	-	-	-	-	-	0.840
Mean	4.250	4.202	3.940	4.178	4.079	4.010	4.140	3.953	4.075
S.D.	0.844	0.831	0.955	0.849	0.873	0.891	0.815	0.941	0.863
Cumulative Percent of Variance	0.602	0.643	0.677	0.722	0.750	0.775	0.800	0.812	0.827
Cronbach's α	0.917	0.819	0.832	0.845	0.919	0.951	0.850	0.889	0.909

Note: please see Table 3 for the meaning of the abbreviations in Table 2.

5.3. Confirmatory Study

Confirmatory factor analysis (CFA) was used to verify the model. It can be seen from Table 3 that standardized load coefficients were all greater than 0.8, and were significant when the confidence was greater than 95%, and R2 was greater than 0.5. Through the confirmatory factor analysis test, it showed that the measurement item structure was good and that the model could be accepted completely (χ^2 = 216.144, df = 114, χ^2/df = 1.896, RMSEA = 0.061, CFI = 0.939, GFI = 0.868, AGFI = 0.835, NFI = 0.880, TLI = 0.931).

Table 3. Parameter estimates of the confirmatory factor analysis.

Factor	Item	Standardized Factor Loading	*t*-Value	R^2
Factor1: Green strategy (GS)	GS1	0.869	-	0.594
	GS2	0.875	8.319	0.602
	GS3	0.803	8.407	0.584
	GS4	0.876	10.325	0.630
	GS5	0.865	11.337	0.624
	GS6	0.893	16.204	0.580
Factor 2: Value Co-creation with supplier (VCS)	VCS1	0.890	-	0.517
	VCS2	0.862	15.089	0.720
	VCS3	0.819	13.730	0.690
	VCS4	0.849	18.427	0.823

Table 3. *Cont.*

Factor	Item	Standardized Factor Loading	*t*-Value	R^2
Factor 3: Value co-creation with competitor (VCC)	VCC1	0.831	-	0.785
	VCC2	0.907	16.082	0.723
	VCC3	0.886	17.656	0.779
	VCC4	0.850	21.299	0.712
Factor 4: Value co-creation with retailer (VCR)	VCR1	0.883	-	0.747
	VCR2	0.844	8.692	0.596
	VCR3	0.864	8.670	0.598
	VCR4	0.804	9.407	0.548
Factor 5: External environment pressure (EEP)	LE1	0.806	-	0.519
	LE2	0.874	10.873	0.566
	LE3	0.865	11.940	0.575
	LE4	0.882	10.684	0.581
	LE5	0.889	10.467	0.843
	LE6	0.893	10.556	0.849
	LE7	0.918	8.769	0.701
	CE1	0.921	8.692	0.940
	CE2	0.837	8.670	0.898
	CE3	0.970	9.407	0.917
	CE4	0.948	9.949	0.787
	CE5	0.957	10.873	0.716
	IE1	0.887	9.548	0.832
	IE2	0.846	8.470	0.765
	IE3	0.912	8.484	0.946
	PS1	0.875	8.687	0.892
	PS2	0.973	30.232	0.838
	PS3	0.944	9.465	0.740
	PS4	0.916	9.653	0.710
	PS5	0.815	9.428	0.858
Factor 6: Internal environment support (IES)	MA1	0.824	-	0.782
	MA2	0.815	19.425	0.536
	MA3	0.899	17.367	0.527
	MA4	0.858	18.739	0.525
	SCR1	0.819	20.846	0.516
	SCR2	0.898	17.436	0.607
	SCR3	0.803	16.239	0.572
	SCR4	0.827	13.285	0.755
	SCR5	0.860	12.459	0.562
	SCR6	0.843	19.048	0.584
	DC1	0.926	18.475	0.567
	DC2	0.884	22.532	0.572
	DC3	0.661	20.604	0.858
	DC4	0.654	21.204	0.892
	DC5	0.652	20.791	0.736
	DC6	0.818	22.841	0.965
	DC7	0.879	20.873	0.890
	DC8	0.856	19.023	0.901
Factor 7: Operational performance (OP)	OP1	0.869	-	0.754
	OP2	0.880	8.697	0.722
	OP3	0.982	9.732	0.830
Factor 8: Financial performance (FP)	FP1	0.943	-	0.772
	FP2	0.949	20.620	0.961
	FP3	0.868	21.703	0.871
	FP4	0.849	24.339	0.812
	FP5	0.911	21.661	0.578
Factor 9: Innovative performance (IP)	IP1	0.879	-	0.524
	IP2	0.980	30.314	0.511
	IP3	0.933	27.233	0.639
	IP4	0.901	22.583	0.575
	IP5	0.843	21.490	0.670
	IP6	0.926	21.963	0.637
	IP7	0.884	18.806	0.594

The results show that the t-value of each item was higher than the critical value, the significance level was 0.05, and the R2 value of each variable was greater than 0.5. This is sufficient evidence of convergence efficiency. The discriminant validity was tested by comparing the mean variance (AVE) extracted and the square correlation between structures. The results show that the AVE was greater than 0.5, and the correlation coefficient between factors was less than AVE, indicating that there is good discrimination validity between factors (see Table 4).

Table 4. Discriminant validity.

	AVE	GS	VCS	VCC	VCR	EEP	ITE	OP	FP	IP
GS	0.625	-	-	-	-	-	-	-	-	-
VCS	0.661	0.004	-	-	-	-	-	-	-	-
VCC	0.725	0.021	0.021	-	-	-	-	-	-	-
VCR	0.682	0.003	0.152	0.160	-	-	-	-	-	-
EEP	0.816	0.038	0.199	0.256	0.065	-	-	-	-	-
IES	0.809	0.010	0.002	0.011	0.242	0.087	-	-	-	-
OP	0.811	0.025	0.116	0.106	0.007	0.039	0.145	-	-	-
FP	0.719	0.087	0.179	0.198	0.076	0.059	0.096	0.310	-	-
IP	0.842	0.203	0.214	0.236	0.157	0.095	0.208	0.178	-	-

5.4. Hypotheses Testing and Results

A hierarchical regression method was used to test the proposed hypothesis and passed the multicollinearity test. To avoid the threat of multi-collinearity, the variance inflation factors were computed, and the value of 1.37 revealed that the dataset was suitable for regression analysis [92], thus dataset was suited for the regression analysis. Additionally, the value of the adjusted R^2 of each model confirmed that the output of the regression models was accepted.

As shown in Table 5, the significant coefficient of model 1 ($\beta = 0.832$, $p < 0.001$) indicates that GS is positively correlated with VCS, thus supporting H1a; the significant coefficient of model 3 ($\beta = 0.870$, $p < 0.001$) indicates that GS is positively correlated with VCC, supporting H1b; and the significant coefficient of model 5 ($\beta = 0.778$, $p < 0.001$) indicates that GS is positively correlated with VCS, supporting H1c. Therefore, it shows that the green strategy is positively related to the value co-creation behavior. Model 8 shows that value co-creation behaviors (VCS, VCC, VCR) have a positive and significant effect on OP ($p < 0.001$), so H2 is accepted. The results of model 11 show that VCC is positively correlated with FP ($\beta = 0.520$, $p < 0.001$) and VCR is positively correlated with FP ($\beta = 0.231$, $p < 0.05$), but the relationship between VCC and FP is not significant, so H3b and H3c are accepted and H3 is rejected. Similarly, the results of model 14 show that VCC is positively correlated with IP ($\beta = 0.321$, $p < 0.001$) and VCR is positively correlated with IP ($\beta = 0.368$, $p < 0.001$), but VCC is not significantly related to IP, thus accepting H4b and H4c and rejecting H4a.

The research results show that models 2, 4, and 6 test the moderating effect of ETE and ITE on the GS and value co-creation behavior (VCS, VCC, VCR), indicating that ETE has a positive moderating effect on GS and value co-creation behavior (VCS, VCC, VCR) ($p < 0.05$), while ITE has no moderating effect on GS and value co-creation behavior (VCS, VCC, VCR), so it accepts H5a and rejects H6a. Model 9 tests the moderating effect of EEP and IES on the value co-creation behavior (VCS, VCC, VCR) and OP, indicating that EEP and IES have no moderating effect on the relationship between value co-creation (VCS, VCC, VCR) and OP, thus rejecting H5b and H6b. Model 12 tests the moderating effect of EEP and IES on value co-creation behavior (VCS, VCC, VCR) and FP and shows that EEP has no moderating effect on the relationship between value co-creation behavior (VCS, VCC, VCR) and FP, but IES has a positive moderating effect on the relationship between value co-creation behavior (VCS, VCC, VCR) and FP ($p < 0.05$). Therefore, it rejects H5c and accepts H6c. Model 15 tests the moderating effect of EEP and ITE on value co-creation behavior (VCS, VCC, VCR) and IP, indicating that EEP and IES both have a positive moderating effect on the relationship between value co-creation behavior (VCS, VCC, VCR) and IP ($p < 0.05$), so H5d and H6d is accepted.

In addition, in the study of control variables, firm age has a positive effect on FP, but has no effect on OP and IP, while firm size and firm ownership have no effect on OP, FP, and IP.

Table 5. Result of hierarchical regression analysis.

Model	VCS		VCC		VCR		OP			FP			IP		
	1	2	3	4	5	6	7	8	9	10	11	12	13	14	15
Independent Variable															
GS	0.832 ^{a}	0.355 ^{b}	0.870 ^{a}	0.150	0.778 ^{a}	0.132	-	-	-	-	-	-	-	-	-
VCS	-	-	-	-	-	-	-	0.125 ^{a}	0.849	-	0.177	1.168	-	0.189	1.650 ^{c}
VCC	-	-	-	-	-	-	-	0.173 ^{a}	0.356	-	0.520 ^{a}	1.962 ^{b}	-	0.321 ^{a}	1.086 ^{c}
VCR	-	-	-	-	-	-	-	0.187 ^{a}	0.367	-	0.231 ^{c}	0.773	-	0.368 ^{a}	0.643
Moderator Variable															
EEP * GS	-	0.052 ^{c}	-	0.149 ^{c}		0.024 ^{c}	-	-	-	-	-	-			
IES * GS	-	0.023	-	0.011	-	0.079	-	-	-	-	-	-			
EEP * VCS	-		-	-	-	-	-	-	0.509	-	-	0.146			1.046 ^{b}
EEP * VCC	-		-	-	-	-	-	-	0.044	-	-	0.108			0.087 ^{c}
EEP * VCR	-		-	-	-	-	-	-	0.378	-	-	0.055			0.920 ^{b}
IES * VCS	-		-	-	-	-	-	-	0.274	-	-	0.200 ^{c}			0.582 ^{b}
IES * VCC	-		-	-	-	-	-	-	0.284	-	-	0.333 ^{c}			0.169 ^{c}
IES * VCR	-		-	-	-	-	-	-	0.569	-	-	0.165 ^{c}			0.825 ^{b}
Control Variable															
firm age	-	-	-	-	-	-	0.126	0.024	0.048	0.174 ^{c}	0.055 ^{c}	0.081 ^{c}	0.117	0.011	0.034
firm size	-	-	-	-	-	-	0.103	0.046	0.062	0.088	0.026	0.042	0.085	0.026	0.056
firm ownership	-	-	-	-	-	-	0.084	0.092	0.123	0.070	0.076	0.078	0.028	0.035	0.104
Constant	0.686 ^{a}	1.386 ^{a}	0.243 ^{a}	1.741 ^{a}	0.874 ^{a}	1.819 ^{a}	4.359 ^{a}	0.454 ^{a}	0.957 ^{a}	0.052	0.209	0.372	0.575 ^{b}	0.182	0.518
Adjusted R^2	0.696	0.744	0.539	0.692	0.581	0.689	0.737	0.704	0.776	0.552	0.711	0.817	0.528	0.717	0.812

Note: ^{a} $p < 0.001$; ^{b} $p<0.01$; ^{c} $p<0.05$.

6. Discussion and Implications

6.1. Effects of Green Strategy on Value Co-Creation Behavior

This paper conducted an empirical study on value co-creation behavior in the green supply chain. First of all, this paper studied the impact of green strategy on value co-creation in the supply chain where the results showed that VCS, VCC, and VCR had a positive impact, and the implementation of a green strategy of manufacturers can effectively promote value co-creation among suppliers and other supply chain members. At the same time, both corporate responsibility and public awareness attach great importance to "green", which proves the importance of co-creation under the green supply chain. The establishment of a green strategy by the state and enterprises can promote enterprises to seek more active ways to realize the strategy, and then effectively promote the green cooperation and value creation between enterprises and other enterprises in the supply chain.

6.2. Enablers of Performance

The article found that there was a positive correlation between value co-creation behavior in the supply chain and financial performance and innovation performance. Moreover, the nature of the manufacturer also has an impact on performance, which is rarely mentioned in the literature on green supply chains. Therefore, this study provides a new theoretical perspective.

Empirical results support H5 and H6, indicating that value co-creation behaviors are positively related to manufacturers' financial performance and innovative performance in green supply chains. The results showed that value co-creation with suppliers was positively related to manufacturers' operational performance, which confirmed H4a. The value co-creation of manufacturers and suppliers can deepen the information sharing and cooperation between the two sides, improving the supply efficiency of raw materials and the supply chain flexibility of manufacturing enterprises. However, when it comes to value co-creation between manufacturers and competitors, the value co-creation between manufacturers and retailers was not significant for operational performance, which indicates that hypothesis H4b and H4c are not valid. Green co-creation requires manufacturers to invest more money in technology for suppliers with a relatively weak technology level. Although it helps to improve operational efficiency, it may lead to higher costs. At the same time, due to the low co-creation degree of the sample enterprises in this paper, the innovation practice of suppliers cannot meet the production needs of the market and manufacturing enterprises, so it cannot promote the production of innovative products/services.

Regarding the influence of control variables on manufacturer performance, the results showed that only firm age had a positive effect on the manufacturers' financial performance. This indicates that the age of the company is a major factor affecting corporate performance. Large companies with a long history and good reputation are more likely to win the trust of other firms, thus promoting value co-creation with other firms. However, the influence of firm size and firm ownership on performance was not significant, and the results did not show that company size and ownership had an impact on performance.

6.3. Effects of Moderators

This paper analyzed the moderating effect of external and internal contextual variables on the relationship between value co-creation behavior and enterprise performance in a green supply chain.

The results shows that external environment pressure (EEP) had a positive regulating effect on the relationship between green strategy (GS), value co-creation with supplier (VCS), value co-creation with competitor (VCC), and value co-creation with retailer (VCR), while internal environmental support (IES) had no significant regulating effect on the relationship between green strategy (GS), value co-creation with supplier (VCS), value co-creation with competitor (VCC), and value co-creation with retailer (VCR). This indicates that external contextual variables (legal environment, consumption environment, industry environment) have positive effects on promoting value co-creation in the green supply chain. Effective policies such as green subsidies and the attention of suppliers, consumers, and competitors to the green strategy can encourage enterprises to actively seek the realization channels of green products, and then promote the co-creation of green value with upstream and downstream enterprises. However, internal contextual variables (manager's attitude, supply chain relationship, digitization capability) have little effect on promoting value co-creation in the green supply chain.

At the same time, external environment pressure (EEP) had a positive moderating effect on the relationship between co-creation behavior (value co-creation with supplier, value co-creation with competitor, and value co-creation with retailer) and innovative performance (IP), while external environment pressure (EEP) had no significant moderating effect on co-creation behavior (value co-creation with supplier, value co-creation with competitor, and value co-creation with retailer) and operational performance (OP) and financial performance(FP), indicating that good external contextual factors can promote the improvement of enterprise innovation performance, but not for operational performance and economic performance. This shows that good legal, production, and consumption environments can promote enterprises to increase the degree of innovation in the development and production of green products to design products more in line with the requirements, but at the same time, it often needs to spend more money and manpower, resulting in its impact on financial performance, and operational performance is not significant. Internal environmental support (IES) has no significant moderating effect on co-creation behavior (value co-creation with supplier, value co-creation with competitor, and value co-creation with retailer) and operational performance (OP) relationship, while internal environmental support (IES) has a positive regulating effect on co-creation behavior and the relationship between financial performance (FP) and innovative performance (IP), indicating that a good internal enterprise environment has a positive effect on improving enterprise financial performance and innovation performance, but has little effect on operational performance. This shows that managers' attention to green products can promote enterprises to invest more innovation costs in the research and development of green products to improve the innovation performance of the enterprise. At the same time, the higher the degree of digitalization the enterprise has, the less the cost of replacing green products, so the financial performance is higher.

6.4. Theoretical Implications

This paper empirically analyzed the influence of different value co-creation modes on the enterprise performance of the supply chain and contributes to green supply chains and value co-creation literature in many ways. First, this paper studied the green supply chain from the perspective of value co-creation, while most of the current studies on value co-creation are based on the general supply chain. Second, this paper selected different types of value co-creation behaviors between manufacturers, suppliers, competitors, and retailers on the factor selection of value co-creation behaviors. Third, this paper considered the regulating effect of the internal environment and external environment. In the internal environment, this paper innovatively analyzed the impact of digital capability on the performance of the green supply chain of enterprises, which is more comprehensive than the related studies in the previous literature. Fourth, unlike the previous literature, which focused more on the influence of financial performance, this paper analyzed it from the three aspects of operational performance, financial performance, and innovation performance in order to evaluate the influence of value co-creation in green supply chains on enterprise performance from multiple aspects. Finally, the empirical research of this paper paid more attention to the interaction between the model of value co-creation and other influencing factors (such as internal and external environmental factors), which is unlike many studies in previous research that only focused on the single influence of the model of value co-creation.

6.5. Managerial Implications

This study has some valuable management implications for manufacturers and governments. First, manufacturers should actively adopt a green strategy, which has a positive effect on promoting value co-creation behavior between manufacturers and their suppliers, competitors, and retailers. Second, manufacturers should actively seek friendly cooperation with other firms, which can help improve the performance of firms such as financial performance, innovative performance, and operational performance. Finally, the internal and external contextual factors of firms play a regulating role in the mechanism of co-creation in the green supply chain. Therefore, the government should increase the publicity of environmental protection awareness, guide the demand of firms and consumers for green products, and create a good policy and consumption environment to reduce the pressure of supply chain enterprises. Firms should strengthen their managers' understanding of green production, improve the level of digital management, establish friendly cooperative relations with upstream and downstream enterprises in the supply chain, and improve the internal environment to promote green value co-creation behavior in the manufacturing supply chain and improve the performance of firms.

7. Limitations and Future Research Lines

This article has some contributions in theory and practice, but still has some limitations. First of all, the research object of this paper was domestic manufacturing firms, and there was no further subdivision of enterprise type. Different types of manufacturing firms may face different situations. Second, this paper only considered the co-creation behavior of two subjects in the supply chain, while in actual operation, there may be more possibilities for the value co-creation behavior of firms. In addition, from the perspective of organizational behavior and psychology, there is a certain gap between the subjective perception of managers and the actual situation of the company during the investigation process, which may lead to the conclusion of this article not being objective and unscientific [93].

Based on the main findings and limitations, we propose several questions that can be explored in depth in future studies. First, the research object can be determined in more segmented manufacturing enterprises such as automobiles, electronic parts, electrical appliances, etc. to make the empirical results more accurate. Second, we can consider the impact of value co-creation between multiple entities in the supply chain or between enterprises and the government on enterprise performance. Third, the model can be considered to add coordination variables and control variables such as product diversity and corporate reputation to make the model more fully studied. Fourth, we can consider expanding the sample size and can choose enterprises from different countries or regions for comparison such as comparing the implementation effect of value co-creation between Chinese enterprises and enterprises from developed countries. Perhaps due to the differences in national policies and development environment, we will obtain unexpected research conclusions.

Author Contributions: Conceptualization, G.L. and X.S.; Methodology, X.S.; Formal Analysis, Y.Y. and C.D.; Resources, C.D.; Data Curation, G.L.; Writing—Original Draft Preparation, Y.Y.; Writing—Review & Editing, C.D. and X.S.; Project Administration, X.S.; Funding Acquisition, G.L. and Y.Y. All authors have read and agreed to the published version of the manuscript.

Funding: This research was partially supported by the Fundamental Research Funds for the Central Universities (grant number: 2020JBW002), Beijing Social Science Foundation of China (grant number: 18GLC081) and the National Science Foundation of China (grant number: 71971215, 71601187).

Conflicts of Interest: The authors declare no conflicts of interest.

Appendix A

Table A1. Measurement items.

Factor	Measured Variable	Item
Green strategy (GS)	Managers are very willing to implement the green supply chain strategy.	GS1
	Select environmentally friendly suppliers.	GS2
	Can carry out green design and green production for products.	GS3
	Can pay attention to the principle of resource conservation of environmental protection.	GS4
	Advocate that consumers choose green products that are not polluted or contribute to public health.	GS5
	Try to reduce the generation of harmful substances.	GS6

Table A1. *Cont.*

Factor		Measured Variable	Item
Value co-creation with supplier (VCS)		Can adopt positive attitude and cooperative behavior with suppliers.	VCS1
		Can jointly design products and systems to meet production requirements.	VCS2
		There are norms about the relationship of responsibility, right, interest and some behaviors.	VCS3
		Can establish the mechanism of interest distribution.	VCS4
Value co-creation with competitor (VCC)		Can adopt positive attitude and cooperative behavior with competitors.	VCC1
		Can share information, technology and resources to improve competitive advantage.	VCC2
		There are norms about the relationship of responsibility, right, interest and some behaviors.	VCC3
		Can establish the mechanism of interest distribution.	VCC4
Value co-creation with retailer (VCR)		Can adopt positive attitude and cooperative behavior with retailers.	VCR1
		Can receive retailers' feedback in time and let retailers participate in product design and development.	VCR2
		There are norms about the relationship of responsibility, right, interest and some behaviors.	VCR3
		Can establish the mechanism of interest distribution.	VCR4
External Environmental Pressure (EEP)	Legal Pressure (LP)	There are national laws, regulations and standards on environmental protection of manufacturing firms.	LP1
		There are local laws, regulations and standards on environmental protection of manufacturing firms.	LP2
		There are laws, regulations, and standards related to environmental protection in market.	LP3
		Relevant departments strengthen the supervision and management of firms' compliance with laws, regulations and standards related to environmental protection.	LP4
		Law enforcement personnel increase the punishment for firms that violate relevant laws, regulations and standards.	LP5
		The government has increased financial subsidies to reduce the pollution and damage of firms to environmental resources.	LP6
		Relevant policies issued by the government strongly support the sustainable development.	LP7
	Consumers' Pressure(CP)	Consumers' social responsibility and awareness of environmental protection are enhanced.	CP1
		Consumers' expectations and requirements for green products and product safety increase.	CP2
		The export of products shall meet the environmental protection standards and safety standards.	CP3
		Consumers require us to have a good social and environmental image.	CP4
		Non-governmental organizations advocate social responsibility and environmental protection.	CP5
	Competitors' Pressure (CmP)	Leading firms in the industry have begun to implement sustainable supply chain management.	CmP1
		Firms in the industry that have implemented sustainable supply chain management perform well in terms of environment, society and economy.	CmP2
		Most firms in the industry have begun to implement sustainable supply chain management.	CmP3
	Suppliers' Pressure (SP)	Suppliers pay more attention to environmental protection.	SP1
		Suppliers put forward environmental protection requirements for the firms.	SP2
		The firm has reached certain environmental partnership with suppliers.	SP3
		Suppliers can provide environmentally friendly or recyclable raw materials.	SP4
		Suppliers can provide environmentally friendly or recyclable raw material packaging.	SP5
Internal Environmental Support (IES)	Managers' attitude (MA)	Top management believes that environmental protection is an important part of corporate strategy.	MA1
		Senior managers support activities to improve social responsibility and environmental performance.	MA2
		Top managers pay attention to corporate social responsibility and require corporate behavior to meet environmental requirements.	MA3
		Senior managers are directly responsible for the work of firm environment.	MA4
	Supply chain relationship (SCR)	Firms and government departments have established friendly relations at various levels.	SCR1
		Firms and major suppliers have established friendly relations at multiple levels.	SCR2
		Different departments of the firms have established friendly relations at multiple levels.	SCR3
		Firms and distributors have established friendly relations at many levels.	SCR4
		Firms and retailers have established friendly relations at many levels.	SCR5
		Firms and suppliers have established friendly relations at multiple levels.	SCR6

Table A1. *Cont.*

Factor	Measured Variable	Item
Digitalization capability (DC)	Managers have digital leadership.	DC1
	Firms have more information professionals.	DC2
	Digital facilities and information systems are widely used.	DC3
	The business object of the enterprise has been digitized by means of mobile technology, etc.	DC4
	Can conduct digital contact analysis on the business of the enterprise.	DC5
	Business events and business decisions are driven by data analysis and algorithms.	DC6
	Firms have digital based business model innovation.	DC7
	Firms' IT capability can support digital transformation.	DC8
Operational performance (OP)	Customer service level of the company has been improved.	OP1
	The company's responsiveness has been improved.	OP2
	Decrease in customer complaints.	OP3
Financial performance (FP)	Market share expansion.	FP1
	Increase in revenue.	FP2
	Decrease in total operating costs.	FP3
	The return on assets of has been improved.	FP4
	The inventory turnover cycle is optimized.	FP5
Innovative performance (IP)	Often take the lead in launching new products/services in the industry.	IP1
	Often take the lead in applying new technologies in the industry.	IP2
	Cost increase of the manufacturer's investment in product transformation and innovation.	IP3
	Increase in the manufacturer's intellectual property (including application for patent, trademark and design, etc.	IP4
	The manufacturer's innovative products have a good market response.	IP5
	Increase in sales revenue of innovative products.	IP6
	The manufacturer's innovative products are highly irreplaceable.	IP7

References

1. Bai, C.; Kusi-Sarpong, S.; Sarkis, J. An implementation path for green information technology systems in the Ghanaian mining industry. *J. Clean. Prod.* **2017**, *164*, 1105–1123. [CrossRef]
2. Hong, Z.; Guo, X. Green product supply chain contracts considering environmental responsibilities. *Omega* **2018**, *83*, 155–166. [CrossRef]
3. Kusi-Sarpong, S.; Gupta, H.; Sarkis, J. A supply chain sustainability innovation framework and evaluation methodology. *Int. J. Prod. Res.* **2019**, *57*, 1990–2008. [CrossRef]
4. Lacoste, S. Sustainable value co-creation in business networks. *Ind. Mark. Manag.* **2016**, *52*, 151–162. [CrossRef]
5. Isaksson, R.; Johansson, P.; Fischer, K. Detecting Supply Chain Innovation Potential for Sustainable Development. *J. Bus. Ethics* **2010**, *97*, 425–442. [CrossRef]
6. Li, S.; Haney, M.H.; Lee, G.; Kang, M.; Ko, C. The effect of task conflict on outsourcers' long-term orientation toward suppliers: The moderating role of formal control and Chinese guanxi. *J. Bus. Ind. Mark.* **2019**, *35*, 260–269. [CrossRef]
7. Lo, S.M.; Zhang, S.; Wang, Z.; Zhao, X. The impact of relationship quality and supplier development on green supply chain integration: A mediation and moderation analysis. *J. Clean. Prod.* **2018**, *202*, 524–535. [CrossRef]
8. Tariq, A.; Yuosre, F.; Badir, W.T.; Bhutta, U.S. Drivers and consequences of green product and process innovation: A systematic review, conceptual framework, and future outlook. *Technol. Soc.* **2017**, *51*, 8–23. [CrossRef]
9. Liu, Y.; Luo, Y.; Liu, T. Governing buyer–supplier relationships through transactional and relational mechanisms: Evidence from China. *J. Oper. Manag.* **2009**, *27*, 294–309. [CrossRef]
10. Somsuk, N.; Laosirihongthong, T. Prioritization of applicable drivers for green supply chain management implementation toward sustainability in Thailand. *Taylor Fr.* **2016**, *24*, 175–191. [CrossRef]
11. Shi, X.; Chan, H.L.; Dong, C. Value of bargaining contract in a supply chain system with sustainability investment: An incentive analysis. *IEEE Trans. Syst. Man Cybern. Syst.* **2020**, *50*, 1622–1634. [CrossRef]
12. Shi, X.; Dong, C.; Zhang, C.; Zhang, X. Who should invest in clean technologies in a supply chain with competition? *J. Clean. Prod.* **2019**, *215*, 689–700. [CrossRef]
13. Dong, C.; Liu, Q.; Shen, B. To be or not to be green? Strategic investment for green product development in a supply chain. *Transp. Res. Part E Logist. Transp. Rev.* **2019**, *131*, 193–227. [CrossRef]

14. Huang, X.X.; Hu, Z.P.; Liu, C.S.; Yu, D.J.; Yu, L.F. The relationships between regulatory and customer pressure, green organizational responses, and green innovation performance. *J. Clean. Prod.* **2016**, *112*, 3423–3433. [CrossRef]
15. Schiederig, T.; Tietze, F.; Herstatt, C. Green innovation in technology and innovation management—An exploratory literature review. *R&D Management* **2012**, *42*, 180–192.
16. Swami, S.; Shah, J. Channel coordination in green supply chain management. *J. Oper. Res. Soc.* **2013**, *64*, 339–351. [CrossRef]
17. Wei, J.; Zhao, J. Pricing Decisions for Substitutable Products with Horizontal and Vertical Competition in Fuzzy Environments. *Ann. Oper. Res.* **2016**, *242*, 505–528. [CrossRef]
18. Zu-Jun, M.; Zhang, N.; Dai, Y.; Hu, S. Managing channel profits of different cooperative models in closed-loop supply chains. *Omega* **2016**, *59*, 251–262. [CrossRef]
19. Chen, R.R.; Roma, P. Group Buying of Competing Retailers. *Prod. Oper. Manag.* **2011**, *20*, 181–197. [CrossRef]
20. Wei, J.; Zhao, J.; Li, Y. Price and warranty period decisions for complementary products with horizontal firms' cooperation/noncooperation strategies. *J. Clean. Prod.* **2015**, *105*, 86–102. [CrossRef]
21. Yang, L.; Zhang, Q.; Ji, J. Pricing and carbon emission reduction decisions in supply chains with vertical and horizontal cooperation. *Int. J. Prod. Econ.* **2017**, *191*, 286–297. [CrossRef]
22. Li, H.; Wang, C.; Shang, M.; Ou, W. Pricing, Carbon Emission Reduction, Low-Carbon Promotion and Returning Decision in a Closed-Loop Supply Chain under Vertical and Horizontal Cooperation. *Int. J. Environ. Res. Public Health* **2017**, *14*, 1332. [CrossRef] [PubMed]
23. Dai, R.; Zhang, J.; Tang, W. Cartelization or Cost-sharing? Comparison of cooperation modes in a green supply chain. *J. Clean. Prod.* **2017**, *156*, 159–173.
24. Ge, Z.; Hu, Q. Collaboration in R&D activities: Firm-specific decisions. *Eur. J. Oper. Res.* **2007**, *185*, 864–883.
25. Gupta, S. Research Note—Channel Structure with Knowledge Spillovers. *Mark. Sci.* **2008**, *27*, 247–261. [CrossRef]
26. Ge, Z.; Hu, Q.; Xia, Y. Firms' R&D Cooperation Behavior in a Supply Chain. *Prod. Oper. Manag.* **2014**, *23*, 599–609.
27. Yenipazarli, A. To collaborate or not to collaborate: Prompting upstream eco-efficient innovation in a supply chain. *Eur. J. Oper. Res.* **2017**, *260*, 571–587. [CrossRef]
28. Song, H.; Gao, X. Green supply chain game model and analysis under revenue-sharing contract. *J. Clean. Prod.* **2018**, *170*, 183–192. [CrossRef]
29. Dong, C.; Shen, B.; Chow, P.S.; Yang, L.; Ng, C.T. Sustainability investment under cap-and-trade regulation. *Ann. Oper. Res.* **2016**, *240*, 509–531. [CrossRef]
30. Heydari, J.; Govindan, K.; Aslani, A. Pricing and greening decisions in a three-tier dual channel supply chain. *Int. J. Prod. Econ.* **2019**, *217*, 185–196. [CrossRef]
31. Porter, M.E. *Competitive Strategy: Techniques for Analyzing Industries and Competitors*; Social Science Electronic Publishing: New York, NY, USA, 1980; pp. 86–87.
32. Huemer, L. Supply Management: Value Creation, Coordination and Positioning in Supply Relationships. *Long Range Plan.* **2006**, *39*, 133–153. [CrossRef]
33. Ketchen, D.J.; Hult, G.T.M. Bridging Organization Theory and Supply Chain Management: The Case of Best Value Supply Chains. *J. Oper. Manag.* **2007**, *25*, 573–580. [CrossRef]
34. Prahalad, C.K.; Ramaswamy, V. Co-creation experiences: The next practice in value creation. *J. Interact. Mark.* **2010**, *18*, 5–14. [CrossRef]
35. Freeman, R.E. Managing for Stakeholders: Trade-offs or Value Creation. *J. Bus. Ethics* **2010**, *96*, 7–9. [CrossRef]
36. Freudenreich, B.; Lüdeke-Freund, F.; Schaltegger, S. A Stakeholder Theory Perspective on Business Models: Value Creation for Sustainability. *J. Bus. Ethics* **2019**, *8*, 1–6. [CrossRef]
37. Donaldson, T.; Preston, L.E. The Stakeholder Theory of the Corporation: Concepts, Evidence, and Implications. *Acad. Manag. Rev.* **1995**, *20*, 65–91. [CrossRef]
38. Haslam, C.; Tsitsianis, N.; Andersson, T.; Gleadle, P. Accounting for business models: Increasing the visibility of stakeholders. *J. Bus. Models* **2015**, *3*, 62–80.
39. Breuer, H.; Lüdeke-Freund, F. Values-based network and business model innovation. *Int. J. Innov. Manag.* **2017**, *21*, 1750028. [CrossRef]
40. Dentoni, D.; Bitzer, V.; Pascucci, S. Cross-sector partnerships and the co-creation of dynamic capabilities for stakeholder orientation. *J. Bus. Ethics* **2016**, *135*, 35–53. [CrossRef]

41. Vargo, S.L.; Lusch, R.F. Service-dominant logic: Continuing the evolution. *J. Acad. Mark. Sci.* **2008**, *36*, 1–10. [CrossRef]
42. Cachon, G.P. Supply Chain Coordination with Contracts. *Handb. Oper. Res. Manag. Sci.* **2003**, *11*, 227–339.
43. Anni-Kaisa, K.; Katrina, L. The Underlying Potential of Supply Management in Value Creation. *J. Purch. Supply Manag.* **2012**, *18*, 68–75.
44. FitzPatrick, M.; Varey, R.J.; Grönroos, C.; Davey, J. Relationality in the service logic of value creation. *J. Serv. Mark.* **2015**, *29*, 463–471. [CrossRef]
45. Yngfalk, F. 'It's not us, it's them!'—Rethinking value co-creation among multiple actors. *J. Mark. Manag.* **2013**, *29*, 1163–1181. [CrossRef]
46. Lei, M.; Liu, H.; Deng, H.; Huang, T.; Leong, G.K. Demand Information Sharing and Channel Choice in A Dual-channel Supply Chain with Multiple Retailers. *Int. J. Prod. Res.* **2014**, *52*, 6792–6818. [CrossRef]
47. Mukhopadhyay, S.K.; Yao, D.Q.; Yue, X. Information Sharing of Value-adding Retailer in A Mixed Channel Hi-tech Supply Chain. *J. Bus. Res.* **2008**, *61*, 950–958. [CrossRef]
48. D'Aspremont, C.; Jacquemin, A. Cooperative and Noncooperative R&D in Duopoly with Spillovers: Erratum. *Am. Econ. Rev.* **1990**, *80*, 641–642.
49. Kamien, M.I.; Zang, M.I. Research Joint Ventures and R&D Cartels. *Am. Econ. Rev.* **1992**, *82*, 1293–1306.
50. Luo, Z.; Chen, X.; Wang, X. The role of co-opetition in low carbon manufacturing. *Eur. J. Oper. Res.* **2016**, *253*, 392–403. [CrossRef]
51. Gnyawali, D.R.; Park, B.R. Co-opetition Between Giants: Collaboration with Competitors for Technological Innovatio. *Res. Policy* **2011**, *40*, 650–663. [CrossRef]
52. Yim, C.K.; Chan, K.W.; Lam, S.S.K. Do Customers and Employees Enjoy Service Participation? Synergistic Ef. *J. Mark.* **2012**, *76*, 121–140. [CrossRef]
53. Xie, X.; Tsai, N.; Xu, S.; Zhang, B. Does customer co-creation value lead to electronic word-of-mouth? An empirical study on the short-video platform industry. *Soc. Sci. J.* **2019**, *56*, 401–416. [CrossRef]
54. Lusch, R.F.; Vargo, S.L.; O'Brien, M. Competing through service: Insights from service-dominant logic. *J. Retail.* **2007**, *83*, 5–18. [CrossRef]
55. Payne, A.F.; Storbacka, K.; Frow, P. Managing the co-creation of value. *J. Acad. Mark. Sci.* **2008**, *36*, 83–96. [CrossRef]
56. Merz, M.A.; Zarantonello, L.; Grappi, S. How valuable are your customers in the brand value co-creation process? The development of a Customer Co-Creation Value (CCCV) scale. *J. Bus. Res.* **2018**, *82*, 79–89. [CrossRef]
57. Navarro, S.; Llinares, C.; Garzon, D. Exploring the relationship between co-creation and satisfaction using qca. *J. Bus. Res.* **2016**, *69*, 1336–1339. [CrossRef]
58. Franke, N.; Schreier, M. Why Customers Value Self-Designed Products: The Importance of Process Effort and Enjoyment. *J. Prod. Innov. Manag.* **2010**, *27*, 1020–1031. [CrossRef]
59. Glassman, M.; Mcafee, R.B. Enthusiasm: The missing link in leadership. *Sam Adv. Manag. J.* **1990**, *55*, 4–6.
60. Zhang, M.; Guo, L.; Hu, M.; Liu, W. Influence of customer engagement with company social networks on stickiness: Mediating effect of customer value creation. *Int. J. Inf. Manag.* **2017**, *37*, 229–240. [CrossRef]
61. Tuli, K.R.; Kohli, A.K.; Bharadwaj, S.G. Rethinking Customer Solutions: From Product Bundles to Relational processes. *J. Mark.* **2007**, *71*, 1–17. [CrossRef]
62. Kim, H.-S.; Kim, Y.-G. A CRM performance measurement framework: Its development process and application. *Ind. Mark. Manag.* **2007**, *38*, 477–489. [CrossRef]
63. Butler, B.; Batt, P.J. Re-assessing value (co)-creation and cooperative advantage in international networks. *Ind. Mark. Manag.* **2014**, *43*, 538–542. [CrossRef]
64. Kowalkowski, C.; Witell, L.; Gustafsson, A. Any way goes: Identifying value constellations for service infusion in SMEs. *Ind. Mark. Manag.* **2012**, *42*, 18–30. [CrossRef]
65. Gummesson, E.; Nenonen, S.; Storbacka, K. Business model design: Conceptualizing networked value co-creation. *Int. J. Qual. Serv. Sci.* **2010**, *2*, 43–59.
66. Ulaga, W.; Reinartz, W. Hybrid Offerings: How Manufacturing Firms Combine Goods and Services Successfully. *J. Mark. A Q. Publ. Am. Mark. Assoc.* **2011**, *75*, 5–23. [CrossRef]
67. Zhu, Q.; Geng, Y.; Lai, K. Circular economy practices among Chinese manufacturers varying in environmental-oriented supply chain cooperation and the performance implications. *Environ. Manag.* **2010**, *91*, 1324–1331. [CrossRef]
68. Green, K.W.; Zelbst, P.J.; Bhadauria, V.S.; Meacham, J. Do environmental collaboration and monitoring enhance organizational performance? *Ind. Manag. Data Syst.* **2012**, *112*, 186–205. [CrossRef]

69. Ghosh, D.; Shah, J. A comparative analysis of greening policies across supply chain structures. *Prod. Econ.* **2012**, *135*, 568–583. [CrossRef]
70. Grekova, K.; Calantone, R.J.; Bremmers, H.J.; Trienekens, J.H.; Omta, S.W.F. How environmental collaboration with suppliers and customers influences firm performance: Evidence from Dutch food and beverage processors. *Clean. Prod.* **2016**, *112*, 1861–1871. [CrossRef]
71. Xu, X.; He, P.; Xu, H.; Zhang, Q. Supply chain coordination with green technology under cap-and-trade regulation. *Int. J. Prod. Econ.* **2017**, *183*, 433–442. [CrossRef]
72. Hou, Y.; Zhang, S.; Hong, J. Study on the selection of enterprise suppliers under the green strategic alliance. *China Bus. Theory* **2017**, *17*, 73–75.
73. Qiu, E. *Research on Enterprise Green Management System*; Harbin Engineering University: Harbin, China, 2006.
74. Wang, M.; Gao, Q.; Liang, D. Stackelberg game model between retailers and two competitive manufacturers. *Mod. Bus.* **2014**, *23*, 17–18.
75. Lee, H.L.; Padmanabhan, V.; Whang, S. Information distortion in a supply chain. *Bullwhip Eff. Manag. Sci.* **1997**, *43*, 546–558.
76. Kang, Y.; Chen, J.; Tong, S. Service strategy and service performance: The regulatory effect of value co-creation. *Soft Sci.* **2016**, *30*, 103–107.
77. Holweg, M.; Pil, F.K. Theoretical perspectives on the co-ordination of supply chains. *J. Oper. Manag.* **2008**, *26*, 389–406. [CrossRef]
78. Liu, H.; Zhang, X. Influence of governance mechanism of online retail platform on opportunistic behavior of sellers-Taking perceived uncertainty as moderating variable. *Bus. Econ. Manag.* **2017**, *04*, 17–28.
79. Zhu, Q.; Sun, Y.; Zhou, L. Research on the relationship between platform empowerment, value co-creation and enterprise performance. *Sci. Res.* **2019**, *37*, 2026–2033.
80. Shi, H. *Research on the Impact of Value Co-Creation among Firms on Innovation Cooperation Based on Resource Theory*; Harbin University of Technology: Harbin, China, 2012.
81. Roper, S.; Du, J.; Love, J.H. Modelling the innovation value chain. *Res. Policy* **2008**, *37*, 961–977. [CrossRef]
82. Bansal, P.; Roth, K. Why Companies Go Green: A Model of Ecological Responsiveness. *Acad. Manag. J.* **2000**, *43*, 717–736.
83. Zhu, Q.H. An inter-sectoral comparison of GSC management in China: Drivers and practices. *J. Clean. Prod.* **2006**, *14*, 472–486. [CrossRef]
84. Wu, F.; Gu, F.; Zhang, J. Factors influencing financial performance of Listed Companies. *J. Syst. Manag.* **2013**, *022*, 715–719.
85. Jiang, Q.; Tan, Q. Research on the configuration path of enterprise innovation performance based on fuzzy qualitative analysis. *China Sci. Technol. Forum* **2019**, *08*, 67–76.
86. Yang, J. The determinants of supply chain alliance performance: An empirical study. *Int. J. Prod. Res.* **2009**, *47*, 1055–1069. [CrossRef]
87. Carter, C.R.; Carter, J.R. Interorganizational determinants of environmental purchasing: Initial evidence from the consumer products industries. *Decis. Sci.* **1998**, *29*, 659–685. [CrossRef]
88. Ehrgott, M.; Reimann, F.; Kaufmann, L.; Carter, C.R. Social sustainability in selecting emerging economy suppliers. *J. Bus. Ethics* **2011**, *98*, 99–119. [CrossRef]
89. Lee, J.N. The impact of knowledge sharing, organizational capability and partnership quality on IS outsourcing success. *Inf. Manag.* **2001**, *38*, 323–335. [CrossRef]
90. Ngo, L.V.; O'Cass, A. Creating value offerings via operant resource-based capabilities. *Ind. Mark. Manag.* **2009**, *38*, 45–59. [CrossRef]
91. Ren, S.J.; Hu, C.; Ngai, E.W.T.; Zhou, M. An empirical analysis of inter-organisational value co-creation in a supply chain: A process perspective. *Prod. Plan. Control* **2015**, *26*, 969–980. [CrossRef]
92. Flynn, B.B.; Huo, B.; Zhao, X. The impact of supply chain integration on performance: A contingency and configuration approach. *J. Oper. Manag.* **2010**, *28*, 58–71. [CrossRef]
93. Schruijer, S.G.; Curseu, P.L. Looking at the gap between social psychological and psychodynamic perspectives on group dynamics historically. *J. Organ. Chang. Manag.* **2014**, *27*, 232–245. [CrossRef]

energies

Article

Policy Implications for the Clean Energy Transition: The Case of the Boston Area

Almona Tani and Piergiuseppe Morone *

Bioeconomy in Transition Research Group, Unitelma Sapienza University of Rome, Viale Regina Elena, 291, 00161 Rome, Italy; almona.tani@unitelmasapienza.it
* Correspondence: piergiuseppe.morone@unitelmasapienza.it

Received: 26 March 2020; Accepted: 20 May 2020; Published: 21 May 2020

Abstract: In this paper, we investigate the transition to clean energy technologies in the Boston area, as perceived through the lens of strategic niche management. The main goal of the study was to assess the role of policy in fostering/hindering the development of the clean energy niche and the complete deployment of clean energy technologies in this area. Using argumentative discourse analysis, our research showed that the clean energy niche in the Boston area is generally perceived as strong and dynamic. However, the public de-legitimizing narrative identified gaps at the policy level that include, among others, the limited engagement of the local and federal government in breaking through well-established practices and regulatory frameworks, funding, and infrastructure. These gaps are likely to delay the market uptake of clean energies in this area.

Keywords: socio-technical transition; strategic niche management; clean energy technologies; argumentative discourse analysis; innovation policy

1. Introduction

Clean technologies are defined as "all the techniques, processes, and products that are of importance in preventing or reducing the burden on the environment" [1]. They compete directly with the older and more traditional technologies employed in existing infrastructures, which are often less expensive (due to scale economies), more stable, and more widespread in the market. These rivalry aspects are detrimental to the market uptake of clean technologies and must therefore be mitigated by policies that incentivize the shift to a more sustainable society. Moreover, many older infrastructures are associated with significant sunk costs that cannot be recovered by incentives alone. For this reason, infrastructural restoration for clean technologies requires public support [2,3]. Furthermore, as clean technologies lack long-term research support, they are frequently considered risky and unattractive to private investors. Finally, public investment in sustainable innovations is a strategic policy decision that might be undermined by short-sighted political actions [4].

Against this background, the main goal of the present study was to map policies and assess their role in fostering or hindering the emergence of the clean energy niche and the deployment of clean energy technologies in the Boston area. (Note: The Boston area comprises the city of Boston, the city of Cambridge, the city of Somerville, neighboring cities and surrounding suburbs—all of which are home to numerous universities, research centers and firms. The Boston area is neither a statistical nor an administrative unit; rather, it is a delimitation of eastern Massachusetts (US) that is commonly used in scientific papers and thus suitable for the present study. In accordance with Berry et al. [5] and Owen-Smith and Powell [6], we define the Boston area as a functional economic area with a certain gravitational and commuting influence on its surrounding areas. With this geographical focus in mind, we attempt here to illustrate the historical evolution of the development of the area's clean energy niche.)We selected the Boston area because it is considered a leading region in research and innovation relating to clean energy technologies (classified as second in the US) [7]. Nonetheless, this area generates only 10.5% of its net electricity from renewable energy resources; this is less than the US average (15.9%) [7], showing a mismatch (or incomplete transition) between technological development and deeper societal changes.

In our study, we investigated the transition process to clean energy technologies in the Boston area using the lens of strategic niche management (SNM). This perspective links the emergence and empowerment of technological niches to interactive learning processes and institutional changes [8,9] (Note: Bearing in mind the regional dimension of the case study, the Boston area clean energy transition could be equally understood through the framework of regional innovation systems (RIS), as this area possesses "significant supralocal governance capacity and cohesiveness," which differentiates it from the national context and that of other regions [10] (p. 480). RIS "opened up the way to exploring the extent to which innovation processes at regional level could be defined as systemic" [10] (p. 489) and, in this sense, is associated with both innovation research and regional science. However, RIS falls short in capturing the broader perspective of transition from an incumbent system to a more sustainable one by means of socio-technological change [11,12]. In fact, innovation systems provide only a narrow definition of socio-economic factors, disregarding the fulfillment of societal functions as a driver of innovation processes [11,13]). In socio-technical transitions, the emergence of a technological niche is not conceived as a technology push process, but one that is triggered by the interactions between technology, user practices, societal needs, and regulatory structures [14]. In the successful evolution of a niche, SNM involves three internal mechanisms [9,14,15]:

(i) the convergence of future expectations, in order to overcome the initial lack of confidence in innovation and to attract attention, thus legitimating niche development;
(ii) learning processes, which reduce uncertainties around innovations by increasing formal and informal knowledge on several dimensions (i.e., technical, social and environmental, cultural, economic, regulatory); and
(iii) networking, in order to build a community of supporters, facilitate stakeholder interactions and bring in (financial, human, physical) resources for niche development.

The destabilization of incumbent socio-technical systems and associated institutional structures is generated by the emergence of innovative and sustainable socio-technical configurations that receive increasing political support [16]. Furthermore, the process of creating legitimacy plays a pivotal role in boosting the maturity of socio-technical niches [17,18] by expanding the network of supportive actors.

Socio-technical transitions are not simple and linear, but complex and long-term transformations of socio-technical systems influenced by numerous dimensions at different levels of development [19], guided by sustainability goals and policies [16]. In these uncertain transitional environments, policy plays an important role in providing direct infrastructural support and building economic and regulatory framing conditions for the development and diffusion of sustainability innovations [17]. The transition to sustainable socio-technical systems can only be accomplished with the support of innovation policy (see, e.g., [20–22]). However, policy interventions maintain a conflicting position: on the one hand, they are crucial for building favorable conditions for niche maturity by boosting niche internal mechanisms; on the other hand, they are affected by path-dependent institutions and incumbent lobbies.

According to Markard et al. [16] policy affects socio-technical systems and their sustainability transitions in different ways. First, policy contributes to the development of innovations and technologies by means of knowledge generation and diffusion. From a deployment and diffusion perspective, policy plays a crucial role in market formation, regulation and the up-scaling of emergent socio-technical systems. Second, policy can contribute to destabilizing established socio-technical systems by removing subsidies and/or increasing taxes on traditional technologies.

In a study examining the role of policy in supporting energy democracy in the US, Burke and Stephens [23] identified four policy categories pertaining to socio-technical transitions. The first category, *regulatory context*, lays the ground for the implementation of further ad hoc policies boosting the diffusion of renewable energies. This category includes renewable energy standards, green public procurement and community benefit agreements, among other regulatory instruments. An additional policy category concerns *financial inclusion measures*, or financial instruments and monetary incentives that promote energy system changes. The most important instrument in this category is the feed-in tariff (FIT), which guarantees a long-term minimum fixed purchasing price for renewable energy. Other instruments may relate to green subsidies, on-bill financing and repayment programs, public bonds, carbon tax-and-invest programs and cooperative financing. A third policy category pertains to *economic institutions*, in the form of new socio-economic institutions and economic opportunities for communities. This category focuses on communities' and/or public actors' ownership of renewable energy systems. Policy instruments in this category include renewable energy cooperatives and the re-municipalization of public assets, such as water, sewage and electricity systems. Finally, the fourth policy category consists of *new energy system institutions*, which support and facilitate institutional change in the energy system. This category includes policy instruments concerning microgrids and democratized grid management, energy regions and sustainable energy utilities.

The necessity of and commitment to a transition to sustainable innovation has dominated the discourses of global and local actors; this is particularly true in the US, where industrial production still depends to a very large extent on non-renewable energy resources. Socio-technical transitions entail political negotiation [24,25] between stakeholders with conflicting positions and opinions; such negotiation depends on the framing and definition of the institutionalizing process of innovation [26]. Indeed, the sustainability transition is shaped by social values and political discourse, and further nuanced by stakeholders' differing perceptions of sustainability issues, goals and policies. Accordingly, in the present study, we used argumentative discourse analysis (ADA) to examine the discourse of key actors concerning the role of policy in the development and deployment of clean energy technologies in the Boston area.

The research investigated the way in which constellations of actors legitimized or delegitimized innovations within transition episodes in terms of multi-dimensional discursive interactions. The aim was not to provide a deterministic view of the transition dynamics, but to explain the differing perspectives of the conflicting actors and to link narratives at the micro level to discourse at the macro level, which influences collective knowledge and discursive events. To this end, our investigation: scrutinized dominant discourses relating to the clean energy transition in the Boston area; framed innovations within discrete narratives (with a particular focus on narratives about the role of policy interventions); and assessed any enduring narratives that could obstruct the sustainable transition to clean energy technologies in the Boston area.

2. Materials and Methods

Moving within this framework, we investigated the transition process to clean technologies in the Boston area by applying argumentative discourse analysis (ADA), as proposed by Hajer [26] and subsequently developed by Hajer and Versteeg [27]. This analysis of discourse and narratives has been applied by many scholars to describe a problem, identify solutions and mediate between actors' positions in the transition process [28–35].

ADA is a valuable methodology for critically examining the environmental discourse embedded in the analysis of energy policies. For instance, by examining the discourses of key actors, ADA exposes contradictory narratives and conflicts formed around particular opinions. The main component of ADA is the storyline, which is a narrative sustained by a socio-political coalition that plays a crucial role in "clustering of knowledge, positioning of actors, and ultimately, in the creation of coalitions amongst the actors of a given domain" [26]. Storylines within an environmental discourse are characterized by specific emblems or "issues that dominate the perception of the ecological dilemma in a specified period" [26]. Since storylines emerge between and within political boundaries and do not conform to specific political and institutional settings, they are very helpful in investigating the influence of policy on niche maturation, by revealing the hegemonic ways in which environmental conflicts are argued.

Hajer [26] defines 10 tasks of ADA, which are generally summarized in three main steps. The first step consists of making a preliminary assessment of the context and its development by analyzing written documents and official communications. This provides an overview of the developmental process, which is later enriched with further information or reframed by the interviews conducted in the second step [26]. In the present study, to explore the context of clean energy technologies in the Boston area, we reviewed reports, industry roadmaps, empirical studies and analyses [36–44], as well as official websites [45–55]. This enabled us to identify concepts and ideas that structured the discourse.

The second step of ADA consists of interviewing key players in order to collect more information on specific events. In our case, the relevant event involved the sustainability transition towards a clean energy sector. As a result of the preliminary assessment, we identified five key actors involved in clean energy technologies in the Boston area. Each actor was asked to provide at least two names of other relevant actors. After examining the suggested actors' profiles, we identified seven additional actors. In this way, we built a final list of 12 key actors, with whom we conducted formal interviews using a qualitative, semi-structured questionnaire (see Appendix A). The interviews were conducted between June and July 2016. Figure 1 presents a graphic representation of the types of actors interviewed in this step, differentiated between core and peripheral actors. Table 1 provides the full list of the interviewed actors.

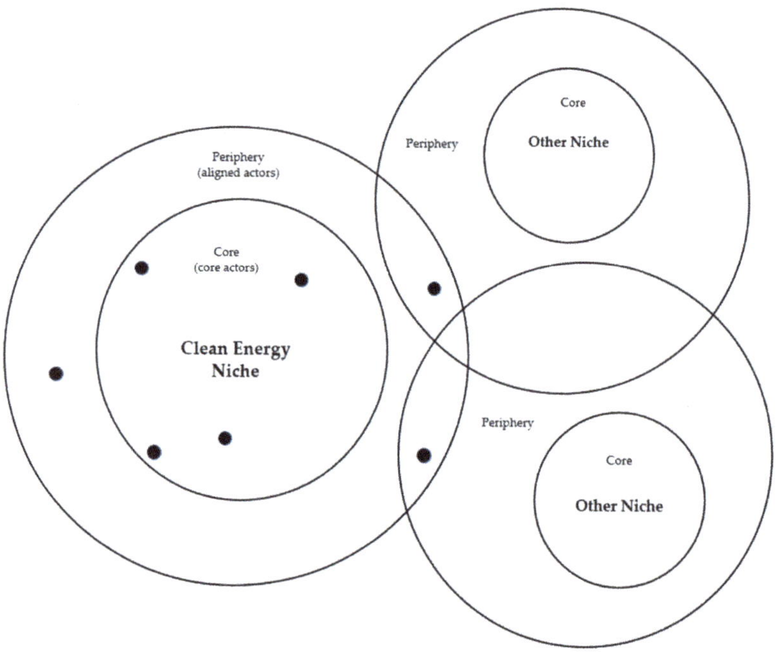

Figure 1. Core and peripheral actors. *Source:* Authors, based on Rosenbloom et al. [28].

Table 1. List of key actors' organizations.

Niche Actors	Peripheral Actors	Policy Actors
Clean-Tech Officer—Technology Licensing Office, MIT	Clean-Tech Officer—Massachusetts Technology Transfer Center	Manager of Grant Programs—Massachusetts Clean Energy Center
Clean-Tech Officer—Center for Research Innovation, Northeastern University	Board Member—Venture Café Foundation	Senior Cluster Development Specialist—EPA Office of Research and Development
Clean-Tech Officer—Harvard Office of Technology Development	Board Member—New England Water Innovation Network	Program and Research Analyst—Innovation Institute (Massachusetts Technology Collaborative)
Manager of Projects in Innovation and Industry Support—Greentown Labs		Director of OTA—Executive Office of Energy and Environmental Affairs Director of Real Estate—Boston Redevelopment Authority

Source: Authors.

As shown in both Figure 1 and Table 1, four of the selected actors were classified as part of the "core" of the clean energy niche, given their role in developing and diffusing the innovative technology. This group was composed of actors from three technology transfer offices (Harvard University, MIT, Northeastern University) and one clean technology business accelerator (Greentown Labs). The second column of Table 1 indicates three peripheral actors at the "fringe" of the niche [28] (p. 1279); these actors participated in initiatives promoting the development and deployment of clean technologies, but were not directly involved in either the niche or relevant policy (New England Water Innovation Network, Venture Café Foundation, Massachusetts Technology Transfer Centre). For instance, some of these actors engaged in technology transfer and the promotion of new technologies and start-ups in support of other niches. The third and final group of interviewees, who are not reflected in Figure 1 because they operate outside of the niche, was composed of five actors at the policy level (Massachusetts Clean Energy Centre, Environmental Protection Agency (EPA), Massachusetts Technology Collaborative: Innovation Institute, Executive Office of Energy and Environmental Affairs: Office of Technical Assistance and Technology (OTA); Boston Redevelopment Authority); these actors are grouped in the third column of Table 1. Within each organization, the selection of interviewees was made on the basis of interviewees' strategic role (e.g., manager of grant programs, innovation projects, industry support, clean technology, etc.).

To complete the ADA, a third and final step was undertaken, involving the analysis of particular events or incidents that emerged from the reviews and interviews. The aim of this step was to validate the reliability of the storylines, given the controversial opinions and experiences they contained. To this end, we analyzed the interviews and documents, querying all relevant discourse elements and events.

By following these three steps, we identified a dominant storyline concerning the flourishing dynamics of the clean energy niche. We also identified two struggling storylines: one legitimizing the crucial role and commitment of public intervention in the development of the clean energy niche, and another delegitimizing the engagement of public bodies, showing a lack of niche power to break through the incumbent market. In Section 3, we present the identified storylines along with illustrative quotes from the interviews.

3. Results

Building on the theoretical and empirical framework depicted in the previous sections, we will now illustrate the identified discourse surrounding clean energy technologies in the Boston area. The storylines that emerged in the research highlighted two specific trajectories:

1. How actors recognize the presence of a clean technology niche; and
2. How actors frame the policy context in which this niche innovation is developed and deployed.

We placed the main storyline within the first trajectory. Within the second discourse trajectory, we placed two other storylines: one legitimizing the role of public intervention and another delegitimizing its commitment. In the following subsections, we discuss these three storylines in some detail.

3.1. Dominant Storyline

Overall, most actors expressed the need to shift to a cleaner energy system, and thereby commit to more sustainable production. As emerged from the discourse analysis, the adoption of the Green Community Act in 2008 was considered the breaking point with traditional energy production. This event was found to significantly influence the development of this dominant storyline, regarding the current development of clean technologies.

STORYLINE 1: Clean energy technologies are central to a thick network that exchanges knowledge and engages for a cleaner common future.

In this storyline, we identified all three mechanisms characterizing the development of the clean energy niche in the Boston area (Table 2). According to SNM, one of the main mechanisms for the development of an innovative niche is building a common vision through shared expectations. On the one hand, universities use the commercialization of research and patents to encourage social use of their inventions; on the other hand, the state adopts environmentally friendly laws and incentives for more energy efficient and cleaner production (e.g., the Commonwealth's 2016–2018 Three-Year Energy Efficiency Plan, the Affordable Access to Clean and Efficient Energy Initiative, etc.) [46]. The annual Industry Report of the Massachusetts Clean Energy Centre (MassCEC) [36] mentions specific goals: "In August 2008, Massachusetts required all economic sectors to reach a 25% reduction in GHG emissions by 2020 and an 80% reduction by 2050 under the Global Warming Solutions Act," becoming "one of the first States in the Nation to move forward with a comprehensive regulatory program to address climate change." Moreover, the private sector "continues the trend of becoming more 'pure-play', meaning that all of their activities are clean energy related" [55].

The achievement of this goal at all levels is driven by knowledge creation and sharing, which consists of local learning processes and investment in human capital and a specialized labor force. The interviewed actors focused particularly on research and innovation, for two reasons: first, the development of clean energy technologies requires intense research to generates radical innovation; and second, as mentioned above, the Boston area is characterized by a high number of academic research institutions conducting theoretical and applied clean energy research. More specifically, since the US government adopted the Bayh-Dole Act in 1980, universities have become key actors of innovation by commercializing their research to firms operating in the market. MIT's Technology Licensing Office files approximately 200 patents each year, but only a quarter of these relate to clean technology. Both MIT and Northeastern University license half of their filed patents to existing companies, and 20–25% of these result in spin-offs. These institutions have engaged in an intense transfer of mature technology, amounting to 147 patents awarded to 36 companies working with pre-commercial products, 229 patents awarded to firms focusing on energy efficiency and 25 patents awarded to firms working exclusively with energy goods and services [36]. Grentown Labs, a clean technology accelerator, has filed 32 innovative patents and supported nine university-born companies. On the other hand, according to Venture Café, there remains a dire need for technology transfer: "Clean energy in the Boston area is composed by small companies; there are no large companies yet. Therefore, technology transfer is really important because all the research and innovation comes from universities. Small companies don't have budget for research."

Indeed, knowledge sharing and technology transfer in the Boston area represent a regional competitive advantage that is disseminated through networked actors. Collaboration and networking are quoted as incentives, either for the location of clean energy companies in Boston or the success of clean energy development. Indeed, one-third of the patents filed at MIT are developed in collaboration with other universities or companies at which their students are likely to find employment. Also, according to the respondent at Greentown Labs, their success depends on their dense network, composed of 102 host start-ups and mature and specialized companies, with expertise in topics ranging from IP to tax filings, fundraising and clean energy technology.

Table 2. Overview of the dominant storyline.

Dominant Storyline	**Clean Energy Technologies Are Central to a Thick Network that Exchanges Knowledge and Engages for a Cleaner Common Future**		
Key narratives	Universities, industry and public bodies share common expectations and work for a cleaner future	Knowledge creation, local learning processes and technology transfer are unique to the Boston area	Networking is an asset for the location of clean energy companies in the Boston area

Source: Authors.

3.2. *Legitimizing and Delegitimizing the Role of Policy*

The core message of the dominant storyline pertains to the existence of a dynamic and sufficiently mature clean energy niche with a thick network of actors engaged in learning processes and knowledge exchanges. Moreover, a common thread throughout this storyline is the need for public institutions and actors to proactively exert pressure upon the incumbent energy regime, prompting the deployment of clean energy technologies. We investigated this aspect further by identifying two additional storylines about the role of policy in niche development.

STORYLINE 2: Public intervention is everywhere.

The second storyline is based on three key narratives relating to policy intervention areas in support of the clean energy niche: the regulatory framework, funding for knowledge creation and sharing, and infrastructures for a collaborative environment. Such policy interventions, which aim at developing an environmentally friendly regulatory framework in the Boston area, account for a significant number of laws and regulations concerning air quality, toxic and hazardous substances, waste and recycling. One particular example is the Toxics Use Reduction Act (TURA), which receives support from both industry and environmental groups. This Act requires companies to analyze and report their annual chemical use, quantifying the chemical content in the final product, the chemical content released into the environment at the end of the production process and any waste treatment performed. Furthermore, it requires companies to draw up long-term pollution reduction plans.

The TUR Institute, located on the Lowell Campus of the University of Massachusetts, is tasked with meeting TURA's goal of developing innovative techniques for cleaner production. As stated by the OTA interviewee: "the Institute is engaged in alternative assessments, in research developing alternative chemistries and in evaluation of alternative technologies in order to complete their program of technology transfer."

Moreover, following the Green Communities Act of 2008, the Massachusetts government now promotes the development of renewable energy, energy efficiency and conservation, "green communities" and the implementation of Regional Greenhouse Gas Initiatives. This Act aims at reducing the costs of renewable energies for consumers and "increase[ing] generation from low or zero-carbon resources within Massachusetts" [43]. The Green Communities Program, instituted by this Act, provides incentives to municipalities that engage in energy efficient and renewable technology. Furthermore, the Pioneer Valley Planning Commission Sustainability Toolkit mentions that the Green Communities Division of the Department of Energy Resources shall provide "up to 10 million dollars per year state-wide in technical and financial help to the communities involved." A portion of this money is raised by the Regional Greenhouse Gas Initiative (enacted under the same Act), which draws on market-based cap-and-trade CO_2 emissions. The emission allowances issued under this initiative are auctioned, and the generated funds are used for zero-interest loans for municipal energy-efficiency projects.

With respect to knowledge creation and sharing, the Green Jobs Act, adopted in 2008, designated "125 million dollars to train about 30,000 people in green collar jobs" [47]. In order to achieve this goal, the Act instituted the MassCEC, which became "the most important quasi-public agency in job creation and for the economic development of the clean energy industry."(Interviewee from the MTTC.) Indeed, both the MassCEC and the Innovation Institute have programs in place to support students who are transitioning to the labor market, "because Boston has a high rate of international students who would chose to work in the Boston area only if there is a good work environment."(Interviewee from the Innovation Institute.) Similarly, the MassCEC Workforce Development Programs aim at increasing awareness of clean energy job opportunities and placing job seekers in clean energy jobs in Massachusetts [36]. Since 2011, this organization has placed "more than 1300 interns at over 250 companies which have received a reimbursement of $14/h for the internships." (Interviewee from the MassCEC.)

Moreover, according to technology officers from MIT, Harvard University and Northeastern University, 80–90% of the research at these universities is publicly funded, with most of the funds granted by the NSF and NIH for basic research. However, the MassCEC and SBIR are key actors in placing their students in green collar jobs, commercializing their staff's clean patents and funding their inventors' spin-offs.

Considering that Massachusetts's basic research at universities, research institutes and hospitals is worth $4.5 billion, in 2003, the state legislature created and funded the Massachusetts Technology Transfer Centre (MTTC). The MTTC aims at improving knowledge generation and the diffusion of innovative technologies. While it does not provide funding, it supports universities' technology transfer offices, assists institutions that do not have their own technology transfer office and develops programs in collaboration with the MassCEC and the Massachusetts Technology Collaborative (MassTech)—both quasi-public agencies that provide financial support for MTTC programs.

The MassTech provides capital funding grants with a co-participation of one-third (up to $5 million) on projects relating to innovation development. It generally funds non-profit institutions—especially universities—that collaborate with private companies or for-profit institutions (Interviewee from the Innovation Institute). To confirm its role in the up-scaling of the emergent clean energy system, the MassCEC implements many grant programs, targeting university innovators and start-ups as well as growing companies, and provides pilot test areas for GTL companies. Table 3 summarizes the MassCEC grant programs that support the development of new clean energy technologies.

Table 3. MassCEC grant programs.

Grant Program	Target	Grant Amount
Catalyst	To help researchers and young companies develop prototypes and proof-of-concept studies	$2.1 million to 55 companies
AmplifyMass	To support Massachusetts-based awardees of ARPA-E (Advanced Research Projects Agency-Energy)	$3 million to 14 awardees
AccelerateMass	To support graduates in transitioning out of accelerator programs	$50,000 in phase 1, $100,000 in phase 2
InnovateMass	To help young clean energy and water companies overcome financial barriers to commercializing products and technologies	$2.2 million to 19 companies
DeployMass	To help companies seeking a first or early customer to validate the commercial readiness of their technology	up to $160,000
Direct Equity Convertible Debt Investment	To help early stage companies	average investment of $500,000
Venture Debt Investment Program	To fill funding gaps for clean-tech companies seeking venture debts but unable to attract private venture capital	$100,000 to $1 million

Source: Authors, based on data from MassCEC [54].

At the federal level, Small Business Innovation Research (SBIR) is a special program of the Small Business Association (SBA), which supports the creation of spin-offs. In particular, the SBA funds small companies that collaborate with universities, with funding also provided for the universities.

Public intervention also targets infrastructure in support of a collaborative environment. At the local level, the city of Somerville is building fabrication laboratories that are small makerspaces in high schools. Also, to stimulate innovation, Somerville is providing physical connections, in addition to economic ones. As highlighted by MIT and stated by Harvard representatives in Cambridge: "the State of Massachusetts and the municipalities of Boston and Cambridge have an economic development project for the creation of an innovation district and their engagement focusses particularly in infrastructural works of connecting Kendall Square with the rest of the area by the bridge and the red line."

Incubators and accelerators represent important infrastructure for start-ups and innovative companies in the Boston area. Some examples include the Cambridge Innovation Center (hosting 600 new companies), the Roxbury Innovation Center (situated in a marginalized neighborhood), Venture Café, MassChallenge and the Boston Innovation Center, which are all supported by the state of Massachusetts as well as local municipalities, as part of their innovation policy initiatives. All of these co-working spaces support start-ups in preparing business plans and networking with strategic partners, mainly for mentorship and resources. Furthermore, in 2015, the city of Boston launched an initiative involving a "startup czar." This czar, named by the municipality, has a background in planning and is tasked with analyzing the city's opportunities for welcoming start-ups and providing them with, in particular, physical connections (Interviewee from the Greentown Labs).

Table 4 summarizes the three key narratives of Storyline 2 related to policy intervention areas in support of the clean energy niche.

Table 4. Overview of Storyline 2 on legitimizing public engagement.

Storyline 2	**Public Intervention Is Everywhere**		
Key narratives	Adoption of significant number of environmentally friendly laws and regulations	Availability of (public) funds for knowledge creation and sharing in clean energy	Public intervention also targets infrastructures for a collaborative environment

Source: Authors.

Most actors perceived the general framework of public engagement in the development of innovative technologies as rather positive and supportive of the emergence of the clean energy niche. However, the discourse analysis uncovered some hurdles in the maturation of the clean energy niche and the deployment of these technologies in the Boston area, as reflected in the third storyline.

STORYLINE 3: diffusion of clean energy technologies in the Boston area is limited by incumbent barriers.

This final storyline represents the interviewees' discourse in response to the questions: "In your opinion, which is the level of diffusion of clean energy technologies stemming from research in this field, as used by companies and households, respectively? For what reason(s)?" The concepts and ideas that emerged in the discourse mainly concerned the limited policy commitment to break through well-established practices and regulatory frameworks, the lack of public programs and policy support for attracting venture capital to up-scale innovations and the lack of infrastructural interventions for the commercialization of clean energy technologies (Table 5). Accordingly, this storyline significantly reframes the role of policy in the clean energy transition in the Boston area.

Table 5. Overview of Storyline 3 on delegitimizing public engagement.

Storyline 3	Diffusion of Clean Energy Technologies in the Boston Area Is Limited by Incumbent Barriers		
Key narratives	Scarce commitment of policy to break through well-established practices and regulatory framework	Lack of public programs and policy support to attract venture capital for the market uptake of clean energy technologies	Lack of infrastructural interventions for the commercialization of clean energy technologies

Source: Authors.

The storyline rightly assumes that clean energy technologies are based on radical innovations that demand change in well-established practices and regulatory frameworks. The interviewed actors perceived limited policy commitment to increasing awareness and improving regulations to support a trustworthy and long-term transition in the Boston area. Indeed, the adoption of the Energy Policy Act in 2005 by the US government intended to extend federal production tax credit to renewable energies. The creation of this policy involved varying public and private stakeholders concerned with environmental regulation and incentives. However, although initially promising, the Act had no real effects (regulatory or otherwise) on the energy sector. In fact, according to the Union of Concerned Scientists [56], the $14.5 billion provided for the implementation of the bill was spent mainly on nuclear and fossil fuel; only 9% was spent on renewable energy and 21% on energy efficiency. Moreover, community-driven innovations such as Greenovate (which promotes bike sharing and solar panels) are not necessarily attractive and struggle to establish shared practices. Consequently, these innovations are unlikely to become established in the market, unless policy intervenes with regulations aimed at raising awareness and forming new markets, for example through State purchasing and procurement. This is the main reason why only 30–50% of university patents are commercialized (Interviewee from the OTA).

Lack of resources also limits the up scaling of clean energy technologies, thereby hindering their ability to gain purchase in the market. Policy has failed to build long-lasting public–private partnerships and to financially plan innovative activities: "15 years ago, there was a program called the Strategic Technology Environmental Partnership (STEP). Its purpose was to take new clean technologies and make the proof-of-concept for commercialization. Once the technology was ready to be deployed, we proposed it to the companies. The program no longer exists, mainly because of lack of resources and change of administration." (Interviewee from the OTA.) Indeed, incentives that support the clean energy sector in the Boston area depend heavily on the governing party. However, there is a mismatch between the horizon of clean energy development and the horizon of political turnover; this increases the uncertainty of clean energy, making it less attractive for venture capitalists (Technology officer from MIT). Although venture capitalists are key for the flourishing of innovation activities in the US private sector, they are not particularly attracted to and active in the clean energy market due to its risky nature [2]. US states (with the exception of California) tend to be conservative with venture capital, due to the long-time horizons involved. In particular, the transition to clean energy technologies cannot occur in the short term (Interviewee from the EPA). This is one of the reasons why companies are afraid of investing in clean energy technologies, as they are unsure whether such technologies will prove sound investments over the long term.

With respect to infrastructural interventions for the commercialization of clean energy, the interviewed actors recognized a lack of policy. Massachusetts is highly dependent on external sources, considering that the state's energy consumption exceeds its production. The state does not produce any petroleum, coal or natural gas, yet its net electricity generation is 73.3% dependent on natural gas, 0.3% dependent on petroleum, 3.7% dependent on coal, 14.5% dependent on nuclear energy, and only 5.7% dependent on renewables (mainly hydroelectric and biomass) [7]. Moreover, Boston—the largest city in the state—has the oldest active port in the US; this port has petroleum product terminals and the only liquefied natural gas (LNG) import terminals. In addition, Massachusetts is home to the largest coal-fired power plant in New England. This shows that state policy has mainly engaged in infrastructure for economic development (see storyline 2), rather than infrastructure that could boost the deployment of clean energy technologies (e.g., the MassCEC funding programs neither supply testing areas nor build infrastructural facilities for start-ups in the clean energy technologies niche).

All of these elements, which emerged from the discourse analysis, substantially undermine the belief that policy has embarked on a flourishing path to a clean energy transition in the Boston area.

Overall, storylines 2 and 3 seem to confirm what emerged in storyline 1—that the clean energy niche in the Boston area is dynamic and potentially able to achieve maturity. However, these two storylines provide two rather contrasting perspectives on this potential. On the one hand, storyline 2 stresses the role of public policy in supporting the emergence of the niche. On the other hand, storyline 3 suggests that policy has failed to support the deployment of clean energy technologies in the Boston area due to a lack of pressure on the incumbent energy regime; thus, breakthrough of the clean energy niche has been impeded.

4. Discussion and Concluding Remarks

Overall, our investigation showed that the clean energy niche in the Boston area is characterized by a fast growing and dynamic innovation environment. In particular, important actors in the quickly evolving clean energy niche are: (i) local research institutions and universities that provide innovation, a specialized workforce, laboratories, and equipment; and (ii) technology business incubators (e.g., Greentown Labs) and clean technology business accelerators (e.g., Cleantech Open Northeast, Venture Café, MassChallenge), which support start-ups in developing innovative business plans and networking with strategic partners, mainly for mentorship and resources. The activity of these actors lays the groundwork for the emergence of an innovative niche, by means of knowledge generation and networking. However, these actors are failing to up-scale clean energy technologies into innovative systems and to fully penetrate the market. In addition, they are lacking in long-term financial resources, as they are dependent on private funds and/or affected by instable political support.

Given the importance that the literature attributes to policy intervention for the development of clean technologies, we further tested the engagement of public agencies in the Boston area in relation to the three niche mechanisms determined through SNM. We identified a legitimizing public engagement storyline (storyline 2) grounded on three key narratives, each reflecting a policy intervention affecting niche mechanisms. The first policy intervention emerged from interviewees' discourse concerning the adoption of a significant number of environmentally friendly laws and regulations, paving the way to a common expectation of clean energy technologies development. The second policy regarded the availability of public funding for knowledge creation and sharing, in order to trigger actors' learning processes. Finally, the third policy targeted infrastructures for a collaborative environment to support networking among actors—both within and beyond the niche.

However, the narratives in the public delegitimizing storyline (storyline 3) identified gaps at the policy level that, according to the respondents, hindered the upscale of clean energy technologies, and thus the development of the clean energy niche in the Boston area. These gaps related generally to three major pitfalls in the public strategy for developing the clean energy niche: (i) state policy engaged more in adopting new laws rather than harmonizing regulation to destabilize well-established practices; (ii) policy intervention had not yet succeeded in attracting private investment to the clean energy sector, resulting in a dependent relation between (unstable) public funding and the market uptake of clean energy technologies; and (iii) by only occasionally providing key infrastructures (e.g., facilities for market-level technology deployment), policy had not yet succeeded in building an effective commercialization program, resulting in the limited deployment of clean energy technologies in the Boston area and hindering the clean energy niche breakthrough.

Overall, until policy designs and implements focused interventions, the emerging clean energy niche in the Boston area, though significantly dynamic, will struggle to overturn the incumbent regime.

This explains why, although the government has invested significantly in clean technologies research and development, the Boston area remains at an early stage of deployment of clean energy technologies.

As a final remark, we shall suggest a possible action to overcome these pitfalls and speed up the transition process. As noted, the clean energy sector in the Boston area is largely composed of small companies—particularly start-ups—that are dispersed across the state. This industrial structure has struggled to develop and, for this reason, is not attractive for venture capital investment, which is "focused on some of the safer bets rather than on the radical innovation that is required to allow the sector to transform society so as to meet the double objective of promoting economic growth and mitigating climate change" [2] (p. 136). Therefore, larger amounts of public resources must be invested to stimulate the growth of small companies. This will attract private investments, facilitating a fruitful entrepreneurial environment and stimulating the development of dedicated infrastructures for the deployment of innovative clean energy technologies.

One limitation of the methodological approach we have applied in this work regards the inclusion of only local actors in the niche, while according to the socio technical transition literature, niches can be defined as networks of both local and global actors [57–60]. In that respect, it could be important to investigate learning processes that aggregate niches' actors in networks generating global structures. However, exploring global (knowledge) spillovers of the Boston area clean energy niche goes beyond the scope of this paper—whose purpose is to assess the role of policy in fostering/hindering the emergence of the clean energy niche and the deployment of clean energy technologies in the Boston area—and would represent a very fertile line for future research.

Furthermore, although the policy gaps and shortfalls identified in this paper are context-specific—grounded in the discourse of local actors—the analysis has highlighted divergent policy needs in the different phases of niche emergence and maturation. Considering the role of actors' perceptions and needs in transition studies, ADA could be a suitable methodology for pinpointing context-specific needs in the energy transition process. For this reason, as an interesting and important further line of research, we suggest that ADA be applied in developing countries that have only recently begun to engage in energy transition pathways.

Author Contributions: Conceptualization: A.T. and P.M.; methodology: A.T.; validation: A.T. and P.M.; formal analysis: A.T.; investigation: A.T.; resources: A.T.; data curation: A.T.; writing—original draft preparation: A.T.; writing—review and editing: A.T. and P.M. All authors have read and agreed to the published version of the manuscript.

Funding: This research received no external funding.

Conflicts of Interest: The authors declare no conflict of interest.

Appendix A Outline of the Semi-Structured Interviews

- Describe your organization and the role it plays in the clean energy sector.
- What are the main drivers of your presence in the Boston area (MA)?
- Have you introduced innovative technologies and/or processes in the market, and/or filed any patent related to clean energy technologies?
    - Did you develop the innovation/patent on your own or in collaboration?
    - What percentage of your R&D/innovation activity has been funded by public financial support/private organizations?
    - Have you received any other support from private or public organizations?
- Has your organization developed any kind of collaboration with other public or private organizations?
    - What kind of collaboration?
- Has public policy supported the development of clean energy technologies?

- Through which correlated or direct activities?

- In your opinion, what is the level of development of clean energy technologies in the Boston area (as compared to other places)?

  - For what reason(s)?

- In your opinion, what is the level of diffusion of clean energy technologies in the Boston area (compared to the R&D in this field)?

  - For what reason(s)?

- Please rate the following statements on a scale from 1 (strongly disagree) to 5 (strongly agree):

  (a) The legislation supports innovative activities.
  (b) Special aid is available from the government for innovations.
  (c) Starting up one's own business is encouraged in Massachusetts.

References

1. Schot, J.W. Constructive technology assessment and technology dynamics: The case of clean technologies. *Sci. Technol. Hum. Val* **1992**, *17*, 36–56. [CrossRef]
2. Mazzucato, M. *The Entrepreneurial State: Debunking Public vs. Private Sector Myths*, 3rd ed.; Public Affairs: New York, NY, USA, 2015.
3. Hopkins, M.; Lazonick, W. Soaking up the Sun and Blowing in the Wind: Renewable Energy Needs Patient Capital. Airnet Working Paper. 2012. Available online: http://www.theairnet.org/files/research/Hopkins/CleanTech_PatientCapital_20121129a.pdf (accessed on 20 May 2020).
4. David, P.A.; Hall, B.H.; Toole, A.A. Is public R&D a complement or substitute for private R&D? A review of the econometric evidence. *Res. Policy* **2000**, *29*, 497–529.
5. Berry, B.J.L.; Goheen, P.G.; Goldstein, H. *Metropolitan Area Definition: A Re-Evaluation of Concept and Statistical Practice (Rev.)*; US Bureau of the Census Working Paper No. 28; US Bureau of the Census: Washington, DC, USA, 1969.
6. Owen-Smith, J.; Powell, W.W. Knowledge networks as channels and conduits: The effects of spillovers in the Boston biotechnology community. *Organ. Sci* **2004**, *15*, 5–21. [CrossRef]
7. US Energy Information Administration (eia). Available online: https://www.eia.gov/state/?sid=US (accessed on 25 February 2017).
8. Schot, J.; Hoogma, R.; Elzen, B. Strategies for shifting technological systems: The case of the automobile system. *Futures* **1994**, *26*, 1060–1076. [CrossRef]
9. Kemp, R.; Schot, J.; Hoogma, R. Regime shifts to sustainability through processes of niche formation: The approach of strategic niche management. *Technol. Anal. Strat.* **1998**, *10*, 175–198. [CrossRef]
10. Cooke, P.; Uranga, M.G.; Etxebarria, G. Regional innovation systems: Institutional and organisational dimensions. *Res. Policy* **1997**, *26*, 475–491. [CrossRef]
11. Geels, F.W. From sectoral systems of innovation to socio-technical systems: Insights about dynamics and change from sociology and institutional theory. *Res. Policy* **2004**, *33*, 897–920. [CrossRef]
12. Markard, J.; Truffer, B. Technological innovation systems and the multi-level perspective: Towards an integrated framework. *Res. Policy* **2008**, *37*, 596–615. [CrossRef]
13. Smith, A.; Voß, J.P.; Grin, J. Innovation studies and sustainability transitions: The allure of the multi-level perspective and its challenges. *Res. Policy* **2010**, *39*, 435–448. [CrossRef]
14. Schot, J.; Geels, F.W. Strategic niche management and sustainable innovation journeys: Theory, findings, research agenda, and policy. *Technol Anal. Strat.* **2008**, *20*, 537–554. [CrossRef]
15. Hoogma, R.; Kemp, R.; Schot, J.; Truffer, B. *Experimenting for Sustainable Transport. The Approach of Strategic Niche Management*, 1st ed.; EF&N Spon: London, UK, 2002.
16. Markard, J.; Suter, M.; Ingold, K. Socio-technical transitions and policy change: Advocacy coalitions in Swiss energy policy. *Environ. Innov. Soc. Transit.* **2016**, *18*, 215–237. [CrossRef]

17. Turnheim, B.; Geels, F.W. The destabilisation of existing regimes: Confronting a multi-dimensional framework with a case study of the British coal industry (1913–1967). *Res. Policy* **2013**, *42*, 1749–1767. [CrossRef]
18. Geels, F.W. The multi-level perspective on sustainability transitions: Responses to seven criticisms. *Environ. Innov. Soc. Transit.* **2011**, *1*, 24–40. [CrossRef]
19. Smith, A.; Kern, F.; Raven, R.; Verhees, B. Spaces for sustainable innovation: Solar photovoltaic electricity in the UK. *Technol. Soc.* **2014**, *81*, 115–130. [CrossRef]
20. Kivimaa, P.; Kern, F. Creative destruction or mere niche support? Innovation policy mixes for sustainability transitions. *Res. Policy* **2016**, *45*, 205–217. [CrossRef]
21. Kivimaa, P.; Virkamäki, V. Policy mixes, policy interplay and low carbon transitions: The case of passenger transport in Finland. *EPG* **2014**, *24*, 28–41. [CrossRef]
22. Weber, K.M.; Rohracher, H. Legitimizing research, technology and innovation policies for transformative change: Combining insights from innovation systems and multi-level perspective in a comprehensive failures framework. *Res. Policy* **2012**, *41*, 1037–1047. [CrossRef]
23. Burke, M.; Stephens, J.C. Energy democracy: Goals and policy instruments for sociotechnical transitions. *Energy Res. Soc.* **2017**, *33*, 35–48. [CrossRef]
24. Smith, A.; Raven, R. What is protective space? Reconsidering niches in transitions to sustainability. *Res. Policy* **2012**, *41*, 1025–1036. [CrossRef]
25. Hermwille, L. Climate change as a transformation challenge: A new climate policy paradigm? *GAIA* **2016**, *25*, 19–22. [CrossRef]
26. Hajer, M.A. *The Politics of Environmental Discourse: Ecological Modernization and the Policy Process*; Oxford University Press Scholarship Online: Oxford, UK, 2003. [CrossRef]
27. Hajer, M.; Versteeg, W. A decade of discourse analysis of environmental politics: Achievements, challenges, perspectives. *J. Environ. Pol. Plan.* **2005**, *7*, 175–184. [CrossRef]
28. Rosenbloom, D.; Berton, H.; Meadowcroft, J. Framing the sun: A discursive approach to understanding multi-dimensional interactions within socio-technical transitions through the case of solar electricity in Ontario, Canada. *Res. Policy* **2016**, *45*, 1275–1290. [CrossRef]
29. Cotton, M.; Rattle, I.; Van Alstine, J. Shale gas policy in the United Kingdom: An argumentative discourse analysis. *Energy Policy* **2014**, *73*, 427–438. [CrossRef]
30. Usher, M. Defending and transcending local identity through environmental discourse. *Environ. Politics* **2013**, *22*, 811–831. [CrossRef]
31. Bern, M.R.; Winkel, G. Nuclear reaction to climate change? Comparing discourses on nuclear energy in France and Germany. In *Methodologie und Praxis der Wissenssoziologischen Diskursanalyse: Theorie und Praxis der Diskursforschung*; Keller, R., Truschkat, I., Eds.; Springer VS: Wiesbaden, Germany, 2013.
32. Hunold, C.; Leitner, S. "Hasta la vista, baby!" The solar grand plan, environmentalism, and social constructions of the Mojave Desert. *Environ. Politics* **2011**, *20*, 687–704. [CrossRef]
33. Jessup, B. Plural and hybrid environmental values: A discourse analysis of the wind energy conflict in Australia and the United Kingdom. *Environ. Politics* **2010**, *19*, 21–44. [CrossRef]
34. Mander, S. The role of discourse coalitions in planning for renewable energy: A case study of wind-energy deployment. *Environ. Plann C Politics Space* **2008**, *26*, 583–600. [CrossRef]
35. Szarka, J. Wind power, discourse coalitions and climate change: Breaking the stalemate? *Eur. Environ.* **2004**, *14*, 317–330. [CrossRef]
36. Massachusetts Clean Energy Center. *Massachusetts Clean Energy Industry Report*; BW Research Partnership: Boston, MA, USA, 2015; Available online: https://s3.amazonaws.com/images.masscec.com/2015MassCleanEnergyIndustryReport.pdf (accessed on 20 May 2020).
37. Massachusetts Clean Energy Center. *Massachusetts Clean Energy Industry Report*; BW Research Partnership: Boston, MA, USA, 2016; Available online: https://files.masscec.com/2016%20MassCEC_IndustryReport_Full_Web.pdf (accessed on 20 May 2020).
38. Battelle Memorial Institute; Redwood Innovation Partners. *Massachusetts Water Technology Industry Roadmap*; Massachusetts Clean Energy Center: Boston, MA, USA, 2015; Available online: https://files.masscec.com/research/MassachusettsWaterTechnologyIndustryRoadmap.pdf (accessed on 20 May 2020).
39. Office of Technical Assistance and Technology (OTA). *TURA 25th Anniversary Leaders Demonstrate Product Innovation, Quality and Safety*; OTA: Boston, MA, USA, 2014.
40. International Monetary Fund (IMF). *World Economic Outlook*; IMF: Washington, DC, USA, 2016.

41. Innovation Institute at the Massachusetts Technology Collaborative. *The Index of the Massachusetts Innovation Economy*; Innovation Institute at the Massachusetts Technology Collaborative: Boston, MA, USA, 2016; Available online: https://masstech.org/sites/mtc/files/documents/InnovationInstitute/Index/MAInnovationEconomy_2015.pdf (accessed on 20 May 2020).
42. United Nations. *World Population Prospects: The 2015 Revision*; United Nations: New York, NY, USA, 2015.
43. Hibbard, P.J.; Tierney, S.F.; Darling, P.G. *The Impacts of the Green Communities Act. on the Massachusetts Economy: A Review of the First Six Years of the Act's Implementation*; Analysis Group: Boston, MA, USA, 2014; Available online: https://www.analysisgroup.com/globalassets/content/insights/publishing/analysis_group_gca_study.pdf (accessed on 20 May 2020).
44. International Energy Agency. World Energy Outlook 2004–2016. Available online: https://www.iea.org/ (accessed on 10 March 2017).
45. Energy Storage Association. Available online: http://energystorage.org/ (accessed on 10 March 2017).
46. Executive Office of Energy and Environmental Affairs (EEA). Available online: http://www.mass.gov/eea/ (accessed on 10 March 2017).
47. Green for All Initiative of Dream Crops. Available online: https://www.thedreamcorps.org/our_programs#green (accessed on 10 March 2017).
48. Centre for Energy Efficiency and Renewable Energy (CEERE). Available online: http://www.ceere.org/ (accessed on 7 March 2017).
49. Massachusetts Institute of Technology (MIT). Available online: http://global.mit.edu/collaborations (accessed on 7 March 2017).
50. Mission Innovation. Available online: http://mission-innovation.net/ (accessed on 10 March 2017).
51. Northeast Clean Energy Council (NECEC). Available online: http://www.necec.org/ (accessed on 10 March 2017).
52. Water Innovation Network for Sustainable Small Systems (WINSS). Available online: http://www.umass.edu/winsss/ (accessed on 7 March 2017).
53. World Resources Institute. Available online: http://www.wri.org/ (accessed on 10 March 2017).
54. Massachusetts Clean Energy Center (MassCEC). Available online: http://www.masscec.com/financial-information (accessed on 7 March 2017).
55. Massachusetts Executive Office of Labor and Workforce Development. Available online: https://www.mass.gov/orgs/executive-office-of-labor-and-workforce-development (accessed on 10 March 2017).
56. Carley, S. Historical analysis of U.S. electricity markets: Reassessing carbon lock-in. *Energy Policy* **2011**, *39*, 720–732. [CrossRef]
57. Raven, R.; Schot, J.; Berkhout, F. Space and scale in socio-technical transition. *Environ. Innov. Soc. Transit.* **2012**, *4*, 63–78. [CrossRef]
58. Coenen, L.; Benneworth, P.; Truffer, B. Toward a spatial perspective on sustainability transitions. *Res. Policy* **2012**, *41*, 968–979. [CrossRef]
59. Truffer, B.; Murphy, J.T.; Raven, R. The geography of sustainability transitions: Contours of an emerging theme. *Environ. Innov. Soc. Transit.* **2015**, *17*, 63–77. [CrossRef]
60. Boschma, R.; Coenen, L.; Frenken, K.; Truffer, B. Towards a theory of regional diversification: Combining insights from Evolutionary Economic Geography and Transition Studies. *Reg. Stud.* **2017**, *51*, 31–45. [CrossRef]

energies

MDPI

Article

100% Renewable Energy Grid for Rural Electrification of Remote Areas: A Case Study in Jordan

Loiy Al-Ghussain [1,*], Mohammad Abujubbeh [2], Adnan Darwish Ahmad [3,*], Ahmad M. Abubaker [3], Onur Taylan [4], Murat Fahrioglu [5] and Nelson K. Akafuah [3]

1 Mechanical Engineering Department, University of Kentucky, Lexington, KY 40506, USA
2 Department of Electrical and Computer Engineering, Kansas State University, Manhattan, KS 66506, USA; abujubbeh@k-state.edu
3 Institute of Research for Technology Development (IR4TD), University of Kentucky, Lexington, KY 40506, USA; ahmad.abubaker@uky.edu (A.M.A.); nelson.akafuah@uky.edu (N.K.A.)
4 Mechanical Engineering Program, Middle East Technical University Northern Cyprus Campus, Kalkanli, Guzelyurt via Mersin NCC 99738, Turkey; ontaylan@metu.edu.tr
5 Electrical and Electronics Engineering Program, Middle East Technical University Northern Cyprus Campus, Kalkanli, Guzelyurt via Mersin NCC 99738, Turkey; fmurat@metu.edu.tr
* Correspondence: loiy.al-ghussain@uky.edu (L.A.-G.); adnandarwish@uky.edu (A.D.A.)

Received: 28 August 2020; Accepted: 17 September 2020; Published: 18 September 2020

Abstract: Many developing countries suffer from high energy-import dependency and inadequate electrification of rural areas, which aggravates the poverty problem. In this work, Al-Tafilah in Jordan was considered as a case study, where the technical, economic, and environmental benefits of a decentralized hybrid renewable energy system that can match 100% of the city demand were investigated. A tri-hybrid system of wind, solar, and hydropower was integrated with an energy storage system and optimized to maximize the match between the energy demand and production profiles. The optimization aimed at maximizing the renewable energy system (RES) fraction while keeping the levelized cost of electricity (LCOE) equal to the electricity purchase tariff. The techno-economic analysis showed that the optimal system in Al-Tafilah comprises a 28 MW wind system, 75.4 MW PV, and 1 MW hydropower, with a 259 MWh energy storage system, for which a RES fraction of 99% can be achieved, and 47,160 MtCO_2 are avoided yearly. This study can be easily extended to other rural cities in Jordan, as they have higher renewable energy system (RES) potential. The presented findings are essential not only for Jordan's planning and economy-boosting but also for contributing to the ongoing force against climate change.

Keywords: Energy hybridization; hydropower; wind energy; solar energy; rural areas; 100% Renewable grid

1. Introduction

Energy is the most important ingredient in both developing and developed countries that contributes to the overall economic and social development [1–3]. There is a strong relationship between energy sources availability and economic growth and hence social development. Studies revealed that energy demand is rapidly increasing globally. It is expected to increase by about 65% in the developing countries and approximately about 34% on the global scale by 2040 [4]. This is due to several factors, including fast-paced prosperity, economic expansion, and continuous population growth. However, during the last 25 years, the reality and impact of environmental degradation have become visible as a result of a combination of factors such as increased energy demand and industrialization, which both imposed stress on energy production. Nowadays, conventional energy production systems—Namely fossil fuel-based systems—Are considered to be the primary cause of the increased environmental degradation due to their carbon dioxide (CO_2) emissions [5]. According to Timmerberg et al. [6], the estimated rate of CO_2 emissions in the Middle East region is 0.396–0.682 kg CO_2 kWh^{-1}, which means a considerable amount of CO_2 is released into the atmosphere. However, as they indicated, if the target for 2030 of the renewable energy share is met, the electricity-production CO_2 emissions are expected to drop to 0.341–0.514 kg CO_2 kWh^{-1}. Attaining solutions to such environmental problems that we face today, long-term planning and actions become vital for sustainable development [1]. Also, a large portion of the industry's research is heavily focused on advancements that reduce operational [7,8] or mending energy [9,10]. Besides, more importantly, renewable energy sources play a significant role in mitigating environment-related problems as they are environmentally friendly with affordable and competitive costs relative to conventional energy systems [11–13].

Jordan is one of the developing countries in the Middle East, with more than 97% of its energy being imported. This has a significant impact on the country's economy [14]. Moreover, the energy security of the country is threatened by the fluctuations of fossil fuel prices as well as the disruption of fossil fuel supplies, which happened in 2011 [15]. The most affected zones are usually rural areas, where infrastructure is not quite robust, and the electrical service coverage is weak. According to the economic research service of the United States Department of Agriculture [16], the delineation introduced in 2000 defines countryside with 500 people per square miles or less as rural areas. Developing countries' poverty problem is aggravated by inadequate electrification of these remote sites. In Malaysia, for example, around 809 schools do not have a 24-hour electricity supply [17]. Modern energy forms are considered an economic good that can lift up the life of billions worldwide, predominantly in developing countries that lack service availability [18,19]. That said, from an economic standpoint, grid transmission through the impenetrable terrains and dense forests to supply a small village is not viable. Therefore, because of the high distribution cost and accompanying transmission losses, serving rural areas become unfeasible [17]. To avoid these high costs accompanying transmission via national grids, decentralized energy systems can be used [18]. Small-scale hydropower, wind, photovoltaic (PV), and diesel-engine generator are amongst the common off-grid electrification systems used in the developing regions; for instance in, Africa, the Caribbean, Latin America, and Asia [18]; nonetheless, most of the population in rural areas (>2.4 billion individuals worldwide), still rely on traditional fuels for cooking.

As discussed in [18], consideration of renewable energy systems (RES) as part of the remedy for rural electrification is of particular importance. This is because most of the previous attempts for conventional electrification of remote areas in developing countries were not successful. If the right policy is adopted, sustainable-development technology comprises an effective means for reducing energy poverty [17]. Solar, wind, and hydro technology are feasible and reliable energy sources. They can be used to generate off-grid electricity and assist in the rural-electrification capacity-expansion; nevertheless, the technical-, installation-, and social issues should be accounted for to ensure the project success [17]. Additionally, the importance of utilizing the Internet of Things (IoT) paradigm when integrating renewable energy sources into smart grids was discussed in the literature [20,21]. This provides a powerful tool for simultaneous monitoring and control of the resources. For instance, [22] provides a novel framework for state estimation in wind-turbine integrated grids.

Along with conventional Supervisory Control and Data Acquisition (SCADA) measurement methods, the IoT provides information about the resources in very short time frames, which can be utilized for control or monitoring purposes. Other literature investigated the use of IoT in energy-aware residential buildings [23]. Here, the IoT paradigm helps not only in appliance automation but also in monitoring household energy demand and remote control of appliances during peak and off-peak hours, which has tremendous economic value on the monthly rates of the household.

The shortcoming of remote areas being far from the capital cities can be utilized for our advantage, as it offers a large area of uninhabited land that can be used to build renewable systems. The government in Jordan should focus on maximizing the utilization of the available resources to decrease the dependency on fossil fuel. Jordan has a great potential of RES that can cover the country's energy needs if implemented wisely, especially wind and solar resources [14,15,24–26]. However, solar and wind energies fluctuate with time. They do not match the demand profile, which forms a challenging issue related to energy security, power grid quality, and reliability of such systems [27]. The hybridization of solar and wind resources increases the reliability of the system and enhances the matching between demand and supply. Few studies investigated the potential of RES in Jordan. For instance, Essalaimeh et al. [28] found that the electricity generated by a combination of solar and wind energy grid-connected systems can be utilized for various types of applications, including space heating and cooling. Halasa and Asumadu [29] demonstrated the feasibility of having large scale wind and solar hybrid power systems in Jordan, where wind energy was placed in the mountains of Northern Jordan and Solar energy in the Eastern Desert. The installation costs were approximated to be US$290 million for the 100 – 150 MW wind farm and US$560 million for the 100 MW solar. Another study [30] proposed an off-grid hybrid solar PV and wind energy systems in two locations in Jordan, Alhasan industrial estate, and Alayn Albayda, taking economic, reliability, and sustainability measures into consideration.

Furthermore, a grid-connected hybrid system consisting of solar PV and wind turbines was proposed in [31]. The system was investigated at four different locations in Jordan, and results showed that it could power a small village in Jordan. However, in the latter study, no-optimization of the system's size or economic feasibility was performed. Also, the mismatch between the demand and supply was not considered. The kingdom is rich in solar and wind resources, and also has a considerable potential of hydropower in some regions that can be utilized to supply a significant part of the baseload demand [32]. As such, a tri-hybridization of solar PV, wind, and hydropower forms an effective and environmentally friendly solution that ensures energy security and, at the same time, contributes to the mitigation of greenhouse gases (GHGs) [33,34].

Al-Tafilah governorate, located about 112 miles to the southwest of Amman, the capital of Jordan, spans an area of 853 mi^2. The population of the governorate is estimated to 106,000 people, which is about 1% of Jordan's population. This means a population density of 125 people/mi^2 compared to 6827 people/mi^2 in the capital city, Amman [35,36], with a total annual electricity demand of 137 GWh. As presented in Kiwan et al. [37], the transition towards a 100% renewable energy grid in Jordan shall be gradual, and about 1/3 of the 2050 total capacity should be installed by 2030. This transition will help Jordan overcome its 94% energy-import dependency. Since Jordan is amongst the signatories to the Paris agreement, it will also help the country adhere to the set emissions limit by reducing its carbon dioxide production associated with electricity generation. In this work, the techno-economic feasibility of installing PV/wind-hydropower systems to supply and match 100% of the electrical demand in Al-Tafilah forecasted for the year 2030 was investigated. Multiple scenarios were considered, including the integration of an energy storage system. Sensitivity analysis was conducted to examine the effect of resources and demand variation, RES costs, and annual discount rate on the techno/economic performance of the system. It's also worth noting that the expansion of this work to other similar areas in Jordan is possible, as Kiwan et al. [37] showed other rural cities, like AlMafraq and Maan, represent sites of highest hybrid RES potential. Hence, Al-Tafilah signifies the more challenging case towards a 100% renewable energy grid and is therefore considered here. As such, the presented study will not only set forth a decentralized system that will suffice the city of its energy needs while mitigating the issues mentioned above of centralized grids. It will also open the doors for consideration of rural renewable decentralized systems as part of the 2030 projection for the 100% renewable transition presented by Kiwan et al.

2. System Description and Methodology

2.1. RES Description

2.1.1. Solar Energy System

The ambient conditions of the location where the PV plant is installed affect the efficiency of the PV module and hence the energy production. As the module temperature increase, the PV efficiency decreases, and so does the amount of produced energy. By neglecting the effects of relative humidity and wind speeds and considering the effect of the ambient temperature, the efficiency of the PV module can be estimated using Equation (A1) shown in Appendix A.

The specifications of the PV modules are vital for the estimation of the energy output from the PV system. In this study, a Canadian Solar company type CS6K-285M modules were used. Obtaining the global insolation on the PV module in addition to the module efficiency is necessary for the estimation of the energy produced. The estimation of the global insolation was achieved using the methodology presented by Duffie and Beckman [38], which was not included here for brevity. The hourly energy generated from the PV power plant, E_p, can be calculated, as shown in A.2. The hourly solar resources, as well as the hourly ambient temperature for Al-Tafilah, were obtained using Meteonorm software, which provides the data based on Typical Metrological Year (TMY).

2.1.2. Wind Energy System

The wind speed at hub height varies depending on several factors, such as the wind speed at ground level, the hub height, time (hour, day, season), the nature of the terrain, and the ambient temperature. These variables can be represented by one factor known as the wind profile exponent or wind shear coefficient (α). In the absence of the site-specific data, can be taken as 1/7 [39,40]. The wind speed at hub height (u_Z) can be extrapolated using equation A.3. By assuming that the energy generated is constant during the hour and each turbine generates the same amount in the case of having multiple turbines, the total hourly electrical energy generated by a wind turbine(s), E_w, can be estimated by A.4 in App. An under wind energy model section. The Weibull distribution shape parameter of the available wind speed, K can be calculated based on Justus theory using Equation (A5). A wind turbine with 2 MW of rated power from GAMESA company (G114-2.0) was used in this study. It should be noted the hourly TMY wind speeds at ground level were obtained from Meteonorm software.

2.1.3. Hydropower System

Al-Tafilah has the potential to construct two small hydropower systems with a total capacity (P_h) of 1 MW, as reported in [32]. In this study, a 1 MW hydropower system with capacity factor (CF_h) of 80% [41] is designed to provide continuous energy generation as part of the baseload of Al-Tafilah, where the annual energy production from the hydropower system, E_h, can be estimated as shown in A.6.

2.2. *System Modelling and Energy Flow: With and Without an Energy Storage System*

Energy Storage System is introduced into the power system as a solution for the intermittent behavior of renewable energy resources. This way, is it possible to reach a match between supply and demand. In this work, two scenarios consisting of and without an Energy Storage System (ESS) are adopted and discussed for the fitness of a hybrid off-grid system to supply the electricity demand of Al-Tafilah. Flow chart for with and without ESS are introduced in Figure 1. In comparison with other batteries for the ESS, this study considered Zinc-Bromine batteries (ZBB) as they show better performance and lower cost. In this study, the round-trip efficiency of the ZBB was taken as 72% and the depth of discharge was considered as 80% as reported in [42,43].

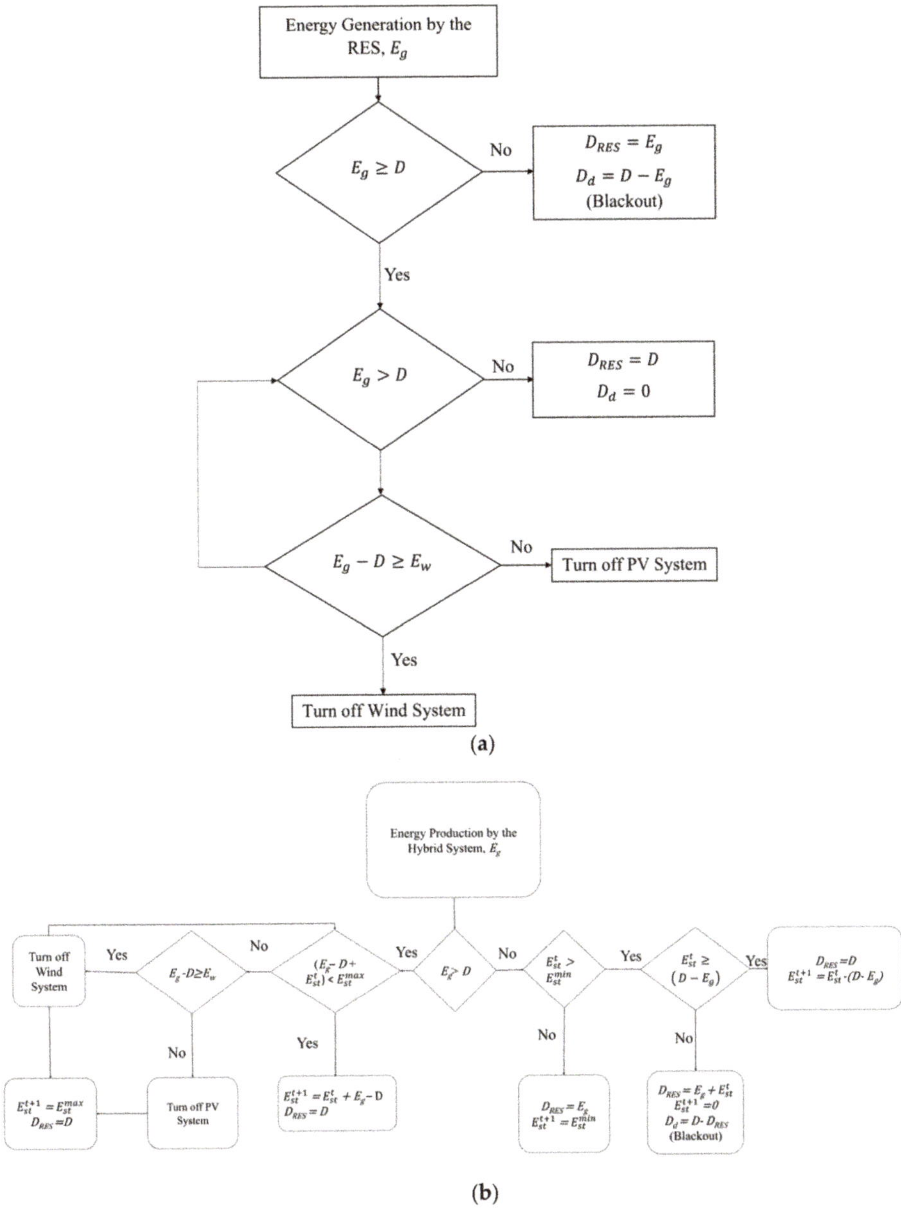

Figure 1. Energy flowchart of the photovoltaic (PV)/wind/hydro hybrid system: (**a**) Without an energy storage system (ESS) and (**b**) with ESS.

To inspect the matching between the energy production from the hybrid system and the demand, the RES Fraction (F_H), which represents the annual fraction of demand met by the RES, was used. F_H can be calculated by A.7. The forecasted demand of 2030 was used in this study, where the average hourly demand of Al-Tafilah in 2010 obtained from [44] was used for estimating the 2030 demand. The population of 2030 was forecasted using the Piecewise Cubic Hermite Interpolating Polynomial method, where 1994, 2004, 2015, and 2019 populations were used for this forecast [35,36].

2.3. *Optimization Procedure*

As the hybrid system operating variables have a nonlinear relation, it is required to find a nonlinear solving algorithm to catch the optimal design. Therefore, the Generalized Reduced Gradient (GRG) algorithm was used. This algorithm can figure out the capacities of the hybrid systems. The adoption of the optimization procedure is depicted in Figure 2. Moreover, the optimization objective, constraints, and decision variables are given in Table 1.

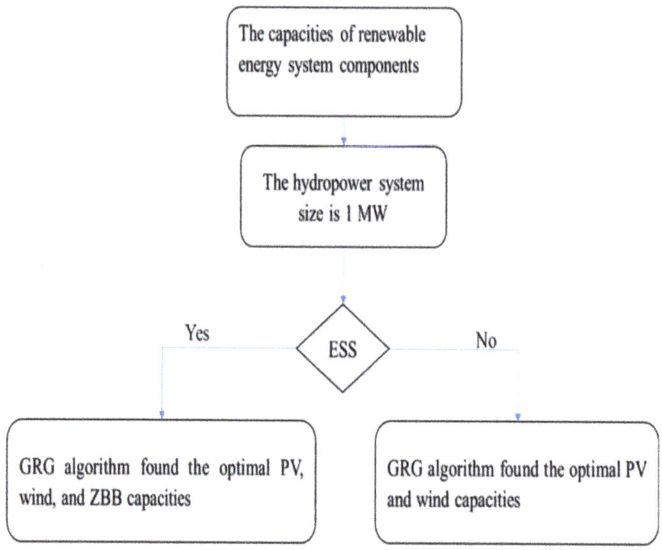

Figure 2. The optimization procedure of the hybrid system components.

Table 1. The components of the optimization problem.

Decision Variables		Objective Functions	Constraints
Without ESS	**With ZBB**		
PV and wind capacities	PV, wind and ZBB capacities	Maximizing F_R	$LCOE \leq 0.12$

2.4. *Economic Assessment and System's Feasibility*

The economic feasibility of the energy system is vital as the technical feasibility wherein most of the cases, the choice of decision-makers is based on the economic feasibility of the project. The levelized cost of electricity (*LCOE*) is one of the economic parameters that is used to determine the economic feasibility of the energy systems. *LCOE* for the hybrid system can be calculated using Equation (A8) to incorporate the effect of the mismatching between the energy generation and the demand, the demand met by the hybrid system was used instead of the energy generation [45,46]. Table 2 shows all the economic parameters used in this study.

Table 2. The economic parameters of the PV, the wind, and the hydropower systems in addition to the electricity purchase tariff and the annual discount rate for Al-Tafilah, Jordan.

Parameter	Value	Reference
PV system capital cost ($/kW)	1533	[47]
Wind system capital cost ($/kW)	1516	[47]
Hydropower capital cost ($/kW)	3000	[41]
Zinc-Bromine capital cost ($/kW)	195	[48]
PV system maintenance cost ($/kW)	24.68	[49]
Wind system maintenance cost ($/kW)	39.53	[50]
Hydro power maintenance cost ($/kW)	75	[41]
System expected lifetime (Years)	25	[41,48,51]
Electricity purchase tariff ($/MWh)	120	[52]
Annual discount rate (%)	5	[37]

3. Results and Discussion

Al-Tafilah has a hydropower potential of 1 MW and a high potential of solar and wind resources that can achieve both economic and technical feasibilities. The RES sizing in this study was made based on maximizing the RES fraction and constraining the *LCOE* to the electricity purchase tariff in Al-Tafilah. This will maximize the environmental benefits of the system by meeting the highest demand possible, at the same time guaranteeing the economic feasibility of the system. The technical and economic parameters of the optimal sized PV/wind-hydro without ESS are shown in Table 3. Without an energy storage system, the tri-hybrid system comprises a 29.4 MW PV system and a 56 MW wind system. This system can supply the demand with a RES fraction equal to 71.5%, as seen in Table 3. In the case where no constraints were forced on the *LCOE*, the RES fraction reached a value of 87.9%; however, the system was very large and not feasible, hence, unrealistic for consideration.

Table 3. Technical and economic performance parameters of the PV/wind-hydro system without ESS.

Parameter	Value
PV Capacity (MW)	29.37
Wind Capacity (MW)	56
Hydropower Capacity (MW)	1
Capacity Factor (%)	26
RES Fraction (%)	71.5
LCOE ($/kWh)	0.12
NPV (M$)	167.53
PBP (Years)	6.235

Figure 3 shows the demand-met energy profile of the hybrid system components and how their summation compares with the demand profile. It is clearly seen from the superposition of the demand met by the system components that these RES can work synergistically as a hybrid system, as the hybridization mode could achieve higher RES fractions. However, although the amount of energy generated is enough, the profiles of the energy generation and energy demand do not match; a result of the fluctuation of the solar and wind resources and the limited hydropower capacity of the system. Consequently, the system is not capable of meeting the demand at 100%. This is seen in Figure 4, where the mismatch is represented by the deficit portion of the bar chart. Figure 4 shows that although the quantity of energy can be more than enough, the mismatch between the supply and demand profiles could still cause energy deficiency. Therefore, when a RES is to be assessed, not only energy quantity should be taken into account, but rather the time it is needed in as well. Since 100% could not be reached solely by the RES, an ESS must be integrated into the system to supply the deficit.

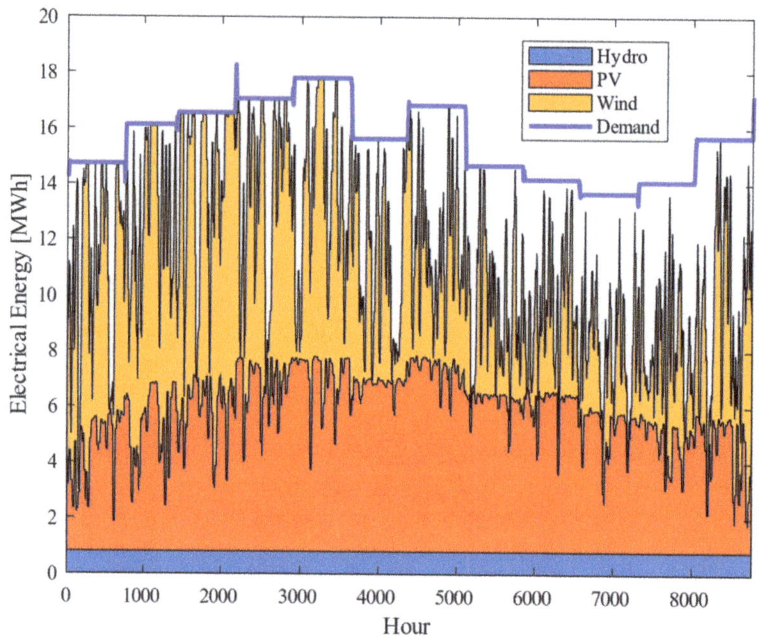

Figure 3. Demand met by the renewable energy system (RES) components without ESS (levelized cost 4 of electricity (LCOE) of 0.12 $/kWh).

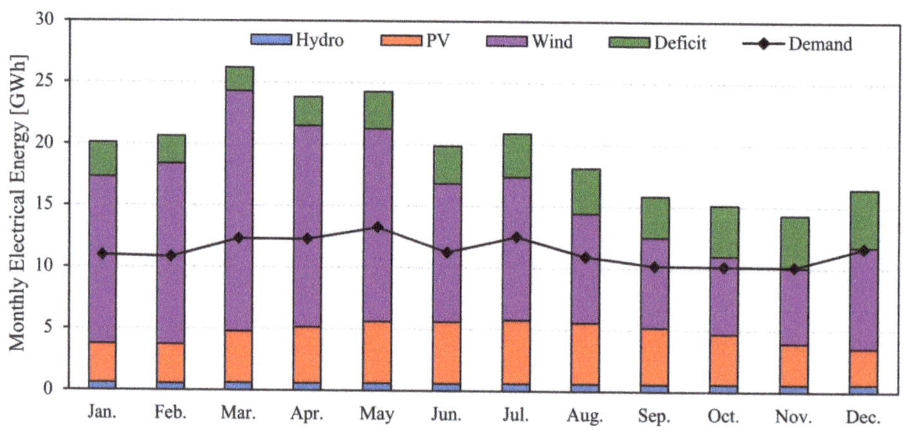

Figure 4. Monthly energy generation by the RES components without ESS.

To ensure Al-Tafilah could base its decentralized grid on the hybrid RES proposed, 100% of their demand must be met and matched. Therefore, an ESS that can supply the deficit must be integrated. The technical and economic parameters of the optimal sized PV/wind-hydro with ESS are shown in Table 4. The table examines two scenarios for the ESS-integrated system; a constrained and unconstrained LCOE. In the first scenario, the LCCOE was again fixed to the electricity purchase tariff at Al-Tafilah. Results showed that the system would comprise of 75.4 MW PV system and 28 MW wind system. It worth noting that the size of the PV system is larger than in the case where no ESS was integrated, while the size of the wind system decreased. The size increase could be attributed to the utilization of more energy to supply the storage system as well as the demand in the time when solar energy is available. The result is a larger system, but with the great advantage of meeting the demand at 98.8%, RES fraction compared to only 71.5% previous value. If RES size was optimized to maximize the RES without any constraints on the LCOE, the system could meet 99.9% of the demand, with an acceptable LCOE of 0.165 $/kWh compared with the current value of 0.12 $/kWh. It is worth mentioning that the presented calculation of the LCOE was based on current prices of the RES technologies, which means these values will be less in the future as the prices of these technologies are expected to drop significantly in the next decades [53]. Figure 5 shows the demand-met energy profile of the hybrid system components when integrated with an ESS, and how their summation compares with the demand profile. The difference between this and Figure 3 resembles the power of the integration of the ESS in solving the mismatch problem. It is apparent in the figure that the demand-met and the actual demand curves coincide almost all the time. This is further clarified by Figure 6, with a significant reduction in the deficit portion when compared to Figure 4 that had no ESS backup. As explained earlier, there is still a 1% deficit due to the constraint forced on the LCOE. However, as this constraint is removed, Figure 7 shows that the demand and demand-met profiles coincide at all times, and no deficit portion is seen, which is further confirmed in Figure 8; i.e., the demand was met at 100%, but with a slightly higher LCOE compared to the current value.

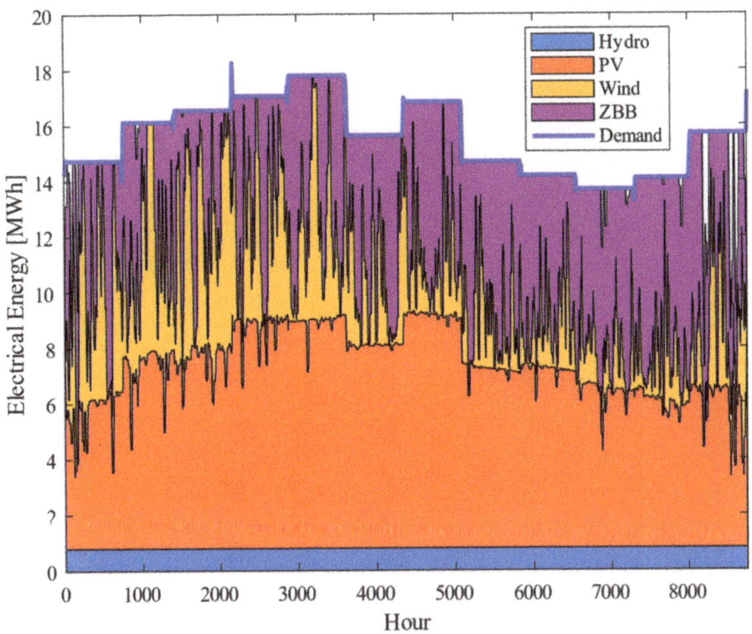

Figure 5. Demand met by the RES components with ESS (LCOE of 0.12 $/kWh).

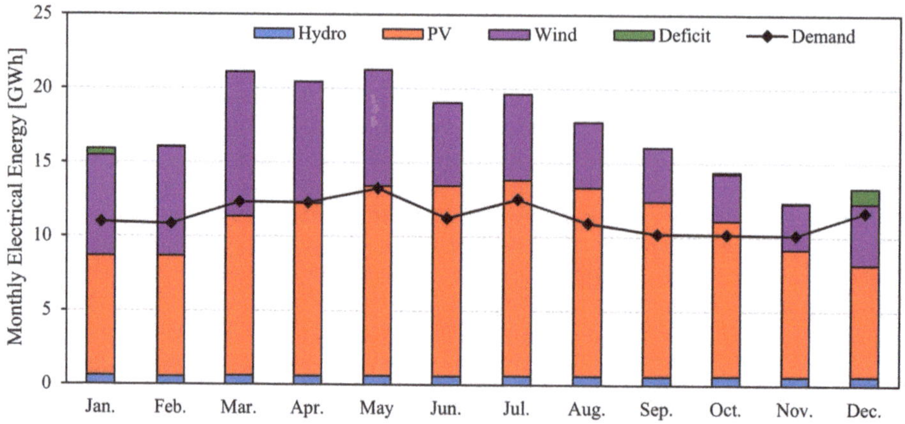

Figure 6. Monthly energy generation by the RES components with ESS (LCOE of 0.12 $/kWh).

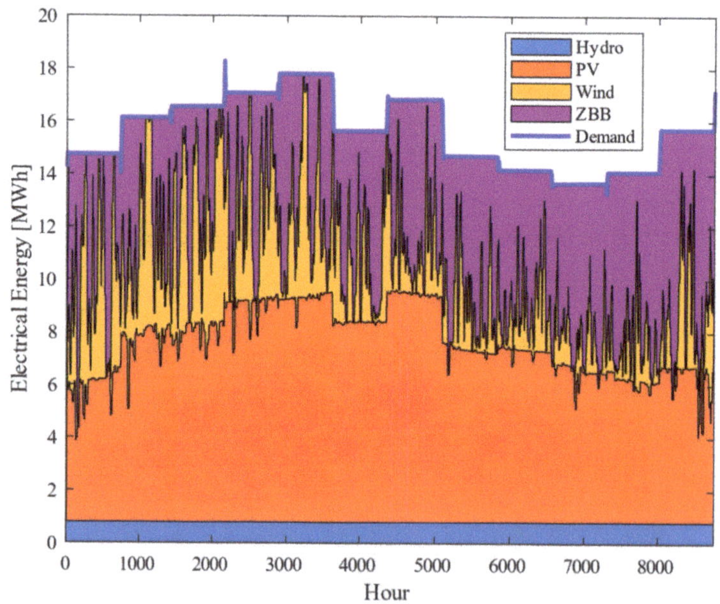

Figure 7. Demand met by the RES components with ESS (LCOE of 0.165 $/kWh).

Table 4. Technical and economic performance parameters of the PV/wind-hydro system with ESS.

Parameter	Value	
	Constrained	**Unconstrained**
PV Capacity (MW)	75.4	116.39
Wind Capacity (MW)	28	26
Hydropower Capacity (MW)	1	1
ZBB (GWh)	0.259	0.415
Capacity Factor (%)	22.48	21.55
RES Fraction (%)	98.79	99.93
LCOE ($/kWh)	0.12	0.165
NPV (M$)	119.55	140.45
PBP (Years)	9	9.62

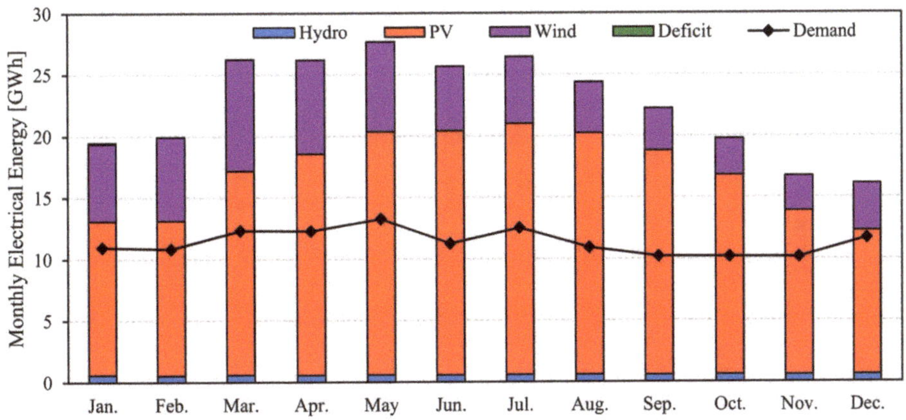

Figure 8. Monthly energy generation by the RES components with ESS (LCOE of 0.165 $/kWh).

The RES system with an ESS not only supply Al-Tafilah with decentralized electricity and help Jordan take a step further towards reducing energy-import dependency, but it also comes with economic and environmental benefits. In Jordan, power plants operated on combined cycles and fired with natural gas are used to supply part of the demand. When the proposed system provides the energy demand, the annual saving of fuel from a typical Jordanian plant, for example, the Al-Qatrana power plant, is equal to 20,365 tons-saved, which is equivalent to a reduction of 47,160 MtCO_2 [3,54]. The fuel-saving and reduction in CO_2 calculations were based on a previously developed code for the simulation of the power plant presented in those latter references, where detailed analysis of fuel consumption and CO_2 emission calculations for the power plant can be found. Based on estimations by the United States Environmental Protection Agency, this CO_2 reduction is equivalent to the carbon sequestered annually by US forest spreading an area of 61,589 acres. Therefore, the presented system also progresses Jordan's adherence to the greenhouse gas limit set by the Paris agreement. For a developing country like Jordan, a transition towards 100% is crucial, as a study by Mathiesen et al. [55] associated such transition with large earnings on export potential, creating jobs, and economic growth. So, this work, which demonstrated a step further towards a 100% renewable energy grid, will support a more robust economy, at the same time, a greener Jordan.

The fluctuations in the renewable energy resources (solar and wind resources), as well as the variation in the electrical demand, significantly affect the performance of RESs where the amount of variations depends on the RES configurations and the existence of ESS, as shown in Figure 9. It can be depicted from Figure 9 that the RES fraction of the hybrid PV/wind/hydro system without ESS is the most sensitive one to the variation in the resources and the electrical demand, which is expected because the ESS acts as a backup system to cover the deficit caused by the drop in the resources or the increase in the demand. Moreover, it can be seen that the increase in the resources and the decrease in the demand do not change the RES fraction of the hybrid system with ESS (specifically for the unconstraint scenario) since they reached almost 100% RES fraction.

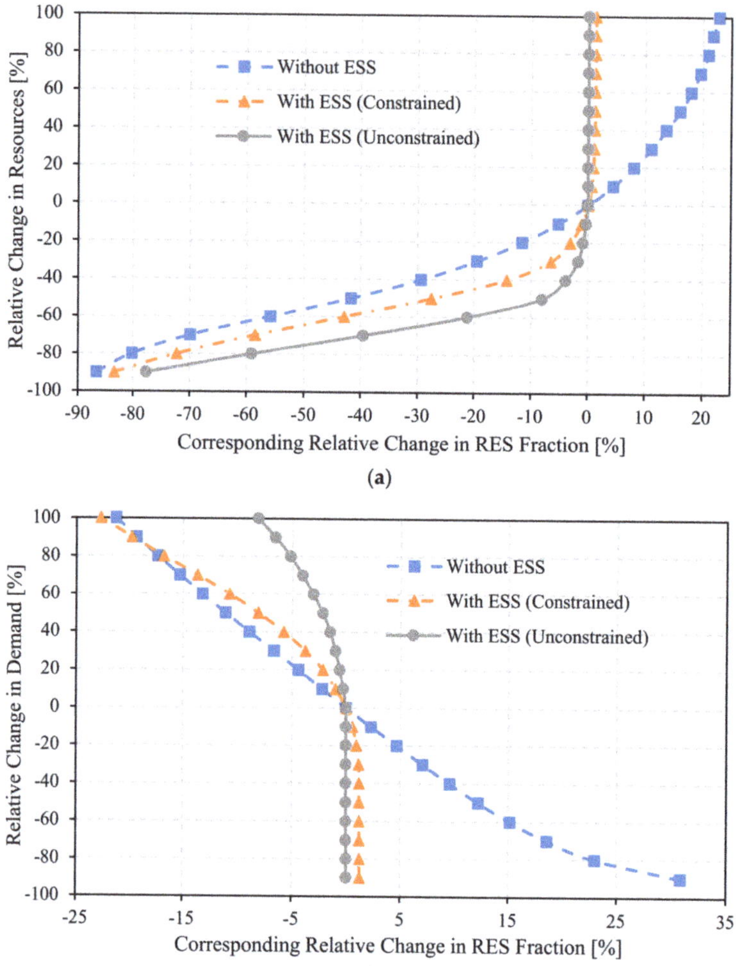

Figure 9. The sensitivity of the RES fraction to the variations in the: (**a**) Solar and wind resources and (**b**) demand.

RES fraction Likewise, the LCOE of the system is sensitive to the variations in the renewable energy resources and the demand and also to the variations in the RES's costs (capital and maintain ace costs) as well as the annual discount rate where the largest variations in the LCOE is caused by the change in the resources and demand as shown in Figure 10 since the amount of energy met by the RES is controlling the LCOE which is in return affected by the change in the resources and the demand. It can be depicted from Figure 10c that all the LCOE of the hybrid system configurations have almost the same sensitivity to the variation in the RES costs. At the same time, it can be seen in Figure 10d that the LCOE of the system without ESS is less sensitive to the variation in the annual discount rate, unlike the systems with ESS.

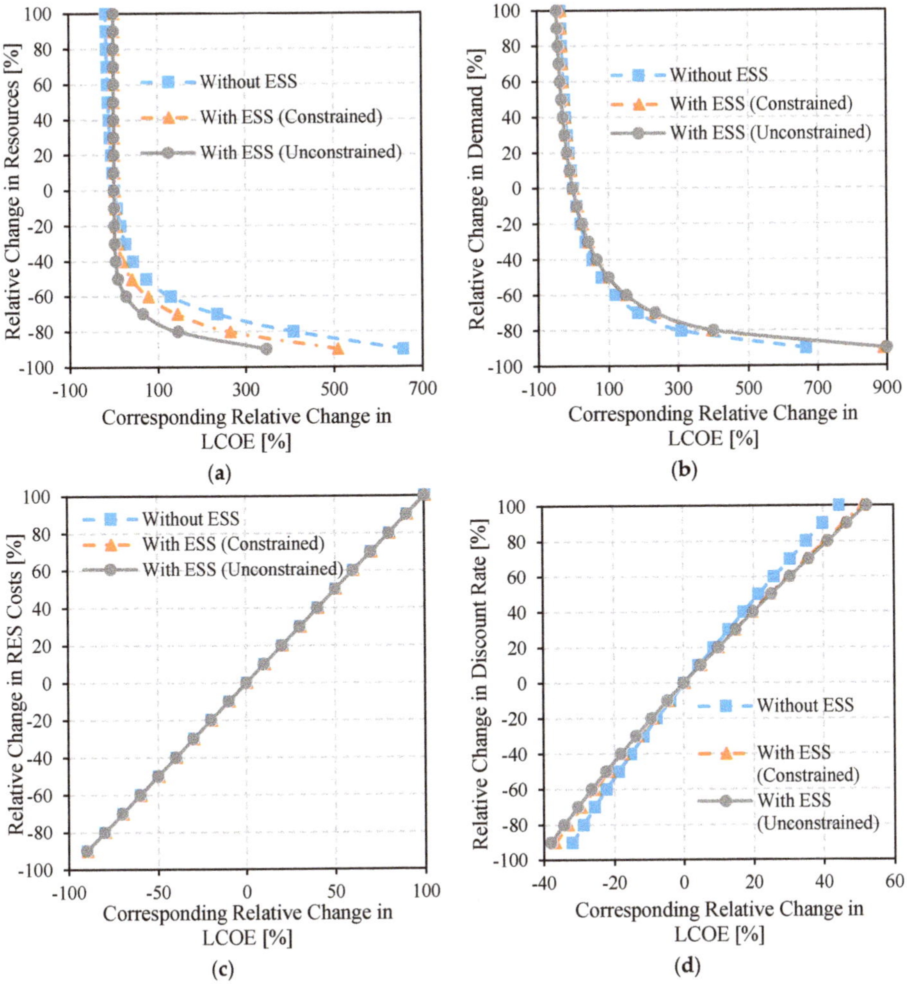

Figure 10. The sensitivity of the LCOE to the variations in the: (**a**) Solar and wind resources, (**b**) demand, (**c**) RES costs, and (**d**) annual discount rate.

4. Conclusions

Many developing countries across the globe rely heavily on imported sources to cover their energy needs. As such, the development of a reliable yet cost-efficient means of energy production is vital for advancing the movement toward sustainable societies and fastening the economic growth in the desired country. Jordan, with no exception, relies on imported oil from neighboring countries to cover their domestic demand. Additionally, the lack of services and poor electrification of rural areas in the kingdom worsens the poverty problem. In this work, Al-Tafilah of Jordan was presented as a case study, and the technical, economic, and environmental benefits of a decentralized renewable energy system that can supply and match 100% of the city energy demand were investigated. Other rural cities in Jordan have higher RES potential; therefore, this study can be expanded to other parts of Jordan. A tri-hybrid system of wind, solar, and hydropower, and an energy storage system were modeled and optimized to maximize the matching between the energy demand and production profiles. Three main scenarios were considered, first a tri-hybrid RES without an energy storage system, with a constraint on LCOE (a maximum of LCOE equal to the electricity purchase tariff was set). The second and third scenarios were investigations of the RES system integrated with an ESS, with and without constraints on the LCOE. Results showed that without an ESS, the hybrid system could only reach up to 71.5% RES fraction. However, the techno-economic analysis of the PV/wind-hydropower system with an ESS showed that the optimal system in Al-Tafilah comprising a 28 MW wind turbine, 75.4 MW PV, 1 MW hydropower, and a 259 MWh energy storage facility could achieve a 99% RES fraction. It offers an attractive LCOE of 0.12 $/kWh (equal to the purchase tariff) and a payback period of 9 years. Results also indicated that with the installation of this system, an equivalent of 47,160 $MtCO_2$ emissions could be avoided yearly, which demonstrates the environmental benefits of the proposed work. Therefore, these finding are essential not only for future renewable energy planning in the country and improving its economy but also for contributing to the ongoing force against climate change. When the constraint on the LCOE was removed, the RES fraction achieved was equal to 100%, with a slightly higher LCOE of 0.165 $/kWh. Since the prices of RES technologies are expected to drop dramatically in the next decades, the last scenario can also be adopted as the LCOE drops significantly with the RES price reduction. Finally, sensitivity analysis showed that the RES fraction of the hybrid PV/wind/hydro system without ESS is the most sensitive configuration to the variation in the resources and electrical demand. In contrast, the LCOE of the three configurations showed the largest sensitivity to the variation in the resources and demand compared to its sensitivity to the RES costs and annual discount rate variations.

Author Contributions: Conceptualization, L.A.-G., A.D.A. and M.A.; methodology, L.A.-G.; software, L.A.-G.; formal analysis, L.A.-G. and A.D.A.; investigation, L.A.-G., A.D.A. and A.M.A.; data curation, L.A.-G. and A.D.A.; writing—Original draft preparation, L.A.-G., A.D.A., A.M.A. and M.A.; writing—Review and editing, L.A.-G., A.D.A., M.A. and N.K.A.; supervision, O.T., M.F. and N.K.A.; project administration, O.T., M.F. and N.K.A.; funding acquisition, N.K.A. All authors have read and agreed to the published version of the manuscript.

Funding: This research was funded internally by the Institute of Research for Technology Development, University of Kentucky.

Acknowledgments: The authors would like to thank Prof. Derek Baker from Middle East Technical University for providing the TMY data of Al-Tafilah.

Conflicts of Interest: The authors declare no conflict of interest.

Nomenclature

- E_h Annual energy production from the hydropower system, kWh
- E_P Hourly energy generated from the PV power plant, kWh
- E_w generated by a wind turbine(s), kWh
- E_{st}^t Energy stored at time t, kWh
- E_{st}^{max} Battery capacity, kWh
- F_H RES Fraction, %
- I_R Reference insolation at nominal conditions, Wh/m^2
- I_T Global insolation on a tilted surface, Wh/m^2
- K Shape parameter of the Weibull distribution
- L Lifetime of the system, years
- *LCOE* Levelized Cost of Electricity, USD/kWh
- M_t Yearly fixed maintenance cost of the RES, USD
- N_m Number of modules in the PV power plant
- NOCT Nominal operating cell temperature, °C
- P_h Hydropower capacity, kW
- P_R Rated electrical power of the wind turbine, kW
- r Annual discount rate, %
- T_a Ambient temperature, °C
- $T_{R,NOCT}$ Reference module temperature at nominal conditions, °C
- $T_{Ref,STC}$ Reference module temperature at standard conditions, °C
- u_C Cut-in wind speed of the wind turbine, m/s
- u_F Cut-out wind speed, m/s
- u_g Average wind speed at ground level, m/s
- u_R Rated wind speed, m/s
- u_Z Wind speed at hub height, m/s
- $\bar{u}$ Mean wind speed at hub height, m/s
- Z Hub height, m
- Z_g Height of the ground level, m

Acronyms and Abbreviations

- ESS Energy storage system
- GHGs Greenhouse gases
- GRG Generalized reduced gradient
- LCOE Levelized cost of electricity
- NPV Net present value
- PBP Payback period
- PV Photovoltaic
- RES Renewable energy system
- ZBB Zinc-Bromine battery

Greek letters

- α Wind shear coefficient
- β_R Temperature coefficient, 1/°C
- η_l System losses, %
- $\eta_{PV,R}$ Reference efficiency of the PV module, %
- η_{PV} PV module efficiency
- σ Standard deviation of the wind speeds sample, m/s

Appendix A

Table A1. System Modelling Equations.

Quantity	Equation	Key Points	Equation Number
	PV Energy model		
The PV module efficiency	$\eta_{PV} = \eta_{PV,\,R} \times \left[1 - \beta_R \times \left(T_a + \left(NOCT - T_{R,NOCT}\right) \times \frac{I_T}{I_R} - T_{Ref,STC}\right)\right]$	Neglecting the effects of relative humidity and wind speeds, while considering the effect of the ambient temperature [56,57].	(A1)
The hourly energy generated from the PV power plant	$E_P = \eta_{PV} \times I_T \times A_m \times N_m \times \eta_l$	η_l was taken 0.85 based on [58,59]	(A2)
	Wind Energy model		
The wind speed at hub height	$u_Z = u_g \times \left(\frac{Z}{Z_g}\right)^{\alpha}$	α can be taken as 1/7 [39], Z_g is the height of the ground level [m] at which speed is measured and it is equal 10 m.	(A3)
total hourly electrical energy generated by a wind turbine(s)	$E_w = \begin{cases} 0 & , u_Z < u_C \; or \; u_Z > U_F \\ N \times P_R \times \frac{(u_C)^K - (u_Z)^K}{(u_C)^K - (u_R)^K} & , u_c \leq u_Z \leq u_R \\ N \times P_R & , u_R < u_Z \leq u_F \end{cases}$	By [39]	(A4)
Shape Parameter	$K = \{(\sigma / \; u \;)^{-1.086}$	$1 \leq K \leq 10$	(A5)
	Hydropower Model		
Annual energy production from the hydropower system	$E_h = P_h \times CF_h \times 24 \times 365$	(CF_h) of 80% [41]	(A6)
	Performance assessment of RES		
RES Fraction	$F_H = \frac{D_{RES}}{D}$	D is the hourly demand of Al-Tafilah [kWh] which was obtained from [44]	(A7)
	Economic Assessment of the RES		
The Levelized Cost of electricity	$LCOE = \frac{C_i + \sum_{t=1}^{L} \frac{M_t}{(1+r)^t}}{\sum_{t=1}^{L} \frac{D_{RES}}{(1+r)^t}}$		(A8)

References

1. Jaber, J.; Badran, O.O.; Abu-Shikhah, N. Sustainable energy and environmental impact: Role of renewables as clean and secure source of energy for the 21st century in Jordan. *Clean Technol. Environ. Policy* **2004**, *6*, 174–186. [CrossRef]
2. Hrayshat, E.S. Analysis of renewable energy situation in Jordan. *Renew. Sustain. Energy Rev.* **2007**, *11*, 1873–1887. [CrossRef]
3. Ahmad, A.D.; Abubaker, A.M.; Najjar, Y.S.; Manaserh, Y.M.A. Power boosting of a combined cycle power plant in Jordan: An integration of hybrid inlet cooling & solar systems. *Energy Convers. Manag.* **2020**, *214*, 112894. [CrossRef]
4. Bortolini, M.; Gamberi, M.; Graziani, A. Technical and economic design of photovoltaic and battery energy storage system. *Energy Convers. Manag.* **2014**, *86*, 81–92. [CrossRef]
5. Al-Ghussain, L. Global warming: Review on driving forces and mitigation. *Environ. Prog. Sustain. Energy* **2019**, *38*, 13–21. [CrossRef]
6. Timmerberg, S.; Sanna, A.; Kaltschmitt, M.; Finkbeiner, M. Renewable electricity targets in selected MENA countries—Assessment of available resources, generation costs and GHG emissions. *Energy Rep.* **2019**, *5*, 1470–1487. [CrossRef]
7. Abubaker, A.M.; Magableh, M.N.A.; Darwish Ahmad, A.; Najjar, Y.S.H. Efficiency boosting and steam saving for a steam-injected gas turbine engine: An optimization study of the running conditions. *J. Energy Eng.* [CrossRef]
8. Ahmad, A.D.; Abubaker, A.M.; Salaimeh, A.A.; Akafuah, N.K. Schlieren visualization of shaping air during operation of an electrostatic rotary bell sprayer: Impact of shaping air on droplet atomization and transport. *Coatings* **2018**, *8*, 279. [CrossRef]
9. Ahmad, A.D.; Singh, B.B.; Doerre, M.; Abubaker, A.M.; Arabghahestani, M.; Salaimeh, A.A.; Akafuah, N.K. Spatial positioning and operating parameters of a rotary bell sprayer: 3D mapping of droplet size distributions. *Fluids* **2019**, *4*, 165. [CrossRef]
10. Sharadga, H.; Dawahdeh, A.; Al-Nimr, M.A. A hybrid PV/T and Kalina cycle for power generation. *Int. J. Energy Res.* **2018**, *42*, 4817–4829. [CrossRef]
11. Harris, C.B.; Meyers, J.P.; Webber, M.E. A unit commitment study of the application of energy storage toward the integration of renewable generation. *J. Renew. Sustain. Energy* **2012**, *4*, 013120. [CrossRef]
12. Al-Ghussain, L.; Samu, R.; Taylan, O.; Fahrioglu, M. Sizing renewable energy systems with energy storage systems in microgrids for maximum cost-efficient utilization of renewable energy resources. *Sustain. Cities Soc.* **2020**, *55*, 102059. [CrossRef]
13. Najjar, Y.S.; Abubaker, A.M. Thermoeconomic analysis and optimization of a novel inlet air cooling system with gas turbine engines using cascaded waste-heat recovery. *Energy* **2017**, *128*, 421–434. [CrossRef]
14. Al-Ghussain, L.; Taylan, O.; Fahrioglu, M. Sizing of a photovoltaic-wind-oil shale hybrid system: Case analysis in Jordan. *J. Sol. Energy Eng.* **2018**, *140*, 011002. [CrossRef]
15. Anagreh, Y.N.; Bataineh, A.S. Renewable energy potential assessment in Jordan. *Renew. Sustain. Energy Rev.* **2011**, *15*, 2232–2239. [CrossRef]
16. Cromartie, J. What is Rural? Available online: https://www.ers.usda.gov/topics/rural-economy-population/rural-classifications/what-is-rural/#:~{}:text=According%20to%20this%20system%2C%20rural,with%20fewer%20than%202%2C500%20people (accessed on 17 September 2020).
17. Azad, H.B.; Mekhilef, S.; Saidur, R.; Boroumandjazi, G. Potential application of renewable energy for rural electrification in Malaysia. *Renew. Energy* **2013**, *59*, 210–219. [CrossRef]
18. Alazraque-Cherni, J. Renewable energy for rural sustainability in developing countries. *Bull. Sci. Technol. Soc.* **2008**, *28*, 105–114. [CrossRef]
19. Anderson, D. Energy and economic prosperity. In *World Energy Assessment: Energy and the Challenges of Sustainability*; Goldemberg, J., Ed.; United Nations Development Programme: New York, NY, USA, 2000; pp. 393–414.
20. Abujubbeh, M.; Al-Turjman, F.; Fahrioglu, M. Software-defined wireless sensor networks in smart grids: An overview. *Sustain. Cities Soc.* **2019**, *51*, 101754. [CrossRef]
21. Al-Turjman, F.; Abujubbeh, M. IoT-enabled smart grid via SM: An overview. *Futur. Gener. Comput. Syst.* **2019**, *96*, 579–590. [CrossRef]

22. Noor-A-Rahim, M.; Khyam, M.O.; Li, X.; Pesch, D. Sensor fusion and state estimation of IoT enabled wind energy conversion system. *Sensors* **2019**, *19*, 1566. [CrossRef]
23. Pan, J.; Jain, R.; Paul, S.; Vu, T.; Saifullah, A.; Sha, M. An internet of things framework for smart energy in buildings: Designs, prototype, and experiments. *IEEE Internet of Things J.* **2015**, *2*, 527–537. [CrossRef]
24. Al-Ghussain, L.; Ahmed, H.; Haneef, F. Optimization of hybrid PV-wind system: Case study Al-Tafilah cement factory, Jordan. *Sustain. Energy Technol. Assess.* **2018**, *30*, 24–36. [CrossRef]
25. Hrayshat, E.S. Wind resource assessment of the Jordanian southern region. *Renew. Energy* **2007**, *32*, 1948–1960. [CrossRef]
26. Alsaad, M.A. Wind energy potential in selected areas in Jordan. *Energy Convers. Manag.* **2013**, *65*, 704–708. [CrossRef]
27. Deshmukh, M.; Deshmukh, S. Modeling of hybrid renewable energy systems. *Renew. Sustain. Energy Rev.* **2008**, *12*, 235–249. [CrossRef]
28. Essalaimeh, S.; Al-Salaymeh, A.; Abdullat, Y. Electrical production for domestic and industrial applications using hybrid PV-wind system. *Energy Convers. Manag.* **2013**, *65*, 736–743. [CrossRef]
29. Halasa, G.; Asumadu, J.A. Wind-solar hybrid electrical power production to support national grid: Case study-Jordan. In Proceedings of the IEEE 6th International Power Electronics and Motion Control Conference, Wuhan, China, 17–20 May 2009; pp. 903–909.
30. Aiad, M.; Badran, A.; Shihabi, S. Optimal selection of hybrid PV/wind systems for jordanian conditions. In Proceedings of the GCREEDER, Amman, Jordan, 10–12 September 2013; pp. 1–9.
31. Dihrab, S.S.; Alghoul, M.; Sopian, K.; Sulaiman, M. Potential of hybrid PV/wind turbine system in Jordan. In Proceedings of the Global Conference on Renewables and Energy Efficiency for Desert Regions, Amman, Jordan, 31 March–2 April 2009.
32. Jaber, J.O. Prospects and challenges of small hydropower development in Jordan. *Jordan J. Mech. Ind. Eng.* **2012**, *6*, 110–118.
33. Bekele, G.; Tadesse, G. Feasibility study of small Hydro/PV/Wind hybrid system for off-grid rural electrification in Ethiopia. *Appl. Energy* **2012**, *97*, 5–15. [CrossRef]
34. Delucchi, M.A.; Jacobson, M.Z. Providing all global energy with wind, water, and solar power, Part II: Reliability, system and transmission costs, and policies. *Energy Policy* **2011**, *39*, 1170–1190. [CrossRef]
35. Al-Momani, A. *Estimated Population 2019 and Some Selected Data*; Department of Statistics: Amman, Jordan, 2020.
36. City Population. Available online: https://www.citypopulation.de/en/jordan/tafilah/?fbclid=IwAR0v2OxWg_BsQFkG4sbAj_2Ub_68xmfV_zvNO7T5uczIPPapasmrHOlk0HY (accessed on 17 September 2020).
37. Kiwan, S.; Al-Gharibeh, E. Jordan toward a 100% renewable electricity system. *Renew. Energy* **2020**, *147*, 423–436. [CrossRef]
38. Duffle-John, A.; Beckman-William, A. *Solar Engineering of Thermal Processes*; John Wiley & Sons: New York, NY, USA, 2006.
39. Manwell, J.F.; McGowan, J.G.; Rogers, A.L. *Wind Energy Explained:Theory, Design and Application*; Wiley: Hoboken, NJ, USA, 2009.
40. Al-Ghussain, L.; Taylan, O.; Samu, R.; Fahrioglu, M. Techno-Economic Analysis of Photovoltaic-Hydrogen Fuel Cell/Pumped Hydro Storage System for Micro Grid Applications: Case Study in Cyprus. In Proceedings of the 2018 International Conference on Photovoltaic Science and Technologies (PVCon), Ankara, Turkey, 4–6 July 2018; Institute of Electrical and Electronics Engineers (IEEE): Piscataway, NJ, USA, 2018; pp. 1–6.
41. IRENA. *Renewable Energy Cost Analysis: Hydropower*; IRENA Press: Abu Dhabi, UAE, 2012.
42. Sadati, S.S.; Jahani, E.; Taylan, O. Technical and economic analyses for sizing PV power plant with storage system for METU NCC. In Proceedings of the ASME International Mechanical Engineering Congress and Exposition, Houston TX, USA, 13–15 November 2015; American Society of Mechanical Engineers: New York, NY, USA, 2015.
43. Al-Ghussain, L.; Samu, R.; Taylan, O.; Fahrioglu, M. Techno-economic comparative analysis of renewable energy systems: Case study in Zimbabwe. *Inventions* **2020**, *5*, 27. [CrossRef]
44. El-Tous, Y.; Al-Battat, S.; Hafith, S.A. Hybrid wind-PV grid connected power station case study: Al Tafila, Jordan. *Int. J. Energy Environ.* **2012**, *3*, 605–616.
45. Al-Ghussain, L.; Taylan, O. Sizing methodology of a PV/wind hybrid system: Case study in Cyprus. *Environ. Prog. Sustain. Energy* **2018**, *38*, e13052. [CrossRef]

46. Al-Ghussain, L.; Taylan, O.; Baker, D.K. An investigation of optimum PV and wind energy system capacities for alternate short and long-term energy storage sizing methodologies. *Int. J. Energy Res.* **2019**, *43*, 204–218. [CrossRef]
47. Fichter, T.; Trieb, F.; Moser, M.; Kern, J. Optimized integration of renewable energies into existing power plant portfolios. *Energy Procedia* **2014**, *49*, 1858–1868. [CrossRef]
48. Yamchi, H.B.; Shahsavari, H.; Kalantari, N.T.; Safari, A.; Farrokhifar, M. A cost-efficient application of different battery energy storage technologies in microgrids considering load uncertainty. *J. Energy Storage* **2019**, *22*, 17–26. [CrossRef]
49. Sangster, A.J. Solar photovoltaics. In *Electromagnetic Foundations of Solar Radiation Collection*; Springer International Publishing: Cham, Switzerland, 2014; pp. 145–172.
50. Breeze, P. *Wind Power Generation*; Academic Press: Cambridge, MA, USA, 2016.
51. Yang, H.; Wei, Z.; Chengzhi, L. Optimal design and techno-economic analysis of a hybrid solar–wind power generation system. *Appl. Energy* **2009**, *86*, 163–169. [CrossRef]
52. Company, J.E.P. Available online: http://www.jepco.com.jo/jepco/index.php?option=com_content&view=frontpage&Itemid=1&lang=en (accessed on 17 September 2020).
53. IRENA. *Future of Solar Photovoltaic: Deployment, Investment, Technology, Grid Integration and Socio-Economic Aspects (A Global Energy Transformation: Paper)*; International Renewable Energy Agency: Abu Dhabi, UAE, 2019.
54. Najjar, Y.S.; Manaserh, Y.M.A. Aligning combined cycle power plant performance with field measurements. *Arab. J. Sci. Eng.* **2019**, *44*, 1657–1669. [CrossRef]
55. Mathiesen, B.V.; Lund, H.; Karlsson, K.B. 100% Renewable energy systems, climate mitigation and economic growth. *Appl. Energy* **2011**, *88*, 488–501. [CrossRef]
56. Dubey, S.; Sarvaiya, J.N.; Seshadri, B. Temperature dependent photovoltaic (PV) efficiency and its effect on pv production in the world—A review. *Energy Procedia* **2013**, *33*, 311–321. [CrossRef]
57. Al-Nimr, M.A.; Kiwan, S.; Sharadga, H. Simulation of a novel hybrid solar photovoltaic/wind system to maintain the cell surface temperature and to generate electricity. *Int. J. Energy Res.* **2018**, *42*, 985–998. [CrossRef]
58. Reich, N.H.; Mueller, B.; Armbruster, A.; van Sark, W.; Kiefer, K.; Reise, C. Performance ratio revisited: Is PR > 90% realistic? *Prog. Photovoltaics: Res. Appl.* **2012**, *20*, 717–726. [CrossRef]
59. Ueda, Y.; Kurokawa, K.; Kitamura, K.; Yokota, M.; Akanuma, K.; Sugihara, H. Performance analysis of various system configurations on grid-connected residential PV systems. *Sol. Energy Mater. Sol. Cells* **2009**, *93*, 945–949. [CrossRef]

energies

MDPI

Article

A Preliminary Assessment of the Potential of Low Percentage Green Hydrogen Blending in the Italian Natural Gas Network

Marco Pellegrini [1,2,*], Alessandro Guzzini [2] and Cesare Saccani [3]

1 Department of Industrial Engineering (DIN), University of Bologna, 47121 Forlì, Italy
2 Interdepartmental Centre for Industrial Research in Renewable Resources, Environment, Sea and Energy, 48123 Ravenna, Italy; alessandro.guzzini2@unibo.it
3 Department of Industrial Engineering (DIN), University of Bologna, 40136 Bologna, Italy; cesare.saccani@unibo.it
* Correspondence: marco.pellegrini3@unibo.it

Received: 15 September 2020; Accepted: 21 October 2020; Published: 23 October 2020

Abstract: The growing rate of electricity generation from renewables is leading to new operational and management issues on the power grid because the electricity generated exceeds local requirements and the transportation or storage capacities are inadequate. An interesting option that is under investigation by several years is the opportunity to use the renewable electricity surplus to power electrolyzers that split water into its component parts, with the hydrogen being directly injected into natural gas pipelines for both storage and transportation. This innovative approach merges together the concepts of (i) renewable power-to-hydrogen (P2H) and of (ii) hydrogen blending into natural gas networks. The combination of renewable P2H and hydrogen blending into natural gas networks has a huge potential in terms of environmental and social benefits, but it is still facing several barriers that are technological, economic, legislative. In the framework of the new hydrogen strategy for a climate-neutral Europe, Member States should design a roadmap moving towards a hydrogen ecosystem by 2050. The blending of "green hydrogen", that is hydrogen produced by renewable sources, in the natural gas network at a limited percentage is a key element to enable hydrogen production in a preliminary and transitional phase. Therefore, it is urgent to evaluate at the same time (i) the potential of green hydrogen blending at low percentage (up to 10%) and (ii) the maximum P2H capacity compatible with low percentage blending. The paper aims to preliminary assess the green hydrogen blending potential into the Italian natural gas network as a tool for policy makers, grid and networks managers and energy planners.

Keywords: hydrogen blending; natural gas networks; power-to-hydrogen; hydrogen and compressed natural gas; renewable energy; hydrogen strategy

1. Introduction

Are the existing infrastructures ready for the decarbonized energy systems of the future, or do they need to be adapted through design and development of new solutions? In the last years investments in renewable power plants have grown rapidly worldwide moving towards a "renewable electrical networks scenario" [1,2]. However, due to the production unpredictability of some renewable power sources (i.e., wind and solar) and the possible mismatch between production and demand, energy storage solutions are essential to avoid grids' instability [3].

Among the available technological solutions, power-to-gas (P2G), based on chemical energy storage concept, is considered as one of the most interesting for energy system decarbonization [4]. In fact, through P2G the power surplus is stored as renewable fuel, i.e., a fuel produced by converting renewable energy sources into chemical molecules for use in various applications with minimum greenhouse emissions or without adding net CO_2 to the atmosphere [5], that can be used for different purposes, like feedstock for in industrial processes [6], energy carrier [7], fuel in residential/district heating and cooling [8] and in the transport sector [9].

ENTSOG, i.e., the European Network of Transmission System Operators for Gas, proposed the "2050 roadmap for gas grids" in which several recommendations and actions are suggested to implement a European P2G strategy [10]. Particularly, three configurations are proposed for the energy grid of the future, i.e., the grids towards a close to carbon neutral gas system: (i) the use of biomethane and synthetic natural gas (SNG) that ensure no adaptation of end-user applications; (ii) an increasing hydrogen blending percentage into the existing natural gas networks; and (iii) the retrofitting of the natural gas networks to transport only hydrogen. From an environmental point of view, the first option should be preferred to the other two, since a neutral or a net negative balance of CO_2 could be obtained in the production of biomethane and SNG, respectively. However, several resources and long times could be required to put in place such approach: (i) plants for CO_2 capture from flue gases emissions should be realized, and (ii) infrastructures dedicated to CO_2 storage and transport to the final users should be implemented. For these reasons, the first configuration is considered for a long-term energy strategy. The third option seems as well to be not feasible in the short-medium term for the same reasons of the first option, i.e., high infrastructural costs. Therefore, hydrogen blending into the natural gas grids appears to be the most viable solutions in the short-medium terms [10].

Among possible P2G configurations, power-to-hydrogen (P2H) is the simplest, the most reliable and energy efficient. In addition, renewable hydrogen production, i.e., "green hydrogen", produced from renewable or nuclear sources [11], is an essential topic in the recent "Hydrogen strategy for a climate-neutral Europe" promoted in 2020 by the European Commission [12]. The strategy includes the natural gas sector as a key driver for the effective implementation of a hydrogen economy, since the existing natural gas infrastructures can play a relevant role in the early stage of the hydrogen strategy development as a way to transport and store green hydrogen [13].

Nevertheless, in the literature several technological limitations have been identified to hydrogen blending in the existing natural gas networks. First of all, safety concerns have to be considered since metallic pipelines shows a higher risk of failure in case of operation with hydrogen and compressed natural gas (HCNG) blend. Several authors investigated the interaction of high and low pressure hydrogen in metallic and plastic pipelines [14]. Assuring the highest safety condition in gas infrastructures should be the first aim of gas operators [15,16]. Particularly, higher leakage rate, i.e., a greater hazardous distance in case of failure, is expected for HCNG for high pressure systems [17] even if they are comparable for low pressure distribution systems [18]. Secondly, HCNG quality, i.e., the energy content, supplied to final end-users has to be controlled and correctly measured. In fact, since hydrogen concentration could change with time, smart metering is crucial to monitor the hydrogen percentage and to measure the effective energy content of the HCNG flow [19]. In addition to metering issues, the hydrogen concentration in HCNG is limited by existing end-users' devices and equipment that are designed and certified only for NG supply. Based on a literature review, [20] reported a maximum concentration up to 20% for vehicle engines, burners and boilers while higher concentration, i.e., up to 50%, could be considered for gas cookers and CHP application. HCNG quality also affects the performances of equipment installed in the transportation and distribution networks. Particularly, a maximum hydrogen concentration of 10% is suggested for the operation of existing compressors installed along the natural gas network [21]. Furthermore, since hydrogen percentage increasing causes a reduction of the low heating value (LHV) of the HCNG [22], higher mass flowrates, and so possible congestion, are expected in the network to convey the same quantity of energy.

Among the non-technological barriers, it is relevant that only qualitative evaluations have been carried out about the potential of green hydrogen blending in the existing natural gas networks [23]. In particular, a fundamental question for the development of a long-term strategy is "how much green hydrogen could be yearly produced and blended in the existing natural gas networks without any relevant impact on the infrastructure and the end-users?" In fact, without the assessment of the nominal capability of the network to transport HCNG, insufficient information would be available also for the proper localization, planning and design of P2H plants.

The aim of the paper is to propose a methodology for the quantitative estimation of the Italian natural gas network capacity to accept green hydrogen and transport HCNG with low hydrogen concentration. Moreover, the paper includes a first assessment of the Italian P2H plants capacity and location.

2. Methodology

The following section reports the description of the methodology followed by the Authors to quantitatively estimate the HCNG transportation potential of Italian natural gas infrastructure in the case of low percentage blending of hydrogen. The Italian natural gas network and the main technical operative conditions are firstly presented. Then, the main concepts of the paper's methodology are introduced. After that, the assumptions for the following calculation are described, discussed and justified.

2.1. The Italian Natural Gas Networks

Two different kind of networks are operated in Italy: the transportation and the distribution networks. More than 90% of natural gas is imported by foreign countries. The Italian natural gas network is characterized by the presence of seven "Import Points", which are connected to the Italian transportation system for natural gas supply [24]:

- Five import points connected to foreign pipelines: located in Mazara del Vallo (Trapani–Sicily), Gela (Caltanissetta–Sicily), Passo Gries (Verbano Cusio Ossola–Piedmont), Tarvisio (Udine–Friuli Venezia Giulia) and Gorizia (Friuli Venezia Giulia);
- Three import points connected to liquefied natural gas (LNG) gasification plants: located in Panigaglia (La Spezia–Liguria), Porto Viro (Rovigo–Veneto) and Livorno (Toscana).

Two further connections should be considered, that are between the Italian natural gas transportation system and national natural gas storages, which are located in Campo Collalto (Treviso–Veneto) and Montalfano (Chieti–Abruzzo).

From the operative point of view, the Italian transportation system is operated at a pressure between 24 and 75 bar g, even if submarine pipelines are operated at a pressure up to 115 bar g. Figure 1 shows through different colors and thickness how the transportation system is indeed divided into two networks, the National Transportation (NT) system and the Regional Transportation (RT) system (in blue and light blue, respectively, in Figure 1). Figure 1 includes only the Transportation system managed by SNAM (in Italian "Società Nazionale Metanodotti), that is the most important of the nine Italian Transportation System Operator (TSO) that controls more than the 93.2% of the Italian system [25].

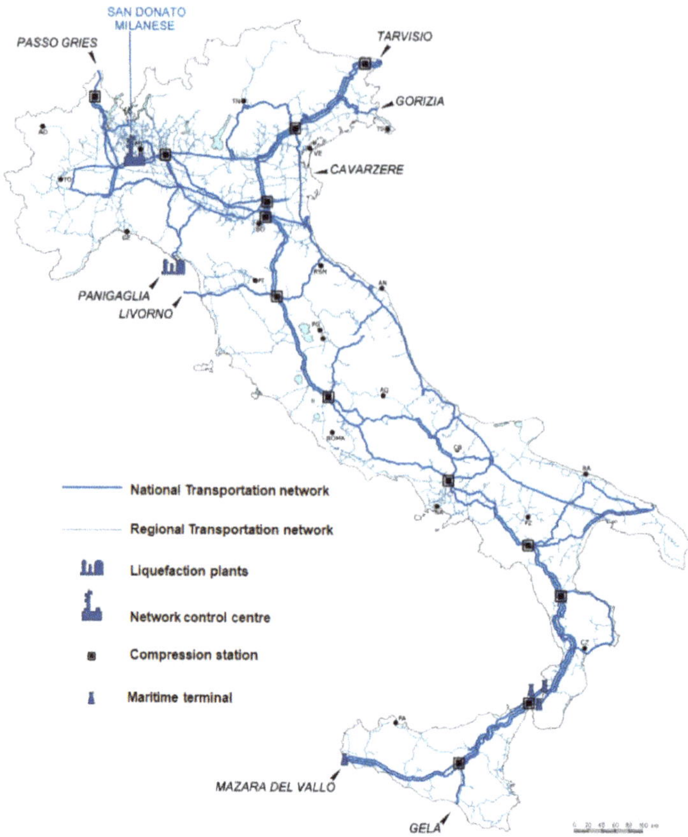

Figure 1. The Italian National and Regional Transmission Systems [26].

In accordance to the Decree of the Ministry of Industry and Economic activities 22/12/2000 [27], the NT system consists of networks with a total length of 10,272 km that connects the North with the South of Italy conveying the natural gas from the Import Points to the Interconnection Points with the RT systems and the two storage plants. Thirteen gas compression plants, with a total load of 961 MW el, are installed to compensate the pressure drops along the TN system [26]. Particularly, centrifugal gas compressors are installed. However, due to the high flowrate elaborated, i.e., up to 1,500,000 Sm^3/h, a maximum compression ratio up to 1.4–1.5 is available in gas compression plants. Therefore, since gas compression plants have to be able to restore the downstream pressure up to 75 bar g in case of a national peak demand, a minimum upstream pressure of 50 bar g (=75/1.5) is allowed by gas transmission code [28]. In accordance to the Decree of the Economic Ministry 29/9/2005 [29], the RT system, with a total length of 24.700 km and 20 interconnection points with the NT, accounts for the distribution of natural gas though the national territory and, particularly, to power plants and to local distribution networks that are connected through 567 ReMi (Regolazione and Misurazione in Italian) stations at a minimum pressure up to 24 bar g. An updated list of TN and RT networks is available at [30].

The Italian Distribution system is responsible for natural gas supply to final customers. Almost 30 GSm^3 of natural gas, equivalent to almost 300 TWh, are annually supplied by more than 200 Distribution System Operators (DSO) to more than 23 million final Italian customers through more than 260.000 km of local networks that are mainly in the Northern of Italy, wherein 70% of the Italian natural gas consumption is concentrated [31]. Respect to the NT and RT systems, gas pressures lower than 5 bar g are operated in the Distribution networks [32]. Due to the lower nominal pressures than TN and RT systems, in addition to steel also polyethylene, iron and copper have been used [33]. Although distribution network is considered as a possible short-term storage for syngas produced in local P2G applications, concerns exist about the implementation of hydrogen blending along the Distribution network. First of all, the presence of multiple hydrogen injection point would be responsible for very different concentrations of the HCNG along the local networks that could impede to DSOs an effective and reliable control of the network operation. Secondly, the high number of DSOs connected to the transportation system could create difficulties in terms of management of the energy fluxes with the transmission system. Particularly, more than 500 "connection points" between distribution and transmission networks are present in Italian gas system [26]. Each "connection point" would become a hydrogen blending point into the transmission system. Therefore, the resulting hydrogen concentration of the transportation system depends on the hydrogen concentrations and on the flowrates entering from each connection point. A very complex coordination between DSOs would be therefore required to not exceed the hydrogen concentration threshold. Therefore, hydrogen blending is assumed only in Italian transmission gas systems while distribution gas networks are not considered as an option in the following sections of the papers for location of P2H plants. Nevertheless, the result of the preliminary assessment in terms of quantification of low percentage hydrogen blending potential is not affected by this choice.

2.2. Premises and Main Hypothesis for Hydrogen Blending Potential Estimation in Italian Natural Gas System

In general, based on research to date [14], only minor or no issues should arise with limited percentage of hydrogen blends, i.e., less than 5–15% hydrogen by volume. More significant problems would be addressed for higher blends, in the range of 15–50%, such as conversion of household appliances, an increase in compression capacity along distribution mains serving industrial users, and the development of a complex control strategies to monitor hydrogen injection and hydrogen percentage blend into the network. Hydrogen blending above 50% is expected to be possible on through challenging actions across multiple areas, including pipeline materials, safety, and substantial modifications required for end-use appliances or other uses. Nevertheless, up to now the limits for hydrogen blending into the natural gas networks have been usually kept very low, varying between 0.2% up to 6% [34]. Even if Italian regulation allows hydrogen concentration for blending only up to 1.0% [35], as defined for biomethane injection, experimental activities have been already performed in Italy to evaluate the impact of higher concentrations in existing networks: 5% blending has been already tested in a small closed network near the southern city of Salerno [36], while new tests have been planned with the aim of testing 10% hydrogen injection [37].

Moving towards a hydrogen economy will require the design and implementation of a complex and long-term national strategy. While potential targets and techno-economic impact by 2050 of the hydrogen economy in Italy have been already estimated [38], a national strategy is still far from being clearly defined. A fundamental part of the EU hydrogen strategy is the "first step", i.e., the public and private investments to be planned in the next 4 years, targeting 2024. Accordingly, short term actions must be planned to stimulate the growth of the hydrogen market and to start the hydrogen penetration in the Italian energy sector. How to approach the opportunity of hydrogen blending into the natural gas network by 2024 is crucial since Italy has one of the largest natural gas network infrastructures in Europe [39], connected with several foreign and strategic areas like Northern Africa and East Europe. Furthermore, Italy also has a huge potential for renewable power generation via wind and solar: [40] identifies in 18.4 GW the wind potential that can be installed by 2030, which would correspond to an annual electricity production of 40.1 TWh, while [41] estimates in about 127 TWh per year the power production from photovoltaics (PV) integrated in buildings.

From a practical point of view, P2H plants will be needed to blend green hydrogen into the Italian natural gas network. So, the design of a strategy moving towards a growing percentage of green hydrogen injected into the natural gas network requires to plan the design, installation and simultaneous operation of an increasing number of P2H plants year by year. Furthermore, since renewable power is needed to produce green hydrogen, the planning of new P2H plants cannot be realized without taking into consideration the current location of renewable power plants as well as the setting up of new ones, if needed. Table 1 shows the current installed power capacity of PV and wind turbine power plants in Italy by Regions [42].

Table 1. Current installed power capacity of photovoltaic (PV) and wind turbine powerplants in Italy by Regions (data updated to June 2020, from [42]).

Region	PV Installed Power (MW)	Wind Turbine Installed Power (MW)
Piedmont	1662	24
Valle d'Aosta	25	3
Lombardy	2458	0
Trentino Alto-Adige	448	0
Veneto	2039	13
Friuli-Venezia Giulia	550	0
Liguria	116	66
Emilia-Romagna	2128	45
Tuscany	848	143
Umbria	498	2
Marche	1110	19
Lazio	1403	71
Abruzzo	748	264
Molise	177	376
Campania	852	1743
Apulia	2848	2575
Basilicata	373	1301
Calabria	545	1150
Sicily	1458	1906
Sardinia	913	1105
Total	21,197	10,806

Therefore, the complexity of the hydrogen economy development will increase with the increasing of green hydrogen percentage injected into the natural gas network due to (i) the impact of hydrogen blending into the existing infrastructures and end-users, and (ii) the interactions between renewable power generation and hydrogen production. However, in a first phase these issues can be minimized if (i) the percentage of green hydrogen is kept relatively low and (ii) the installation of P2H is optimized by taking into account the current locations of both natural gas network and renewable power plants.

The aim of the paper is to identify what is the total amount of green hydrogen that could be produced and injected right now in the Italian natural gas network without compromising its integrity and with no relevant drawbacks for the end-users. The quantification of such a target is fundamental to calculate the P2H installations needed and to evaluate in a first assessment the geographical distribution and the required budget for the realization of these new P2H plants in relation with natural gas network characteristics and current regional distribution of renewable power plants.

2.3. Analytical Description of the Methodological Approach

The evaluation of the maximum green hydrogen blending capacity to be injected in the Italian natural gas network with no relevant impacts has been done accordingly to the following considerations. The maximum blending threshold (BT), defined as in Equation (1), is the limit to hydrogen blending beyond which many actions are needed to guarantee infrastructure integrity, end-users safety and an effective control of hydrogen percentage flowing in the natural gas network. BT, calculated in (Sm^3/h), can be computed if (i) the minimum natural gas (MNG) flowrate in (Sm^3/h) measured in the natural gas network is known, and if (ii) the allowed blending percentage (ABP) is fixed. ABP can be defined as the upper limit of hydrogen blending percentage in volume in the natural gas grid under which modifications on the network and its auxiliaries and on the end-users are not required. ABP is defined in ($\%_{vol}$) Natural gas and hydrogen density are respectively defined as ρ_{NG} and ρ_{H2}, both in (kg/Nm^3). A safety factor (SF) in (%) and lower than 1 is also introduced in Equation (1) to take into account of the available data quality.

$$BT = SF \times \frac{ABP \times \rho_{H2}}{(1 - ABP) \times \rho_{NG}} \times MNG \tag{1}$$

An energy density correction factor (EDF) is also needed to take into account the reduction of the Lower heating value (LHV) of the HCNG volumetric flowrate (Q_{HCNG}) respect to the pure natural gas case. This reduction depends on the energy density of natural gas and hydrogen, in accordance to the respective higher heating Values (HHV_{NG} = 9.70–12.58 kWh/Sm^3 [43]) and HHV_{H2} = 3.36 kWh/Sm^3). In fact, since the total energy demand by the end-users (E_{Demand}) does not change, an increase of HCNG flowrate is required proportionally to the reduction of the energy content of the gas mixture resulting from the hydrogen blending. The HCNG volumetric flowrate is the sum of the natural gas (Q_{NG}) and hydrogen (Q_{H2}) volumetric flowrates (Sm^3/h) as reported in Equation (2):

$$Q_{HCNG} = Q_{NG} + Q_{H2} \tag{2}$$

where, considering w_{NG} and w_{H2} as the volumetric concentrations of natural gas and hydrogen in the HCNG, Equations (3)–(5) apply:

$$Q_{H2} = Q_{HCNG} \times w_{H2} \tag{3}$$

$$Q_{NG} = Q_{HCNG} \times w_{NG} \tag{4}$$

$$w_{NG} + w_{H2} = 1 \tag{5}$$

Since end-users' energy demand does not depend on the composition of the gas supplied, the same amount of energy in case of pure natural gas flowrate has to be delivered through HCNG. Particularly, if Q_{NG}' is the natural gas flowrate when no hydrogen is blended in (Sm^3/h), the existing energy demand E_{Demand} (kWh) of the end-users can be calculated as in Equation (6):

$$E_{Demand} = Q_{NG'} \times LHV_{NG} \tag{6}$$

where LHV_{NG} is the lower heating value of the natural gas in (kWh/Sm^3). The same amount of energy has to be transported by HCNG mixture. Defining the energy delivered by the HCNG flowrate as E_{HCNG} (kWh), Equation (7) has to be considered:

$$E_{HCNG} = E_{Demand} \tag{7}$$

The energy transported by the HCNG flowrate can be calculated as in Equation (8):

$$E_{HCNG} = Q_{HCNG} \times LHV_{HCNG} \tag{8}$$

Where LHV_{HCNG} (kWh/Sm^3) is the lower heating value of the HCNG flowrate and it is calculated as in Equation (9):

$$LHV_{HCNG} = LHV_{NG}w_{NG} + LHV_{H2}w_{H2} \quad (9)$$

From Equation (7) and by the use of Equations (6), (8) and (9), the HCNG flowrate required to supply the same amount of energy that end-users require is calculated as in Equation (10):

$$Q_{HCNG} = Q_{NG'}\frac{LHV_{NG}}{LHV_{NG}w_{NG} + LHV_{H2}w_{H2}} \quad (10)$$

In accordance to Equation (10), the HCNG flowrate increases as the hydrogen concentration in the HCNG mixture rises due to the lower volumetric energy density of hydrogen respect to natural gas. Therefore, EDF, which is greater than 1 and defined as in Equation (11), is introduced in Equation (2) to calculate an energy corrected blending threshold (BT_{corr}) in accordance to Equation (12):

$$EDF = \frac{LHV_{NG}}{LHV_{NG}w_{NG} + LHV_{H2}w_{H2}} \quad (11)$$

$$BT_{corr} = SF \times EDF \times \frac{ABP \times \rho_{H2}}{(1 - ABP) \times \rho_{NG}} \times MNG \quad (12)$$

Even if different operative conditions in terms of operative mixture pressure and temperature could verify during the years, it should be noted that the density ratio (ρ_{H2}/ρ_{CH4}) can calculated as follow. In fact, in accordance to the real gas law, Equations (13) and (14) apply:

$$\frac{p_{NG}}{\rho_{NG}} = Z_{NG}\frac{R_0}{M_{NG}}T_{NG} \quad (13)$$

$$\frac{p_{H2}}{\rho_{H2}} = Z_{H2}\frac{R_0}{M_{H2}}T_{H2} \quad (14)$$

where p_{NG} and p_{H2} are natural gas and hydrogen pressures [Pa], Z_{NG} and Z_{H2} are natural gas and hydrogen compressibility factors in [#], R_0 is the universal gas constant in [kJ/kmol K], M_{NG} and M_{H2} are natural gas and hydrogen molecular weights (kg/kmol) and T_{NG} and T_{H2} are the natural gas and hydrogen operative temperatures [K]. Even if operative annual temperature of natural gas conveyed in buried pipelines changes during the year [44], the variation can be considered negligible for the purpose of the following evaluations. However, the same consideration is not valid for pressure that depends on the specific point of the network. However, in the reported pressure range, i.e., [25 bar g, 75 bar g], the ratio between the hydrogen and methane compressibility factor can be considered almost constant. In fact, assuming an operative temperature of 285.15 K, the reduced temperature of hydrogen is equal to 6.8, resulting in a compressibility factor Z_{H2} equal almost to 1, independently from the reduced pressure. Concerning natural gas, assuming the same properties of methane, a reduced temperature of 1.49 and a reduced pressure between [0.04, 0.13] is calculated. A compressibility factor Z_{NG} between 1 and 0.96 is obtained from available diagrams [45]. Therefore, compressibility factors are neglected in following evaluations. Equations (13) and (14) can be elaborated as reported in Equations (15) and (16):

$$\rho_{H2} = p_{H2} \times \left(\frac{R_0}{M_{H2}}T_{H2}\right)^{-1} \quad (15)$$

$$\rho_{NG} = p_{NG} \times \left(\frac{R_0}{M_{NG}}T_{NG}\right)^{-1} \quad (16)$$

Therefore, the density ratio is calculated as in Equation (17) based on Equations (15) and (16):

$$\frac{\rho_{H2}}{\rho_{NG}} = \left(\frac{p_{NG}}{p_{H2}} \times \frac{M_{CH4}}{M_{H2}}\right)^{-1} \quad (17)$$

where also T_{NG} is assumed equal to T_{H2} since natural gas and hydrogen are in the same mixture.

But why the authors define the MNG as natural gas flowrate reference for hydrogen blending? The hypothesis is that if the P2H blending capacity is calculated starting from the lowest capacity of the current natural gas flowrate, i.e., when the natural gas flowrate delivered by the national transportation system is at the minimum level, some important benefits occur:

1. No control is needed on each P2H plant: each P2H plant can produce hydrogen at the maximum capacity at any time with no risk for the natural gas network since real time natural gas flowrate will be always higher than MNG;
2. No control is needed to compensate P2H plants production: since the sum of the maximum capacity of all the P2H plants will be always lower than BT, it is not necessary to design and realize an effective general control system able to monitor and to control in real-time the hydrogen flow rate injected and the percentage in volume;
3. No control is needed on the real hydrogen percentage in volume in the natural gas network: it is not necessary to measure the hydrogen content in the natural gas network, since it will be always lower than the ABP; hydrogen concentration monitoring would be required only for energy billing purposes;
4. Infrastructures and auxiliaries as well as end-users' equipment are not subjected to adaptation or revamping since ABP will not be overcome.
5. No roll-out plan is required to substitute existing smart meters at end-users to take into account of hydrogen concentration during energy bills calculation. In fact, if gas chromatographs could be installed at REMI stations to calculate the concentrations of the HCNG delivered to the distribution networks, all the end-users will handle the same HCNG mixture. Figure 2 shows the natural gas flowrate hourly variation respect to the daily average flowrate in the transmission system. Particularly, four days randomly taken in different seasons of the years 2019–2020 have been considered. As shown, a very slow variation in natural gas flowrate occurred during the day. Therefore, assuming an interval up to 1 h, i.e., the frequency at which natural gas flowrate can be measured by the smart meters, the assumption that the hydrogen concentration measured at REMI station is constant in each hour would result in an error. For example, the greatest variation occurs the 23/24 March 2019 between 12:00 and 14:00 when a variation in natural gas flowrate up to 6% (=6% − 0%) occurs in the period. In such case, a reduction of hydrogen concentration up to the 5.7% of the initial value measured at 12:00 would result. However, the acceptability of such error in billing procedures is out of the scope of the paper.

After the BT has been identified, it is part of the strategy to define how much fast the threshold should be reached, i.e., how many MW of P2H plants are planned to be realized every year up to 2024 as schematically shown in Figure 3. The cumulative green hydrogen blending capacity is influenced by policy makers and energy planners' decisions, since the slope of the cumulative curve may allow to reach the threshold before ($\alpha 1$ in Figure 3) or close to the deadline ($\alpha 2$ in Figure 3). The second step of the strategy will start once the BT has been overcome, and will require relevant actions, as synthetized in Figure 3, as well as the practical implementation of actions over the time (curve slope $\beta 1$ or $\beta 2$ in Figure 3). Therefore, it is crucial to properly set the first phase timing to not reach too early the blending threshold, thus avoiding the risk of dead time waiting for the revamping/adaptation needed to increase the hydrogen blending percentage.

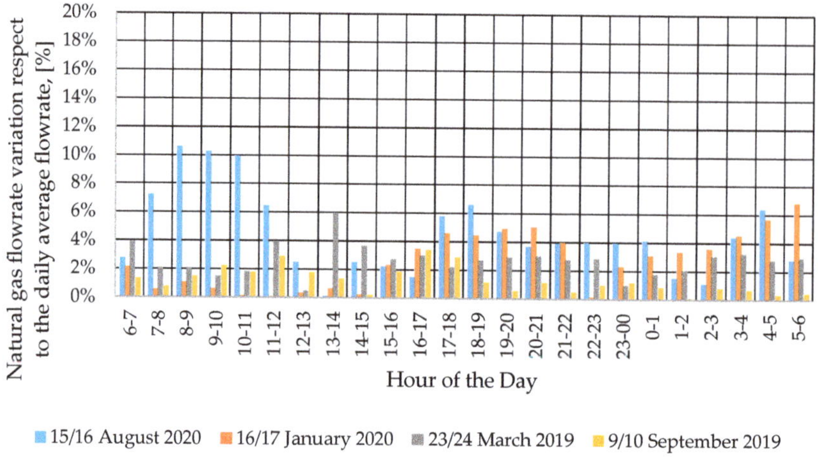

Figure 2. Natural gas flowrate variation respect to the daily average flowrate. Original elaboration based on data from [46].

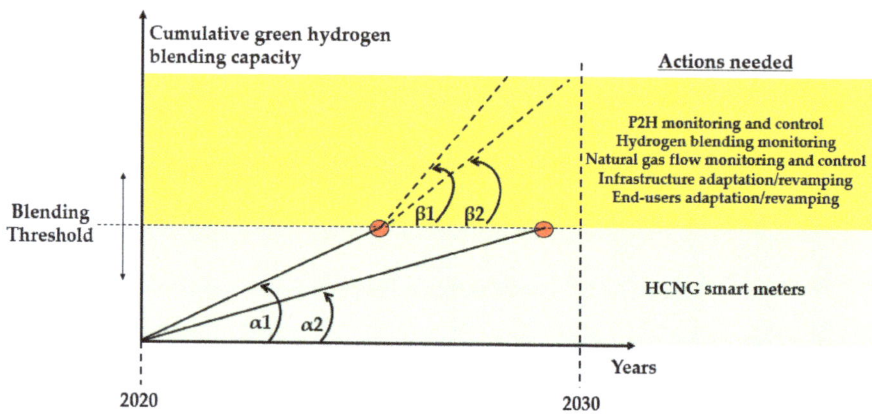

Figure 3. Design of the first phase of implementation of the hydrogen blending strategy.

2.4. Natural Gas Thermodynamic Parameters

Several parameters have to be evaluated to assess the Italian hydrogen blending threshold. Table 2 shows the natural gas composition that is conveyed by the Italian natural gas networks [43]. A natural gas density equal to 0.904 kg/m^3 (= 0.7 × 1.292) at 0 °C and 101,325 Pa is conservatively assumed. In fact, to assume the highest possible density for the natural gas density signifies to evaluate the minimum BT in accordance to Equation (1). In addition, applying the ideal gas law a molar mass of 20.3 g/mol results in the case study. For hydrogen a density equal to 0.0899 kg/m^3 is assumed in the same thermodynamic conditions.

Table 2. Mean natural gas composition conveyed by the Italian natural gas networks [43].

Parameter	Acceptability Limit	Unit of Measure
Methane	*	
Ethane	*	
Propane	*	
Iso-butane	*	
Normal-butane	*	
Iso-pentane	*	
Normal-pentane	*	
Hexanes	*	
Nitrogen	*	
Oxygen	≤0.6	% mol
Carbon dioxide	≤2.5	% mol
Hydrogen sulfide	≤5	mg/Sm^3
Sulfur from mercaptans	≤6	mg/Sm^3
Sulfur, total	≤20	mg/Sm^3
Higher heating value	34.95–45.28	MJ/Sm^3
Wobbe Index	47.31–52.33	MJ/Sm^3
Relative density	0.555–0.7	#

(*) Values are limited by the respect of gas mixture's Wobbe Index.

The values reported in Table 3 are conservatively assumed for the hydrogen blending threshold evaluation.

Table 3. Values assumed for the evaluation of hydrogen blending threshold.

Parameter	Value	Unit of Measure
Hydrogen concentration, w_{H2}	10	%
Methane concentration, w_{CH4}	90	%
LHV_{CH4} (*)	11.86	kWh/Sm^3
LHV_{H2}	2.83	kJ/Sm^3
ρ_{NG}	0.904	kg/m^3 at 0 °C and 101.325 kPa
ρ_{H2}	0.0899	kg/m^3 at 0 °C and 101.325 kPa

(*) Since gas natural mixture composition varies, an average value was considered.

2.5. Considerations about Reliability of Available Data on Natural Gas Flowrates

Concerning the safety factor SF, the value was defined by the Authors in accordance to the available data about MNG. In particular, Snam provides the data of hourly gas imports, storages and exports that enter the national transmission system. In a preliminary evaluation, it can be assumed that the sum of the natural gas imports, the national production and eventually of the natural gas from the storage fields correspond to the flowrate that is going to be conveyed through the transportation system. In Figure 4 the average daily natural gas imports per months for the period 2017–2019 is reported, and also the maximum and the minimum daily gas imports are shown by the error bars that show the maximum positive and negative deviations of the daily average values from the average. As expected, higher energy is delivered in the winter season. On the other hand, despite of the winter months, natural gas flow rate varies a little during the summer months regardless of the year. Minimum values equal to 877.3 kWh/day, 840.4 kWh/day and 850.1 kWh/day were calculated in August respectively for the year 2017, 2018 and 2019.

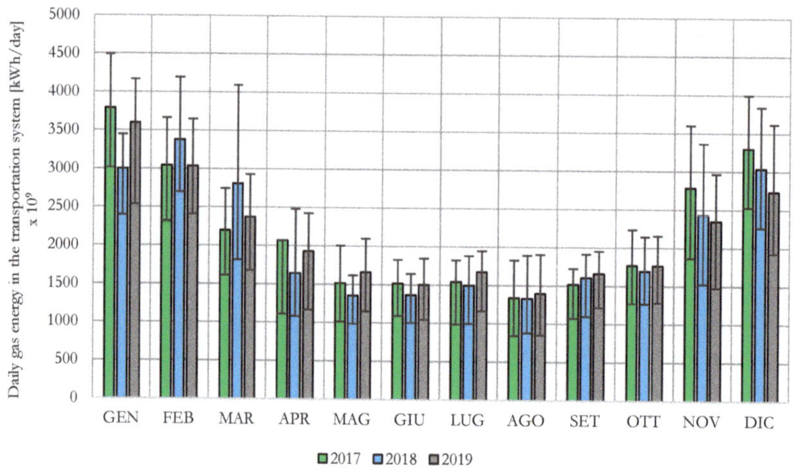

Figure 4. Average daily gas energy in the transportation system during 2017, 2018 and 2019. Data elaborated from [47].

Since instantaneous natural gas flowrate could change during the day, the hourly flowrates of fifty days randomly selected were analysed for a statistical evaluation: 17 days were selected in both 2019 and 2018, 16 days in 2017.

2.6. Identification of P2H Plants Size and Location in the Italian Territory

Based on the hydrogen blending threshold calculated in the previous section, the P2H plants' total capacity (P_{P2H}) in (kW) can be calculated from Equation (18):

$$P_{P2H} = BT_{corr} \times LHV_{H2} \tag{18}$$

Furthermore, electrical needs have to be calculated. Three main electrical equipment are considered in the following analysis: (i) the electrolyzers, (ii) the compressor units and (iii) other auxiliaries. Since compressed physical storage is considered as the preferred option for its maturity level, no additional energy consumption due to storage section is considered. Particularly, the P2H total electric power capacity in (kW), $P_{EL,P2H2}$, is calculated in accordance to Equations (19) as the sum of the electric capacity of the components that are implemented in the plant.

$$P_{EL,P2H} = P_{EL,ELECTROLYSER} + P_{EL,COMPRESSOR} + P_{EL,AUXILIARIES} \tag{19}$$

Water electrolyzer electric capacity is calculated in Equation (20) as the ratio between the P2H plants' total capacity and electrolyzer efficiency ($\eta_{ELECTROLYSER}$) in [%]. It should be noted that electrolyzer' efficiency is calculated as the ratio between the hydrogen energy production (based on LHV_{H2}) and electrical power consumption.

$$P_{EL,ELECTROLYSER} = \frac{P_{P2H}}{\eta_{ELECTROLYSER}} \tag{20}$$

Hydrogen compressors' electric capacity is calculated in Equation (21) as the ratio between the isentropic compression power and compressors' total efficiency. It should be noted that $L_{is,COMPRESSOR}$ is the isentropic work of compression [kJ/kg]. An isentropic ($\eta_{is,COMPRESSOR}$) and electric ($\eta_{el,COMPRESSOR}$) efficiencies are also introduced [%]:

$$P_{EL,COMPRESSOR} = \frac{BT_{corr} \times L_{is,\ COMPRESSOR} \times \rho_{H2}}{\eta_{is,\ COMPRESSOR} \times \eta_{el,\ COMPRESSOR}} \quad (21)$$

Auxiliaries' electric capacity are calculated as in Equation (22). For the purpose a safety factor (SF') is introduced:

$$P_{EL,AUXILIARIES} = SF' \times (P_{EL,ELECTROLYSER} + P_{EL,COMPRESSOR}) \quad (22)$$

Concerning the electrolysis section, due to the higher maturity level and the lower Capital Expenditures respect to alternative solutions, alkaline electrolyzers are assumed as the preferred option for P2H Italian plants. Based on state of the art [4], an average efficiency between [62%, 82%] is recognized for alkaline electrolyzers. A conservative efficiency of 65% is considered in the paper. Concerning the compression section, a downstream pressure up to 70 bar is considered as appropriate for hydrogen blending into the transmission system. Based on available review in the literature [48], reciprocating, linear and diaphragm compressors can be considered for the purpose. Particularly, diaphragm compressors are considered for the following analysis. In this case efficiencies in the range [80%, 85%] can be considered. For a conservative approach an efficiency value of 80% is considered. Therefore, based on Equation (21) and data reported in Table 4, the compression isentropic work between 1 bar and 70 bar is calculated equal to 9940 kJ/kg, i.e., a real work of 12.4 MW. Considering the total hydrogen flowrate (2326 kg/h), a total installed electrical power supply up to 8 MW is required to operate hydrogen compressors.

Table 4. Data used for the calculation of the isentropic compression work.

Parameter	Value	Unit of Measure
Isentropic coefficient, k	1.407	#
Upstream pressure, p_0	101,325	Pa
Downstream pressure, p_{out}	70 × 101,325	Pa
Hydrogen density at upstream compressor inlet, ρ_0	0.085	kg/m^3
Isentropic efficiency, η_{is}	80	%

Once total P2H plants' capacity is calculated, the localization of each plant should be performed. However, in this preliminary estimation, it was assumed that P2H plants are located at the eight Italian transportation system import points. Based on this conceptualization, a distribution of P2H plants in Italy would be possible. P2H plants' capacity localization in the national territory is affected by both current renewable power plants distribution and concentration as well as renewable power potential. Since the paper aim is to firstly assess the compatibility of the proposed P2H power plants distribution and the existing renewable power plants, only a qualitative comparison will be performed between available and needed power on a regional basis.

2.7. Estimation of the Economic Investment Required

Based on the quantitative evaluation of P2H Italian potential, a preliminary assessment of the capital expenditure (CAPEX) is performed. The realization of new renewable power plants is not considered. For the purpose, referenced data available in the literature were used. Particularly, CAPEX is the investment required for P2H plant design, realization and tests. As shown in Table 5, the voice regarding the hydrogen storage volume is considered. For the purpose, a storage volume able to store up to 1 h of the nominal hydrogen capacity is considered. Concerning hydrogen compression, as reported in the available literature, estimates for compressors' investment vary widely from 144 €/kW to 18,500 €/kW [49]. Therefore, an average value equal to 10,000 €/kW was assumed as reported in Table 5. Since several assumptions were made, a safety factor for the purpose was defined. Particularly, since P2H plant's design strictly depends on the specific boundary conditions, a value equal to 25% was selected to take into account all the expenditures that were not included in electrolyzers, compressors and storage tanks, such as for example, engineering activities, ATEX certification, the purchase of interconnecting and bulk materials, the purchase of the land, etc.

Table 5. Capital expenditure (CAPEX) parameters.

Parameter	Value	Unit of Measure
Electrolyzers' section (instrumentation and tubing are included) [50]	700	€/kW
Compressors' section	10,000	€/kW
Hydrogen storage volume [51]	8300	€/m^3
Safety factor to take into account of engineering, tests and other activities (of the total CAPEX)	25	%

(*) CAPEX costs for electrolyzers with a size of 5 MW.

Based on the specific costs reported in Table 1, the following CAPEXs (€) are calculated in accordance to Equations (23) and (24):

$$CAPEX_{ELECROLYSER} = c_{electrolyser} \times P_{EL,ELECTROLYSER} \quad (23)$$

$$CAPEX_{COMPRESSOR} = c_{compressor} \times P_{EL,COMPRESSOR} \quad (24)$$

$$CAPEX_{STORAGE} = c_{storage} \times V_{storage} \quad (25)$$

$$CAPEX_{OTHER} = (CAPEX_{ELECROLYSER} + CAPEX_{COMPRESSOR} + CAPEX_{STORAGE}) \times SF'' \quad (26)$$

where $c_{electrolyser}$, $c_{compressor}$ and $c_{storage}$ are the specific cost of water electrolyzers and compressors in [€/kW] as reported in Table 5. $V_{storage}$ is the storage volume [m^3]. SF'' is the safety factor to take into account other costs that are necessary to design, install and operate a P2H plant.

3. Results and Discussion

3.1. Computation of the Energy Density Factor EDF

In accordance to Equation (11), the energy density factor results equal to 1.085. Therefore, to deliver the same amount of energy, an increase up to the 8.5% of the HCNG flowrate is required respect to the existing natural gas flowrate. Moreover, in accordance to Equation (17), a density ratio (ρ_{H2}/ρ_{NG}) of 0.099 results.

3.2. Evaluations about the Safety Factor SF

Figure 5 represents the hourly gas flowrates from the 6:00 of the analysed day to the 6:00 of the following day for the 2019. Different values of MNG are obtained depending on the days, i.e., the lowest flowrate is calculated during the weekends and during public holidays.

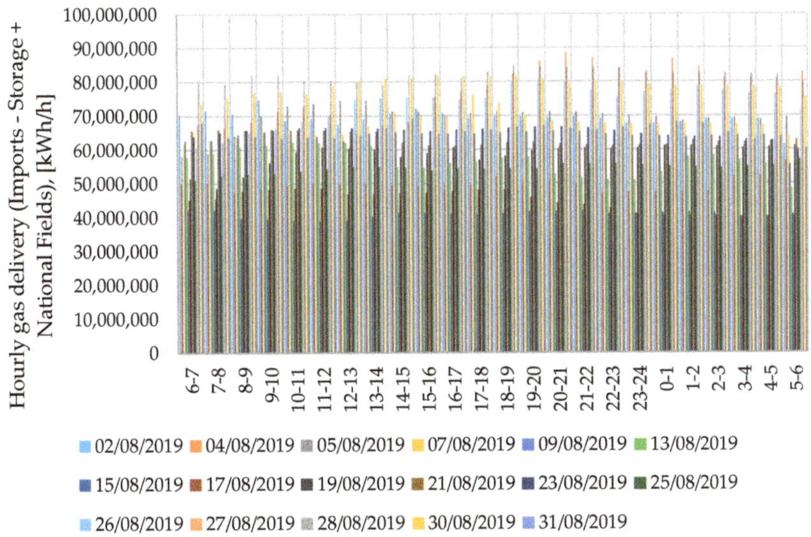

Figure 5. Hourly gas delivery for the 17 days analyzed in August 2019. Data from [45].

Figure 5 seems to show similar values for the lower flowrates. This trend has been further investigated. The results of the statistical data analysis for the three years are summarized in Table 6. As shown, the maximum and average flowrates are almost the same for the three years, while a difference occurs for the minimum value that in 2018 reached the smallest value, i.e., 16,965,404 kWh/h in the time period 02:00–03:00 of the 15th of August 2018. However, it should be noted that, considering a Gaussian statistical distribution for 2018 data, from a probabilistic approach the 99.7% of the values should be within the range [19,750,649 kWh, 93,321,690] (=$\mu - 3\sigma$). Applying the Le Chauvenet's criterion the minimum natural gas flowrate calculated in August 2018 was discarded resulting as a statistical outlier. This value, in fact, could be justified by different reasons, such as for example a maintenance activity. Based on remaining data, the MNG was assumed equal to 30,000,000 kWh/h, i.e., an hourly natural gas flowrate of 2,529,511 Sm^3/h based on the LHV reported in Table 4.

Table 6. Results from the data analysis performed concerning hourly gas flowrate in August 2017, 2018 and 2019.

Parameter	2017	2018	2018
Maximum hourly gas flowrate, (kWh/h)	79,914,957	80,479,252	88,454,383
Average hourly gas flowrate (μ), (kWh/h)	56,286,215	56,536,170	63,661,102
Minimum hourly gas flowrate, (kWh/h)	30,806,987	16,965,404 *	38,865,590
Standard deviation (σ), (kWh/h)	14,466,853	12,261,840	11,989,874

(*) The minimum value calculated in the 15th of August can be considered as an outliner and not considered for the analysis. Applying the Chauvenet's criterion a minimum value equal to 30,442,626 kWh/h results for 2018.

Due to the assumptions and simplifications performed for the calculation of the MNG, a safety factor SF equal to 0.9 was considered appropriate for the hydrogen threshold calculation.

3.3. Computation of the Hydrogen Blending Threshold

Applying Equation (12) and the values reported in Table 7 a corrected blending threshold equal to 27,293.4 Sm^3/h, that are equivalent to 2326 kg/h of hydrogen, can be computed. Based on an annual timeframe of 3500 h/year, typical for a P2H plant for electric grid stability service [52], a total hydrogen production up to 8141 ton/year can be estimated. In Table 8 the results of the preliminary assessment are summarized.

Table 7. Parameters used in the calculation of corrected hydrogen blending threshold.

Parameter	Value	Unit of Measure
Allowed blending percentage (ABP)	10	%
Energy density factor (EDF)	1.085	#
Density ratio, (ρ_{H2}/ρ_{NG})	0.099	#
Safety factor (SF)	0.9	#

Table 8. Results of the preliminary assessment about hydrogen blending threshold.

Parameter	Value	Unit of Measure
Corrected hourly hydrogen blending threshold (BT_{corr})	2326	kg/h
Annual P2H plants' working hours	3500	h/year
Annual blended hydrogen	8141	ton/year

3.4. Preliminary Estimation of P2H Plants Size

Based on the preliminary assessment of the hydrogen blending threshold, an estimation of the P2H size and distribution can be performed: a total installed P2H plant output capacity of 77.5 MW can be considered for the Italian framework in the first green hydrogen development phase. To produce such amount of energy, in accordance to Section 2.3, P2H plants should be designed with (i) an electrolysis section and (ii) a compression section and (iii) other auxiliaries. By using Equations (19)–(22) the total electrical power supply for electrolyzers and compressors is calculated equal to 127.2 MW. Assuming the safety factor equal to 5% to take into account other auxiliaries' consumption a total power supply of 133.6 MW can be assumed. From an energy point of view, this would result in an annual electrical consumption of 467.6 GWh/year. Results are summarized in Table 9.

Table 9. Preliminary estimation of Italian P2H plants' total size.

Parameter	Value	Unit of Measure
Total installed output capacity (green hydrogen)	77.5	MW
Electrical installed capacity for electrolyzers	119.2	MW
Electrical installed capacity for compressors	8.0	MW
Electrical installed capacity for other auxiliaries in the plant	6.4	MW
Total electrical installed capacity	133.6	MW
Total P2H plant efficiency estimation	58.0	%
Annual electricity consumption	467.6	GWh/year

3.5. Preliminary Estimation of P2H Plants Distribution

The total installed capacity of Italian P2H plants was distributed in the Italian territory based on the following consideration. Particularly, it was assumed that natural gas flowrate at the import points is the minimum when the total Italian gas flowrate is the minimum. Therefore, it is assumed in this preliminary evaluation that the control of the gas flowrate at each import point is proportional to the total gas flowrate in the network. A detailed analysis of natural gas flowrates for each import points will be performed in a following paper.

Then, P2H plants capacity was divided per each location as reported in Table 10 are calculated. In Table 10 the natural gas flowrate from national fields is not shown. In Figure 6 the size and the location of the P2H plants are shown in the Italian territory in comparison with already installed PV and wind turbine power plants capacity.

Table 10. Preliminary estimation of Italian P2H plants' total size.

Parameter	Tarvisio	Gorizia	Passo Gries	Mazara del Vallo	Gela	GNL Cavarzere	GNL Livorno	GNL Panigaglia
Natural gas imports in 2017, (Sm^3/year) × 10^9	30.2	0.03	18.88	7.25	4.64	6.85	0.91	0.62
Percentage of the total, (%)	43.5	0.04	27.2	10.4	6.7	9.9	1.3	0.9
Total installed capacity, (MW)	34	0.03	21.1	8.1	5.2	7.7	1.0	0.7
Total electrical installed capacity, (MW)	58.2	0.06	36.4	14.0	8.9	13.2	1.8	1.2
Annual electricity consumption, (GWh/year)	204	0.2	127.3	48.9	31.3	46.2	6.1	4.2

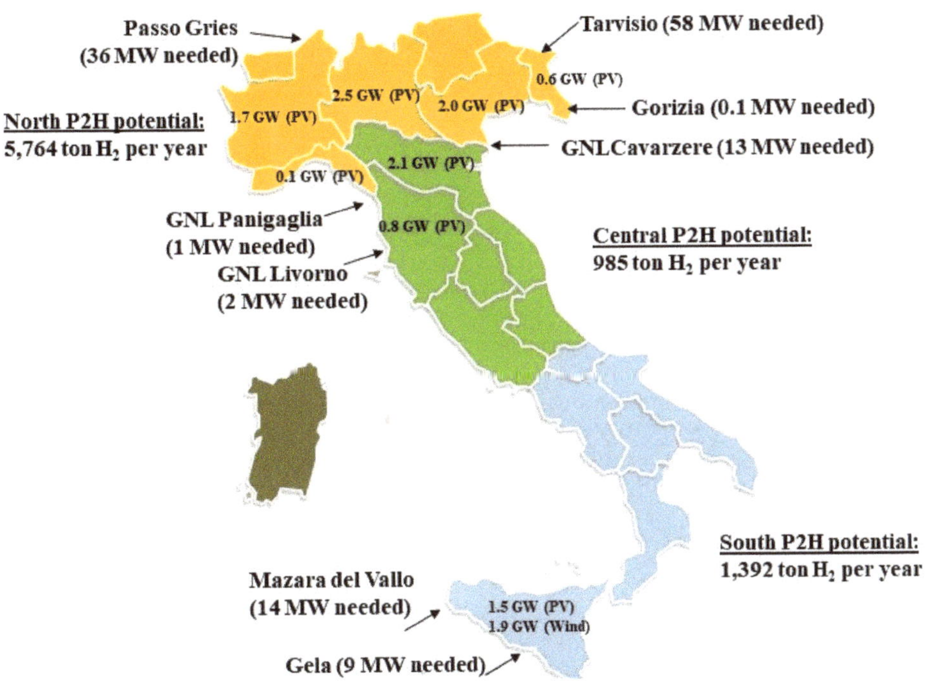

Figure 6. Location and size of the P2H plants in Italy and relation with regional renewable power availability.

3.6. Preliminary Economic Assessment for Italian P2H Plants

Based on Equations (23)–(26) and values introduced by Table 11, a CAPEX equal to 487.5 M€ can be estimated. As shown in Table 10, the physical storage volume is responsible for the highest voice of cost. In fact, ATEX certified components, instrumentation and high-pressure stainless steel storage tank are required for safety reason.

Table 11. Capital expenditure (CAPEX) parameters.

Parameter	Value	Unit of Measure
Electrolyzers' section (instrumentation and tubing are included)	83.4	M€
Compressors' section	80.0	M€
Hydrogen storage volume	226.6	M€
Safety factor to take into account of engineering, tests and other activities (of the total CAPEX)	97.5	M€
Total	487.5	M€

3.7. Discussion and Next Steps

Figure 6 defines the optimum location of the first kind of P2H plants for low percentage green hydrogen blending in Italy. The final results have been achieved by considering only (i) the definition of hydrogen blending threshold to be not overcome, and (ii) the hydrogen injection close to the principal connections of the natural gas network. It is interesting to note how the final result is coherent with energy consumption. In fact, 71% of hydrogen produced by the P2H plants would be blended in the north area of the natural gas network, wherein energy demand is the highest. Therefore, hydrogen generation and blending are optimized in relation with energy demand. Nevertheless, a relevant component of the P2H plants location strategy should consider not only consumers' location, but also renewable power availability and/or potential. Figure 6 shows that, due to the conservative hypothesis related to the maximum hydrogen blending threshold identified in the paper, the installed power required for each P2H plant related with a principal connection of the natural gas network is much lower than the installed renewable power (PV plus wind) already available in the Regions potentially involved.

Further steps are needed to identify in detail if and which existing renewable power plants can be directly connected to the new P2H plants in each region. An economic evaluation will be performed starting from the available GIS database about natural gas network and renewable power plants. Nevertheless, Apulia region will be taken into consideration as a relevant case study too: in fact, while Apulia currently has no relevant natural gas network connection, it has the highest power production from renewables. Therefore, Apulia is the most interesting Italian region to be studied as a reference for the second step of the national hydrogen strategy, i.e., hydrogen blending over the threshold as defined in the paper.

4. Conclusions

The paper aims to assess the actual green hydrogen potential in Italy based on natural gas network characteristics. The estimation has been performed by considering some relevant hypothesis and limitations to minimize the impact on natural gas infrastructure and end-users, thus allowing a short-term implementation at local level of P2H installations for green hydrogen blending. The paper shows how up to 8100 ton/year of green hydrogen blending, i.e., 715,000 Sm^3/year could be injected right now in the existing natural gas network with a proper location and sizing of P2H plants. This green hydrogen potential corresponds to an installed capacity of about 78 MW of electrolyzers and about 488 M€ of investment. Further analysis is needed to better evaluate the geographical positioning of P2H plants, including also integration with existing renewable power plants.

The objective of the EU hydrogen strategy is to install in the first phase of its development, from 2020 up to 2024, at least 6 GW of renewable hydrogen electrolyzers and the production of up to 1 million ton of renewable hydrogen. Therefore, Italy could give at least a 1% contribution to this strategy by immediately implementing the realization of P2H plants for green hydrogen blending as described in the paper. Nevertheless, the results of the preliminary assessment show how the design and development of more complex strategies, including natural gas network revamping and end-users adaptation, are necessary if the ambitious goals of the European strategy want to be reached by Italy and also by the Member States. That is why the Italian government is urgently called to agree on a national hydrogen strategy to boost energy transition and stimulate the technological innovation of the Italian hydrogen industry.

Author Contributions: M.P. conducted the conceptualization and was in charge for the revision of the document; A.G. prepared the methodology and elaborated data; C.S. provided comments and supported designing the study. All authors have read and agreed to the published version of the manuscript.

Funding: This research was funded by the Italian Minister of University and Research (MUR) in the framework of the "*SuperP2G: Synergies Utilizing Renewable Power Regionally by Means of Power To Gas*" project under the joint programming initiative "*ERA-Net Smart Energy Systems—Focus initiative integrated, Regional Energy Systems*".

Conflicts of Interest: The authors declare no conflict of interest.

Nomenclature

ABP	Allowed Blending Percentage, ($\%_{vol}$)
BT	Blending threshold, (Sm^3/h)
BT_{corr}	Corrected blending threshold, (Sm^3/h)
$CAPEX_{ELECTROLYSER}$	Water electrolyzers' CAPEX, (€)
$CAPEX_{COMPRESSOR}$	Compressors' CAPEX, (€)
$CAPEX_{STORAGE}$	Storages' CAPEX, (€)
$c_{electrolyser}$	Water electrolyzer's specific cost, (€/kW)
$c_{compressor}$	Compressor's specific cost, (€/kW)
$c_{storager}$	Storage's specific cost, (€/kW)
E_{demand}	End-users' energy demand, (kWh)
E_{HCNG}	Energy delivered by the HCNG mixture, (kWh)
EDF	Energy density factor, (%)
HCNG	Hydrogen and compressed natural gas
$L_{is,COMPRESSOR}$	Compressors' isentropic work (kJ/kg)
LHV_{NG}	Lower heating value of natural gas, (kWh/Sm^3)
LHV_{HCNG}	Lower heating value of the natural gas and hydrogen mixture, (kWh/Sm^3)
LHV_{H2}	Lower heating value of hydrogen, (kWh/Sm^3)
MNG	Minimum Natural Gas flowrate, (Sm^3/h)
M_{NG}	Natural gas molecular weight, (kg/kmol)
M_{H2}	Hydrogen molecular weight, (kg/kmol)
P_{P2H}	P2H plants' capacity, (kW)
$P_{EL,P2H2}$	P2H Total electric power capacity, (kW)
$P_{EL,ELECTROLYSER}$	Electric power capacity of water electrolyzers, (kW)
$P_{EL,COMPRESSOR}$	Electric power capacity of hydrogen compressors, (kW)
$P_{EL,AUXILIARIES}$	Electric power capacity of P2H plant auxiliaries, (kW)
p_{HCNG}	HCNG pressure, (Pa)
p_{NG}	Natural gas pressure, (Pa)
p_{H2}	Hydrogen pressure, (Pa)
Q_{H2}	Hydrogen volumetric flowrate, (Sm^3/h)
Q_{NG}	Natural gas volumetric flowrate, (Sm^3/h)
$Q_{NG'}$	Natural gas volumetric flowrate when no hydrogen is blended into the network, (Sm^3/h)

Q_{HCNG}	HCNG volumetric flowrate, (Sm^3/h)
R_0	Universal gas constant, (kJ/kmolK)
SF	Safety factor, (%)
T_{NG}	Natural gas temperature, (K)
T_{H2}	Hydrogen temperature, (K)
$V_{STORAGE}$	Storages' volume, (m^3)
$\eta_{el,ELECTROLYSER}$	Water electrolyzers' electric efficiency, (%)
$\eta_{el,COMPRESSOR}$	Compressors' electric efficiency, (%)
$\eta_{is,COMPRESSOR}$	Compressors' isentropic efficiency, (%)
ρ_{H2}	Hydrogen density, (kg/m^3)
ρ_{NG}	Natural gas density, (kg/m^3)
w_{H2}	Volumetric concentrations of hydrogen in the HCNG, ($\%_{vol}$)
w_{NG}	Volumetric concentrations of natural gas in the HCNG, ($\%_{vol}$)
Z_{NG}	Natural gas compressibility factor
Z_{H2}	Hydrogen compressibility factor

References

1. IRENA. *Renewable Energy Statistics 2020*; The International Renewable Energy Agency: Abu Dhabi, UAE, 2020; ISBN 9789292601379.
2. BloombergNEF. *Global Trends in Renewable Energy Investment 2019*; Frankfurt School of Finance & Management gGmbH: Frankfurt am Main, Germany, 2019.
3. Denholm, P.; Ela, E.; Kirby, B.; Milligan, M. *The Role of Energy Storage with Renewable Electricity Generation*; National Renewable Energy Laboratory: Golden, CO, USA, 2010.
4. Götz, M.; Lefebvre, J.; Mörs, F.; McDaniel Koch, A.; Graf, F.; Bajohr, S.; Reimert, R.; Kolb, T. Renewable Power-to-Gas: A technological and economic review. *Renew. Energy* **2016**, *85*, 1371–1390. [CrossRef]
5. Dincer, I.; Bicer, Y. *Integrated Energy Systems for Multigeneration*, 1st ed.; Elsevier Science: Amsterdam, The Netherlands; Oxford, UK; Cambridge, MA, USA, 2019; ISBN 9780128131756.
6. IEA. *The Future of Hydrogen*; IEA: Paris, France, 2019.
7. Mazloomi, K.; Gomes, C. Hydrogen as an energy carrier: Prospects and challenges. *Renew. Sustain. Energy Rev.* **2012**, *16*, 3024–3033. [CrossRef]
8. Dodds, P.E.; Staffell, I.; Hawkes, A.D.; Li, F.; Grünewald, P.; McDowall, W.; Ekins, P. Hydrogen and fuel cell technologies for heating: A review. *Int. J. Hydrog. Energy* **2015**, *40*, 2065–2083. [CrossRef]
9. Wulf, C.; Kaltschmitt, M. Hydrogen supply chains for mobility-Environmental and economic assessment. *Sustainability* **2018**, *10*, 1699. [CrossRef]
10. ENTSOG. *2050 Roadmap for Gas Grids*; ENTSOG: Brussels, Belgium, 2020.
11. CertifHy—Developing a European Framework for the Generation of Guarantees of Origin for Green Hydrogen Definition of Green Hydrogen. Available online: https://ec.europa.eu/jrc/sites/jrcsh/files/Vanhoudt%20Definition%20of%20Green%20Hydrogen%20SFEM.pdf (accessed on 12 August 2020).
12. European Commission. *A Hydrogen Strategy for a Climate-Neutral Europe*; European Commission: Brussels, Belgium, 2020.
13. van Wijk, A.; Chatzimarkakis, J. *Green Hydrogen for a European Green Deal A 2x40 GW Initiative*; Hydrogen Europe: Brussels, Belgium, 2020.
14. Melaina, M.W.; Antonia, O.; Penev, M. *Blending Hydrogen into Natural Gas Pipeline Networks: A Review of Key Issues*; National Renewable Energy Laboratory: Golden, CO, USA, 2013.
15. Bianchini, A.; Guzzini, A.; Pellegrini, M.; Saccani, C. Natural Gas Distribution Networks: How Failures' Databases Can Improve Existing Safety Performances. In Proceedings of the XXIV Summer School Francesco Turco, Brescia, Italy, 11–13 September 2019; pp. 430–436.
16. Bianchini, A.; Guzzini, A.; Pellegrini, M.; Saccani, C. Natural gas distribution system: A statistical analysis of accidents data. *Int. J. Press. Vessel. Pip.* **2018**, *168*, 24–38. [CrossRef]
17. Messaoudani, Z.L.; Rigas, F.; Hamid, M.D.; Hassan, C.R. Hazards, safety and knowledge gaps on hydrogen transmission via natural gas grid: A critical review. *Int. J. Hydrog. Energy* **2016**, *41*, 17511–17525. [CrossRef]
18. Hormaza Mejia, A.; Brouwer, J.; Mac Kinnon, M. Hydrogen leaks at the same rate as natural gas in typical low-pressure gas infrastructure. *Int. J. Hydrog. Energy* **2020**, *45*, 8810–8826. [CrossRef]

19. Hydrogen Europe. *Hydrogen Europe Vision on the Role of Hydrogen and Gas Infrastructure on the Road toward a Climate Neutral Economy-A Contribution to the Transition of the Gas Market*; Hydrogen Europe: Brussels, Belgium, 2019.
20. Gondal, I.A. Hydrogen integration in power-to-gas networks. *Int. J. Hydrogen Energy* **2019**, *44*, 1803–1815. [CrossRef]
21. SNAM. Snam Testa Con Baker Hughes la Prima Turbina 'Ibrida' a Idrogeno Al Mondo Per Una Rete Gas. Available online: https://www.snam.it/it/media/comunicati-stampa/2020/Snam_Baker_Hughes_prima_turbina_ibrida_a_idrogeno.html (accessed on 15 September 2020).
22. Abeysekera, M.; Wu, J.; Jenkins, N.; Rees, M. Steady state analysis of gas networks with distributed injection of alternative gas. *Appl. Energy* **2016**, *164*, 991–1002. [CrossRef]
23. Saccani, C.; Pellegrini, M.; Guzzini, A. Analysis of the Existing Barriers for the Market Development of Power to Hydrogen (P2H) in Italy. *Energies* **2020**, *13*, 4835. [CrossRef]
24. SNAM. Piano Decennale Di Sviluppo Delle Reti Di Trasporto Di Gas Naturale. Available online: http://pianodecennale.snamretegas.it/it/infrastrutture-del-gas-in-italia-ed-europa/rete-di-trasporto-di-snam-rete-gas.html (accessed on 15 September 2020).
25. ARERA. Reti Delle Società Di Trasporto. Available online: https://www.arera.it/it/dati/gm58.htm (accessed on 15 September 2020).
26. SNAM. Piano Decennale Di Sviluppo Della Rete Di Trasporto Del Gas Naturale 2018–2027. Available online: http://pianodecennale.snamretegas.it/includes/doc/2/2019012208362018-decennale_web.pdf (accessed on 17 October 2020).
27. Ministero dell'Industria, del Commercio e dell'Artigianato. *Decreto 22 dicembre 2000: Individuazione della rete nazionale dei gasdotti ai sensi dell'art. 9 del decreto legislativo 23 maggio 2000, n. 164*; Ministero dell'Industria, del Commercio e dell'Artigianato: Rome, Italy, 2000.
28. SNAM. Codice di Rete. Available online: https://www.snam.it/it/trasporto/codice-rete-tariffe/Codice_di_rete/Aree/codice_rete.html (accessed on 17 October 2020).
29. Ministero delle Attività Produttive. *Decreto 29 settembre 2005: Indirizzi e criteri per la classificazione delle reti regionali di trasporto e per l'allacciamento diretto di clienti finali alle stesse reti*; Ministero delle Attività Produttive: Rome, Italy, 2005.
30. MISE. Gas Naturale Reti di Trasporto Nazionale e Regionale. Available online: https://www.mise.gov.it/index.php/it/energia/gas-naturale-e-petrolio/gas-naturale/trasporto (accessed on 15 September 2020).
31. ARERA. Attività di Distribuzione Per Regione. Available online: https://www.arera.it/it/dati/gm53.htm (accessed on 15 September 2020).
32. Bianchini, A.; Donini, F.; Guzzini, A.; Pellegrini, M.; Saccani, C. Natural gas pipelines distribution: Analysis of risk, design and maintenance to improve the safety performance. In Proceedings of the XX Summer School 'Francesco Turco' Industrial Systems Engineering, Naples, Italy, 16–18 September 2015; pp. 243–248.
33. ARERA. *Documento per la Consultazione 338/2019/R/GAS*; ARERA: Milan, Italy, 2019.
34. IEA. Limits on Hydrogen Blending in Natural Gas Networks. Available online: https://www.iea.org/data-and-statistics/charts/limits-on-hydrogen-blending-in-natural-gas-networks-2018 (accessed on 15 September 2020).
35. ARERA. *Allegato A Deliberazione 64/2020/R/gas*; ARERA: Milan, Italy, 2020.
36. SNAM. Snam: Per la Prima Volta in Europa Fornitura di Idrogeno Misto a Gas Naturale Su Rete di Trasmissione a Utenti Industriali. Available online: https://www.snam.it/it/media/comunicati-stampa/2019/Snam_prima_volta_Europa_fornitura_idrogeno_misto_gas_naturale.html (accessed on 15 September 2020).
37. SNAM. Snam: Immissione Sperimentale di Idrogeno a Contursi Raddoppiata al 10%. Available online: https://www.snam.it/it/media/news_eventi/2020/Snam_immissione_sperimentale_idrogeno_Contursi_raddoppiata.html (accessed on 15 September 2020).
38. The European House—Ambrosetti. *SNAM H2 ITALY 2050 Una Filiera Nazionale Dell'idrogeno e la Decarbonizzazione Dell'Italia*; SNAM: Milan, Italy, 2020.
39. Bianchini, A.; Guzzini, A.; Pellegrini, M.; Saccani, C. Earthquake and Earth Movement Monitoring: The Possibility to Use Natural Gas Distribution Infrastructure. In Proceedings of the XXII Summer School 'Francesco Turco' Industrial Systems Engineering, Palermo, Italy, 13–15 September 2017; pp. 143–149.
40. ANEV. *Il Potenziale Eolico Installabile*; ANEV: Rome, Italy, 2019.
41. IEA. *Potential for Building Integrated Photovoltaics*; IEA: Paris, France, 2002.
42. Terna, S.P.A. Gaudì. Available online: https://www.terna.it/it/sistema-elettrico/gaudi (accessed on 19 October 2020).

43. MISE. *Decreto Ministeriale 18/5/2018*; MISE: Rome, Italy, 2018.
44. Oosterkamp, A. Heat transfer modelling of natural gas pipe flow-effect of yearly ambient temperature cycles. In Proceedings of the 26th International Ocean and Polar Engineering Conference, International Society of Offshore and Polar Engineers, Rhodes, Greece, 26 June–2 July 2016.
45. Compressibility Chart|McGraw-Hill Education Access Engineering. Available online: https://www.accessengineeringlibrary.com/content/book/9780071830829/back-matter/appendix8 (accessed on 17 October 2020).
46. SNAM. Operating Data—Physical Flows on the National Network. Available online: https://www.snam.it/en/transportation/operational-data-business/0-Phisical_Flows_on_the_national_network/index.html (accessed on 15 September 2020).
47. SNAM. Trend dal 2005. Available online: https://www.snam.it/it/trasporto/dati-operativi-business/2_Andamento_dal_2005/ (accessed on 15 September 2020).
48. Sdanghi, G.; Maranzana, G.; Celzard, A.; Fierro, V. Review of the current technologies and performances of hydrogen compression for stationary and automotive applications. *Renew. Sustain. Energy Rev.* **2019**, *102*, 150–170. [CrossRef]
49. STORE & GO. *Innovative Large-scale Energy Storage Technologies and Power-to-Gas Concepts after Optimisation Report on the Costs Involved with PtG Technologies and Their Potentials across the EU*; STORE&GO Project: Karlsruhe, Germany, 2018.
50. Proost, J. State-of-the art CAPEX data for water electrolysers, and their impact on renewable hydrogen price settings. *Int. J. Hydrog. Energy* **2019**, *44*, 4406–4413. [CrossRef]
51. Wang, Y.; Kowal, J.; Leuthold, M.; Sauer, D.U. Storage system of renewable energy generated hydrogen for chemical industry Selection and/or peer-review under responsibility of Canadian Hydrogen and Fuel Cell Association. *Energy Procedia* **2012**, *29*, 657–667. [CrossRef]
52. Guzzini, A.; Bianchini, A.; Pellegrini, M.; Saccani, C. Analysis of the existing barriers and of the suggested solutions for the implementation of Power to Gas (P2G) in Italy. In Proceedings of the 5th International Conference on Smart Energy Systems, Copenhagen, Denmark, 10–11 September 2019; Available online: https://smartenergysystems.eu/wp-content/uploads/2019/09/7-2_AlessandroGuzziniSESAAU2019.pdf (accessed on 16 August 2020).

energies

MDPI

Article

Success Factors for the Implementation of Community Renewable Energy in Thailand †

Nilubon Luangchosiri *, **Takaya Ogawa**, **Hideyuki Okumura** and **Keiichi N. Ishihara**

Graduate School of Energy Science, Kyoto University, Kyoto 6068501, Japan; ogawa.takaya.8s@kyoto-u.ac.jp (T.O.); okumura@energy.kyoto-u.ac.jp (H.O.); ishihara.keiichi.6w@kyoto-u.ac.jp (K.N.I.)

* Correspondence: luangchosiri.nilubon.t72@kyoto-u.jp; Tel.: +81-080-8471-6522

† This paper is an extended and revised article presented at the International Conference on Sustainable Energy and Green Technology 2019 (SEGT 2019), Bangkok, Thailand, 11–14 December 2019.

Abstract: Community renewable energy (CRE) has recently been proposed as one of the effective policy tools to make a community sustainable and to strengthen this association. CRE is, however, not widespread yet in Thailand, and evidence-based research on CRE in the country is rather scarce. This paper aims to investigate the characteristics of CRE in Thailand to identify the key factors affecting its implementation. Data were collected through semi-structured interviews and document analysis for 26 active CRE projects in Thailand. Results show that the characteristics of CRE in Thailand are rooted in the geographical location of a community. A legal structure was created from a pre-existing group to implement a CRE project. The primary motivation for implementing CRE is sustainable development of a community. We examined essential factors separated into two groups—internal and external factors—in reference to CRE in developed nations. We highlight several issues concerning the further development of CRE in Thailand. First, a strong group leader and a formal structure are essential to drive and manage a project. Second, networks of CRE projects and long-term revolving funds are crucial external support to implement CRE.

Keywords: small-scale; success factor; community renewable energy; sustainable energy

Citation: Luangchosiri, N.; Ogawa, T.; Okumura, H.; Ishihara, K.N. Success Factors for the Implementation of Community Renewable Energy in Thailand. *Energies* **2021**, *14*, 4203. https://doi.org/10.3390/en14144203

Academic Editors: Vincenzo Costanzo and Tapas Mallick

Received: 23 April 2021
Accepted: 7 July 2021
Published: 12 July 2021

1. Introduction

In the modern world, the dominant paradigm of energy supply and production consists of a centralized, usually fossil-fuel-based energy production that generates energy remotely and supplies it to consumers over a vast region. As an alternative to this paradigm, in recent years, there have been efforts to establish decentralized, community-owned, small-scale facilities that produce energy from renewable resources. One of the movements following this paradigm is community renewable energy (CRE).

CRE offers communities the ability to tackle poverty and raise the standard of living [1–3]. Many studies have argued that various socioeconomic benefits are to be gained from the production and use of renewable energy (RE) resources at the community level [4–10]. These benefits include job creation, income generation, self-sufficient energy supply, social cohesion, and human capacity development. CRE projects in agricultural communities help increase farm productivity and rural sustainability [6]. However, establishing a CRE project involves many issues, such as financial, organizational, technical, and extensive cooperation [11]. Additionally, CRE requires contribution (e.g., capital, labor, commitment, trust) from community members to drive a project to run [12–14]. Using locally available RE resources requires collective actions and agreements among community members [15]. The crucial factors for a CRE project's success might differ in the different contexts of communities [13,16–19], which then require contrasting support structures for CRE implementation.

Researchers have intensively investigated the key characteristics and factors affecting CRE development in developed countries/regions, such as the UK, Denmark, Scotland,

Germany, and Australia [13,17,20–22]. G. Seyfang et al. examined the strengths and weaknesses of the UK's CRE [17]. They conducted a web-based survey of the CRE projects undertaken between June and October in 2011. Their results highlighted that CRE is a diverse sector: their multiple objectives need joined-up thinking among governmental departments. Networking among groups and monetary policy support are essential factors driving CRE development. Another research investigated the effect of financial incentives on CRE development in Germany, Denmark, the UK, and Ontario [20]. The study argued that appropriate monetary policy support at each stage of CRE development helps communities overcome project setbacks. Network support among CREs is also vital for sector development, as in the case of the Samso renewable energy island in Denmark [13].

In contrast, there are a limited number of papers that explore the factors affecting the deployment of CRE projects in developing countries. A study assessed the impact and identified drivers of and barriers to CRE success by analyzing two community micro-hydroprojects in Indonesia [18]. The results showed positive socioeconomic impacts, and the essential factor is the intermediary organization's role in supporting project development. E. Macabebe et al. investigated key factors to achieve community-based SHS (solar home system) adoption and sustainability in the Philippines [23]. The results showed that giving priority to the economic value of the technology, capacity building of locals on the technical aspect, management perspectives of the project, and creation of a supply chain for replacement parts are crucial factors for the sustainability of programs. Another study identified factors that influence the success of a rural electric cooperative case in Costa Rica [24]. They argued that six factors affecting success are community governance, capacity building and engagement, ownership structure, technical design, operation and management, and system and project sustainability measures.

Evidence-based research on a community-led initiatives strategy is limited in Thailand as a developing country. A few studies have focused on Thai CRE development. One study involves exploring various aspects of best management practices (BMPs) for sustainable CRE [25]. Results indicated that the key concept of BMPs for sustainability in small-scale CRE is the "actual problem-based solving approach", which will help CRE avoid the creation of unintended negative impacts and develop effective RE implementation. Another study identified how CRE contributes to sustainable energy transitions in Thailand and the Philippines [26]. The article highlighted that valuable insights can be generated from rural- and community-driven renewable energy initiatives and underscored their power to reimagine the future of energy systems in the Global South. Previously, there was a study [9] analyzing financial issues in developing CRE in Thailand in 26 projects, which are also used in this study. The results showed the benefits of CRE projects to the local economy to be income generation, expense savings, and environmental protection; however, they did not focus on factor analysis. Despite the benefits identified, there is still a lack of a comprehensive view on the characteristics and critical factors of implementing CRE projects in Thailand. Thus, there is no basis on which support structures can be built by either communities themselves or the academe or government. Moreover, there are a limited number of studies in developing countries: two cases in Indonesia [18], three cases in the Philippines [23], and one case in Costa Rica [24]. Therefore, it is essential to identify the barriers and drivers of community-led initiatives for energy production in Thailand.

This study, hence, aims to identify factors that lead to successful implementation of Thailand's CRE. We also discuss the factors via a comparison with those in developed nations and identify potential issues and solutions. The results will contribute to the knowledge necessary for informed policymaking, practical knowledge exchange, and research collaboration between CRE stakeholders.

2. Data Collection

2.1. CRE Subsidy Program in Thailand

In 2013–2016, the Ministry of Energy (MOEN) of Thailand launched a CRE subsidy program (total of THB 60 M (million) or approximately USD 1.7 M budget) to support the development of CRE using a variety of renewable energy technologies. The objectives of the projects were to identify the necessary financial and technological support for CRE. The MOEN subsidized the project investment to reduce the payback period for a CRE project to less than 3 years.

In this subsidy program, "CRE" was defined as "renewable energy (RE) utilization in the community", where:

- Community members invest either purely financially or with in-kind contributions.
- Resources in the community are used for energy production.
- The energy produced should be traded or self-utilized.
- An entity is formed within the community to govern the project.

Communities that meet the criteria can submit a proposal to garner financial support from the program. Forty-eight CRE projects out of 75 proposals passed the MOEN criteria, while 27 projects were rejected due to lack of technical or economic feasibility. The rejected proposals could be revised and resubmitted to the MOEN before the deadline of proposal submission. Finally, 26 CRE projects were implemented, and the remaining 22 projects canceled their contracts due to several reasons, for example, an unstable market of the product of the project (25%), lack of financial resources (19%), uncertainty of project success (19%), inability to form a legal entity (13%), and others (25%). The overall schedule of the CRE subsidy program in Thailand is shown in Table 1.

Table 1. Schedule of the CRE subsidy program in Thailand.

Timeline	Activities	Community Participation
July 2013	Program dissemination and call for proposal in 6 regions	632 persons
August 2013–September 2016	Consultation in the communities Preliminary assessment of RE resources, technology, and economics	147 communities
	Proposal development and submission	75 communities
August 2013–October 2016	Screening and approval of the proposal	48 communities
	Project success installation and implementation	26 communities

2.2. After-Action Review (AAR)

Successful implementation of CRE is highly dependent on the local context of communities and the vast societal structures and institutional arrangements of a project itself. Seyfang and Haxeltine emphasized the importance of resolving internal and external factors in grassroots innovations [27]. Wirth employed an institutional theory to analyze the emergence of a community-based energy project in the concrete local, internal context of a community in Italy [28]. He also suggested the importance of external factors along with his analysis. These factors and their inter-relations should be understood in detail if we want to realize the specific reasons why particular community energy projects succeed or fail. Thus, in this study, we follow Wirth's suggestion to study the concrete internal context and combine it with the external context of CRE in Thailand.

Our research approach involves qualitative content analysis [29] using document analysis and lesson-learned discussions with key actors determined via an after-action review (AAR) technique [30] and semi-structured interviews (SSIs) [31]. Permission to access 26 unpublished proposals was given by the MOEN of Thailand. The data included information about technology type and design, scale, objective, ownership, and mode of organization. Background knowledge allowed us to more deeply understand when, where, and why a community created a CRE project, which provided a sound basis for subsequent data collection.

The extraction of success factors involved the AAR technique and SSIs. Although SWOT (strengths, weaknesses, opportunities, and threats) analysis is widely used to identify key factors that have an impact on CRE development in the UK [17] and to evaluate renewable energy projects in Thailand [32], it is not suitable for our objective analysis. SWOT analysis provides a strategic planning framework to evaluate an organization, a plan, a project, or a business activity [33]. On the other hand, AAR is a systematic method that helps individuals learn significant lessons from successful experiences and apply them to others [30]. AARs have three functions for learning from experiences: (i) self-explanation, (ii) data verification, and (iii) feedback. The self-explanation function enables individuals to analyze their behaviors and to suggest explanations for their success and/or failed actions. The data validation function enables learners to cross-validate the information they hold before changing or correcting their mental models. The feedback function enables learners to confirm, add to, overwrite, tune, or restructure information. The main advantage of feedback is that it concentrates not only on performance outcomes but also on the process of task performance [30]. Thus, the AAR approach is appropriate for our analysis objective.

AAR was conducted twice. The first AAR was conducted in Chiang Mai, Thailand, on 24 October 2016 for a preliminary survey. Twenty-one representatives from 10 CRE cases participated in discussions. The authors and the MOEN's staff participated in and facilitated the AAR. The second AAR was conducted in Bangkok, Thailand, on 16 November 2016 to identify key factors influencing CRE implementation. Thirty-seven representatives from 26 CRE cases participated in the AAR. The MOEN's staff facilitated the AAR, and an author was a participant, observing the activity. The AARs were recorded and then subsequently transcribed. Each AAR meeting took about 3 h. All the data collected through documents and interviews were analyzed qualitatively or semi-quantitatively. The overall flow of the AAR process is shown in Table 2.

Table 2. The AAR process.

Steps	Activities
Introduction (40 min)	The main facilitator explains the objective and process of the AAR meeting and questionnaires and divides the participants into 4 groups, namely, community enterprise, cooperative, local authority, and foundation. Each community answers the questionnaire.
Brainstorming (40 min)	Brainstorming in the group based on the answer.
Presenting the results of brainstorming (15 min)	The participants present the ideas generated during the brainstorming session.
Summarizing the AAR (40 min)	The main facilitator consolidates the ideas obtained from the brainstorming with the help of all the participants.

3. Analysis of Collected Data

3.1. Characteristics of CRE Projects in Thailand

3.1.1. Detail of Community Energy Groups and Projects

The 26 pilot CRE projects are located in 15 provinces across Thailand, as shown in Figure 1. Overall, most CRE projects are in the northern and southern parts of the country. This is mostly due to the strong network support from academic institutions in the areas. Few projects are located in the central, eastern, and northeastern regions. It seems that CRE's network support from an academic institution strongly affects the number of CRE projects implemented in an area. The average networking of CRE was 1.15 partners per project.

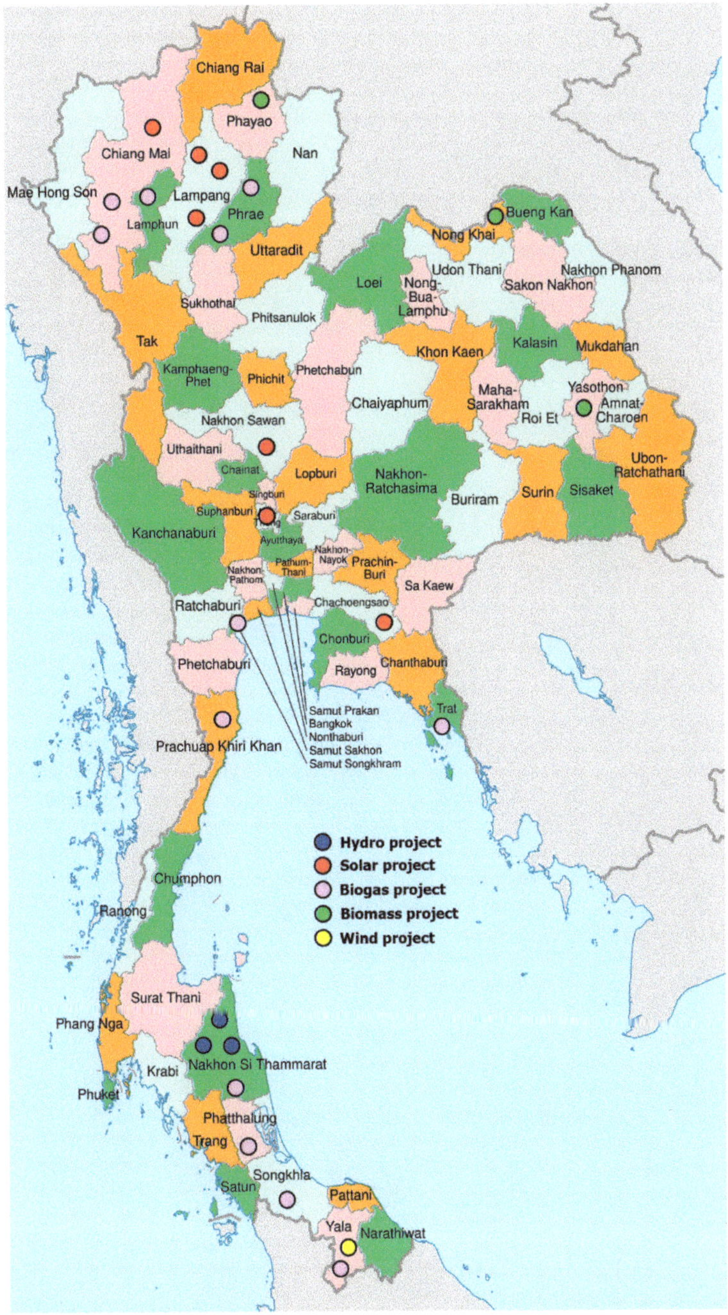

Figure 1. Map showing the geological location of the 26 CRE projects in Thailand: Adapted from [34].

Figure 2a shows the energy generation activities of the 26 CRE projects. Biogas thermal and solar thermal technologies are the most commonly installed technologies due to the fitting of their thermal demand with cost-effectiveness. Most of the CRE projects are implemented in agricultural communities. Thermal energy is needed for agronomic activities. Few projects are involved in electricity generation mostly because of communal accesses to the national grid. CRE that generates electricity is only located in an off-grid area or regions with unstable electricity supply.

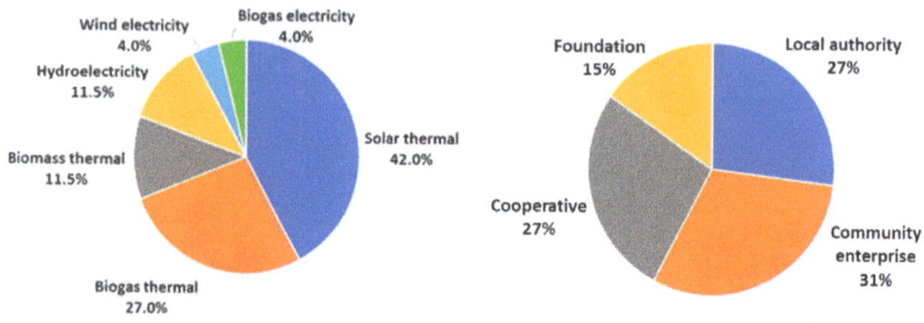

Figure 2. Detail of community energy projects: (**a**) Energy generation activities and (**b**) group types of the 26 CRE projects.

Figure 2b shows the types of legal entities that implemented the CRE projects in Thailand. All the CRE projects had a formal structure before implementing the project. We found that the origin of most of the groups is rooted in civil society. The legal entities of the CRE projects were established and invested by local people and existing groups coming together.

3.1.2. Objectives of CRE Projects

Figure 3 shows the objectives of CRE project development. Three groups are categorized according to aims: economic, environmental, and social. The most common goal was saving money on energy bills, accounting for 85% of the groups. For example, biogas projects used gas to substitute LPG or wood for cooking or agricultural activities. The second common goal was improving the local environment, accounting for 50%. For instance, biogas projects focused on reducing odor pollution and underground water contamination in the area. As regards the third objective, 31% implemented CRE for education in the community. In brief, the economic and environmental dimensions are the main objectives for implementing CRE.

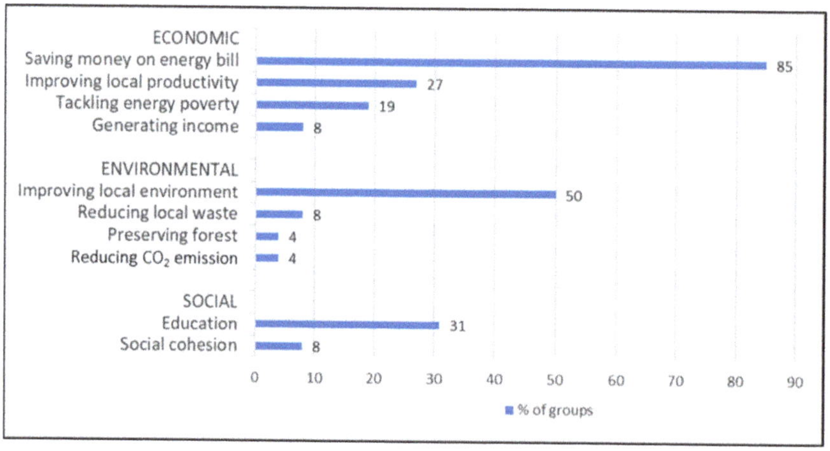

Figure 3. Objectives of Thai CRE groups.

3.2. Key Success Factors for the Implementation of CRE Projects

The key actors identified nine factors that could have influenced CRE implementation. The nine factors can be then categorized into two groups: internal and external. Upon considering the CRE development step, each internal factor is allocated to each step that requires it inside the community, as shown in Figure 4. On the other hand, the external factors mainly are supports from outside the community, which are also assigned to the required step, as shown in Figure 4. Details of the internal and external factors are introduced in the following sections.

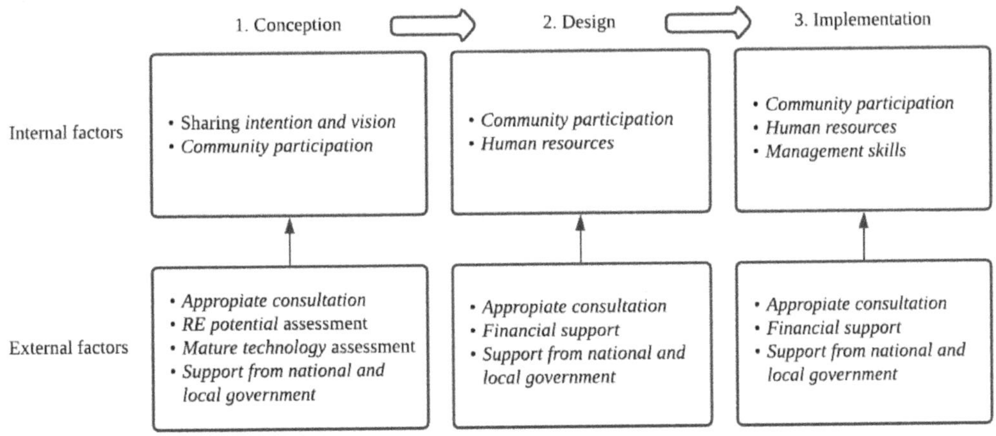

Figure 4. Summary of internal and external conditions and their interaction through a path toward a successful implementation of CRE in Thailand.

3.2.1. Internal Factors

A clear "intention and vision" in a community (for both a leader and team members) creates cooperation and unity, accounting for 84% of the groups. For instance, a cooperative group representative mentioned that "our group has very clear intention and vision on CRE development for improving the local environment and reducing fuel cost of our rubber processing. Therefore, the team members work hard to achieve successful implementation of the project".

"Human resources" is essential in the implementation, accounting for 83% of the groups. A cooperative group representative, for instance, said that "despite the cooperative having many members and staff, we lack the personnel with expertise in renewable energy technologies. We have to ask the help of experts from outside, which leads to some delay in the implementation".

"Management skills" helps drive the project smoothly, accounting for 94% of the groups. A community enterprise representative elaborated that "the project must be expertly managed on time, on budget, with knowledge and integration that organizations need". He added that "without this skill, our project would not be successful in terms of the objective, budget, and time".

"Community participation" exerts a significant influence on project implementation because the participation gains social acceptance, accounting for 50% of the groups. For instance, a biogas network project involved 100 households. The gas produced is supplied through a piping system to the houses of community members. In such a project, community participation plays an essential role because the number of participating households critically affects gas utilization and income.

3.2.2. External Factors

"RE potential" and "mature technology" are significant factors for project development, accounting for 100% and 92% of the groups, respectively. Each group mentioned that "without RE resources and technology, the community would not implement CRE. Furthermore, the lack of RE assessment and technology evaluation causes the project to fail".

"Financial support" is a significant factor in a project's implementation, accounting for 84% of the groups. A grant from the MOEN is the primary source of financial support. The MOEN fund accounts for 30% to 70% of the total budget. Subsequently, the community bears the rest. We found that 85% of the projects raised funds by collecting money from their group members or using the organization's budget. Three cooperative groups and one community enterprise, which represent only 15% of the projects, borrowed money from banks. Many of them indicated that it was challenging to get a loan for a project with a reasonable interest rate from a commercial bank. Furthermore, a representative of a nonprofit foundation group elaborated that a "foundation organization has no choice but to use the foundation's capital because it is quite difficult for the foundation to borrow money from the bank". Some of them mentioned that "it would be easier for us to implement the project if the government could contact some banks to approve the community loans". Therefore, a financial institution's support is vital for CRE implementation to overcome the initial investment cost burden.

"Appropriate consultation" is necessary to design CRE, accounting for 83% of the groups. All the groups mentioned that "the projects would not succeed without the support from the expert or academic institution". In these CRE projects, the MOEN provided the community people with technical support from academic advisors. This factor plays an essential role in the initial stage of CRE project development. Community members cannot design concrete CRE systems by themselves due to limited knowledge and skills.

"Support from national and local government" is needed to authorize a project, accounting for 61% of the groups. Most of the respondents mentioned that "when we create the legal entity in the early development stage, we require support from the relevant government".

4. Discussion

The results show that the formation of a CRE's group and the existence of external support contribute to the successful implementation of CRE in Thailand. The characteristics of Thai communities make it easier to formulate groups due to their collectivist mentality and commitment to the traditional roles of each member. However, the formation of a formal organization is necessary as it provides the structure and rules that enable the community to manage the project as a business. Support from the academe helps less experienced CREs by providing technical and financial knowledge. On the other hand, the government provides financial support to implement the projects since most communities have low income.

4.1. Characteristics of the Community

Thai farmers with a collectivist culture have a tendency to build long-term relationships and trust, which leads to a shared intention and vision. Their commitment to traditional roles leads to everyone having their own rightful place in society, which fosters community participation in maintaining society. Hofstede reported that Thai cultures exhibit a high power distance and tend to be low on individualism [35,36]. Furthermore, these two cultural dimensions (power distance and individualism) relate to the culture of agricultural communities in Thailand because of the kinship principles, which serve as the basic foundation for working together (e.g., rice planting and harvesting are organized around kinship and close friends). Although some traditional agricultural societies are gradually shifting to the agroindustry, collectivist farming vastly remains in Thailand. Thus, Thai agricultural communities can still be considered collectivistic and traditional. Hence, these characteristics lead to (1) effortless formation of groups, (2) willingness to accept help from others, and (3) intention to take leadership roles. However, a strong group leader is needed to drive the project's success.

In the UK, internal and vision (4%) and community participation (5%) were also identified as success factors, but the percentage was relatively low, suggesting a low impact [17]. In contrast to Thailand's collectivism and commitment to traditional roles, the UK's society is more individualistic and competitive [35,36], which explains the low importance placed on a group's effort. Community spirit and sense of locality and responsibility were also identified as success factors for the community-level project in the Samso Renewable Energy Island, Denmark [13]. Moreover, the longevity of the local people's group was identified as a successful social factor in Scotland [21]. This shows that community participation and intention and vision are important despite the difference in culture.

4.2. Characteristics of the Group Structure

Since Thai culture is fundamentally rooted in a kinship society [37], it tends to have unclear business practices, which necessitate a formal structure. On the other hand, individualism in the UK also requires a formal structure to initiate a business. We found that a legitimate organizational structure leads to a successful implementation of CRE, as mentioned in Section 3.1.1. The activity of forming a legal structure sometimes strengthens a group as it provides the opportunity to (1) gather the necessary people with skills needed by a project, (2) establish the rules of an organization, and (3) clarify the duty of members. The legal organization establishes good governance to manage its activities, such as account auditing, annual tax payment, and collective decision making. The results showed that 69% of the total CRE projects (18 projects) had a formal structure before implementing CRE, as shown in Figure 2b. A study in the UK also reported that 65% of the total respondents (354 projects) had a formal structure [17]. In the Thai case, the MOEN required a community to form a legal entity in a project's final application. As a result of forming a formal organization, CRE gained human resources and management skills, which are necessary to manage a project.

4.3. Knowledge Support

CREs in Thailand need technical and financial consultation since this kind of project only started in 2013 with support from the MOEN. Communities still do not have the necessary knowledge to implement a project. The MOEN has just implemented a trial of a CRE subsidy program from 2013 to 2016, which suggests that the program is still young. As the MOEN aims at helping agricultural communities to develop CRE, it recognizes that these communities do not have the necessary knowledge about renewable energy. With this in mind, the MOEN has dispatched academic consultants to help communities to develop a CRE project. Initially, the MOEN only contacted Chiang Mai University, and the university used its network to establish a consultation network to support communities in other regions in Thailand.

Despite the efforts of the consultation network, only five universities supported communities across the country due to the limited number of academic organizations in Thailand. It will be tough to support communities once the number of CREs increases in the future. Therefore, there is a need to increase network support. In the UK and Denmark cases, people involved in a project can support each other due to the strong network among CREs and other organizations, such as local authorities, business entities, academic institutions, and nonprofit organizations [13,17]. The study in the UK reported that the average networking of CRE was 2.7 partners per operative project in partnership with other organizations [17], while in the Thai case, it was only 1.15 partners per operative project, as mentioned in Section 3.1.1. In the case of the Samso Renewable Energy Island, Denmark, a strong network with entrepreneurial individuals was identified to effectively support the island's search for new opportunities for CRE development [13]. Therefore, to promote CRE in Thailand, we suggest that the Thai government or academia help CRE groups create network support among CRE groups. For example, organizing more AAR activities can create a network among CRE projects.

4.4. Financial Support

The participants in the projects are from agricultural sectors that have low income, and most of them need financial support. Actually, 26 CRE projects were implemented with financial support from the government. The project size ranged from USD 3000 to USD 160,000 and USD 40,000 on average per project [9]. With 30%–70% subsidy, each project got around USD 28,000. With a farmer's household net income of about USD 2128 per year in 2017 [38], it is insufficient for members to implement a project only with their incomes. This shows the importance of financial support for the success of a project.

Despite the success of these projects, financial support is limited to that of the MOEN, since only 15% of the participants were successful in borrowing money from financial institutions, as explained in Section 3.2.2. These factors show the limited financial institutes that provide loan for CRE implementation in the country. The Thai government is not entirely decentralized, and as a result, there is minimal local financial autonomy to support CRE. There is a limited grant from the MOEN available for CRE implementation. Moreover, a few soft loans are available for a community to implement a project.

In the UK case, however, since the local government is decentralized, it can directly help CRE projects. Moreover, the UK government created a novel funding mechanism called "Big Society Capital" in 2012 to provide the social sector with affordable financing and build a market for a social investment of social purpose enterprises that expect a social or financial return on investment [17,39]. At the same time, it has a sufficient number of financial institutions that support local activities, with the number of banks totaling about 300 [40]. Thus, the chance to obtain financial support from a bank in the UK is high, which could have led to a significant number of CRE projects in the UK. In contrast, there are only 30 banks in Thailand [41]. Among these small financial institutions, only a few of them provide loans to communities because projects require a low-interest rate. Therefore, the government should establish long-term financial support for CRE. For example, a long-term revolving fund should be created for CRE implementation in Thailand. It is feasible because Thailand already has a revolving fund for the private sector to invest in renewable energy and energy efficiency projects [42].

5. Conclusions

The characteristics and success factors for the implementation of the selected 26 CRE projects in Thailand are analyzed. Results reveal that (i) CRE implementation is rooted in geographically bound communities; (ii) a pre-existing group and its legal structure in communities play an essential role in carrying out CRE; (iii) the primary motivation for CRE employment is sustainable development, which focuses on the economic, environmental, and social dimensions; (iv) the internal success factors are intention and vision, human resources, management skills, and community participation, while the external success factors are RE potential, mature technology, financial support, appropriate consultation, and support from the national and local government. It is strongly suggested that external factors, such as financial policy and network support, should be promoted to further facilitate CRE implementation.

Our finding addresses the lack of a comprehensive view on implementing CRE projects in Thailand. Further research is necessary to understand the significant factors influencing CRE implementation in developing countries.

Author Contributions: Conceptualization, N.L. and K.N.I.; methodology, N.L., T.O., H.O., and K.N.I.; software, N.L., T.O., and H.O.; validation, N.L., T.O., H.O., and K.N.I.; formal analysis, N.L.; investigation, N.L. and K.N.I.; resources, K.N.I.; data curation, N.L.; writing—original draft preparation, N.L.; writing—review and editing, N.L., T.O., H.O., and K.N.I.; visualization, N.L.; supervision, K.N.I.; project administration, N.L.; funding acquisition, K.N.I. All authors have read and agreed to the published version of the manuscript.

Funding: This research received no external funding.

Institutional Review Board Statement: Our study is not applicable.

Informed Consent Statement: Our study is not applicable.

Data Availability Statement: This study did not report any data.

Acknowledgments: The authors are grateful to the communities' leaders and members for sincerely and kindly providing the information required in this study, and especially to the Ministry of Energy and Energy Technology for Environment Research Center, Chiang Mai University, for the great collaboration to get more data for the analysis. Special thanks to Samuel Matthew G. Dumlao for the fruitful discussion, support, and help with English proofreading.

Conflicts of Interest: The authors declare no conflict of interest.

References

1. Sovacool, B.K. A qualitative factor analysis of renewable energy and Sustainable Energy for All (SE4ALL) in the Asia-Pacific. *Energy Policy* **2013**, *59*, 393–403. [CrossRef]
2. Kirubi, C.; Jacobson, A.; Kammen, D.M.; Mills, A. Community-Based Electric Micro-Grids Can Contribute to Rural Development: Evidence from Kenya. *World Dev.* **2009**, *37*, 1208–1221. [CrossRef]
3. Hicks, J.; Ison, N. Community-owned RE rural Australia. *Rural. Soc.* **2011**, *20*, 244–255. [CrossRef]
4. Del Río, P.; Burguillo, M. Assessing the impact of renewable energy deployment on local sustainability: Towards a theoretical framework. *Renew. Sustain. Energy Rev.* **2008**, *12*, 1325–1344. [CrossRef]
5. IRENA. *Energy Benefits: Measuring the Economics*; International Renewable Energy Agency: Abu Dhabi, United Arab Emirates, 2016. Available online: http://www.irena.org/publications/2016/Jan/Renewable-Energy-Benefits-Measuring-the-Economics (accessed on 16 January 2021).
6. Etcheverry, J. *Renewable Energy for Productive Uses: Strategies to Enhance Environmental Protection and the Quality of Rural Life*; University of Toronto: Toronto, ON, Canada, 2003; pp. 1–50. Available online: http://citeseerx.ist.psu.edu/viewdoc/summary?doi=10.1.1.614.1572 (accessed on 18 December 2020).
7. Shoaib, A.; Ariaratnam, S. A Study of Socioeconomic Impacts of Renewable Energy Projects in Afghanistan. *Procedia Eng.* **2016**, *145*, 995–1003. [CrossRef]
8. Cebotari, S.; Benedek, J. Renewable energy project as a source of innovation in rural communities: Lessons from the periphery. *Sustainability* **2017**, *9*, 509. [CrossRef]
9. Chaichana, C.; Wongsapai, W.; Damrongsak, D.; Ishihara, K.N.; Luangchosiri, N. Promoting Community Renewable Energy as a tool for Sustainable Development in Rural Areas of Thailand. *Energy Procedia* **2017**, *141*, 114–118. [CrossRef]
10. Fujimoto, T.; Kagohashi, K. Community-led micro-hydropower development and landcare: A case study of networking activities of local residents and farmers in the gokase township (Japan). *Energies* **2019**, *12*, 1033. [CrossRef]
11. Walker, G. What are the barriers and incentives for community-owned means of energy production and use? *Energy Policy* **2008**, *36*, 4401–4405. [CrossRef]
12. Walker, G.; Hunter, S.; Devine-Wright, P.; Evans, B.; Fay, H. Harnessing Community Energies Harnessing Community Energies: Explaining and Evaluating Community-Based Localism in Renewable Energy Policy in the UK. *Glob. Environ. Polit.* **2007**, *7*, 64–82. [CrossRef]
13. Sperling, K. How does a pioneer community energy project succeed in practice? The case of the Samsø Renewable Energy Island. *Renew. Sustain. Energy Rev.* **2017**, *71*, 884–897. [CrossRef]
14. Walker, G.; Devine-Wright, P.; Hunter, S.; High, H.; Evans, B. Trust and community: Exploring the meanings, contexts and dynamics of community renewable energy. *Energy Policy* **2010**, *38*, 2655–2663. [CrossRef]
15. Hoffman, S.M.; High-Pippert, A. From private lives to collective action: Recruitment and participation incentives for a community energy program. *Energy Policy* **2010**, *38*, 7567–7574. [CrossRef]
16. Madriz-vargas, R.; Bruce, A.; Watt, M. A Review of Factors Influencing the Success of Community Renewable Energy Minigrids in developing countries. In *Proceedings of the Asia Pacific Solar Research Conference 2015*; Egan, R., Passey, R., Eds.; Australian PV Institute: Redfern, Australia, 2015; pp. 1–11. Available online: http://apvi.org.au/solar-research-conference/wp-content/uploads/2015/12/R-Madriz-Vargos_Peer-Reviewed_FINAL.pdf (accessed on 1 December 2020).
17. Seyfang, G.; Park, J.J.; Smith, A. A thousand flowers blooming? An examination of community energy in the UK. *Energy Policy* **2013**, *61*, 977–989. [CrossRef]
18. Guerreiro, S.; Botetzagias, I. Empowering communities–the role of intermediary organisations in community renewable energy projects in Indonesia. *Local Environ.* **2018**, *23*, 158–177. [CrossRef]
19. Dragoman, M.C. Factors Influencing Local Renewable Energy Initiatives in Different Contexts: Comparative Analysis: Italy, Romania and the Netherlands. Master's Thesis, University of Twente, Twente, The Netherlands, 2014.
20. Curtin, J.; McInerney, C.; Johannsdottir, L. How can financial incentives promote local ownership of onshore wind and solar projects? Case study evidence from Germany, Denmark, the UK and Ontario. *Local Econ.* **2018**, *33*, 40–62. [CrossRef]

21. Haggett, C.; Creamer, E.; Harnmeijer, J.; Parsons, M.; Bomberg, E. *Community Energy in Scotland: The Social Factors for Success*; CXC; University of Edinburgh: Edinburgh, UK, 2013; pp. 1–25.
22. Wüste, A.; Schmuck, P. Bioenergy villages and regions in Germany: An interview study with initiators of communal bioenergy projects on the success factors for restructuring the energy supply of the community. *Sustainability* **2012**, *4*, 244–256. [CrossRef]
23. Macabebe, E.Q.B.; Guerrero, R.C.; Domdom, A.C.; Garcia, A.S.; Porio, E.E.; Dumlao, S.M.G.; Perez, T.R. A review of community-based solar home system projects in the Philippines. *MATEC Web Conf.* **2016**, *70*, 12002. [CrossRef]
24. Rolando, M.-V.; Anna, B.; Muriel, W.; Yuri Alvarado, R. Energy with Development: 50 years' Experience of Community-driven Rural Electrification and Future Challenges for COOPEGUANACASTE in Costa Rica. In *Proceedings of the Asia Pacific Solar Research Conference 2016*; Egan, R., Passey, R., Eds.; Australian PV Institute: Redfern, Australia, 2016; pp. 1–12. Available online: http://apvi.org.au/solar-research-conference/wp-content/uploads/2017/02/R-Madriz-Vargas-A-Bruce-M-Watt-and-Y-Alvarado-Rojas-Energy-with-Development-in-Costa-Rica.pdf (accessed on 1 December 2020).
25. Kitisittichai, P.; Aruninta, A. Best Management Practices for Sustainability in a Small-scale Community Renewable Energy: Sathya Sai School, Thailand. *Appl. Environ. Res.* **2015**, *37*, 93–105. [CrossRef]
26. Marquardt, J.; Delina, L.L. Reimagining energy futures: Contributions from community sustainable energy transitions in Thailand and the Philippines. *Energy Res. Soc. Sci.* **2019**, *49*, 91–102. [CrossRef]
27. Seyfang, G.; Haxeltine, A. Growing grassroots innovations: Exploring the role of community-based initiatives in governing sustainable energy transitions. *Environ. Plan. C Gov. Policy* **2012**, *30*, 381–400. [CrossRef]
28. Wirth, S. Communities matter: Institutional preconditions for community renewable energy. *Energy Policy* **2014**, *70*, 236–246. [CrossRef]
29. Elo, S.; Kyngäs, H. The qualitative content analysis process. *J. Adv. Nurs.* **2008**, *62*, 107–115. [CrossRef] [PubMed]
30. Ellis, S.; Davidi, I. After-event reviews: Drawing lessons from successful and failed experience. *J. Appl. Psychol.* **2005**, *90*, 857–871. [CrossRef] [PubMed]
31. Adams, W.C. Conducting Semi-Structured Interviews. In *Handbook of Practical Program Evaluation*; Newcomer, K.E., Hatry, H.P., Wholey, J.S., Eds.; John Wiley & Sons, Inc.: Hoboken, NJ, USA, 2015; ISBN 9781119171386.
32. Iemsomboon, P.; Tangtham, N. Problems and barriers on SHSs management in Thailand's rural areas based on SWOT analysis. *Energy Procedia* **2014**, *56*, 598–603. [CrossRef]
33. GÜREL, E.; TAT, M. Swot Analysis: A Theoretical review. *J. Int. Soc. Res.* **2017**, *10*, 994–1006. [CrossRef]
34. NordNordWest. Paul_012 Thailand Provinces Map en. Available online: https://en.wikipedia.org/wiki/File:Thailand_provinces_en.svg (accessed on 23 April 2021).
35. Country Comparison—Hofstede Insights. Available online: https://www.hofstede-insights.com/country-comparison/thailand,the-uk/ (accessed on 20 April 2021).
36. Srikes, M.; Louvieris, P.; Collins, C. The impact of culture on mobile phone purchasing: A comparison between thai and british consumers. In Proceedings of the 17th European Conference on Information Systems, ECIS 2009, Verona, Italy, 8–10 June 2009. Available online: https://aisel.aisnet.org/ecis2009/408 (accessed on 20 April 2021).
37. Kuwinpant, P. *Thai Society and Culture*; Nagoya University: Nagoya, Japan, 2002.
38. Office of Agricultural Economics. Indicators of Socio-Economic Conditions of Agricultural Households. Available online: https://www.oae.go.th/assets/portals/1/files/socio/socio10june63.pdf (accessed on 29 April 2021).
39. Policy Guide to Scaling Social Innovation 2013—Reports—World Economic Forum. Available online: https://reports.weforum.org/social-innovation-2013/big-society-capital-united-kingdom/?doing_wp_cron=1618670113.8847661018371582031250 (accessed on 17 April 2021).
40. List of Banks in the UK—Top UK Banks. 2018. Available online: https://www.relbanks.com/europe/uk (accessed on 18 December 2020).
41. Number of All Branches and Service Point of Commercial Bank in Thailand. 2020. Available online: https://www.bot.or.th/App/BTWS_STAT/statistics/BOTWEBSTAT.aspx?reportID=904&language=ENG (accessed on 18 December 2020).
42. Energy For Environment Foundation. ESCO Revolving Fund. Available online: http://www.efe.or.th/escofund.php?task=&sessid=&lang=en (accessed on 30 March 2021).

MDPI
St. Alban-Anlage 66
4052 Basel
Switzerland
Tel. +41 61 683 77 34
Fax +41 61 302 89 18
www.mdpi.com

Energies Editorial Office
E-mail: energies@mdpi.com
www.mdpi.com/journal/energies

www.ingramcontent.com/pod-product-compliance
Lightning Source LLC
Chambersburg PA
CBHW061619220926
51850CB00005B/545
9783036523965